Topic	Symbol	Meaning		ge
GRAPHS (continued)	e'	edge of network with		373
	G'	the dual graph of G		318
	K	sink of a network		370
	$K_{m,n}$	bipartite graph joining m vertices with n		321
	K_n	complete graph on n vertices		320
	$P_1(v), P_2(v)$	dual payoff functions		361
	$P \approx Q$	P connects to Q in the graph G		285
	V	set of vertices in a graph		279
	$x \mid y$	flow x through edge of capacity y		370
LANGUAGES, AUTOMATA, AND TURING MACHINES	$\uparrow$	cursor in a Turing machine		591
	$\ominus$	label for the initial state of an automaton		560
	$\oplus$	label for a terminal state of an automaton		560
	$\ominus\!\!\!\oplus$	label for an initial/terminal automaton state		560
	A^*	star operation on A		568
	$A + B$	set union operation		568
	AB	set concatenation operation		568
	$\alpha \Rightarrow \beta$	a production		552
	B	blank symbol in a Turing machine		590
	$c(S)$	code number of a Turing machine symbol S		615
	$C(\mathbf{T})$	code number of a Turing machine $\mathbf{T}$		616
	Γ	$\Sigma \cup V$, the extended alphabet of a grammer		551
	Γ^+	the nonempty strings of Γ		551
	Q	states of a Turing machine		590
	Σ	the alphabet of a language		551
	S	starter symbol in a type-0 language		551
	S_i	elements of the alphabet in a Turing machine		590
	V	variable set in a type-0 language		551
LOGIC	$\forall x$	for any x (universal quantifier)		58
	$\exists x$	there exists an x (existential quantifier)		58
	$\neg p$	not p		25
	p'	alternate form of $\neg p$		455
	$p \wedge q$	p and q		25
	pq	alternate form of $p \wedge q$		455
	$p \vee q$	p or q		25
	$p \rightarrow q$	p implies q		25
	$p \leftrightarrow q$	p if and only if q		25
	$p + q$	p XOR q		464
	$p \mid q$	p NAND q		53
	$p \downarrow q$	p NOR q		53
	$P \equiv Q$	P is logically equivalent to Q		44

The Lending Library at Connecticut College

Rent textbooks

For the semester

For FREE

SUSTAINABILITY

All books are due back to the Lending Library by the last day of finals.

Failure to do so will result in a minimum fee of $25, or the used resale price of the book on Amazon.

Questions? Email: sustainability@conncoll.edu

Book

The Lending Library runs off the generous book donations of students like YOU!

 Connecticut College Office of Sustanablity

Discrete Mathematics

Discrete Mathematics

Melvin Hausner

Courant Institute of Mathematical Sciences
New York University

SAUNDERS COLLEGE PUBLISHING

A Harcourt Brace Jovanovich College Publisher

Fort Worth   Philadelphia   San Diego   New York   Orlando   Austin
San Antonio   Toronto   Montreal   London   Sydney   Tokyo

Text Typeface: Caledonia
Compositor: Maryland Composition
Acquisitions Editor: Robert B. Stern
Managing Editor: Carol Field
Project Editor: Martha Brown
Copy Editor: Tom Whipple
Manager of Art and Design: Carol Bleistine
Art Director: Christine Schueler
Text Designer: Circa 86, Inc.
Cover Designer: Lawrence R. Didona
Text Artwork: Grafacon, Inc.
Director of EDP: Tim Frelick
Production Manager: Bob Butler
Product Manager: Monica Wilson

Cover Credit: IMEC and University of Pennsylvania. (Detail) Neural Net: foveated, retina-like sensor, 1989.

Printed in the United States of America

DISCRETE MATHEMATICS

ISBN 0-03-003278-4

Library of Congress Catalog Card Number: 91-050637

1234 039 987654321

This book is dedicated
   to my daughter Carol,
   my wife Frieda,
   and
   to the memory of my son Leonard.

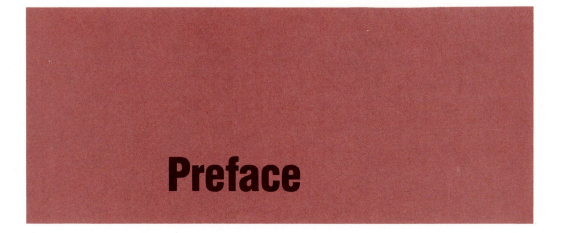

Preface

This book is intended for a first undergraduate course in discrete mathematics. There is enough material for a full-year course, but it may be used for a one-term course as well. Some suggestions for the design of various courses are given in the course outlines section that follows. Calculus is not a prerequisite and in many cases I have avoided the temptation to mention this subject when it may have been useful or informative. The few cases where calculus results appear are clearly indicated.

This text is an outgrowth of a course developed and given by me at New York University. In the early stages, my audience consisted of computer scientists and programmers with very little mathematical background. The purpose was to offer some mathematics relating to computer science, thus preparing these students for some of their more theoretical studies. Gradually, the course proved of interest to the mathematics students (and the computer scientists were becoming more mathematically oriented), and the course developed into the present text.

While it is sometimes felt that this material should be, by its very nature, simpler than the more traditional mathematics taught in college, experience has shown that this is not the case. Throughout the text, care has been taken to illustrate all concepts with examples, usually before the concept is formally introduced. Proofs of results are motivated in almost all cases, and are usually given after the student has had enough experience with the material to make the proofs more transparent.

It is more than coincidence that discrete mathematics today has some of the flavor of computer science. The great breakthroughs in computer technology gave the impetus to this subject. Such topics as logic, number bases, algorithms, trees and graphs, the Chinese Remainder Theorem, and Boolean algebra were all studied before computer technology existed. They were generally thought of as esoteric subjects, of interest only to the mathematically curious. The advent of computers led to the development and better understanding of these subjects. Although I have tried to introduce some computer applications where appropriate, the book is mathematical in its approach. It is safe to say, however, that the contents will be of use to the mathematics as well as the computer science student.

The pseudo code used in the text is introduced in Sections 3.1 and 3.2. It is based on one developed by G. W. Stewart. The choice was not made casually, as it is a compromise between English and a more detailed pseudo code resembling a computer language. The pseudo code used is quite informal, has very few commands, and is fairly easy to convert to the computer language of choice. The result is, I think, quite satisfactory.

The pseudo code has some properties that should be pointed out. It practically forces the user to indent portions of the code involved in a FOR or WHILE loop, or in an IF ... THEN command, since we did not introduce a BEGIN ... END sequence, or an ENDFOR, or END-WHILE instruction. While it is possible to expand the pseudo code in this manner, I have found that these constructs tend to make the code too much like programming, and will therefore inevitably lose a certain portion of the students while perhaps gaining some precision. As it stands, the advantage of this pseudo code is that it forces indentation so that the logical flow of an algorithm can be followed. Also, the number of commands is minimal, and not too strange looking for a student who is not a computer scientist.

The answers to the odd-numbered exercises are given at the back of the text. In the text, the more difficult exercises are introduced with a bold asterisk. These include the more theoretical, the more demanding, and the more time consuming exercises. As a word of caution, an exercise with an asterisk is not meant to be impossible. Occasionally, the student will find that the hardest part of these exercises is to find out why the author decided to asterisk it.

The answers to the even-numbered exercises are in the Instructor's Manual. Many of these exercises are worked out in considerable detail.

If you have any suggestions that you feel may be helpful to other students or instructors, please write to me. I would like to hear from you and to act on your suggestion.

Distinctive Features

The text has a number of features that are unique.

● Many methods and techniques of proof are introduced and analyzed using the tautologies of the propositional calculus. In this way, the traditional methods of proof are not only labeled, but are carried out in depth, within the context of propositional calculus (Section 1.3).

● Formal deductions are carried out in the propositional calculus. Thus, early in the text, students are introduced to a formal deductive system that is quite manageable and is a natural outgrowth of the previous section (Section 1.4).

● An informal introduction to predicate calculus is given, which includes rules for quantification as well as possible pitfalls. Special emphasis is also given on the universe of discourse. In this section the student learns that there is more to logic than truth tables (Section 1.5).

● Permutations on n objects are presented, including cycle decomposition and conjugation. Although group theory is not included, the framework is given. The emphasis is on computation; the effect is to make such traditionally dry subjects as inverse functions, 1-1 functions, and so on, more concrete (Section 2.2).

● A detailed analysis of the box principle, including its abstract version is given with examples. Not only are geometric and numerical applications given, but the underlying theory is analyzed and related to the uniqueness of the cardinality of a set (Section 2.4).

● The pseudo code is introduced and developed in some detail without assuming that the student is familiar with programming techniques. It is gradually introduced, with many examples at each step (Sections 3.1 and 3.2).

● A careful discussion and comparison of various sorting and searching techniques is given. Big and little Oh notation is introduced with many examples and used in the time analysis discussion. Emphasis is also given to the factors contributing to the time analysis, so, in some cases, different times are computed, depending on the presence or absence of these factors (Sections 3.2 and 3.3).

● Linear difference inequalities and equalities are analyzed. The inequality results are dependent on the theory of difference equalities. This bonus gives the tools for a time analysis of various algorithms (Section 4.3).

● The algebra of generating functions is developed informally, with examples. This classical topic is used to solve difference equations and several combinatorial problems (Section 4.4).

● Games of strategy are covered in detail, using directed acyclic graphs. Many specific games are analyzed, including Nim and Wytoff's game. Game trees are introduced as a special case (Section 5.5).

● Networks are introduced as an application of digraphs. The Ford-Fulkerson Algorithm for networks is developed. The emphasis is both algebraic and geometric in character, and shows (without using these terms) how a maximal flow is a linear combination of paths (Section 5.6).

● A modern, algebraic approach to Boolean algebra is given. In the axioms, $x + y$ is interpreted as the exclusive OR of x and y. Within this context, the various Boolean laws, such as De Morgan's Law, properties of complementation and union, and the disjunctive normal form are developed in a traditionally algebraic manner (Section 7.2).

● Properties of congruence are developed, including calculations of quotients mod p, orders of elements, and Fermat's Little Theorem. These are applied to cryptography and random sequences (Section 8.2 and 8.3).

● The Chinese Remainder Theorem is developed, along with different techniques for solving systems of congruences. The theory is applied to error correcting techniques and parallel processing (Section 8.4).

● An introduction to the study of formal languages is given. It is illustrated by developing the syntax of the propositional calculus recursively, and also by using type-0 grammars (Section 9.1).

● Turing machines and recursive function theory are introduced. To facilitate the study of Turing machines, macros are developed (Section 10.1 and 10.2).

● The halting problem is analyzed and various nonrecursive sets are shown to exist. Cantor's diagonal method is introduced and used for this purpose. The significance of these results is discussed in some detail (Section 10.3).

Course Outlines

A one-semester course should include the following key topics.

1. Truth tables and logical reasoning (Sections 1.2 and 1.3)
2. Induction, recursion, and the box principle (Sections 2.3 and 2.4)
3. Algorithms and time analysis, including order of magnitude (Sections 3.1 and 3.2)
4. Some combinatorics (at least Sections 4.1 and 4.2)
5. Graph theory along with some classical problems in this field (Sections 5.1 through 5.4)

A one-semester course could cover Chapters 1 through 5. It may be necessary to eliminate some material because of time considerations. For a one-semester course, possible exclusions are: Sections 1.4 and 1.5; Section 2.2; Section 3.3; Sections 5.5 and 5.6. In addition, the introductory Section 1.1 (Sets and Functions) will be considered as a review at many schools, and should not be emphasized at the expense of the other material.

The book is arranged so that it is possible, in many cases, to cover various topics without necessarily going through all of the previous sections. Thus, according to your taste, you may include games of strategy (Section 5.5) perhaps at the expense of Hamiltonian paths. In a one-semester course, you may include tree theory (Chapter 6) or Boolean algebra and circuit theory (Chapter 7), if time permits.

In addition to the topics given for a one-semester course, a two-semester course should include, at a minimum, the following key topics.

1. Trees and applications (Sections 6.1 through 6.4)
2. Switching circuits and Boolean algebra (Sections 7.1 and 7.2)
3. Some number theory (Sections 8.1 and 8.2)
4. Automata and regular sets (Sections 9.2 and 9.3)

A full-year course could also include much of Chapters 6 through 9, and some of Chapter 10. I have included Chapter 10 (Turing Machines) because of its intrinsic interest and because many of my students requested elaboration on this topic. This chapter may prove difficult for the average student, but I have had success with it. These days, it is not uncommon to have students who can out-program the instructor, and to find students who can program in assembly language. Such students, at least, will not find Turing machines hard to fathom, and will be interested in the theory leading to the discussion of the halting problem.

Once again, variations are possible, depending on the interests of the instructor, the aptitude and interests of the students, departmental requirements, and the inevitable time pressure.

The logical dependence of the chapters is given in the following table. This table should not be interpreted strictly, since much material in a chapter can be used that does not depend on results and methods of allegedly prerequisite chapters. As usual, experimentation is often the way of obtaining the best and most interesting course.

Table of Chapter Dependencies

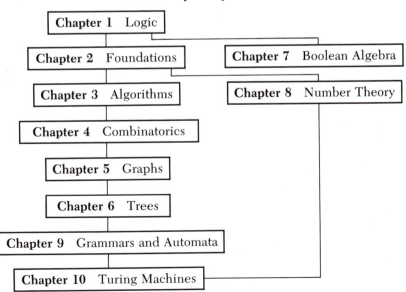

Acknowledgments

The final version of this book benefited from the comments of many mathematicians. These include

Robert A. Chaffer, Central Michigan University
Robert Davis, Southern Methodist University
L. Joyce Donohoe, Southwest Texas State University
D. J. Hartfiel, Texas A & M University
Stephen C. Locke, Florida Atlantic University
Richard K. Molnar, Macalester College
Tom Wineinger, University of Wisconsin-Eau Claire

Without their careful (and sometimes sharp) criticism and suggestions, the book would be considerably less intelligent and more tedious than the final product. I especially want to thank Richard Molnar, who taught me a thing or two, and Stephen Locke, who carefully went through the final manuscript locating miscalculations and typographical errors.

A special thank you to my student and one-time editor, Shelly Langman, who convinced me to put some lectures and notes into a book. Thanks, also, to my many students in several classes at New York University who subjected themselves to the original version of the text.

This book would not have been possible without the assistance of Robert Stern, Jay Ricci, and Martha Brown of Saunders College Publishing. Their care and intelligence in the process of getting the book into print cannot be overstated.

Finally, I would like to thank my wife, Frieda, for her patience, encouragement, and understanding of the time and energy needed to complete the book.

Melvin Hausner
December 1991

Contents

C H A P T E R 1

Logic

Logic, in some form, has been used since history was recorded, and it has been studied for at least 2000 years. Since the nineteenth century, the subject has been made more mathematical, and in the twentieth century much of the work in logic was quite deep and had a major impact on how we think about logic and its power. Logic plays a critical role in computer science, where program and component design use the ideas of this subject.

In this chapter we concentrate on the use of logic in reasoning. Before beginning, we give a brief introduction to two of the basic notions of mathematics: sets and functions.

1.1 Sets and Functions

This section is a short introduction to the concepts of set and function, topics that will be used throughout the text. These topics are by now understood to be part of the working knowledge of today's mathematician and scientist. The two topics are related, and we shall consider sets first.

Sets

A **set** is a collection of objects, called the **elements** of that set. The objects can be arbitrary. They can be numbers, points, lists, programs, words, or book titles. Sets are often denoted by capital letters, such as *A* or *B*, and elements of sets are often designated by small letters, such as *x* or

y. The following are some of the sets of numbers that will often be used in this text.

DEFINITION 1.1.1

$\mathbb{N}$ = The set of **natural numbers** (nonnegative integers)
= $\{0, 1, 2, 3, 4, \dots\}$.

$\mathbb{N}^+$ = The set of **positive integers**
= $\{1, 2, 3, 4, \dots\}$.

$\mathbb{Z}$ = The set of **integers**
= $\{\dots, -3, -2, -1, 0, 1, 2, 3, \dots\}$.

$\mathbb{I}_n$ = The set of integers between 1 and n inclusive
= $\{1, 2, 3, \dots, n\}$. Here n is a positive integer.

$\mathbb{Z}_n$ = The set of integers between 0 and $n - 1$ inclusive
= $\{0, 1, 2, \dots, n - 1\}$. Again, n is a positive integer.

$\mathbb{R}$ = The set of **real numbers**.

$\mathbb{Q}$ = The set of **rational numbers**.

All of these examples are sets of **numbers**. The following are some nonnumerical examples of sets.

EXAMPLE 1.1.2

Consider the set of symbols that can be printed on a particular typewriter. We may call this set TYP. It is a finite set of symbols. The set of capital letters $\{A, B, C, \dots, Z\}$ is a subset of TYP. We call this set ALPH, since it consists of the letters of the alphabet.

EXAMPLE 1.1.3   ASCII Characters

Another example of a finite set is the standard set of symbols that can (usually) be sent to a printer from a personal computer. There are 128 such symbols, and they are called ASCII characters. They include the ordinary symbols of TYP, such as "a", "b", "A", and so on (if the type-writer is not an unusual one), but they also include such items as TAB, BEL, ESC, and others, many of which are used to control a printer. ASCII is an acronym for the American Standard Code for Information Interchange.

There is also an extended version (256 in all) of the ASCII characters. Many printers accept the full range of these characters, using some of the extras for figures, Greek letters, and other symbols. The additional characters are called "high-bit" characters. The explanation for this terminology is postponed to Section 2.5. (See Example 2.5.1.)

EXAMPLE 1.1.4   *Strings*

Let ALPH be the alphabet as introduced in Example 1.1.2. We consider the set of strings in this alphabet ALPH. These are sequences of letters of ALPH given in a specific order. (The term **word** is usually used only for strings that can be found in a specified dictionary.) The set of all strings using the letters of ALPH is an infinite set, since the strings can have arbitrarily large length.

The set of *six-letter strings* using the letters of ALPH is a finite set. For example, the strings ABCDEF, SMILES, and AAAABA are included in this set. Note that the string ABCDEF is different from the string CDAFEB, much as the (three-letter) word TAR is different from RAT, even though both words consist of the same letters. The set of the six-letter strings using ALPH is denoted $ALPH^6$.

Similarly, for any positive integer n, we can form strings of length n on *any* set Σ, regarded as the alphabet. The resulting set of n-letter strings is denoted Σ^n.

EXAMPLE 1.1.5   *Decimal Representation of Integers*

We usually represent positive integers by using a string of **decimal digits** D. These are the special symbols $D = \{0, 1, \ldots, 9\}$. When we say that there are 133 objects in a pile, we are describing a number using a string of three digits, 1 followed by 3 followed by 3—an element of D^3. We shall see in Section 2.5 that there are other systems for representing integers that use more, or less, than ten digits.

For accuracy, particular strings are put in quotes. Thus, the accurate way of writing the beginning of the second paragraph of Example 1.1.4 is as follows:

The set of *six-letter strings* using the letters of ALPH is a finite set. For example, the strings "ABCDEF", "SMILES", and "AAAABA" are included in this set.

For example, this allows us to distinguish the set ALPH from the string "ALPH" of four letters.

EXAMPLE 1.1.6   *Use of Quotes*

The integer 18 is divisible by 3, and the string "18" has length 2, but it is meaningless to say that the integer 18 has length 2, or that the string "18" is divisible by "3" or by 3.

EXAMPLE 1.1.7   Bits and Bytes

A **bit** is, by definition, an element of $\mathbb{Z}_2$. Thus, a bit is one of the numbers 0 or 1. The term "bit" comes from **binary digit,** in analogy with the decimal digits of D. We shall see that bits play a useful role in this text because they can represent many concepts: true or false; low or high; on or off.

A **byte** is an element of $(\mathbb{Z}_2)^8$. That is, a byte is a string of bits of length 8. For example, "00101001" is a byte. (For convenience in what follows, we usually drop the quotes in writing strings of bits.) We shall see that there are exactly 256 different bytes ($256 = 2^8$). It is no coincidence that this is how many extended ASCII characters there are—since each such character is stored internally in a computer as a byte. Bytes have special importance for computers, since it is usually by means of bytes that data is stored in a computer. A typical memory cell of a computer consists of a number of places that can receive low or high voltages, called 0 and 1. The more modern computers have 16- or 32-bit memory cells for faster computation—thus, 2- or 4-byte cells—but the byte is still a basic unit of storage.

EXAMPLE 1.1.8   Strings

If Σ is any set of elements, called the alphabet, we may consider strings of arbitrary length using the alphabet Σ. The notation for the set of all strings on the alphabet Σ is Σ^*. By convention, this includes the **empty string ϵ** consisting of no elements of the alphabet and that has length 0. We also write $\epsilon = $ " ". The strings "AA", "ASDF", and "BEAUTY" are all elements of ALPH*.

Of the sets considered so far, $\mathbb{Z}$, $\mathbb{R}$, $\mathbb{Q}$, and $\mathbb{N}$ are infinite sets, as is Σ^* for any nonempty Σ. The rest are finite, and so each of them contains a certain number of elements. In some cases, the number can be easily found: $\mathbb{I}_n$ has n elements—the numbers from 1 to n. Similarly, ALPH has 26 elements. In other cases, it is more difficult. $ALPH^6$ has roughly 309 million elements (26^6). One of the tasks of Chapter 4 is to find accurate counts for large, relatively complicated, finite sets.

We write $x \in A$ to mean that x is an element of A; similarly, $x \notin A$ means that x is *not* an element of A. Two sets A and B are equal, written $A = B$, if they have the same elements. This means that any element of A is necessarily an element of B, *and* any element of B is also an element of A.

EXAMPLE 1.1.9 *Describing Sets*

Sets can be described in several ways. The simplest is to list the elements **explicitly,** enclosing them with braces "{" and "}" and usually separating them by a comma. Thus, the list

$$D = \{0, 1, 2, 3, 4, 5, 6, 7, 8, 9\}$$

fully describes the set of decimal digits. As usual, we often use "..." to avoid a long list. Thus $\{1, 2, \ldots, 100\}$ is understood to be an explicit, though shorthand, description of $\mathbb{I}_{100}$. This idea is often stretched, as in the description $\mathbb{N} = \{0, 1, 2, \ldots\}$, where "..." is used to describe an infinite set.

Another important method of defining sets is **descriptive** rather than explicit. For example, $\{1, 3, 5, 7, 9\}$ may be defined descriptively as the set of odd, single-digit numbers. Since an odd number is a number of the form $2n + 1$ (n is an integer), and a single-digit number is one between 0 and 9 inclusive, this set can be written as

$$\{2n + 1 \mid n \in \mathbb{Z} \text{ and } 0 \le 2n + 1 \le 9\}$$

or, equivalently, as

$$\{2n + 1 \mid n \in \mathbb{N} \text{ and } 2n + 1 \le 9\}$$

The form of this description is $\{X \mid Y\}$, which can be read as "the set of elements of the form X satisfying the condition Y." The symbol "$\mid$" may also be read "such that."

EXAMPLE 1.1.10 *Descriptive Definition*

The rational numbers $\mathbb{Q}$ are defined as the set of numbers that are the quotients of two integers, with nonzero denominator. A descriptive definition is

$$\mathbb{Q} = \{p/q \mid p, q \in \mathbb{Z} \text{ and } q \ne 0\}$$

Similarly the positive rational numbers $\mathbb{Q}^+$ may be defined by the equation

$$\mathbb{Q}^+ = \{p/q \mid p, q \in \mathbb{N}^+\}$$

A set is determined by its elements only, not by the order in which these elements are listed or by how the set is defined. Thus, for example,

$\{1, 3, 5, 7, 9\} = \{3, 7, 5, 1, 9\}$. Similarly, $\{1, 1, 3\} = \{3, 1\}$, since both sets contain the same elements.

DEFINITION 1.1.11

The number of elements in a finite set A, called the **cardinality** of the set A, is written $|A|$:

$$|A| = \text{the cardinality of } A$$

For example, $|\mathbb{Z}_8| = 8$ and $|\text{ALPH}| = 26$. The cardinality of a set is defined more precisely in Section 2.2, as is the notion of a finite set itself, but for now we take it for granted that for any finite set A there is a unique number $|A|$ of elements in A. The number $|A|$ is a nonnegative integer; that is, it is in $\mathbb{N}$. In fact, the elements of $\mathbb{N}$ are called "natural" numbers because they are the counting numbers— probably the first numbers discovered (or invented).

We now consider some of the elementary properties of sets.

DEFINITION 1.1.12

The **empty set**, or **null set**, $\varnothing$ is the set consisting of no elements. That is, for any choice of x, we have $x \notin \varnothing$.

For example, we can define $\varnothing = \{x \mid x \neq x\}$. It is useful to have such a set. For example, suppose we wish to find all words in the dictionary in which the letter q is not followed by the letter u. We can let Σ be the set of such words. If it turns out that there is no such word in the dictionary, we have $\Sigma = \varnothing$. Thus $\Sigma = \varnothing$ is a way of saying that q is always followed by u in the dictionary. (We leave open the question of whether Σ is actually empty.) Because of the null set, we can be sure that a proper description of a set will definitely yield a set.

Set Operations

If A and B are sets, they determine two sets called the *union* and the *intersection* of A and B.

DEFINITION 1.1.13

The **union** of A and B, written $A \cup B$, is the set of elements that are in A or in B (or in both):

$$x \in A \cup B \quad \text{if and only if} \quad x \in A \text{ or } x \in B \tag{1.1.1}$$

Descriptively,

$$A \cup B = \{x \mid x \in A \text{ or } x \in B\}$$

The **intersection** of A and B, written $A \cap B$, is the set of elements that are in both A and B:

$$x \in A \cap B \quad \text{if and only if} \quad x \in A \text{ and } x \in B \qquad (1.1.2)$$

Equivalently,

$$A \cap B = \{x \mid x \in A \text{ and } x \in B\}$$

For example, suppose that

$$A = \{1, 3, 5, 9\} \quad \text{and} \quad B = \{0, 1, 2, 3\}$$

Then $A \cup B = \{0, 1, 2, 3, 5, 9\}$ and $A \cap B = \{1, 3\}$.

The key words in these definitions are *or* for the union, and *and* for the intersection.

EXAMPLE 1.1.14

Suppose that M is the set of names that are ordinarily used for men and, similarly, that W is the set of women's names. Then $M \cup W$ is the set of all names (men's or women's names), and $M \cap W$ is the set of all names that can be used as names for men and women. Times and customs change, but it appears that $M \cap W$ is not empty (Leslie is an element of it), but that $M \cup W \neq M$, since Mary is a member of $M \cup W$ but not of M.

In the same manner, we may define the union and intersection of three or more sets. The union of any number of sets is the set consisting of elements belonging to *at least one* of the sets; the intersection consists of elements belonging to *all* of the sets.

A graphic way of illustrating the union and the intersection of sets is by **Venn diagrams** as in Figure 1.1.1. The set A is regarded as all the points inside curve A, and similarly for the other sets. The intersection and union of sets are indicated by the shaded areas.

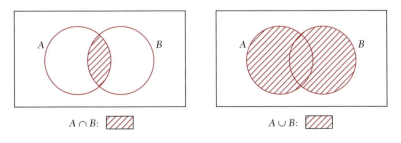

Figure 1.1.1

Addition of natural numbers has always been identified with the *union* of sets. Thus, if there are two apples and three pears, then there are five (2 + 3) pieces of fruit (the union of the apples and pears). But some care has to be taken if this idea is generalized to a union of arbitrary sets. For example, if there are 7 history books and 5 giant-sized books in a library, it would be wrong to assume that the library has 12 books that are either history books or giant-sized. It is possible that there may be some giant-sized history books, and we must take care to see that these sets of books do not overlap in our count. The idea of nonoverlapping sets is formalized in the following definition.

DEFINITION 1.1.15

A and B are said to be **disjoint** or **mutually exclusive** if

$$A \cap B = \varnothing$$

More generally, if $A_1, A_2, \ldots, A_n$ are sets, and

$$A_i \cap A_j = \varnothing \qquad \text{for } i \neq j$$

the sets $A_1, \ldots, A_n$ are said to be **pairwise disjoint.**

The general result on counting unions of sets is as follows:

THEOREM 1.1.16

$$|A \cup B| = |A| + |B|$$

provided A and B are disjoint. More generally,

$$|A_1 \cup \cdots \cup A_n| = |A_1| + \cdots + |A_n|$$

for pairwise disjoint sets A_i. ■

We can write this more succinctly by using the idea of a *partition*.

DEFINITION 1.1.17

Suppose a set U is divided into nonempty subsets $A_1, A_2, \ldots, A_n$ in such a way that

1. $U = A_1 \cup \cdots \cup A_n$.
2. The sets A_i are pairwise disjoint.

Then we say that the sets A_i **partition** the set U.

Remark   Partitioning a set U is a way of classifying its elements. For example, the country is partitioned into two sets—males and females.

Theorem 1.1.16 may then be restated: If A is partitioned into sets A_i, then $|A| = |A_1| + \cdots + |A_n|$.

For example, suppose a library has seven history books and five giant-sized books, and no history book is giant-sized. How many books are either history books or giant-sized? The answer is 12, but let's put it into the context of this theorem. Let H be the set of history books in a library, so $|H| = 7$. Let G be the set of giant-sized books in the library, so $|G| = 5$. Since no history book is giant-sized, G and H are disjoint ($G \cap H = \varnothing$), and so

$$|G \cup H| = |G| + |H| = 12$$

These results are generalized in Chapter 4.

What elements are *not* in $\mathbb{Z}_n$? If you said the integers greater than or equal to n, you gave a reasonable answer, but other answers are possible. For example, -2 is such an element. So is π. So is the letter F. The problem lies with the question itself. It is too broad to ask for *anything* that is not in a given set. What is usually done is to specify one set, called the **universe** or the **universal set**, from which all elements are to be chosen. Thus, if a discussion concerns strings on Σ, it would be stated or understood that the universe is the set Σ^*. With this understanding, the question of what elements are not in some set A depends on the universe. We often use U to designate the choice of universe.

DEFINITION 1.1.18

Let A be a set. The **complement** of A, written A', is defined as the set of all x in U such that $x \notin A$. The set $B - A$ is defined as the set of all elements in B but not in A. This set is also called the **complement** of A in B. Equivalently, $B - A = B \cap A'$.

For example, the complement of $\mathbb{Z}_{10}$ in $\mathbb{N}$ is the set of all integers greater than or equal to 10.

Figure 1.1.2 illustrates these definitions with the help of Venn diagrams.

One way of comparing sets is by *inclusion*.

DEFINITION 1.1.19

If every element of A is also an element of B, we say that A **is included in** B or that A is a **subset** of B, and write $A \subseteq B$.

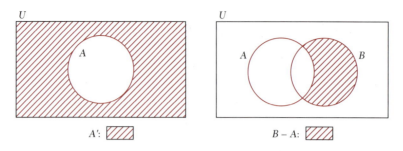

Figure 1.1.2

A basic property of the universal set U is that the sets under discussion are subsets of it: $A \subseteq U$. The definition of equality of sets A and B may now be stated as follows:

$$A = B \quad \text{if and only if} \quad A \subseteq B \text{ and } B \subseteq A$$

It is also true that $\emptyset \subseteq A$. Why is this so? Since there are no elements of $\emptyset$, it might be suspected that it makes no sense to say that all elements of $\emptyset$ are in A. For example, is it true that all words in the dictionary that begin with the letters XXXQRA necessarily end with the letter D? If you look it up, you will find that whenever you find a word beginning with XXXQRA, it will end with D! This is true because no cases have to be checked. In this illustration, $\emptyset$ is the set of words in the dictionary beginning with XXXQRA, and A is the set of words ending with D. It is common to say that the statement $\emptyset \subseteq A$ is **vacuously** true.

Figure 1.1.3 illustrates the relation $A \subseteq B$ with a Venn diagram.

Functions

DEFINITION 1.1.20

For sets A and B, a **function** f from A to B is a unique assignment of an element of B for each element a of A. We write $f: A \to B$ to make explicit the role of the sets A and B. The set A is called the **domain** of f, and B is called the **codomain** of the function. We write

$$b = f(a)$$

to indicate that the element a of A is assigned by f to the element b of B. When $b = f(a)$, a is called the **input** and b is the **output** for this input. The element b is also called the **image** of a. The set of outputs of a function f is called the **range** of f:

$$\text{Range of } f = \{f(x) \mid x \in A\}$$

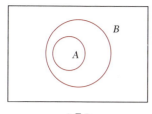

$A \subseteq B$

Figure 1.1.3

The codomain is a set in which $f(x)$ must necessarily land. The range is the set of landing places.

DEFINITION 1.1.21

Equality of functions is defined as follows. Let $f: A \to B$ and $g: C \to D$. Then we say that $f = g$ provided

1. $A = C$ and $B = D$.
2. $f(x) = g(x)$ for all $x \in A$.

Thus, the functions have the same domain and codomain, and the same output for each input.

The following are some examples of functions.

EXAMPLE 1.1.22 Length of a String

The length of a string is the number of letters in it, counting repetitions. This is a function **length:** $\Sigma^* \to \mathbb{N}$. This makes explicit that the length of a string is a function (here called length) from the strings on the non-empty set Σ and having the natural numbers $\mathbb{N}$ as possible values. If we take Σ = ALPH, we see that **length**("SWEETS") = 6. The range of the length function is $\mathbb{N}$.

EXAMPLE 1.1.23 Cardinality of a Set

We have used the notation $|A|$ for the cardinality of a set. To use the functional notation we introduced, let's write $|A|$ as $\#(A)$. What are the domain and codomain of $\#$? We won't say that the domain consists of *all* sets because this is a vague concept and is known to lead to logical difficulties. Instead, we take a finite universal set U and consider only subsets of U. The set of all subsets of U is called the **power set** of U and is denoted $P(U)$. The cardinality of any finite set is in $\mathbb{N}$. Thus, we have $\#: P(U) \to \mathbb{N}$.

If the set U is infinite, # has not been defined. One way of defining it is to adjoin one element, called ∞, to $\mathbb{N}$ and define the cardinality of an infinite set to be ∞. In this way, we have for any set U,

$$\#\colon P(U) \to \mathbb{N} \cup \{\infty\}$$

We shall continue to use the notation $|A|$. Informally, we speak of "the function $|A|$."

In Chapter 10 (Theorem 10.3.10), we show that there are different kinds of infinities and that, for example, the real numbers have more elements than the integers. The preceding method is relatively unsophisticated, since it does not distinguish between different infinite cardinal numbers.

EXAMPLE 1.1.24   ASCII Numbers

We stated that there are 128 standard characters, called ASCII characters. Each one is assigned a number (from 0 through 127), called its ASCII number. For example, "a" has ASCII number 97, "B" has ASCII number 66, and even " has an ASCII number, 34. (We avoided putting " in quotes, although we could have.) Thus, the assignment of ASCII numbers is a function from the ASCII characters to $\mathbb{Z}_{128}$. We may think of this as a simple coding device: Instead of using a symbol, we can replace it by a number. This is often done in programming languages.

EXAMPLE 1.1.25   Explicit Listing

Just as a set can be defined explicitly by listing its elements, a function can be defined explicitly by listing its values. For example, the following table gives an example of a function $f\colon \mathbb{I}_4 \to \mathbb{I}_4$:

a	1	2	3	4
$f(a)$	4	1	2	1

By observation, the range of f is $\{1, 2, 4\}$.

EXAMPLE 1.1.26   Definition by Formula

A function can also be defined by a formula or a description. The equation

$$f(x) = 1 + 3x^3 - x^4$$

tells how to define $f(x)$ for each real x. It therefore defines a function f: $\mathbb{R} \to \mathbb{R}$. In equations like this, the domain and the codomain are understood to be the real numbers $\mathbb{R}$. The calculation of the *range* of this function f is a problem of the calculus, and is not involved with the definition of the function.

DEFINITION 1.1.27 *Greatest Integer Function*

Another example of a function is the **greatest integer function**. If x is any real number, there is a unique integer n such that

$$n \le x < n + 1$$

This integer n is called the integer part of x, and we write $n = [x]$. Thus, $[x] \le x < [x] + 1$. For example, $[\sqrt{5}] = 2$, since $\sqrt{5}$ is between 2 and 3. The function $f\colon \mathbb{R} \to \mathbb{N}$ defined by $f(x) = [x]$ is called the greatest integer function.

EXAMPLE 1.1.28 *The Function n mod m*

If an integer n is divided by a positive integer m, we obtain the quotient q and remainder r, where r is nonnegative and less than m. Thus,

$$n = qm + r, \qquad 0 \le r < m \tag{1.1.3}$$

Here, $q = [m/n]$. To see this, divide this equation by m to obtain $n/m = q + r/m$. Since $0 \le r/m < 1$, we have

$$q \le n/m < q + 1$$

This shows that $q = [m/n]$ by Definition 1.1.27. By Equation (1.1.3) we find $r = n - qm$, and we have the formula for the remainder:

$$r = n - [n/m]m$$

This is often abbreviated $r = n \bmod m$. If we think of m as a fixed positive integer, the function f_m given by the formula

$$f_m(n) = n \bmod m$$

defines a function $f_m\colon \mathbb{Z} \to \mathbb{Z}_m$. The main reason for introducing the set $\mathbb{Z}_m$ is that it is the set of all possible remainders when integers are divided by m. Put another way, it is the range of the function $n \bmod m$.

EXAMPLE 1.1.29   Definition by Cases—Collatz's Function

A more complicated example of a descriptive definition of a function is the function $C: \mathbb{N} \rightarrow \mathbb{N}$ defined as follows:

$$C(n) = \begin{cases} n/2 & \text{if } n \text{ is even} \\ 3n + 1 & \text{if } n \text{ is odd} \end{cases}$$

This function is well defined because it gives an unambiguous value of $C(n)$ for each $n \in \mathbb{N}$.

The function $C(n)$ of Example 1.1.29 is of some interest because it leads to a well-known problem that has not been solved as of this writing. Start with any $n > 0$ in $\mathbb{N}$ and find $C(n)$. Then apply C to the answer, finding $C(C(n))$, and keep applying C in this way. The conjecture is that eventually the answer becomes 1. Thus, starting with 6, we find $C(6) = 3$, $C(3) = 10$, $C(10) = 5$, $C(5) = 16$, and so on through the values 8, 4, 2, 1. This conjecture has been verified for many values of n, but no proof or counterexample has yet been found. The problem is attributed to Collatz, hence the name of the function.

A **function of two variables** is one that takes two inputs and yields one output. A simple example is addition. There is a simple logical device that reduces this kind of function to a function of one variable. The device is to think of the two variables as *one* entity.

When analytic geometry was introduced, Descartes showed how the plane could be regarded as the set of points (x, y), where x and y are real numbers. In this way, *two* independent numbers were treated as *one* object (representing a point in the plane). We now generalize this.

DEFINITION 1.1.30

An ordered couple, or an ordered pair, (a, b), is a listing of two elements in a definite order. The set of *all* ordered couples (a, b), where $a \in A$ and $b \in B$, is called the **Cartesian product** of A and B and it is written $A \times B$.

Thus, in analytic geometry, the plane is $\mathbb{R} \times \mathbb{R}$.

The elements of $A \times B$ can be listed in a table, provided A and B are finite. Merely put the elements of A in the first column and the elements of B in the top row. The body of the table then consists of the elements of $A \times B$. The element (a, b) is located in "row of A, column of B." This is done in Figure 1.1.4, where two points are singled out in the figure.

Definition 1.1.30 can be generalized to a listing of n elements.

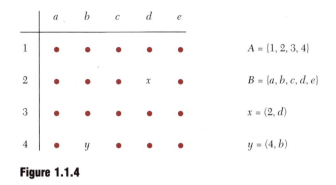

$A = \{1, 2, 3, 4\}$

$B = \{a, b, c, d, e\}$

$x = (2, d)$

$y = (4, b)$

Figure 1.1.4

DEFINITION 1.1.31

An ordered n-tuple $(a_1, a_2, \ldots, a_n)$ is a listing of n elements in a definite order. The set of *all* elements $(a_1, a_2, \ldots, a_n)$, where $a_i \in A_i$, is called the **Cartesian product** of the sets $A_1, \ldots, A_n$ and is written $A_1 \times A_2 \times \cdots \times A_n$.

If $(a_1, a_2, \ldots, a_n)$ is an ordered n-tuple, the element a_i is called its ith component.

Thus, the definition of a function is broad enough to include functions of two or more variables. For example, if $f(x, y) = x^2 + y^2$ (where x and y are real numbers), we may think of f as a function $f \colon \mathbb{R} \times \mathbb{R} \to \mathbb{R}$. The value $f(x, y)$, ordinarily thought of as depending on *two* numbers x and y, can now be regarded as depending on the *one* quantity (x, y) in $\mathbb{R} \times \mathbb{R}$. In general, a function of $f(x, y)$ with values in the set C of two variables is defined as a function $f \colon A \times B \to C$. This definition makes clear where the variables x and y are to be chosen from (A and B, respectively) and in what set the value $f(x, y)$ lies. A function of more than two values may be defined similarly.

An **operation** on a set A is a function $f \colon A \times A \to A$. Some familiar examples are addition, subtraction, and multiplication on $\mathbb{Z}$. Note that division is not strictly an operation on the reals $\mathbb{R}$, since the function given by $f(x, y) = x/y$ is not defined for $y = 0$. Often, operations are not written in strict functional notation. We do not usually use the functional notation $\mathrm{DIFF}(x, y)$ for the difference function. Rather, we use the familiar algebraic notation $x - y$, and similarly for the product and sum.

EXAMPLE 1.1.32   Prefix, Postfix, and Infix Notation

Our usual functional notation is called **prefix** notation, since the operator usually precedes the variables. When the operator is placed between the variables, as in addition and subtraction, the notation is called **infix** notation. Similarly, **postfix** operation is used if the operator is placed after the variables. A familiar example of postfix notation (for a function of one variable) occurs in the use of the percentage symbol %. When we write 50%, % is the operation of dividing by 100, and custom demands that we do not write %50 or %(50) in this case.

Perhaps a less familiar operation is **concatenation** defined on the strings Σ^* on a set Σ. This operation, denoted by $*$, adjoins one string to another. Thus, "NO"$*$"THING" = "NOTHING". Infix notation is generally used for this.

Sequences

A *sequence* is an infinite listing of elements. We can frame the definition in terms of functions, as follows.

DEFINITION 1.1.33

Let A be any set. A **sequence** of elements of A is a function $f: \mathbb{N} \to A$. The element $f(n)$ is called the nth term of the sequence.

The idea is that the nth term of the sequence is a function evaluated at n. Sequences are also called **infinite sequences.**

A sequence is often given informally by listing the first few of its terms. Thus, we might say "Let 0, 1, 4, 9, 16, . . . be the sequence of squares." Since the sequence continues for all n, an explicit listing is not possible, but knowing the first few terms is often helpful. Often the expression for $f(n)$ is given as the "general term," as in the sequence

$$1, 2, 5, 10, \ldots, n^2 + 1, \ldots$$

This is the same as the sequence f given by the function $f(n) = n^2 + 1$, but it gives a better feeling for a sequence if we list its first few elements.

The variable n is often written as a *subscript*. For example, the sequence $a_n = n^2 + 1$ is the same as 1, 2, 5, 10, . . . , $n^2 + 1$, ... Occasionally, it is convenient to start a sequence with $n = 1$ rather than $n = 0$. In this case, we understand that such a sequence is a function $f: \mathbb{N}^+ \to A$, where $\mathbb{N}^+ = \mathbb{N} - \{0\} = \{1, 2, 3, \ldots\}$ = the set of positive integers. In a similar way, we can consider sequences a_n, defined for all integers $n \geq k$, for some fixed k.

DEFINITION 1.1.34

A **finite sequence** of elements of A is a function $f: \mathbb{I}_k \to A$. The number k is called the length of the finite sequence.

Remark   By convention, finite sequences usually start with a first term, not a zeroth term as for infinite sequences.

If k is small, a finite sequence can be listed explicitly. For example, the terms

$$2, 5, 10, 17$$

are a listing of the finite sequence $f: I_4 \to \mathbb{Z}$ of length 4, which can be given by the formula $f(n) = n^2 + 1$. It is not necessary that a formula be given. For example, the listing

$$0, 0, 11, 9, 0, -6, 9, 23$$

is a finite sequence f in which, for example, $f(3) = 11$. (Here, $k = 8$.) In usage, it is indistinguishable from the ordered 8-tuple $(0, 0, 11, 9, 0, -6, 9, 23)$.

Logically speaking, there is no distinction between (a) a finite sequence $f: I_n \to A$, (b) an ordered n-tuple with entries in A, and (c) a string of length n in A^*. Each of these terms is used, depending on the usage desired. For example, strings would usually be used for text manipulation. Sequences of length n are often thought of as the first n terms of an infinite sequence. Ordered n-tuples are a convenient way of considering n objects at one time. In analytic geometry, for example, a **point** is defined as an ordered couple of real numbers; we would ordinarily not consider it as a string of reals of length 2 or a finite sequence of reals of length 2. In some programming languages, finite sequences are called *one-dimensional arrays*.

DEFINITION 1.1.35 *Matrices*

If m and n are positive integers, we define a **matrix** A as a function A from $I_m \times I_n \to \mathbb{R}$. The value $A(i, j)$ is also written a_{ij}. Here, $1 \le i \le m$ and $1 \le j \le n$. The function A is also called an **m by n matrix.**

We can visualize the numbers a_{ij} as a rectangular array of numbers with m rows and n columns. Simply put the number a_{ij} in the ith row and jth column of the rectangle. For example, suppose

$$a_{11} = 0, \quad a_{12} = 3, \quad a_{13} = 7$$
$$a_{21} = 1, \quad a_{22} = 4, \quad a_{23} = 5$$

Then the matrix A is a 2 by 3 matrix, and we can put the numbers a_{ij} in a rectangular array:

$$A = \begin{bmatrix} 0 & 3 & 7 \\ 1 & 4 & 5 \end{bmatrix}$$

Matrices play an important role in mathematics. In the next chapter they are used to represent relations. In Chapter 5, they are used to analyze graphs. In any computing applications of graphs or relations, matrices show up, since they give a numerical representation of these topics that can be input into a computer and subsequently analyzed. The appendix gives a brief summary of those properties of matrices used in this text.

Characteristic Functions

It is possible to code subsets A of a finite set U as strings of bits—that is, as a string of 0's and 1's. We first illustrate with an example.

EXAMPLE 1.1.36

Let the universal set U be $I_5 = \{1, 2, 3, 4, 5\}$. The idea is to use 0 and 1 to denote the status of any element of U, where 0 represents a nonelement of A, and 1 represents an element of A. Thus, the string 01011 is the code name for the set $\{2, 4, 5\}$, since 1 appears in the second, third, and fifth places. In general, we let the nth digit of the string give the status of the integer n: 1 indicates that it is an element of the given set, 0 that it is not. The string 11111 is the code for the universal set I_5, and 00000 the code for the null set. A set with four elements will be coded as a string with exactly four 1's. In brief, the string gives all the information about the subset of U.

To generalize for any subset A of some (possibly infinite) universal set U, we consider any element x of U and give its status relative to A: 1 if it is in A, and 0 if it is not. The result is called the *characteristic function* of the set A.

DEFINITION 1.1.37

Let A be a subset of U. The **characteristic function** of A, denoted f_A, is a function

$$f_A: U \to \mathbb{Z}_2$$

such that $f_A(x) = 1$ if $x \in A$, and $f_A(x) = 0$ when $x \notin A$.

Any function $g: U \to \mathbb{Z}_2$ is the characteristic function of exactly one set $A \subseteq U$. In fact, if $A = \{x \mid g(x) = 1\}$, we have $g = g_A$.

Remark   This is a useful idea in programming and in algorithm constructions. In Example 1.1.36 the set $\{2, 4, 5\}$ was represented by the sequence 0, 1, 0, 1, 1, which is its characteristic function. There are practical advantages to this representation. (1) If n is large and we are representing a subset A of I_n by its characteristic function f_A, we have a sequence of fixed length n. The list of its elements has variable length. (2) To discover whether an element i is in the set A, we need only check whether $f_A(i) = 1$. If we had a list for A, we would have to search through the list.

Another use is as follows. Suppose one wanted to program a prime number tester. This means that a program would take a number n as input and give an output, yes or no, as to whether it is a prime. Now replace yes with 1, and no with 0. The prime number tester is then a *function* from the positive integers to $\mathbb{Z}_2$ with a value 1 if and only if the input is a prime. The prime tester function is thus the characteristic function of the set P of prime numbers.

EXAMPLE 1.1.38

Let the universe consist of all the bit strings of length 3:

$$U = \{000, 001, 010, 011, 100, 101, 110, 111\}$$

Let A be the set consisting of the strings that end with 1. Find the characteristic function f_A, and express that function as a string of length 8.

Method   Simply make a table of values for f_A as indicated. If we order the elements of U as indicated, the associated string for A is the last line of the table, namely 01010101.

x	000	001	010	011	100	101	110	111
$f_A(x)$	0	1	0	1	0	1	0	1

Exercises

1. Let $U = \mathbb{I}_{10}$. Let $A = \{1, 3, 5, 7, 9\}$, $B = \{1, 2, 3\}$, and $C = \{5, 6, 7, 8, 9, 10\}$. List the elements of
   (a) $A \cap C'$ (c) $(A \cup B) \cap C'$
   (b) $A' \cup B$ (d) $A \cap B \cap C'$
2. Let $U = \mathbb{I}_{20}$. Define the sets $A = \{x \mid x \le 10\}$, $B = \{y \mid y < 5\}$, and $C = \{3x \mid x \le 6\}$.
   (a) List the elements of A, B, and C.
   (b) Find $|A \cap B'|$.
   (c) Find $|A \cap B \cap C|$.
   (d) List the elements of $(A \cap B) \cup (B \cap C) \cup (C \cap A)$.
3. Draw a Venn diagram for
   (a) $A \cap B \cap C'$ (b) $A' \cup (B \cap C)$
4. Draw a Venn diagram for
   (a) $A' \cup (B \cap C')$ (b) $(X' \cap Y) \cup (Y' \cap Z)$

5. Let E = the set of positive even integers; let T = the set of all three-digit positive integers (whose first digit is not 0); let S = the set of all integers smaller than 50; and finally let the universe be $\mathbb{Z}$. Describe each of the following sets.
   (a) $S \cap T$ (c) $T \cup S$
   (b) $(E \cup S) \cap T$ (d) $E \cap T'$

6. Let the universe be the set of cars in the showroom. Let C denote the set of sports cars, let R denote the red cars, and let E denote the expensive ones. Describe each of these sets in terms of C, R, and E.
   (a) the inexpensive red sports cars
   (b) cars that are neither red nor expensive
   (c) sports cars that are not red but are expensive

7. Let O denote the odd integers, and let H denote the integers greater than 100. Let the universe be $\mathbb{N}$. Using the set operations defined in the text, describe each of these sets in terms of O and H.
   (a) the even integers
   (b) the odd integers that are 100 or less
   (c) the integers that are either even or greater than 100

*8. Let A and B be finite sets. Suppose $|A \cap B| = 45$, $|A| = 80$, and $|A \cup B| = 120$. Find $|B|$. (*Hint:* Draw a Venn diagram.)

*9. Let A and B be sets in a finite universe U. Suppose that $|U| = 150$, $|A| = 80$, $|B| = 85$, and $|A \cap B| = 30$. How many elements are neither in A nor in B? (*Hint:* Draw a Venn diagram.)

*10. Let $\mathbb{N}_i$ be the integers divisible by i. (For example, $\mathbb{N}_2$ = the even integers.) Give a simple description of the sets $\mathbb{N}_2 \cap \mathbb{N}_3$, $\mathbb{N}_4 \cap \mathbb{N}_8$, and $\mathbb{N}_5' \cup \mathbb{N}_{10}$.

*11. In the previous exercise, what conditions on i and j make $\mathbb{N}_i \subseteq \mathbb{N}_j$?

*12. (a) How many integers are there between 1 and 1000 inclusive that are divisible by 3?
    (b) How many integers are there between 1 and 1000 inclusive that are divisible by 7?
    (c) How many integers are there between 1 and 1000 inclusive that are divisible by both 3 and 7? (This is the same as being divisible by 21.)
    (d) How many integers between 1 and 1000 are neither divisible by 3 nor by 7? [*Hint:* Let A_3 be the set of integers divisible by 3, where $U = \mathbb{I}_{1000}$ is the universal set. Similarly, define A_7. Now draw a Venn diagram and use your results in (a)–(c).]

13. Suppose the alphabet A is chosen as the set $\{1\}$. Describe A^*. If x is a word in this alphabet and length(x) = 5, what is x?

14. Let the universe be $\mathbb{Z}_2^*$. Let $A = \{x \mid \text{length}(x) = 4\}$. Give an explicit listing of A.

*15. Using the $\{X \mid Y\}$ notation, describe each of the following sets. In each case, give the universal set.
   (a) the set of real numbers between 0 and 1, inclusive

     (b) the set of rational numbers that, in lowest terms, have an even denominator

     (c) the set of subsets of U that have an even number of elements (take U to be finite)

16. Let $A = \{1, 2, 3\}$ and $B = \{a, b, c, d\}$. Let the universe U be the set $A \times B$.

     (a) List the elements of $A \times B$.

     (b) Let $C = \{(x, y) \mid x = 3\} \cup \{(x, y) \mid y = b\}$. List the elements of C.

     (c) Draw a diagram of C similar to Figure 1.1.4 and indicate the set C on your drawing.

17. Let $f: \mathbb{N} \to \mathbb{N}$ according to the formula $f(n) = n + 1$. Further, let $g: \mathbb{N} \to \mathbb{N}$ according to the formula $g(n) = 3n$. Compute $f(f(n))$, $f(g(n))$, $g(f(n))$, and $g(g(n))$. What are the ranges of f and g?

18. Let $g: \mathbb{N} \to \mathbb{N}$ with $g(n) = 2n$, and let $h: \mathbb{N} \to \mathbb{N}$ with $h(n) = [n/2]$, where $[x] = $ greatest integer in x. Is $g(h(n)) = n$? Is $h(g(n)) = n$? Explain.

19. Let $f: \mathbb{Z}_7 \to \mathbb{Z}_7$ according to the formula $f(n) = (3n + 2) \bmod 7$. Write a table listing all values of n in $\mathbb{Z}_7$ together with corresponding values $f(n)$.

20. Let $f: \mathbb{Z}_5 \to \mathbb{Z}_5$ according to the formula $f(n) = (n^2 + 3) \bmod 5$. Write a table listing all values of n in $\mathbb{Z}_5$ together with corresponding values $f(n)$.

21. A sequence a_n is given by the formula $a_n = 2^n - 1$.

     (a) Write the first six terms of the sequence.

     (b) Let $b_n = a_{2n}$. Give a formula for b_n and write the first six terms of the sequence b_n.

     (c) Let $c_n = a_{n+2}$. Give a formula for c_n and write the first six terms of the sequence c_n.

     (d) Explain how the sequences defined in parts (b) and (c) are related to the original sequence a_n.

22. The operation $+$ is usually written in infix notation. Thus, we write $a + b$ for the sum of a and b. Suppose we use prefix notation and write $\mathrm{SUM}(a, b)$ for the sum of a and b. Similarly, let us write $\mathrm{PROD}(a, b)$ for the product of a and b.

     (a) Write the associative law $[(a + b) + c = a + (b + c)]$ using prefix notation.

     (b) Write the distributive law $[a(b + c) = ab + ac]$ using prefix notation.

23. The exponential function a^b is written in the infix notation $a \hat{} b$ in some computer languages. Write this function in prefix notation as $\mathrm{EXP}(a, b)$.

     (a) Write the formula $a^b a^c = a^{b+c}$ in prefix notation. (Use the notation introduced in the previous exercise.)

     (b) Repeat part (a) for the formula $(xy)^z = x^z y^z$.

24. Let $E = \mathbb{R} \times \mathbb{R}$ be the Euclidean plane. Define $f: E \to E$ by the formula $f(x, y) = (x, 0)$. Interpret this function geometrically. Show that $f(f(P)) = f(P)$.

25. Let $E = \mathbb{R} \times \mathbb{R}$ be the Euclidean plane. Define $g: E \to E$ by the formula $g(x, y) = (y, x)$. Interpret $g(P)$ geometrically. Show that $g(g(P)) = P$. Which points P satisfy the equation $g(P) = P$?

*26. Let $E = \mathbb{R} \times \mathbb{R}$ be the Euclidean plane. Define $h: E \to E$ by the formula $h(x, y) = (x + y, x - y)$. Find $h(h(x, y))$. If $h(x, y) = (1, 3)$, what is (x, y)?

27. Explain how the power set $P(\mathbb{I}_{10})$ is related to the strings of bits of length 10.

*28. (*For calculus students only.*) The text mentioned the function $f: \mathbb{R} \to \mathbb{R}$ defined by $f(x) = 1 + 3x^3 - x^4$. What is its range?

29. Let C be the Collatz function of Example 1.1.29. If $a_0 = 11$, find $a_1 = C(a_0)$, then $a_2 = C(a_1)$, then $a_3 = C(a_2)$, and so on. Continue until you arrive at the result 1. How many computations of $C(x)$ were made?

30. Repeat the previous exercise for $a_0 = 27$. (Don't give up!)

31. Let A be the integers of I_{20} that are either even or multiples of 3. Find the characteristic function of A, expressing it as either a sequence or a string.

32. Let A be the set $\{1, 4, 9, 16, 25\}$ of squares in $\mathbb{I}_{25}$. Find the characteristic function of A. Express it as a string of length 25.

33. Let $U = (\mathbb{Z}_2)^4$. Let A be the set of strings in U that have exactly two 0's and two 1's. Write the characteristic function of A as a string. (*Hint:* First list the elements of U in a definite order. The standard order is lexicographic: 0 precedes 1. This is similar to alphabetic order. Thus, 1001 precedes 1100.)

*34. Suppose A and B are sets (subsets of U) and f_A and f_B are their characteristic functions.
    (a) If $A \subseteq B$, what can you say about f_A and f_B?
    (b) What is the characteristic function of $A \cap B$ in terms of the characteristic functions f_A and f_B?
    (c) Similarly, what is the characteristic function of $A \cup B$?

35. Let the universal set be $U = \mathbb{I}_{10}$. Suppose the characteristic functions of the sets A, B, and C are given as the following strings: A: 0101010101; B: 1001000010; C: 1111101111.
    (a) Give a descriptive definition of each of these sets.
    (b) Give an explicit description of each of these sets.

36. A 3 by 3 matrix A has entries a_{ij} given by the formula
$$a_{ij} = i^2 - j$$
    What is the matrix A? (Write it out as a square array.)

37. A 4 by 4 matrix has entries $a_{ij} = i + j - 2 \bmod 4$. List this matrix as a square array of numbers.

1.2 The Logical Connectives

Logic is the study of reasoning. This includes the study of statements, how statements are constructed and combined, and whether the statements are true or false. By logic, it is possible to show that some statements are true once other statements are known to be true. Of course, people reasoned before there was a subject called logic. But with the introduction of logic as a subject of study, it was possible to fully analyze and classify the components involved in reasoning and to discover new techniques. In this section, we concentrate on how compound statements are constructed from simpler ones and on how to analyze the truth or falsity of compound statements.

We use the terms "statement" and "proposition" interchangeably. A characteristic of a statement is that it is either true or false, not both. These can be mathematical statements (like $2 + 2 = 4$ or $3 \times 5 = 35$) or nonmathematical ("John is taller than Mary" or "In 1946, there were 48 states"). Here are some further examples.

EXAMPLE 1.2.1

The following are statements.

a. $3 = 4 + 4$.
b. Either 8,131,432 is divisible by 17 or it is not divisible by 19.
c. Alice will get sleepy if she drinks wine.
d. Any integer greater than 100 is not less than 50.
e. Seventeen is a prime number and 15 is not a prime number.
f. If you wash your clothes with SqueekyClean, your clothes will all turn gray or they will disintegrate.

EXAMPLE 1.2.2

The following are *not* statements.

a. $5 =$
b. Go away.
c. Who is there?
d. A rose is a rose is a rose.
e. Good heavens!

The following are not strictly propositions, but can be made so with clarification. We first present them and then discuss what is needed for clarification.

EXAMPLE 1.2.3

a. $x = y$.
b. She is smart in mathematics.
c. If $x = 3$, then $x^2 = 9$.
d. If $ax = ay$, then $x = y$.
e. $x = y$ or $x = 3$.
f. If $x = 0$, then it is not cold outside now.

In Examples 1.2.3a and 1.2.3e we cannot state whether the "statements" are true or false since the values of the variables x and y are unknown to us or have not yet been given. However, once x and y have been assigned a numerical value, we can determine their truth or falsity. In Example 1.2.3b we do not know who "she" is. It is possible that "she" has been "assigned a value." For example, consider the conversation: "Betty Hoskins is in the biology class. She is smart in mathematics." Here "she" has been implicitly assigned a value. All that is now needed to make Example 1.2.3b a proposition is to give a precise criterion for being smart in mathematics.

Example 1.2.3c would appear to be unambiguously true. Still we need to know that x has been assigned a value. Further, can we say that the statement "If Alice = 3, then Alice^2 = 9" is true? What about the statement "If 4 = 3, then $4^2 = 9$"? We shall see in Example 1.2.5 that if x is given any numerical value, Example 1.2.3c will be a true statement. Example 1.2.3d becomes a statement if a, x, and y are given numerical (say real) values. It is not necessarily true in this case. Do you see why? Example 1.2.3f has the same difficulties with unassigned values of a variable. But even if x is assigned a numerical value, and we have a criterion for "coldness outside," is it proper to call "If 5 = 0, then it is not cold outside now" a statement? If so, is it true or is it false? We shall see that this is a true statement, regardless of the criterion for coldness.

We informally take the approach that examples such as 1.2.3a–f are statements. That is, we assume that the variables (such as x, y, she, and he) have been properly assigned and that the various possible ambiguities (coldness, smartness) have been properly defined.

The words "and" and "or" are called **connectives**, since they are used to **connect** statements to form compound statements. Thus, the statement "$x = 2$ and $y = 3$" is formed using the connective "and" to join the two simpler statements "$x = 2$" and "$y = 3$". Another example of a connective is in the statement "If $x = 3$, then $x^2 = 9$". This is an example of an "If . . . then . . ." statement, in which the simple statements "$x = 3$" and "$x^2 = 9$" are connected as a compound statement. The statement $x \neq 3$ may also be regarded as the compound statement not($x = 3$). One of the inherent difficulties in the study of these connectives is that English, the very language we speak, itself uses these connectives freely. (For example, toward the beginning of this paragraph, we spoke of connecting the two statements "$x = 2$" *and* "$y = 3$".) The traditional way out is to use special symbols to analyze these terms. Thus, $\neg$ is used to denote "not," $\vee$ to denote "or," and so on, as we now describe.

The reader should look through Examples 1.2.1 and 1.2.3 to find various examples of these connectives. The "If . . . then . . ." connective appears in Example 1.2.1c and 1.2.1f, as well as 1.2.3c–d. In Example 1.2.1c, the statement "If p then q" is phrased as "q if p." The "or," "and," and "not" connectives similarly appear in these examples.

In our study of mathematical logic (which is called the **propositional calculus**), we use the letters p, q, r, . . . to denote statements (which can be true or false). The symbols $\neg$, $\vee$, $\wedge$, $\rightarrow$, and $\leftrightarrow$ are used to represent *not, or, and, if . . . then . . .*, and *if and only if*, respectively. The symbol $\rightarrow$ is also read *implies*, and $\leftrightarrow$ is also read *is equivalent to*. The usage is described in Table 1.2.1.

Table 1.2.1   The Logical Connectives

Symbol	English	Usage	Read
$\neg$	Not	$\neg p$	Not p
$\wedge$	And	$p \wedge q$	p and q
$\vee$	Or (and/or)	$p \vee q$	p or q
$\rightarrow$	Implies (if . . . then . . .)	$p \rightarrow q$	p implies q (if p then q)
$\leftrightarrow$	If and only if (is equivalent to)	$p \leftrightarrow q$	p if and only if q (p is equivalent to q)

The following example illustrates this notation with ordinary English.

EXAMPLE 1.2.4

Let p and q be the following statements.

p: It is clear outside.

q: The sun is shining.

Here, $p \wedge q$ represents the statement "It is clear outside, and the sun is shining."

Also, $p \rightarrow q$ represents the statement "If it is clear outside, then the sun is shining," or, equivalently, "The sun is shining if it is clear outside."

$q \rightarrow p$ represents the statement "If the sun is shining, then it is clear outside."

$\neg p$ represents the statement "It is not clear outside."

$\neg p \wedge \neg q$ represents the statement "The sun is not shining, and it is not clear outside."

All of the connectives with the exception of ¬ are **binary** connectives because they join two statements. The connective ¬ is called **unary** because it operates on only one statement.

Other notations sometimes used are & for ∧, ⊃ for →, and ~ for ¬.

We can use several of the connectives to build up more and more complicated statements. For example,

$$(\neg p) \rightarrow ((\neg q) \lor r)$$

would be loosely translated: "If not p, then either not q or r." In this formula, the statements p, q, and r are called the **atomic** statements, since they are the building blocks used to form it. The entire statement is called a **compound** statement.

Just as in algebra, care must be taken with parentheses to indicate which operations take precedence. For example, the compound statement $(p \lor q) \land r$ has a different meaning than $p \lor (q \land r)$. A loose translation of the first is "Either p or q, and also r." The second is "Either p, or both q and r."

In algebra, we learned to omit parentheses in order to aid readability. For example, we almost always write $-x + y$ instead of $(-x) + y$. If we intend to do the addition before taking the negative, we would use parentheses: $-(x + y)$. We say that the *negative symbol has priority over the + symbol.* Similarly, $x + yz$ is understood to mean $x + (yz)$ and not $(x + y)z$. Thus *multiplication has priority over addition.*

The following are standard conventions regarding the dropping of parentheses in propositional calculus.

1. Outer parentheses may be dropped. Thus $(p \land q)$ is simply written $p \land q$.
2. The symbol ¬ has first priority. It applies to as little as possible. Thus, $\neg p \rightarrow q$ stands for $(\neg p) \rightarrow q$ rather than $\neg (p \rightarrow q)$.
3. After the priority of ¬, ∧ has next priority over the other connectives. For example,

$$\neg p \lor q \land r$$

stands for

$$(\neg p) \lor (q \land r)$$

4. After the priorities of 2 and 3, ∨ has the next priority over the other connectives. Thus, the expression $p \lor \neg q \rightarrow r$ is an abbreviation for $(p \lor (\neg q)) \rightarrow r$
5. If the only binary connective is ∧, then the first occurrence of this symbol (to the left) receives the first priority, the second occurrence receives the next priority, and so on. For example, $p \land q \land r \land s$ stands for

$$((p \land q) \land r) \land s$$

A similar rule applies for $\vee$ and the other binary connectives. A more formal analysis of the structure of compound statements is given in Section 9.1.

Truth Tables

A truth table is a table that shows how to determine whether a compound statement is true, depending on whether the component parts of it are true. In this chapter, we use T and F to denote "true" and "false." Sometimes, true and false are used without any abbreviations. In later chapters, we use 1 and 0 to good advantage. If a statement p is true, we say that **its truth value is** T. If it is false, we say that **its truth value is** F.

Tables 1.2.2–1.2.6 follow ordinary mathematical and English usage. For example, the only way that "p and q" is true is for both p to be true and q to be true. This is reflected in Table 1.2.2. These tables give every possible combination of truth values for the statements p and q and for the compound statement.

Table 1.2.2

p	q	$p \wedge q$
F	F	F
F	T	F
T	F	F
T	T	T

Table 1.2.3

p	q	$p \vee q$
F	F	F
F	T	T
T	F	T
T	T	T

Table 1.2.4

p	q	$p \rightarrow q$
F	F	T
F	T	T
T	F	F
T	T	T

Although these tables reflect standard English usage, there are some differences that should be pointed out.

In the OR Table 1.2.3, the last line shows that $p \vee q$ is true when both p and q are true. English is sometimes ambiguous on this matter. For example, consider the statement "John is a brilliant student, or he cheated on that exam." In some contexts, this statement might be understood to mean that unless John is brilliant, he cheated. However, by Table 1.2.3, this statement would be true if John were brilliant and if he did cheat on that exam. Whereas English usage is sometimes unclear, logical usage is simply given by the table. All programming languages use Table 1.2.3 to define the truth value of an OR expression. This type

Table 1.2.5

p	q	$p \leftrightarrow q$
F	F	T
F	T	F
T	F	F
T	T	T

Table 1.2.6

p	$\neg p$
F	T
T	F

of OR is called the **inclusive OR**. In English, the term "and/or" is frequently used when the inclusive OR is explicitly desired. An **exclusive OR** is a type of OR in which p or q is understood to be *false* if both p and q are true. The exclusive OR is used in computer languages and is usually called XOR. It is used and discussed in some detail in Section 7.2 (Boolean algebra).

Table 1.2.4 takes us even further from English usage. In English, we usually expect a causal relationship between p and q in a "$p \rightarrow q$" statement. Thus, the statement "If John got 100 on that exam, then he cheated" would be an acceptable statement (possible true or possible false), but a statement such as "If today is sunny, then $4 = 3$" would be questionable as to meaning. In logic, however, any two statements may be connected by an implication sign, and the truth value of the compound statement will be governed by Table 1.2.4. The truth values given in the table are not arbitrary, but are completely consistent with mathematical and computer usage. This table tells us that "If $1 = 4$, then $8 = 8$" is a true statement, since the statements "$1 = 4$" and "$8 = 8$" are false and true respectively. We illustrate with an example.

EXAMPLE 1.2.5

Consider the mathematical statement

$$\text{If } x = 3, \text{ then } x^2 = 9 \qquad (1.2.1)$$

Then we claim that Equation (1.2.1) is true for *all* real values of x. We take three typical values to illustrate the truth of Equation (1.2.1).

$$x = \quad 3: \quad \text{If } \quad 3 = 3, \text{ then } 9 = 9 \qquad (1.2.2)$$

$$x = -3: \quad \text{If } -3 = 3, \text{ then } 9 = 9 \qquad (1.2.3)$$

$$x = \quad 1: \quad \text{If } \quad 1 = 3, \text{ then } 1 = 9 \qquad (1.2.4)$$

Note that Equations (1.2.2)–(1.2.4) are true by lines 4, 2, and 1 respectively in Table 1.2.4. This is a very convenient table, since it makes Equation (1.2.1) true all the time, and we don't have to worry about the value of x.

But the converse of Equation (1.2.1)

$$\text{If } x^2 = 9, \text{ then } x = 3 \qquad (1.2.5)$$

is not true for all values of x. A *counterexample* is the value $x = -3$. For this yields the statement

$$\text{If } 9 = 9, \text{ then } -3 = 3 \qquad (1.2.6)$$

Line 3 of Table 1.2.4 shows that this is a false statement.

Table 1.2.4 shows that each of the following statements is true.

A false hypothesis implies any statement.

Any statement implies a true conclusion.

The implication $p \rightarrow q$ is false only when p is true and q is false.

We can illustrate the truth tables by considering some of the definitions and results about sets in Section 1.1.

EXAMPLE 1.2.6

We stated that the set $A \cup B$ was the set of elements in A or in B. Thus,

$$x \in A \cup B \leftrightarrow (x \in A) \vee (x \in B)$$

The symbol $\leftrightarrow$ is often used in definitions. Here, its use shows that x is in the union *if and only if* x is in A or x is in B. Similarly, the other set definitions may be stated by using logical connectives:

$$x \in A \cap B \leftrightarrow (x \in A) \wedge (x \in B)$$

and

$$x \in A' \leftrightarrow \neg(x \in A)$$

We have

$$A \subseteq B \leftrightarrow [(x \in A) \rightarrow (x \in B)]$$

All of these statements are true for each x in the universe. The inclusion $\emptyset \subseteq A$ can now be restated as

$$x \in \emptyset \rightarrow x \in A$$

This latter statement is true for all x, since the statement $x \in \emptyset$ is always false, by the definition of the null set. In Section 1.1 we said this was "vacuously true." Now we can say that it follows from the definition of inclusion.

Truth Tables of Compound Sentences

Truth tables can be used to determine the truth values of compound statements. This is done by using Tables 1.2.2–1.2.6 over and over. We illustrate in the following example.

EXAMPLE 1.2.7

Find the truth table for the compound statement

$$(\neg p) \to (q \lor p)$$

Method We first assign all possible truth values to p and q, as in the following table.

p	q	$(\neg p) \to (q \lor p)$
F	F	
F	T	
T	F	
T	T	

Now, using Tables 1.2.2–1.2.6, compute the truth values of the quantities in the innermost parentheses. This yields the partially completed table

p	q	$(\neg p)$	$\to$	$(q \lor p)$
F	F	T		F
F	T	T		T
T	F	F		T
T	T	F		T

Finally, based on the truth values so far obtained, fill in the rest of the table:

p	q	$(\neg p)$	$\to$	$(q \lor p)$
F	F	T	F	F
F	T	T	T	T
T	F	F	T	T
T	T	F	T	T

The next example shows how to find the truth table of a compound statement involving more than two atomic statements.

EXAMPLE 1.2.8

Find all truth values of p, q, and r for which the statement $(p \lor \neg q) \to (r \leftrightarrow p \land q)$ *is false*.

Method List all possible truth values of p, q, and r to start the table, and put in the truth values of the statements in the innermost parentheses (the truth value is best placed under the connective):

p	q	r	$(p \lor \neg q) \to (r \leftrightarrow p \land q)$	
F	F	F	T	F
F	F	T	T	F
F	T	F	F	F
F	T	T	F	F
T	F	F	T	F
T	F	T	T	F
T	T	F	T	F
T	T	T	T	F

Proceeding in this way, we progressively fill out the table. We leave it to the reader to work out the final Table 1.2.7.

By noting the F's in the column for the statement, we see that the required truth values for p, q, and r are (F, F, T), (T, F, T), and (T, T, F).

An indirect way of doing such computations is sometimes helpful. The given statement has the form $P \to Q$. Therefore, it can have the truth value F if and only if P and Q have truth values T and F, respectively. Thus, Q must be false and P true. Since Q is the statement $(r \leftrightarrow p \land q)$, we see that

$$r \text{ and } p \land q \text{ must have different truth values} \qquad (1.2.7)$$

On the other hand, P is the statement $(p \lor \neg q)$ and it is true. Therefore the only possible values for p and q are

$$(p, q)\colon (T, F), (T, T), (F, F)$$

Table 1.2.7

p	q	r	$(p$	$\lor$	$\neg q)$	$\to$	$(r$	$\leftrightarrow$	p	$\land$	$q)$
F	F	F		T	T	T		T		F	
F	F	T		T	T	F		F		F	
F	T	F		F	F	T		T		F	
F	T	T		F	F	T		F		F	
T	F	F		T	T	T		T		F	
T	F	T		T	T	F		F		F	
T	T	F		T	F	F		F		T	
T	T	T		T	F	T		T		T	

This fixes the value of r, by Equation (1.2.7). Thus

$$(p, q, r): (T, F, T), (T, T, F), (F, F, T)$$

are the only values that make the given statement false. This agrees with Table 1.2.7.

Exercises

In Exercises 1–8 write the statements in terms of simpler, atomic statements. Use the logical symbols $\wedge, \rightarrow$, and so on. Label your atomic statements p, q, and so on. Try to make the atomic statements as simple as possible.

1. If you take care, you won't get into trouble.
2. If I go, I'll either be sorry or sad, but if I don't go, I'll be sorry.
3. Either that jewelry is fake, or I'm a monkey's uncle.
4. The gelatin won't gel if it is too warm or has too much sugar.
5. If Jane takes the algebra course and it is taught by Mr. Smith, then Jane will fail that course.
6. If Jones is elected president, he will balance the budget provided Smith is not elected.
7. John is always happy, but if Peter watches too much television, he gets unhappy and must go for a long walk.
8. Either that ad is not telling the truth, or if I try this product I will not be able to stay awake at night.

In Exercises 9–14, x and y are understood to be *integers* (positive, negative, or zero). After writing these in terms of simple atomic sentences using the logical connectives, determine whether the statements are always true, always false, or sometimes true and sometimes false, depending on the values of the variables. Give reasons.

9. If $x < y$, then $x^2 < y^2$.
10. If $x < y$, then $x + 1 < y$ or $x + 1 = y$.
11. If $x < 0$ and $x > 0$, then x is even.
12. $x = 0$ or $x^2 = 1$ or $x^2 > 3$.
13. If $x^2 = 5$, then x is even.
*14. If x is even and y is not even, then $x + y$ is even.

In Exercises 15–20, x and y are understood to be *real numbers*. After writing these in terms of simple atomic sentences using the logical connectives, determine whether the statements are always true, always false, or sometimes true and sometimes false, depending on the values of the variables. Give reasons.

15. If $x^2 = 4$, then $x = 2$ or $x = -2$.
16. If $x < y$, then $x + 1 < y$ or $x + 1 = y$.
17. If $x^2 + 1 = 0$, then $x = 2$.
18. If $x^2 = y^2$, then $x = y$.

19. If $x > y$, then $x > 2y$.
20. If $x^2 > 1$, then either $x > 1$ or $-x > 1$.

21. Let p, q, and r be defined as follows.

 > p: This movie is very long.

 > q: This movie is interesting.

 > r: This movie is about insects.

Translate the following into ordinary English.
(a) $p \wedge q$ (b) $r \rightarrow \neg q$ (c) $p \wedge q \wedge \neg r$

22. Let p, q, and r be defined as follows:

 > p: This is a math class.

 > q: This class is boring.

 > r: Logic is taught in this class.

Translate the following into ordinary English.
(a) $p \wedge q$ (b) $r \rightarrow (p \vee q)$ (c) $r \wedge p \wedge \neg q$

23. In this exercise, the variable x is an integer. Let p, q, and r be defined as follows.

 > p: $x^2 = 9$

 > q: $x = 3$

 > r: $x = -3$

Translate the following into mathematical statements and state whether they are always true or always false, or sometimes true and sometimes false, depending on the value of x.
(a) $q \leftrightarrow r \vee p$ (b) $q \rightarrow (\neg r \wedge p)$ (c) $p \wedge q \wedge r$

24. In this exercise, the variables x and y are positive integers. Let p, q, and r be defined as follows.

 > p: $x < y$

 > q: $x + 1 = y$

 > r: $x^2 = 5$

Translate the following into mathematical statements and state whether they are true or false, or sometimes true, depending on the values of x and y.
(a) $p \rightarrow (q \vee \neg r)$ (b) $q \rightarrow p$ (c) $r \rightarrow \neg r$

25. Translate each of the following into smooth English.
(a) (It is raining) $\rightarrow$ (I will stay home)
(b) $\neg$(John walks $\vee$ John sleeps)
(c) (Frederick goes to the movies) $\rightarrow$ (Denise goes to the movies $\wedge$ $\neg$Elise goes to the movies)
(d) (Beth studies math) $\rightarrow$ ($\neg$Beth will watch TV $\vee$ Beth will eat a lot of popcorn)

26. Translate each of the following into smooth English. The variables are understood to represent positive integers.
    (a) $(x^2 = y^2) \rightarrow (x = y)$
    (b) $(x \text{ is even}) \wedge (x > 2) \rightarrow \neg(x \text{ is a prime number})$
27. Translate each of the following into smooth English. The variables are understood to represent positive integers.
    (a) $(x \text{ is divisible by } 6) \rightarrow (x \text{ is even})$
    (b) $(x^2 = 5) \rightarrow (x = 7)$
28. Rewrite the following formulas by dropping as many parentheses as possible according to the notational convention of this section.
    (a) $(p \wedge (q \wedge r))$
    (b) $(p \vee q) \leftrightarrow ((\neg p) \rightarrow r)$
    (c) $((p \vee q) \wedge r)$
    (d) $((p \wedge q) \vee r)$
    (e) $(p \wedge (\neg q)) \rightarrow (r \wedge (q \rightarrow p))$
29. In each of the following, give conditions on the atomic statements p, q, r that are equivalent to the stated conditions.
    (a) $p \vee q$ has truth value F
    (b) $p \wedge \neg q$ has truth value T
30. In each of the following, give conditions on the atomic statements p, q, r that are equivalent to the stated conditions.
    (a) $p \rightarrow (q \rightarrow r)$ has truth value F
    (b) $\neg(\neg p \vee \neg q)$ has truth value T

Find a truth table for each formula in Exercises 31–42.

31. $p \wedge q \rightarrow \neg q$
32. $(p \rightarrow q) \leftrightarrow q \vee p$
33. $p \rightarrow p \rightarrow p \rightarrow q$
34. $(p \leftrightarrow q) \rightarrow (\neg p \vee r)$
35. $(p \rightarrow q) \rightarrow ((q \rightarrow r) \rightarrow (p \rightarrow r))$
36. $(p \rightarrow q \vee r) \rightarrow (p \rightarrow p \vee r)$
37. $\neg(p \rightarrow r) \vee (q \wedge \neg p)$
38. $p \rightarrow (q \vee \neg p)$
39. $p \wedge (q \rightarrow (r \wedge \neg p))$
40. $\neg(d \vee (f \rightarrow \neg e)) \wedge (d \wedge e)$
41. $s \rightarrow \neg(t \rightarrow (s \vee u))$
42. $p \wedge \neg q \wedge (r \rightarrow (p \vee q))$

43. Construct a statement in p and q that is true only when p and q are both false.
*44. Construct a statement in p, q, and r that is true only when p, q, and r are false.
45. Consider the statement $(x < 3) \rightarrow (x < 9)$. Discuss the truth or falsity of this statement when $x = 1$, $x = 5$, and $x = 10$. Is it true for all real x? Why?
46. Let the universe be I_5. Take $A = \{1, 3\}$ and $B = \{1, 3, 4\}$. Show that $A \subseteq B$ by showing that the statement $x \in A \rightarrow x \in B$ is true for all x in the universe.

47. Let the universe be I_5, and let $A = \{1, 3\}$. Suppose B is a set, and the statement $x \in B \leftrightarrow \neg(x \in A)$ is true for all x in the universe. By checking this statement for $x = 1, 2, \ldots, 5$, show that B must be the set $A' = \{2, 4, 5\}$.
48. For what truth values is the statement $p \rightarrow (q \rightarrow (r \rightarrow s))$ false? Explain.
49. For what truth values is the statement $(p \wedge q \wedge \neg r) \rightarrow (r \vee s \vee t)$ false? Explain.
50. For what truth values is the statement $(p \rightarrow q) \wedge (q \rightarrow r)$ true? Explain.

1.3 Tautologies and Reasoning

Certain statements are automatically true because of their structure. For example, the statement "He is guilty or he is not" is true, although perhaps not informative. This statement is a special case of the statement $p \vee \neg p$. It is easily seen that the truth table for this statement is always T for the two choices of truth values for p. Such a statement is called a *tautology*. Much of the logic we use every day is based upon tautologies. In what follows, we continue to let p, q, and so on, denote atomic statements. We use P, Q, and so on, to denote compound statements. If P is a statement involving p and q, we write $P = P(p, q)$ to show this dependence.

DEFINITION 1.3.1

A **tautology** is a statement $P = P(p, q, \ldots)$ that is true for all possible truth values of the atomic statements $p, q, \ldots$.

Thus, $P = P(p) = p \vee \neg p$ is a tautology. It is in theory simple to determine whether a statement is a tautology. We simply construct its truth table and note whether T is its only possible truth value. When people started to reason mathematically, this was not the way it was done. Statements such as $p \vee \neg p$ were stated to be universally true as a "law of logic." Special names were given to some of these tautologies, some of which are still in use. A list of some tautologies and their names is given in Tables 1.3.1 and 1.3.2.

It is convenient to introduce the truth values T and F as special atomic statements in themselves. We may think of these as the constant statements. Thus T is always true, and F is always false. Their use will help simplify some of our statements.

The tautologies in Table 1.3.2 are useful for algebraic manipulations. Many of the names are taken from algebra. Note that the familiar equals sign of algebra is replaced by the symbol $\leftrightarrow$.

All of these tautologies can be easily verified and can be clearly labeled as "logical truths." For example, De Morgan's Law (3) may be para-

Table 1.3.1   Common Tautologies

Tautology	Name
1. $p \vee \neg p$	Law of the Excluded Middle
2. $\neg(p \wedge \neg p)$	Law of Contradiction
3. $\neg(p \vee q) \leftrightarrow \neg p \wedge \neg q$	De Morgan's Law
4. $\neg(p \wedge q) \leftrightarrow \neg p \vee \neg q$	De Morgan's Law
5. $(p \to q) \leftrightarrow (\neg q \to \neg p)$	Contrapositive Law
6. $p \wedge (p \to q) \to q$	Modus ponens
7. $(\neg p \to p) \to p$	Reductio ad absurdum
8. $(\neg p \to F) \to p$	Reductio ad absurdum
9. $(p \to (q \to r)) \leftrightarrow ((p \wedge q) \to r)$	Exportation Law
10. $(p \to q) \wedge (\neg p \to q) \to q$	Proof by cases
11. $p \wedge q \to p \; ; \; p \to p \vee q$	
12. $p \to (q \to p \wedge q)$	
13. $(p \to q) \leftrightarrow (\neg p \vee q)$	
14. $(p \leftrightarrow q) \leftrightarrow (p \to q) \wedge (q \to p)$	
15. $p \to T \; ; \; F \to p$	

phrased as follows: "To say that $(p \vee q)$ is not true is equivalent to saying that both p and q are each not true." A simple illustration follows.

WATSON: "Either the killer was a midget, or the killer was an acrobat."
HOLMES: "Not so, Watson."
WATSON: "Oh, then the killer was not a midget, and also not an acrobat."
HOLMES: "Quite so!"

The dialogue is pure De Morgan. Note that the "or" is changed to an "and," and vice versa, in De Morgan's Law.

Table 1.3.2   Common Tautologies (_continued_)

Tautology	Name
16. $p \wedge q \leftrightarrow q \wedge p$	Commutative Law
$p \vee q \leftrightarrow q \vee p$	Commutative Law
$(p \leftrightarrow q) \leftrightarrow (q \leftrightarrow p)$	Commutative Law
17. $p \wedge (q \wedge r) \leftrightarrow (p \wedge q) \wedge r$	Associative Law
$p \vee (q \vee r) \leftrightarrow (p \vee q) \vee r$	Associative Law
18. $p \wedge (q \vee r) \leftrightarrow (p \wedge q) \vee (p \wedge r)$	Distributive Law
19. $p \vee (q \wedge r) \leftrightarrow (p \vee q) \wedge (p \vee r)$	Distributive Law
20. $p \wedge p \leftrightarrow p \; ; \; p \vee p \leftrightarrow p$	Idempotent laws
21. $\neg\neg p \leftrightarrow p$	Law of the Double Negative
22. $p \wedge T \leftrightarrow p \; ; \; p \vee T \leftrightarrow T$	Property of T
23. $p \wedge F \leftrightarrow F \; ; \; p \vee F \leftrightarrow p$	Property of F
24. $\neg F \leftrightarrow T \; ; \; \neg T \leftrightarrow F$	

THEOREM 1.3.2   De Morgan's Law for Sets

Let A and B be sets in a universe U. Then

$$(A \cup B)' = A' \cap B' \tag{1.3.1}$$

$$(A \cap B)' = A' \cup B' \tag{1.3.2}$$

(In words, the complement of a union is the intersection of the complements; the complement of an intersection is the union of the complements.)

Proof   We prove Equation (1.3.1). For any x in U, we have

$$x \in (A \cup B)' \leftrightarrow x \notin (A \cup B)$$

$$\leftrightarrow \neg(x \in A \cup B)$$

$$\leftrightarrow \neg(x \in A \lor x \in B)\dagger$$

$$\leftrightarrow \neg(x \in A) \land \neg(x \in B)$$

$$\leftrightarrow x \in A' \land x \in B'$$

$$\leftrightarrow x \in A' \cap B'$$

Therefore $(A \cup B)' = A' \cap B'$, since both sets have the same elements. The proof of Equation (1.3.2) is similar and is left to the exercises.   ■

Some Techniques of Reasoning

Some of the classical methods of reasoning are implicitly contained in our table of tautologies. We now discuss them on an informal level, proceeding more formally in the next section.

Reasoning means that we can come to some conclusion if we are given some facts (true statements), called the **hypotheses** or the **premises.** That is, if we know the hypotheses are true, correct reasoning will show that the conclusion must necessarily be true. For example, suppose we were given as facts

John is a good student.

If John is a good student, then he studies.

We may then draw the conclusion:

John studies.

---

† Here is where we use the De Morgan tautology 3.

The form of this deduction is clearly independent of John, of studying, and so on. We are using the following general technique.

If we are given P, and $P \rightarrow Q$, we can deduce Q.

To see why this is so, suppose Q is false. Then since we are given that P is true, it would follow from the truth table for $\rightarrow$ that $P \rightarrow Q$ is false. But we are also given that $P \rightarrow Q$ is true. Therefore, Q cannot be false, and so Q must be true. Thus, the above principle is valid. (Incidentally, the reasoning used in this paragraph is an example of a proof by contradiction, which will be discussed later.)

This form of deduction is very similar to the modus ponens tautology 6 of Table 1.3.1: $p \wedge (p \rightarrow q) \rightarrow q$. The foregoing technique is also called modus ponens.

Modus ponens has a classical history. The following illustration has been used for centuries. The hypotheses are placed over a line and the conclusion placed below it.

If Socrates is a man, then Socrates is mortal;
Socrates is a man.

(Therefore) Socrates is mortal.

EXAMPLE 1.3.3   Proof Using Modus Ponens

Given the premises

1. If John is lazy, he watches TV.
2. If John watches TV, then he eats popcorn.
3. John is lazy.

Prove that John eats popcorn.

Method We let z = John is lazy, t = John watches TV, and p = John eats popcorn. The premises are (1) $z \rightarrow t$; (2) $t \rightarrow p$; (3) z. The conclusion is p.

To show this is a valid conclusion, we have z and $z \rightarrow t$. Therefore, we know that t is true by modus ponens. But we also have $t \rightarrow p$. Therefore, we have p, again by modus ponens.

In this example, we spelled out the atomic sentences z, t, and p. In the following examples, the translation from English to the propositional calculus is left to the reader.

Another method of proof uses the **contrapositive** law (tautology 5 of Table 1.3.1): $(p \rightarrow q) \leftrightarrow (\neg q \rightarrow \neg p)$.

EXAMPLE 1.3.4   *Proof Using Contrapositive*

Given the hypotheses

1. If Max cheated on the exam, Max would pass with flying colors.
2. Max did not pass with flying colors.

Prove that Max didn't cheat on the exam.

Method   The form of the hypothesis is (1) $c \rightarrow p$; (2) $\neg p$. The conclusion we wish to show is $\neg c$.

The contrapositive law shows that $c \rightarrow p$ has the same truth value as $\neg p \rightarrow \neg c$. Since we are given that $c \rightarrow p$, we have $\neg p \rightarrow \neg c$. But we also are given $\neg p$. Therefore we have the conclusion $\neg c$ by modus ponens.

Note that the equivalent of $p \rightarrow q$ is $\neg q \rightarrow \neg p$: the statements p and q get reversed and negated. The statement $\neg q \rightarrow \neg p$ is called the *contrapositive* of the statement $p \rightarrow q$. The contrapositive law, tautology 5 of Table 1.3.1, is often used in proving results when it is easier to work with the negation of a statement rather than the statement itself. For example, suppose we wanted to prove that if $x^2 \neq 4$ then $x \neq 2$. It would be much simpler to prove its equivalent—that if $x = 2$ then $x^2 = 4$— since it is easier to work with equalities than with inequalities.

EXAMPLE 1.3.5   *Use of the Contrapositive*

Consider the statement "If I use a word, it's in the dictionary." This statement is equivalent to its contrapositive "If it's not in the dictionary, I don't use the word." The **converse** of $p \rightarrow q$ is $q \rightarrow p$. The converse is not equivalent to the original statement. In this case, the converse is "If a word is in the dictionary, I use it."

Another important technique for proving results is called a **proof by contradiction.** This is illustrated by the "reduction ad absurdum" tautology 7, $(\neg p \rightarrow p) \rightarrow p$, or 8, $(\neg p \rightarrow F) \rightarrow p$ (where F is any false proposition).

EXAMPLE 1.3.6   *Proof by Contradiction*

Given the hypotheses

1. If $x = 3$, then $y = 4$.
2. If $z < 5$, then $x = 3$.
3. If $y \neq 4$, then $z < 5$.

Prove that $y = 4$.

Method   The hypotheses are (1) $p \rightarrow q$; (2) $r \rightarrow p$; (3) $\neg q \rightarrow r$. We wish to prove q.

Now assume $\neg q$. Using (3), $\neg q \rightarrow r$, we deduce r by modus ponens. Using (2), $r \rightarrow p$, we deduce p, again by modus ponens. Finally, using (1), $p \rightarrow q$, we deduce q.

The assumption of $\neg q$ leads to q. But from this we can say that $\neg q \rightarrow q$. Now we use the reductio ad absurdum tautology: $(\neg q \rightarrow q) \rightarrow q$. Since $(\neg q \rightarrow q)$ is true, it follows by modus ponens that q is true.

In a proof by contradiction, after assuming $\neg q$ and proving q, we immediately say that q is true "by contradiction."

Another method of proof is by a **case analysis.** This is based on tautology 10:

$$(p \rightarrow q) \wedge (\neg p \rightarrow q) \rightarrow q$$

In words, if p implies q and $\neg p$ implies q, then q is true. The idea is that we prove q by cases. Either p is true (case 1) or false (case 2). In either case, we have q. Therefore q is true.

EXAMPLE 1.3.7   Proof by Cases

Given the hypotheses

1. The town will go bankrupt if Jones is elected.
2. If everybody doesn't vote, the town will go bankrupt.
3. If everybody votes, then Jones will be elected.

Prove that the town will go bankrupt.

Proof   The hypotheses are (1) $j \rightarrow b$; (2) $\neg v \rightarrow b$; (3) $v \rightarrow j$. We want to prove b. If j is true, then we have b, using (1) and modus ponens. If $\neg j$ is true (or j is false), then by the contrapositive form of (3), we have $\neg v$. Then using (2) and modus ponens, we have b. Thus, we have b in either case: j or $\neg j$. This proves b by a case analysis.

In life, as well as in a mathematical proof, care must be taken to reject an approach to reasoning that doesn't conform to the methods of logic.

EXAMPLE 1.3.8   Invalid Reasoning

Given the hypotheses

1. If Mary is a millionaire, then Mary can buy this book.
2. Mary can buy this book.

Prove that Mary is a millionaire.

Method   The hypotheses are (1) $m \to b$; (2) b. Our intuition may tell us that Mary might be able to buy this book even though she is not a millionaire. (Nothing was given to the contrary.) Therefore we should not be able to prove that m is true. In fact, we might have m false and b true. Then the hypotheses (1) and (2) would both be true, according to the truth tables. But the proposed conclusion m would not be true. Therefore, no valid line of reasoning can prove m. The alleged deduction is an example of **invalid reasoning.** Of course, Mary might be a millionaire. But we couldn't prove it by using the hypotheses and logical reasoning.

To show that reasoning is invalid, we must show that it is possible for all the hypotheses to be true, and yet the conclusion is false. It is possible to hear invalid reasoning every day, especially around election time. "My opponent is rich. Anyone who steals from the government is rich. (Therefore my opponent steals from the government.)"

We end this section with a real-life mathematical exercise illustrating how some of these ideas can apply in a mathematical situation. We consider the classical proof that $\sqrt{2}$ is irrational. As "hypotheses" we take for granted various facts about numbers. We first start with some lemmas. These are results we intend to use in the theorem we want to prove. All variables m, n, p, q, and so on, are understood to be positive integers.

LEMMA 1.3.9

The square of an odd number is odd. The square of an even number is even.

Proof   The proof is algebraic. If n is odd, this means that n can be written in the form $n = 2p + 1$. By algebra, $n^2 = 4p^2 + 4p + 1 = 2(2p^2 + 2p) + 1$, which is of the form $2q + 1$. Therefore p^2 is odd. Similarly, we can prove that if n is even, then so is n^2. ■

LEMMA 1.3.10

If n^2 is even, then n is even.

Proof (by contradiction)   Assume that n is not even. Then, as an arithmetic fact, n must be odd. By Lemma 1.3.9, n^2 would be odd. But n^2 is even. This is a contradiction, so n must be even.   ■

THEOREM 1.3.11

$\sqrt{2}$ is irrational (not a rational number).

Proof (again, by contradiction)   Assume that $\sqrt{2}$ is rational. This means that $\sqrt{2} = p/q$. As a fact about integers, we may write this in lowest terms. This means that we cannot have both p and q even. Since $\sqrt{2} = p/q$, by algebra this gives $p^2 = 2q^2$. Therefore p^2 is even. By Lemma

1.3.10, p is even. This means that we can write $p = 2r$. Since $p^2 = 2q^2$, we have $(2r)^2 = 2q^2$. By algebra, this yields $q^2 = 2r^2$. Therefore q^2 is even. Again by Lemma 1.3.10, q must be even. Therefore we have shown that both p and q must be even. But we have stated that they can't both be even, since the fraction p/q is in lowest terms. This is a contradiction, so $\sqrt{2}$ is irrational. ■

Exercises

Using truth tables, show that each statement in Exercises 1–5 is a tautology.

1. The contrapositive law
2. The exportation law
3. De Morgan's Laws
4. Reductio ad absurdum
5. Proof by cases

*6. A statement P is called a **contradiction** if it is always false.
    (a) Prove that P is a contradiction if and only if $\neg P$ is a tautology.
    (b) Prove that P is a contradiction if and only if $P \rightarrow Q$ is a tautology for all statements Q.
7. Prove Equation (1.3.2) of Theorem 1.3.2. Use a technique similar to the one used to prove Equation (1.3.1).
8. Prove that if $A \subseteq B$, then $B' \subseteq A'$. Use a technique similar to the one used to prove Equation (1.3.1) in Theorem 1.3.2.

Sometimes, during the course of a deduction, a tautology is needed, so a reasonable one is conjured up. Decide, using truth tables, which of the following statements are tautologies.

9. $(p \rightarrow q \vee r) \leftrightarrow (p \rightarrow q) \vee (p \rightarrow r)$
10. $(p \wedge q \rightarrow r) \rightarrow (p \rightarrow q) \vee (q \rightarrow r)$
11. $((p \wedge q) \rightarrow r) \rightarrow (p \rightarrow r) \wedge (q \rightarrow r)$
12. $(p \rightarrow q) \vee (r \rightarrow \neg q)$
13. $(p \rightarrow p) \rightarrow p$

In Exercises 14–22, prove the required conclusion from the given premises.

14. Hypotheses: If Jill likes the movie *The Logical Sleuths*, John doesn't. John likes the movie *The Logical Sleuths*. Conclusion: Jill doesn't like the movie *The Logical Sleuths*.
15. Hypotheses: Mary is cutting the chemistry lab. If John is not cutting the chemistry lab, then Mary is not cutting that lab. Conclusion: John is cutting the chemistry lab.
16. Hypotheses: If the window is left open, it is cold inside. If it is cold inside, then sweaters are put on. Sweaters are not put on. Conclusion: The window is not left open.

*17. Hypotheses: If the book is interesting, Joan will finish reading it. If the book is not about baseball, then it is interesting. If Joan finishes reading this book, it is about baseball. Conclusion: The book is about baseball.

18. Hypotheses: John cuts class if Mary cuts class. If John doesn't cut class, he does not feel good. John feels good if Mary does not cut class. Conclusion: John cuts class.

19. Hypotheses: Henry will dance all night if he goes to the party. If Henry doesn't go to the party, he will be cross. If Henry dances all night, he will be cross. Conclusion: Henry will be cross.

*20. Hypotheses: If this picture is produced, it will not make money. If this picture stars Warren, it will make money. This picture will make money if it's done as an epic. If this picture is produced, it will either star Warren or be an epic. Conclusion: This picture won't be produced.

21. Hypotheses: If I have homework, I go to the movies. I have homework if my biology class is not easy. If my biology class is easy, I go to the movies. Conclusion: I go to the movies.

*22. Hypotheses: (Based on *Catch-22*, by Joseph Heller) If Yossarian gets out of combat duty, then he is crazy and he asks to get out of combat duty. If Yossarian asks to get out of combat duty, then he's not crazy. Conclusion: Yossarian will not get out of combat duty.

23. It is desired to prove that Zookies are crunchy. The argument goes as follows. (a) If Zookies were not crunchy, then Balzoes would be malleable. (b) If Balzoes were malleable, then Zookies would be crunchy. Therefore, (c) Zookies are crunchy. With (a) and (b) as premises, is the reasoning sound?

24. Is the following reasoning sound? Explain.
Either this book is dull, or I'm tired. This book is not dull, so I must be tired.

*25. Is the following reasoning sound? Explain.
John always gets angry when Mary is late. When Mary is late, she always forgets to wear gloves. Therefore, when Mary forgets to wear gloves, John will get angry.

26. Is the following reasoning sound? Explain.
If John is a Yuppy, then John jogs. John jogs. Therefore, John is a Yuppy.

27. Is the following reasoning sound? Explain.
If Kermit is tired, he reads a book. Kermit is not tired if he rakes leaves. Therefore if Kermit is raking leaves, he is not tired.

*28. Is the following reasoning sound? Explain.
(Based on *Catch-22*, by Joseph Heller) Yossarian will get out of combat duty if he is crazy and he asks to get out of combat duty. If Yossarian asks to get out of combat duty, then he's not crazy. Conclusion: Yossarian will not get out of combat duty.

1.4 Logical Equivalence and Deductions

We stated in the last section that the symbol ↔ replaces the familiar equals sign of algebra in the foregoing tautologies. We now discuss the sense in which this is true. In what follows, $P, Q, \ldots$ denote statements, possible compound statements involving the atomic statements $p, q, \ldots$. For example, P might equal the statement $\neg(p \wedge q)$, and Q might be the statement $\neg p \vee \neg q$. In this case, P and Q have identical truth tables, but they are not the same statements. This fact is equivalent to tautology 4 (De Morgan's Law). In general, we say that statements with the same truth table are logically equivalent.

DEFINITION 1.4.1

Let $P = P(p, q, \ldots)$ and $Q = Q(p, q, \ldots)$ be statements. We say that P and Q are **logically equivalent,** and write

$$P \equiv Q$$

if the truth value of $P(p, q, \ldots)$ is the same as the truth value of $Q(p, q, \ldots)$ for all possible truth values of the atomic statements $p, q, \ldots$, that appear in them.

In particular, P is a tautology if and only if $P \equiv T$.

In any usage in which we are concerned only with the truth value of a statement, we can replace any compound statement with one that is logically equivalent to it. This is also true of any computer program that needs the truth or falsity of a statement P in order to proceed. Thus, the statement

$$\text{IF not(not(X = Y)) THEN Y} := 4$$

can be replaced by the simpler statement

$$\text{IF X = Y THEN Y} := 4$$

Here, we are using the tautological equivalence $\neg\neg P \equiv P$ when P is the statement $X = Y$, and replacing the statement not(not(X = Y)) by its tautological equivalent $X = Y$. (Such IF ... THEN ... statements are discussed in more detail in Sections 3.1 and 3.2.)

The following theorem shows how logical equivalence is related to the logical symbol ↔.

THEOREM 1.4.2

Let P and Q be statements. Then $P \equiv Q$ if and only if $P \leftrightarrow Q$ is a tautology.

Proof   First suppose that $P \leftrightarrow Q$ is a tautology. This means it is always true, regardless of the truth values of $p, q, \ldots$. Referring to Table 1.2.5

(the truth table for $\leftrightarrow$), this means that P and Q are either both true or both false for any choice of truth values for $p, q, \ldots$. Thus, P and Q have the same truth values, so $P \equiv Q$ by definition.

Conversely, suppose $P \equiv Q$. Then the truth values of P and Q are either (T, T), or (F, F). In either case, the truth value of $P \leftrightarrow Q$ is T, again by Table 1.2.5. Thus, $P \leftrightarrow Q$ is a tautology, by definition. ∎

Deductions and Logical Implications

As we saw in the last section, it is possible to use logic in order to prove statements true. A typical proof involves hypotheses, which are given, and a conclusion, which is to be proved. A proof must be such that the conclusion will be true provided all of the hypotheses are true. We now formalize this idea.

DEFINITION 1.4.3

We say that the statements $P_1, P_2, \ldots, P_n$ **logically imply** the statement P, provided that P is true whenever $P_1, \ldots, P_n$ are all true. In this case, we say that P is a **valid conclusion** from the hypotheses $P_1, \ldots, P_n$.

The following theorem relates this idea to the implication symbol $\rightarrow$, and to tautologies.

THEOREM 1.4.4

Statements $P_1, \ldots, P_n$ logically imply P if and only if $P_1 \wedge \cdots \wedge P_n \rightarrow P$ is a tautology.

Proof   First suppose that the P_i logically imply P. The statement $P_1 \wedge \cdots \wedge P_n \rightarrow P$ is false only when $P_1 \wedge \cdots \wedge P_n$ is true and P is false. But if the P_i logically imply P, once $P_1 \wedge \cdots \wedge P_n$ is true, P will necessarily be true. Thus, $P_1 \wedge \cdots \wedge P_n \rightarrow P$ can never be false. Therefore it is a tautology.

Next suppose that $P_1 \wedge \cdots \wedge P_n \rightarrow P$ is a tautology. This means it is always true. We now show that the P_i logically imply P. For suppose each P_i is true. Then $P_1 \wedge \cdots \wedge P_n$ is true as well as $P_1 \wedge \cdots \wedge P_n \rightarrow P$ (since it is a tautology). Hence, P is true by Table 1.2.4. This gives the result. ∎

Warning   Lest the reader think this is a magical way to prove anything, we remark that this result is valid only in the propositional calculus. Thus, the method depends only on properties of the logical connectives (and, or, if . . . then . . . , and so on). For example, Theorem 1.4.4 does not allow us to show that $x = 1$ logically implies $x^2 = 1$. If we let p be the statement $x = 1$, and q the statement $x^2 = 1$, we are stuck with the problem of showing that $p \rightarrow q$ is a tautology. It isn't!

A **proof,** or a **deduction,** in mathematics is a sequence of statements, all of which are true as a consequence of the hypotheses and the rules of logic, and which culminate in the conclusion. It is often very hard to find a proof of a result in mathematics, but in theory, once the proof is presented, it should be an easy matter to verify each step in the proof. We now show this procedure for the propositional calculus.

We introduced the term "modus ponens" in the last section. This is either the statement

$$P, P \to Q \text{ logically imply } Q \tag{1.4.1}$$

or its equivalent, by Theorem 1.4.4,

$$(P \wedge (P \to Q)) \to Q \text{ is a tautology} \tag{1.4.2}$$

Let us say that the statement Q *follows from the statements P and $P \to Q$* by modus ponens. This is traditionally called a *rule of inference.*

DEFINITION 1.4.5

A rule of inference is a way of proceeding from several statements, P_1, $P_2, \ldots, P_k$, to another, P. It is required that these statements P_i logically imply P. We write

$$\frac{P_1, \ldots, P_k}{P}$$

A rule of inference looks suspiciously like a rewording of the definition of logical implication. However, the usage is different. Usually rules of inference start with one or two statements (like modus ponens) to derive a third. They are useful in proofs, allowing us to arrive at a conclusion using a few of the statements at a time. Historically there were many rules of inference, each given a special name. We have retained only the name modus ponens in this discussion.

We can give other valid rules of inference. For example, the statement $(P \to Q) \wedge (Q \to R) \to (P \to R)$ is a tautology, as may be easily verified. Therefore, by Theorem 1.4.4, $P \to Q, Q \to R$ logically imply $P \to R$. Thus, we have the rule of inference

$$\frac{\begin{array}{c} P \to Q \\ Q \to R \end{array}}{P \to R}$$

In the same manner, we can list various other rules of inference. Table 1.4.1 summarizes some of the commonly used ones. See if you can make intuitive sense of them. Then check that they are valid rules of inference, using Theorem 1.4.4.

Table 1.4.1   Some Rules of Inference

(a) P
$\dfrac{P \to Q}{Q}$

(c) P
Q
$\dfrac{}{P \wedge Q}$

(e) $P \to Q$
$\dfrac{\neg Q}{\neg P}$

(b)
$\dfrac{P \wedge Q}{P}$

(d) $P \to Q$
$\dfrac{\neg P \to Q}{Q}$

(f) $P \vee Q$
$\dfrac{\neg P}{Q}$

The reader may recognize part (a) in Table 1.4.1 as modus ponens; part (d) was used in a proof by cases; and part (e) was used in a proof using the contrapositive.

It is now possible to define what is meant by a deduction.

DEFINITION 1.4.6

Let h_1, h_2, ..., h_n be statements (the hypotheses or premises), and let P be a statement (the conclusion). We say that P is a **consequence** of the h_i, or that P **may be deduced from** the h_i, provided there is a sequence of statements $A_1, A_2, \ldots, A_N$ with the following properties.

1. The final statement A_N is the conclusion P.
2. Each of the statements A_i must either
   a. be one of the hypotheses h_j, or
   b. be a tautology, or
   c. follow from some of the previous statements by a valid rule of inference.

The sequence $A_1, \ldots, A_N$ is called a **deduction.**

We illustrate with an example similar to the ones we discussed in the previous section.

EXAMPLE 1.4.7

Given p, $p \to r$, $r \to q$. Deduce q.

Method   We actually give the deduction, together with reasons for each step. (These *must* be any of the items 2a, 2b, or 2c in Definition 1.4.8.)

1. p Hypothesis (Hyp, for short)
2. $p \to r$ Hyp
3. r Modus ponens from 1 and 2 (MP 1, 2 for short)
4. $r \to q$ Hyp
5. q MP 3, 4

Definition 1.4.6 is so general that a deduction can be given by simply listing all of the premises and then the conclusion, provided the premises logically imply the conclusion. However, we use simple rules of inference (such as modus ponens) in which we usually use only one or two statements to obtain a conclusion. We continue to give reasons for each step. When a rule of inference is used, we give "Taut" or "T" as the reason, referencing the lines of the deduction that were used in applying the rule.

EXAMPLE 1.4.8

Given the premises $p \wedge q$, $p \to r$, $q \to s$, deduce the conclusion $r \wedge s$.

Method   The deduction is

1. $p \wedge q$ Hyp
2. $p \to r$ Hyp
3. $q \to s$ Hyp
4. p Taut 1
5. r Taut 2, 4 (or MP 2, 4)
6. q Taut 1
7. s Taut 3, 6
8. $r \wedge s$ Taut 5, 7

Remark   Statements 4 and 6 use the rule of inference (b) of Table 1.4.1. It means "From P and Q, conclude P", so it is fairly natural to use. Conclusion 8 uses the rule of inference (c) from the same table.

The critical fact about a deduction is that the conclusion is logically implied from the hypotheses. That is, if the hypotheses are true, the conclusion must also be true.

THEOREM 1.4.9

Suppose it is possible to deduce P from the hypotheses $P_1, \ldots, P_n$. Then $P_1, \ldots, P_n$ logically imply P.

Proof   It is necessary to show that if each of the hypotheses P_i is true, so is the conclusion P. To do this, we show that if each of the P_i is true, then *every statement in the deduction is true*. But statements are either hypotheses (given true), or tautologies (always true), or follow from preceding statements by a valid rule of inference. If we follow the deduction from beginning to end, we see that each step of the proof must be true. Thus, the conclusion (which is the last statement) must also be true. ■

To paraphrase Theorem 1.4.9, if you can prove something from the premises, and the premises are known to be true, then your conclusion will be true.

Two important proof techniques are not covered by the definition of a deduction. The first is a way of proving an implication $P \rightarrow Q$. The second is the method of proof by contradiction.

THEOREM 1.4.10   The Deduction Theorem

Suppose statements $h_1, h_2, \ldots, h_n$ are a set of hypotheses and that, together with P, they logically imply Q. Then the hypotheses $h_1, \ldots, h_n$ alone logically imply the statement $P \rightarrow Q$.

In short, to prove $P \rightarrow Q$ from a set of hypotheses, assume P also and then prove Q.

Proof We are given that $h_1, h_2, \ldots, h_n, P$ logically imply Q. We wish to show that $h_1, h_2, \ldots, h_n$ logically imply $(P \rightarrow Q)$. Suppose, then, that $h_1, h_2, \ldots, h_n$ are all true. Let us show that $P \rightarrow Q$ is true. We show this by a case analysis. If P is true, then by hypotheses Q is true. Therefore $P \rightarrow Q$ is true. But if P is false, $P \rightarrow Q$ is true by the truth table for $P \rightarrow Q$. Thus, in any case $P \rightarrow Q$ is true. This proves the result. ■

We illustrate Theorem 1.4.10 in the following example.

EXAMPLE 1.4.11

Given the premises $q \rightarrow r$ and $\neg r \vee s$. Deduce $q \rightarrow s$.

Method Since we are to prove the implication $q \rightarrow s$, we assume q and attempt to deduce s. This gives the result by the Deduction Theorem. The method and reasons follow.

1. $q \rightarrow r$ Hyp
2. q Assumption
3. r T 1, 2
4. $\neg r \vee s$ Hyp
5. s T 3, 4
6. $q \rightarrow s$ Discharge Ded 2, 5

The reasons are explained as follows. Statement 2 is simply a new premise, used to apply the Deduction Theorem. Statement 5 is valid, since 3 and 4 logically imply it. (It is a variant of the rule of inference (e) of Table 1.4.1.) Since the hypotheses, along with the additional assumption 2, logically imply 5, the original hypotheses alone logically imply $q \rightarrow s$. We say that the assumption 2 is **discharged** by the Deduction Theorem—the conclusion follows from the hypotheses alone and does not need the additional assumption 2.

The above demonstration is not a deduction according to Definition 1.4.6. But the conclusion is logically implied from the hypotheses, using the Deduction Theorem. We shall extend the definition and call this method a deduction.

Our final technique of proof is the proof by contradiction. In this method, when we wish to prove P, we assume the contrary ($\neg P$) and attempt to arrive at a contradiction F (a false statement). If we do so, this will prove P. Equivalently, if we arrive at the conclusion P, having assumed $\neg P$, this will show that P is logically implied by the hypotheses. We state this as a theorem.

THEOREM 1.4.12   *Proof by Contradiction*

Suppose $h_1, h_2, \ldots, h_n$ are a set of hypotheses, which together with $\neg P$ logically imply F, the false statement. Then the hypotheses $h_1, \ldots, h_n$ logically imply P. Similarly, if $h_1, h_2, \ldots, h_n$, together with $\neg P$, logically imply P, then the hypotheses $h_1, \ldots, h_n$ logically imply P.

Remark   Usually F occurs as a statement $R \wedge \neg R$.

Proof   Suppose $h_1, h_2, \ldots, h_n$ and $\neg P$ logically imply F. Then, by the Deduction Theorem, $h_1, h_2, \ldots, h_n$ logically imply $\neg P \rightarrow$ F. But $\neg P \rightarrow$ F logically implies P, since $(\neg P \rightarrow$ F$) \rightarrow P$ is a tautology. Therefore, $h_1, h_2, \ldots, h_n$ logically implies P.
    The second half is proved in a similar manner.   ■

EXAMPLE 1.4.13

Given the premises $p \rightarrow q$, $\neg q \rightarrow \neg r$, and $\neg p \rightarrow r$, deduce q.

Method   Let's try a proof by contradiction. We want to prove q, so we assume $\neg q$ and try to arrive at a contradiction. The method, with documentation, follows.

1. $\neg q$	Assumption
2. $p \rightarrow q$	Hyp
3. $\neg p$	T 1, 2
4. $\neg p \rightarrow r$	Hyp
5. r	T 3, 4
6. $\neg q \rightarrow \neg r$	Hyp
7. q	T 5, 6
8. q	Discharge Contradiction 1, 7

Remark   Statement 1 is an additional assumption used to prove q by contradiction. Statement 3 uses a form of the contrapositive rule of inference. Statement 5 uses modus ponens. Statement 7 also uses a form of the contrapositive rule of inference. Statement 8 discharges assumption 1 by the method of Theorem 1.4.12, because of the conclusion 7. This means, as in the Deduction Theorem, that the conclusion q follows from the hypotheses alone and does not need the assumption 1.

We conclude with some further illustrations of deductions.

EXAMPLE 1.4.14

Given the hypotheses $p \to (q \to r)$, $s \to \neg r$, and $\neg t \to s$, deduce $q \to (p \to t)$.

Method When we have to prove an implication, it is usually a good idea to assume the first half and use the deduction theorem. So here we assume q and try to prove $p \to t$. How do we prove $p \to t$? Once again, by assuming p and proving t! The deduction is as follows.

1.	$p \to (q \to r)$	Hyp
2.	q	Assumption
3.	p	Assumption
4.	$q \to r$	T 1, 3
5.	r	T 2, 4
6.	$s \to \neg r$	Hyp
7.	$\neg s$	T 5, 6
8.	$\neg t \to s$	Hyp
9.	t	T 7, 8
10.	$p \to t$	Discharge Ded 3, 9
11.	$q \to (p \to t)$	Discharge Ded 2, 10

We introduced assumptions 2 and 3, used them, and discharged them by the Deduction Theorem.

EXAMPLE 1.4.15

Given the hypotheses $(p \to q) \vee (p \wedge s)$, $q \to t$, p, $p \to u$, and $(u \wedge s) \to t$, deduce t.

Method The first hypothesis is an OR statement. If we knew that either half was true, we could prove t. This suggests a case analysis proof. The deduction is as follows.

1.	$p \to q$	Assumption
2.	p	Hyp
3.	q	T 1, 2
4.	$q \to t$	Hyp
5.	t	T 3, 4
6.	$(p \to q) \to t$	Discharge Ded 1, 5
7.	$p \wedge s$	Assumption
8.	$p \to u$	Hyp
9.	u	T 7, 8
10.	$u \wedge s$	T 7, 9
11.	$(u \wedge s) \to t$	Hyp
12.	t	T 10, 11
13.	$(p \wedge s) \to t$	Discharge Ded 7, 12
14.	$(p \to q) \vee (p \wedge s)$	Hyp
15.	t	T 6, 13, 14

The proof by cases variant we used in step 15 follows from the tautology

$$(P \lor Q) \land (P \to R) \land (Q \to R) \to R$$

This yields a rare rule of inference where three statements yield a conclusion.

Finally, we note that it is only statements logically implied by a set of hypotheses that can be deduced from them. No matter how hard you try, you cannot prove the statement p from the hypotheses $q \to p$ and $\neg q$. This can be seen by assigning q and p the truth value F. This assignment makes the hypotheses true but the alleged conclusion false. Thus no proof is possible. When you are trying to prove a result and you seem to be getting nowhere, there is always a possibility that no proof is possible. In this case, try to see if it is possible that all of your hypotheses may be made true and the conclusion false. If this can be done, no proof is possible.

EXAMPLE 1.4.16

Given the hypotheses $p \to q$, $\neg q \to r$, and $\neg r \to (q \to p)$, deduce the conclusion $p \to r$ or show that it cannot be deduced.

Method   A way of showing that no deduction is possible is to use truth tables. We try to make the conclusion false while making each of the hypotheses true. To make $p \to r$ false, we must make p true and r false. We now list the hypotheses and see if we can make them all true. Put truth values under the variables corresponding to p and r.

$$p \to q, \; \neg q \to r, \; \neg r \to (q \to p) \qquad \text{Conclusion } p \to r$$
$$\text{T} \qquad\qquad \text{F T} \qquad\quad \text{T} \qquad\qquad\qquad \text{T F F}$$

By observation, if we assign q the value T, then each of the hypotheses is true. Since the conclusion is false, no deduction is possible.

This method is fairly simple in this case, since we had only one choice in our assignment of truth values for p and r in order to make the conclusion false. If it were impossible in this case to make all of the hypotheses true, then the conclusion would have been valid. Namely, all truth values that make the hypotheses true would make the conclusion true, so the hypotheses would logically imply the conclusion.

Exercises

1. Using Theorem 1.4.2 and tautologies 3–5 of Table 1.3.1, give examples of logically equivalent statements.
2. Using Theorem 1.4.2 and tautologies 9, 13, and 14 of Table 1.3.1, give examples of logically equivalent statements.

3. Using Theorem 1.4.2 and tautologies 17–19 of Table 1.3.2, give examples of logically equivalent statements.
4. Using Theorem 1.4.2 and tautologies 18–19 of Table 1.3.2, give examples of logically equivalent statements.
5. Using Theorem 1.4.2 and tautologies 20–22 of Table 1.3.2, give examples of logically equivalent statements.
6. Using Theorem 1.4.2 and tautologies 23–24 of Table 1.3.2, give examples of logically equivalent statements.
*7. (Reduction to two connectives)
   (a) Find a statement that is logically equivalent to $p \vee q$ but uses only the connectives $\neg$ and $\wedge$. (*Hint:* Use De Morgan's Law.)
   (b) Using part (a), show that any statement P whose only connectives are $\neg$, $\wedge$, and $\vee$ is logically equivalent to a statement whose only connectives are $\neg$ and $\wedge$.
   (c) Using the list of tautologies and part (b), prove that $p \rightarrow q$ and $p \leftrightarrow q$ are each logically equivalent to statements whose only connectives are $\neg$ and $\wedge$.
   (d) Using parts (b) and (c), show that any statement is logically equivalent to one whose only connectives are $\neg$ and $\wedge$.
*8. Find a statement that is logically equivalent to $p \vee q \rightarrow r$ but involves only the logical connectives $\neg$ and $\wedge$. Use the methods outlined in Exercise 7.
*9. Find a statement that is logically equivalent to $p \vee (q \rightarrow r)$ but involves only the logical connectives $\neg$ and $\wedge$. Use the methods outlined in Exercise 7.
*10. The NAND connective $|$ (for NOT AND) and the NOR connective $\downarrow$ (for NOT OR) are defined by the following truth tables.

p	q	$p \mid q$
F	F	T
F	T	T
T	F	T
T	T	F

p	q	$p \downarrow q$
F	F	T
F	T	F
T	F	F
T	T	F

($|$ is called the **Scheffer stroke,** and $\downarrow$ is called the **Peirce arrow.**)
   (a) Prove $p \mid q \equiv \neg(p \wedge q)$.
   (b) Prove $p \mid p \equiv \neg p$.
   (c) Using parts (a) and (b), prove $(p \mid q)\mid(p \mid q) \equiv p \wedge q$.
*11. Using the tables given in Exercise 10,
   (a) prove $p \downarrow q \equiv \neg(p \vee q)$.
   (b) prove $p \downarrow p \equiv \neg p$.
   (c) Using parts (a) and (b), prove $(p \downarrow q) \downarrow (p \downarrow q) \equiv p \vee q$.
*12. Find a simple expression that is logically equivalent to $p \wedge q$ and uses the NOR operation $\downarrow$ only. (*Hint:* Use the previous results and De Morgan's Law.)

*13. Find a simple expression that is logically equivalent to $p \vee q$ and uses the NAND operation $|$ only. (*Hint:* Use the previous results and De Morgan's Law.)

NOTE: Exercises 10 through 13, along with Exercise 7, show that all statements, up to logical equivalence, can be built up using only one logical connective (either NAND or NOR). However, this simplicity leads to statements that most humans would find unreadable. Still, computing machines would have no difficulty with them, and the theory of the structure of statements would be somewhat simpler with their use.

14. Carefully explain the difference between $P \leftrightarrow Q$ and $P \equiv Q$. Explain how these two concepts are related.

15. Carefully explain the difference between $P \rightarrow Q$ and P logically implies Q. Explain how these two concepts are related.

*16. Prove that $P \equiv Q$ if and only if P logically implies Q and Q logically implies P.

*17. Prove that P and Q logically imply R if and only if P logically implies $Q \rightarrow R$.

In Exercises 18–31, use the hypotheses to give a deduction of the conclusion. State reasons for each step, as in the text.

18. Hyp: $p, q \rightarrow \neg r, \neg r \rightarrow \neg p$ Concl: $\neg q$
19. Hyp: $p, p \rightarrow r$ Concl: $r \vee s$
20. Hyp: $p \rightarrow \neg p, \neg p \rightarrow p$ Concl: q
21. Hyp: $p \rightarrow q, q \rightarrow r, r \rightarrow \neg p$ Concl: $\neg p$
22. Hyp: $p \vee q \rightarrow s, \neg s$ Concl: $\neg p$
23. Hyp: $p \rightarrow q \wedge r, r \rightarrow \neg q$ Concl: $\neg p$
24. Hyp: $p \rightarrow q, s \rightarrow \neg q, \neg s \rightarrow p, q$ Concl: $\neg s \wedge p$
25. Hyp: $p \rightarrow q \wedge s, s \rightarrow t, q \rightarrow r$ Concl: $p \rightarrow t \wedge r$
26. Hyp: $p \rightarrow q, \neg p \rightarrow (r \rightarrow q), r$ Concl: q
27. Hyp: $p \vee q \rightarrow r, p \rightarrow s, s \rightarrow \neg p$ Concl: $\neg p \wedge q \rightarrow r$
28. Hyp: $p \rightarrow (r \rightarrow t), r$ Concl: $p \rightarrow t$
29. Hyp: $p \vee q \rightarrow s, s \rightarrow \neg p, \neg r \rightarrow \neg s$ Concl: $q \rightarrow r$
*30. Hyp: $p \rightarrow q \vee r, p \rightarrow t, q \rightarrow \neg p, r \rightarrow \neg t$ Concl: $\neg p$
*31. Hyp: $p \vee q \rightarrow r, t \rightarrow p, \neg t \rightarrow q$ Concl: r

In Exercises 32–40, decide whether the conclusion follows from the hypotheses. If so, give a deduction; if not, explain why with the help of truth tables.

32. Hyp: $p, q \rightarrow \neg r, \neg r \rightarrow \neg p$ Concl: $\neg q \rightarrow p$
33. Hyp: $p \rightarrow q, q \rightarrow r, r \rightarrow \neg p$ Concl: $\neg p \vee q$
34. Hyp: $p \vee q \rightarrow s, \neg s$ Concl: $\neg p \wedge s$
35. Hyp: $p \rightarrow q \wedge r, r \rightarrow \neg q$ Concl: $r \vee p$
36. Hyp: $p \rightarrow q, \neg p \rightarrow (r \rightarrow q), r$ Concl: $r \rightarrow p$
37. Hyp: $p \vee q \rightarrow r, p \rightarrow s, s \rightarrow \neg p$ Concl: r
38. Hyp: $p \rightarrow (r \rightarrow t), r \rightarrow s$ Concl: $p \rightarrow (\neg s \rightarrow r)$
39. Hyp: $p \vee q \rightarrow s, s \rightarrow \neg p, \neg r \rightarrow \neg s$ Concl: $q \vee r$
*40. Hyp: $p \rightarrow q \vee r, p \rightarrow t, q \rightarrow \neg p, r \rightarrow \neg t$ Concl: $t \rightarrow \neg p$

1.5 Predicate Logic

Predicate logic is the logic in which mathematical discourse can be expressed. This logic uses all of the logical symbols of the propositional calculus, as well as other symbols introduced in this section. The atomic statements, however, are no longer simply p, q, and so on, but take the form of meaningful statements of mathematics. We can illustrate with a simple example. Consider the statement

$$\text{If } x < y, \text{ then there is a value } z \text{ such that } x < z < y. \qquad (1.5.1)$$

1. This statement has variables x, y, z in it, and the intention in Equation (1.5.1) is to let x and y be arbitrary. Arbitrary values for x and y can be indicated by the phrase "for all choices of x and y." The symbol $\forall x$ is used for the phrase "for any x," or "for all x," or "for each x." Thus, this statement could be preceded by $\forall x \forall y$.
2. The variable z is not arbitrary, and is preceded by the clause "there is a value." The symbolism for this is $\exists z$, which is interpreted as "there exists a z." The symbols $\forall$ and $\exists$ are called **quantifiers.**
3. The "less than" symbol occurs in several places. Since the expression $x < z < y$ means $x < z$ and $z < y$, this statement has a meaning once the inequality $x < y$ is defined. This inequality is an example of a **predicate** in the two variables x and y. A predicate in several variables has the property that, once the variables are given specific values, it becomes either true or false.
4. Equation (1.5.1) is *false* if it is understood that x, y, and z are integers. (Take $x = 1$ and $y = 2$ to see this.) However, it is *true* if the variables are understood to be real numbers or fractions. (Take $z = (x + y)/2$ to see this.) Thus, we have to know where the variables x, y, and z are coming from in order to ascertain the truth value of this statement. The set of possible values for the variables is called the **universe** U. We can say that Equation (1.5.1) is false if the universe is taken as the integers; it is true if the universe is the reals.

The discussion in 1–3 shows that Equation (1.5.1) may be rewritten as follows.

$$\forall x \forall y ((x < y) \to \exists z (x < z \wedge z < y)) \qquad (1.5.2)$$

Thus, we see that Equation (1.5.1) can be written in terms of the predicate $x < y$, the quantifiers $\forall$ and $\exists$, and some of the logical connectives of the propositional calculus. A similar situation occurs in mathematical usage.

We now consider some of these ideas in more detail.

The Universe U

If we are talking about integers, we choose the universe U as the set of all integers. Once this is done, it will be understood that the variables

$x, y, \ldots$, denote integers, and it will not be necessary to constantly refer to this.

Many programming languages are especially careful about the universe of discourse. For example, variables are "declared" as integers, reals, and so on. Of course, due to the finite nature of computers, we don't expect all positive integers as a possible universe in a programming language. If x is declared to be an integer in Pascal, its size is severely limited, depending on the installation involved. Similarly, there are only finitely many "reals" in computing languages, since computers have a finite (though large) capacity for storage.

Predicates

Predicates are expressions involving variables (which are understood to be in the universe). Predicates take the place of the variable statements involving $p, q, r, \ldots$ of the propositional calculus. They, too, can take on the two truth values T and F, but these values depend on the values of the variables that are substituted in for the variables appearing in them.

EXAMPLE 1.5.1

The following are examples of predicates.

Prime (x): x is a prime number (universe: $\mathbb{N}$) [A prime number is a number greater than 1 whose only divisors are itself and 1.]

$Q(x)$: x is a rational number (universe: The set of reals $\mathbb{R}$)

LT (x, y): $x < y$ (universe: $\mathbb{R}$)

SQ2(x): x is the sum of two squares (universe: $\mathbb{N}$)

Pyth(x, y, z): $x^2 + y^2 = z^2$ (universe: $\mathbb{N}$)

$B(x, y, z)$: y is between x and z (universe: the set of points in the plane)

Thus, we have Prime(13) = T; Prime(12) = F: $Q(\pi)$ = F; $Q(12.567)$ = T; LT(5, -4) = F; SQ2(13) = T: Pyth(5, 12, 13) = T.

Just as in the propositional calculus, it is possible for two differently defined predicates to have the same truth values for all choices of the variables used. In that case, we say that the predicates are **equivalent,** and we write

$$P \equiv Q$$

EXAMPLE 1.5.2   Equivalent Predicates

We have the equivalence

$$(x \neq 2) \vee (y = 3) \equiv (x = 2) \rightarrow (y = 3)$$

by the propositional calculus, since $\neg p \vee q \equiv p \rightarrow q$. But the equivalence

$$x < y \equiv (x + 1 = y) \vee (x + 1 < y)$$

when the universe is $\mathbb{Z}$ does not follow from the propositional calculus but from special properties of $\mathbb{Z}$.

Definitions give equivalences. Thus

$$x \in A \cup B \equiv (x \in A) \vee (x \in B)$$

Since predicates have truth values, we can combine them using logical connectives. For instance, in Example 1.5.1, $\text{Prime}(x) \wedge (100 < x)$ is a new predicate that can be paraphrased as "x is a prime number greater than 100."

Often, in a mathematical theory, some predicates are assumed as given, and other predicates are defined in terms of these. The predicate $x = y$ and its properties are almost always taken as given.

EXAMPLE 1.5.3

Consider the predicate $x < y$. Once this predicate is given, other predicates may be defined in terms of it. Thus

x is positive $\equiv 0 < x$ (universe is $\mathbb{Z}$).

$x > y \equiv y < x$.

x is between y and $z \equiv (y < x) \wedge (x < z) \vee (z < x) \wedge (x < y)$.

x is the largest element in the universe $\equiv \forall y((y < x) \vee (y = x))$.

In this example, $<$ is an **atomic predicate** because the other predicates are defined in terms of it. An atomic predicate is one that cannot be constructed in terms of simpler ones.

Programming languages all have built-in predicates that can be used to define other predicates. The built-ins may be thought of as the atomic predicates. For example, most languages have equality ($=$) as an atomic predicate. This permits the statement ($X = Y$) which is either true or false, depending on whether the variables are equal. Similarly, the predicate $X < Y$ is usually allowed in programming languages. Other predicates in a programming language may be defined as Boolean-valued functions, that is, functions whose value is either true or false. (If Boolean values are not available in a programming language, then 1 and 0 are good alternatives.) For example, suppose we wish to define the predicate $\text{Near}(X, Y)$ to mean that X and Y are within 0.05 of each other: $|X - Y| < 0.05$. This means

$$\text{Near}(X, Y) \equiv (X - Y) < 0.05 \wedge (Y - X) < 0.05$$

The following clause in Pascal could be part of the definition used to define the predicate Near:

$$\text{Near} := ((X - Y) < 0.05) \text{ and } ((Y - X) < 0.05)$$

The Quantifiers ∀ and ∃

The quantifiers ∀ and ∃ are called, respectively, the **universal** and the **existential** quantifiers. The universal quantifier ∀ is used to denote "for any," "for every," "for all," or "for each." The existential quantifier denotes "there exists" or "there is." Thus ∀x is interpreted as "for each x in the universe U," and similarly ∃x is interpreted as "there is an x in the universe such that." Using these symbols, we can write the statement

There is a prime number greater than 100

as

$$\exists x (\text{Prime}(x) \wedge x > 100)$$

Similarly, the statement

Any integer x has a negative y,
that is, a number y such that $x + y = 0$

can be written

$$\forall x \exists y (x + y = 0) \qquad\qquad (1.5.3)$$

The order of writing these quantifiers is crucial. The statement $\exists y \forall x$ $(x + y = 0)$ can be freely translated as "There is a number that is the negative of all numbers."

   Much of our language and reasoning uses quantifiers. A good rule of thumb is to expect ∀ when "any," "every," "each," or "all" occurs, and to expect ∃ when "there is," or "there exists," or "it is possible to find" occurs.

EXAMPLE 1.5.4   Free and Bound Variables

An expression such as $\exists x P(x, y)$ is regarded as a predicate in *one* variable y. For example, the predicate $\exists x(x < y)$ has the meaning "There is some number smaller than y." x has been "quantified out." The variable x is called a **dummy variable** in this formula. We say that x is a **bound variable** in the formula $\exists x(x < y)$, and y is a **free variable** in this predicate. A free variable x is not quantified by ∀x or ∃x, but a bound variable is.

   We can freely substitute a value for the free variable y in $\exists x(x < y)$. For example, setting $y = 2$, we obtain $\exists x(x < 2)$, which is a statement asserting that there is a number in the universe smaller than 2. But setting $x = 2$ gives $\exists 2(2 < y)$, which is meaningless, since quantifiers must occur only before variables. Similarly, $\exists x \forall y (R(x, y, z, w))$ is a predicate in the two variables z and w. The variables x and y are bound. A predicate with no free variables is called a **statement.**

EXAMPLE 1.5.5 Statements and Predicates

$\text{Prime}(x) \wedge (100 < x)$ is a *predicate* with x as its only free variable. It states that x is a prime number greater than 100. However,

$$\exists x (\text{Prime}(x) \wedge (100 < x))$$

has no free variables. It is a statement. It is true (with universe $= \mathbb{N}$), and it asserts that there is prime number greater than 100. The expression

$$\forall y \exists x (\text{Prime}(x) \wedge (y < x))$$

also has no free variables and is therefore a statement. It asserts that there is a prime number larger than any number that may be given. This is a true statement, and is a way of saying that there are infinitely many primes. Again, the universe is taken as $\mathbb{N}$.

If the universe is *finite*, quantifiers can be replaced by ORs and ANDs.

EXAMPLE 1.5.6

Let $P(x)$ be the predicate $x < 3$, with universe $U = I_5 = \{1, 2, 3, 4, 5\}$. Find the truth value of $\forall x P(x)$ and $\exists x P(x)$.

Method $\forall x P(x)$ is true provided $P(x)$ is true for all x in U. This means that the statement $P(1) \wedge P(2) \wedge P(3) \wedge P(4) \wedge P(5)$ is true. Thus $\forall x P(x)$ is equivalent to this statement. Similarly, $\exists x P(x)$ is true provided that $P(x)$ is true for at least one x in U. This means that the statement $P(1) \vee \cdots \vee P(5)$ is true. For this example, $P(2)$ is T and $P(3)$ is F. This is enough to show that the truth value of $\exists x P(x)$ is T, and the truth value of $\forall x P(x)$ is F.

Even for a finite universe, it is clearly useful to use quantifiers, since no one wants to list many statements connected by ORs. In Example 1.5.6, for instance, the simple statement $\exists x (x < 3)$ replaced the compound statement

$$1 < 3 \vee 2 < 3 \vee 3 < 3 \vee 4 < 3 \vee 5 < 3$$

The quantified statement is sometimes preferable. If the universe is large, it becomes a practical necessity.

When the universe is infinite, quantifiers are a way of avoiding an infinite number of ORs and ANDs. For the universe $\mathbb{N}$, the best we could do for a statement $\exists x Q(x)$ is

$$Q(0) \lor Q(1) \lor Q(2) \lor \cdots$$

This is not an allowable statement because it is infinitely long. Compare it to $\exists x Q(x)$, which is allowed in predicate logic, and which is made up of exactly six symbols.

The famous conjecture known as Fermat's Last Theorem can be stated as

$$\forall x \forall y \forall z \forall n((x^n + y^n = z^n) \to (n = 1 \lor n = 2)) \qquad (1.5.4)$$

(The universe is $\mathbb{N}^+$, the set of positive integers.) Nobody knows whether this result is true or false, although Fermat claimed he had a proof. No obvious finitistic approach is available. It is of little use to write this statement using an infinite number of ANDs connecting variable-free statements.

Terms

A **term** is a function of variables in the universe, whose value is also in the universe. Equation (1.5.3) uses the term $x + y$, and Fermat's Last Theorem [Equation (1.5.4)] uses the term x^n as well as $u + v$. Since the value of a term is in the universe when the variables are, it makes sense to substitute a term for a free variable.

In the Pascal example defining $|x - y| < 0.05$, which followed Example 1.5.3, the term $x - y$ was used. Terms are usually built into computer languages. However, many of these are not strictly terms in our sense of the word, since the limitation of size often makes some operations invalid. Thus, we cannot arbitrarily add two integers since the sum may be "out of range." Despite these limitations, it is often convenient, for the purpose of illustration, to regard the built-in functions of a programming language as terms.

We can summarize the discussion so far by stating that predicate logic is the study of statements and predicates built up from atomic predicates, terms, and constants, using variables, quantifiers, and the connectives of the propositional calculus. We now illustrate how the foregoing notions are used in some familiar mathematical situations.

EXAMPLE 1.5.7

In the universe $\mathbb{N}^+$ of positive integers, define the notion of divisibility $(x \mid y$ or x divides $y)$ and define the predicate "x is a prime number".

Use equality ($=$) as the only atomic predicate, and the product xy as the only term.

Method We say that x divides y if y is a multiple of x. This means that $y = zx$ for some z. Thus,

$$x \mid y \equiv \exists z(y = xz)$$

A number x is defined to be a prime number if any factorization of x into positive integers must have one of the factors equal to 1 (and the other equal to x). In addition, a prime must be greater than 1. Thus,

$$\text{Prime}(x) \equiv (x \neq 1) \wedge \forall y \forall z[(x = yz) \to (y = 1) \vee (y = x)]$$

However, since we have already defined the "divides" predicate $x \mid y$, we can restate this as

$$\text{Prime}(x) \equiv (x \neq 1) \wedge \forall y((y \mid x) \to (y = 1) \vee (y = x))$$

Here, $x \neq 1$ is a substitute for $\neg(x = 1)$. We did not use the statement $x > 1$, since we did not assume that $<$ or $>$ was an atomic predicate.

The following example illustrates the use of this notation in a familiar, nonmathematical situation.

EXAMPLE 1.5.8   Relatives

Using a few atomic predicates, it is possible to describe the relationship of various people in a family. We take the universe as the people in a city and introduce the following atomic predicates:

$$M(x) \equiv x \text{ is male;} \qquad F(x) \equiv x \text{ is female;}$$
$$P(x, y) \equiv x \text{ is a parent of } y$$

A *father* is a male parent. We can thus define the new predicate

$$\text{Fa}(x, y) \equiv P(x, y) \wedge M(x)$$

Motherhood is similarly defined:

$$\text{Mo}(x, y) \equiv P(x, y) \wedge F(x)$$

To say that John and Mary are siblings means they have the same mother:

$$\exists x(\text{Mo}(x, \text{John}) \wedge \text{Mo}(x, \text{Mary}))$$

(Here, John and Mary are constants in the universe.) We can similarly define the predicate $\text{Sib}(x, y)$ (x and y are siblings) using this idea. Let's generalize to allow them to have the same parent. Since we don't want John, for example, to be his own sibling, we must stipulate that $x \neq y$:

$$\text{Sib}(x, y) \equiv \exists z(P(z, x) \land P(z, y)) \land (x \neq y)$$

The statement that Adam has no father can be expressed as

$$\forall x \neg \text{Fa}(x, \text{Adam})$$

Freely translated, this says "Each person you name is not the father of Adam." Another way of expressing this is to say that there does not exist a father of Adam. Thus,

$$\neg \exists x \text{Fa}(x, \text{Adam})$$

This is an example of the equivalence of $\forall x \neg$ and $\neg \exists x$, which will be discussed later in this section.

Lest the reader be overly concerned with the biblical interpretation of these last two equivalent statements, we remind the reader that (1) nothing is said about the truth or falsity of this statement and (2) quantified variables must come from the universe U. Assuming we are talking about the population of a *finite* city, it will be an easy matter to locate an individual who has no parent: take the oldest individual in that universe. The reader can continue the family example in the exercises.

EXAMPLE 1.5.9

Discuss whether the statement

$$\forall x(x^2 = 2 \rightarrow x = 5) \tag{1.5.5}$$

is true or false. Take the universe as the integers, and then as the real numbers.

Method   We ask if

$$c^2 = 2 \rightarrow c = 5 \tag{1.5.6}$$

is true for all choices of c in the universe. If the universe is $\mathbb{Z}$, then c must be an integer, and $c^2 = 2$ is false. Therefore Equation (1.5.6) is true for all choices of c (by the propositional calculus), so Equation (1.5.5) is a true statement.

If the universe is $\mathbb{R}$, we can take $c = \sqrt{2}$, so Equation (1.5.6) becomes

$$(\sqrt{2})^2 = 2 \rightarrow \sqrt{2} = 5$$

which is clearly false. Therefore Equation (1.5.5) is false if the universe is chosen as $\mathbb{R}$.

This example once again highlights the point that before the truth value of a statement can be found, it is necessary to specify the universe.

In previous sections we discussed deductions in the propositional calculus. Deduction can also be defined in predicate logic, formalizing a mathematical proof. However, this topic is beyond the scope of this text.

What is the meaning of statements that are preceded by two quantifiers? Suppose $P(x, y)$ is a predicate that has x and y as free variables. The statement $\forall x \forall y P(x, y)$ can be read "For all x and for all y, $P(x, y)$." Its meaning is that P is true for all choices of x and y. It is thus equivalent to the same statement with the order of quantification reversed.

$$\forall x \forall y P(x, y) \equiv \forall y \forall x P(x, y)$$

Similarly $\exists x \exists y P(x, y)$ can be read "There exists an x and a y such that $P(x, y)$." This is equivalent to $\exists y \exists x P(x, y)$, and it means that $P(x, y)$ is true for at least one set of values x and y in the universe. We can abbreviate:

$$\forall x \forall y \equiv \forall y \forall x; \qquad \exists x \exists y \equiv \exists y \exists x \qquad (1.5.7)$$

On the other hand, the statement $\forall x \exists y P(x, y)$ is not equivalent to $\exists y \forall x P(x, y)$. We have already indicated in this section that these are not equivalent.

EXAMPLE 1.5.10

The statement

$$\forall x \exists y (y^2 = x) \qquad (1.5.8)$$

means (for the universe in question) that one can always find a square root of any number. The statement

$$\exists y \forall x (y^2 = x) \qquad (1.5.9)$$

means that there is an element that is the square root of all the others. (We have assumed that y^2 is a term in the language.)

The reader should try interpreting the statements $\forall x \exists y$ and $\exists y \forall x$ on the predicate $\text{Lv}(x, y)$, meaning "x is loved by y". The universe can be taken as all people.

We mentioned that a negation of a universal quantifier yields an existential quantifier. The following are theorems of predicate logic.

THEOREM 1.5.11

$$\neg \forall x P(x) \equiv \exists x \neg P(x) \tag{1.5.10}$$

Similarly, we have

$$\neg \exists x P(x) \equiv \forall x \neg P(x) \tag{1.5.11}$$

We shall not prove these. We can see them for a finite universe, since both of these formulas turn out to be equivalent to De Morgan's Law. Take the elements of the universe to be $a, b, \ldots$. Then

$$\neg \forall x P(x) \equiv \neg(P(a) \wedge P(b) \wedge \cdots)$$
$$\equiv \neg P(a) \vee \neg P(b) \vee \cdots$$
$$\equiv \exists x \neg P(x)$$

This is the equivalence Equation (1.5.10). Equation (1.5.11) is shown in the same way. The equivalence Equations (1.5.10) and (1.5.11) conform to common usage, as the following examples illustrate.

EXAMPLE 1.5.12

To say that "It is not true that all teachers are intelligent" is equivalent to saying "There is a teacher who is not intelligent." Here, the universe may be taken as the teachers, $I(x)$ as the predicate that x is intelligent, and the equivalence is $\neg \forall x I(x) \equiv \exists x \neg I(x)$.

A mathematical example: The statement "It is not true that $f(x) \leq 10$ for all x" can be converted to "$f(x) > 10$ for at least one value of x".

A typical use of these equivalences occurs in a proof by contradiction. Suppose we want to prove "All such x's are rational numbers", and our proof is to be by contradiction. Then we would start our proof with the statement "Suppose not. Then there is such an x which is not rational." Similarly, to deny that "there is an x satisfying $x < 10$" is to say that "for each x, $x \geq 10$".

We can abbreviate Equations (1.5.10) and (1.5.11) as follows.

$$\neg \forall x \equiv \exists x \neg; \qquad \neg \exists x \equiv \forall x \neg \tag{1.5.12}$$

Taking negations, we obtain the equivalences

$$\forall x \equiv \neg \exists x \neg; \qquad \exists x \equiv \neg \forall x \neg$$

These equivalences can be used for a succession of quantifiers. Thus,

$$\neg\forall x\exists y \equiv \exists x\forall y\neg \qquad \neg\exists x\forall y \equiv \forall x\exists y\neg \qquad (1.5.13)$$

EXAMPLE 1.5.13

The statement "Regardless of how tall you are, there is someone taller than you" is claimed. You wish to deny this. Using the methods of Equation (1.5.13), write this statement and its negative in the language of predicate logic.

Method Introduce the predicate $T(x, y)$ to mean that x is taller than y. The given statement becomes (if "you" is taken as x and "someone" as y) $\forall x\exists y T(y, x)$. Its negation is $\neg\forall x\exists y T(y, x)$. By (1.5.13) this is equivalent to $\exists x\forall y\neg T(y, x)$. This asserts that there is (a person called) x, such that for any person y, y is not taller than x. Freely translated, "There is a certain person such that no person is taller than that person." More freely, "There is a tallest person." This is the commonsense negation of the original assertion.

Exercises

In the "relatives" Example 1.5.8, define the predicates in Exercises 1–10 using quantification and the symbols of the propositional calculus. You may use predicates that have been defined in this example.

1. Bro(x, y): (x and y are brothers)
2. Son(x, y): (x is a son of y)
3. Gramma(x, y): (x is a grandmother of y)
4. OnlyCh(x): (x is an only child)
5. Cou(x, y): (x and y are cousins)
6. Cou2(x, y): (x and y are second cousins)
7. Mary and Alice have the same grandmother.
8. Hazel has only one child.
9. Robert is the nephew of Matilda, and has no children.
10. John is married to a cousin of Beverly. (Use a new atomic predicate Ma(x, y): x is married to y.)

In Exercises 11–14, the universe is taken to be $\mathbb{R}$, and the only atomic predicates are taken to be $<$ and $=$. The predicate $\leq$ may be used once it is defined in terms of these, as in Exercise 11. The usual arithmetic operations are allowable terms. Using these predicates only, define each of the following using quantification and the symbols of the propositional calculus.

11. $x \leq y$.
12. x is larger than both y and z.
13. $y = |x|$.
*14. x is closer to y than it is to z.

In Exercises 15–19, the universe is taken to be $\mathbb{R}$, and the only atomic predicates are taken to be $<$, $=$, and the predicate $\text{Int}(x)$, meaning x is an integer. The usual arithmetic operations are allowable terms. Using these only, define each of the following predicates using quantification and the symbols of the propositional calculus.

15. There is a positive square root of 2.
16. x is a rational number.
17. x is irrational.
*18. Any cubic equation has a real root.
*19. There is a rational number between any two distinct real numbers.

In Exercises 20–24, the universe is taken to be $\mathbb{Z}$, and the only atomic predicates are taken to be $<$, $=$. The usual arithmetic operations are allowable terms. Using these only, define each of the following predicates using quantification and the symbols of the propositional calculus.

20. There is a positive square root of 2.
21. x is an even number.
22. x is a perfect square.
23. Any positive integer is the sum of four squares.
*24. The product of two odd numbers is odd.

In Exercises 25–34, write each of the following statements in the language of predicate logic. Then write its negation, using the techniques discussed in this section. Finally, translate the negation into ordinary English. You may simplify any statement according to the rules of the propositional calculus. Use the ordinary arithmetic operations and the predicates $x = y$ and $x < y$.

25. There is a number whose square is -1.
*26. Any number has a next largest number.
27. One person in this family is the parent of all the rest.
28. If the square of a number is 4, then that number is 2.
29. There is a number that is greater than or equal to all other numbers.
30. No number is larger than all the others.
31. There are two different numbers whose square is 9.
32. Any even number greater than 2 is the sum of two prime numbers. [You may use the $\text{Prime}(x)$ predicate.]
*33. There is a quadratic equation with no solution.
34. No person in this family is the parent of all the rest.

35. Discuss the truth value of the statements $\forall x(1 + 1 = 2)$ and $\exists x(1 + 1 = 2)$. Take $\mathbb{N}$ as the universe.
*36. Take as three possible universes the sets $\mathbb{N}$ of natural numbers, $\mathbb{Z}$ of integers, and $\mathbb{R}$ of reals. Using $=$ and $<$ as the atomic predicates, and using no terms, find a statement of predicate logic that is true for $\mathbb{N}$ and false for $\mathbb{Z}$ and $\mathbb{R}$. Similarly find another statement that is true for $\mathbb{Z}$ but false for the other two, and another statement that is true for $\mathbb{R}$ but not for the other two.

*37. Take as three possible universes the sets $\mathbb{Q}$ of rational numbers, $\mathbb{Z}$ of integers, and $\mathbb{R}$ of reals. Using $=$ as the only atomic predicate, and using the term $x \cdot y$, find a statement of predicate logic that is true for $\mathbb{Z}$ and false for $\mathbb{Q}$ and $\mathbb{R}$. Similarly find another statement that is true for $\mathbb{Q}$ but false for the other two, and another statement that is true for $\mathbb{R}$ but not for the other two.

Exercises 38–42 contain some definitions used in the study of the calculus. Express each of these definitions using the language of predicate logic.

38. A function $f(x)$ defined over a set A is *bounded by* M provided $f(x) \leq M$ for all x in the set A.
39. $f(x)$ is said to be *bounded* provided it is bounded for some M. (See the previous exercise.)
40. If $f(x)$ is not bounded, it is said to be *unbounded*. (See the previous two exercises.)
*41. $f(x)$ is said to *attain its maximum* if it is bounded by some M and $f(x) = M$ for some value of x in A. (See Exercise 38.)
*42. If $f(x)$ attains its maximum M, as in the previous exercise, then M is said to be the *maximum* of $f(x)$ in A.

43. The predicate $SQ2(x)$ was defined in Example 1.5.1 as meaning that x is the sum of two squares. The universe is $\mathbb{N}$. Define the predicate $SQ2(x)$ using the symbols of predicate logic, the terms $x + y$ and $x \cdot y$ and the predicate $=$.

In Exercises 44–47, the universe is $\mathbb{I}_n$. Let $P(x, y)$ be a predicate, and let a_{ij} be the matrix of truth values $P(i, j)$. Thus, the ith row, jth column of a_{ij} is either T or F, depending on the truth value of $P(i, j)$. Describe when each of these statements is true in terms of the matrix a_{ij}.

44. $\forall x \forall y P(x, y)$
45. $\exists x \exists y P(x, y)$
*46. $\forall x \exists y P(x, y)$
*47. $\exists x \forall y P(x, y)$

By naming the appropriate predicates in the statements of Exercises 48–53, write each one using the language of predicate logic. In each case, write the negative of the statement; simplify, if possible; and convert into ordinary English.

48. If there were a Martian on Earth, it would be smaller than 4 feet tall.
49. John is smarter than everyone.
50. Someone is smarter than anyone else.
51. No one is smarter than Jane.
*52. For any positive real number ϵ there is positive real number δ such that if $|x| < \delta$, then $\sqrt{x} < \epsilon$.
53. Judy likes anyone who likes her, unless that person also likes Karen.

54. Simplify
    (a) $\neg \forall x \exists y (Q(x) \lor R(y))$
    (b) $\neg \forall x \forall y \exists z (z < x \land z < y)$
    (c) $\neg \exists u \exists v (u^2 + v^2 = 1003)$
55. Simplify
    (a) $\neg \forall r \forall s \exists t (P(r, s, t) \lor Q(r, s, t))$
    (b) $\neg \forall x \exists y \forall z (x < y \lor y < z)$
    (c) $\neg \exists x \exists y \exists z (x^4 + y^4 = z^4)$

In Exercises 56–61, the universe is $\mathbb{N}^+$, the positive integers. Determine whether each statement is true or false. In each case, explain your answer.

56. $\forall x \exists y (x \mid y)$
57. $\exists y \forall x (x \mid y)$
58. $\forall y \exists x (x \mid y)$
59. $\exists x \forall y (x \mid y)$
60. $\exists x \exists y (x \mid y)$
61. $\forall x \forall y (x \mid y)$

Summary Outline

- $\mathbb{N}$ = the set of **natural numbers** (nonnegative integers).
  = $\{0, 1, 2, 3, 4, \ldots\}$.

- $\mathbb{Z}$ = the set of **integers**.
  = $\{\ldots, -3, -2, -1, 0, 1, 2, 3, \ldots\}$.

- $\mathbb{I}_n$ = the set of integers between 1 and n inclusive.
  = $\{1, 2, 3, \ldots, n\}$. Here n is a positive integer.

- $\mathbb{Z}_n$ = the set of integers between 0 and $n - 1$ inclusive.
  = $\{0, 1, 2, \ldots, n - 1\}$. Again, n is a positive integer.

- A **bit** is an element of $\mathbb{Z}_2$, that is, either the number 0 or 1.

- A **byte** is an element of $(\mathbb{Z}_2)^8$, that is, a string of bits of length 8.

- **ASCII characters** are a standard set of symbols that can usually be sent to a printer from a computer.

- The **cardinality** of a finite set A is the number of elements in A. It is written $|A|$.

- The **empty set**, or the **null set**, $\varnothing$, is the set consisting of no elements. That is, for any choice of x, we have $x \notin \varnothing$.

- **Set operations**
  The **union** of A and B, written $A \cup B$, is the set of elements that are in A or in B (or in both).

The **intersection** of A and B, written $A \cap B$, is the set of elements that are in both A and B.

The **complement** of A, written A', is the set of all $x \notin A$. x must be in the universe U.

The **difference set** $B - A$ is the set of all elements in B but not in A.

- $A \subseteq B$ (**A is included in B**) if and only if every element of A is included in B.

- A **Venn diagram** is a diagram used to illustrate various set operations.

- A and B are **disjoint,** or **mutually exclusive,** if $A \cap B = \emptyset$.

- The sets $A_1, \ldots, A_n$ are **pairwise disjoint** if A_i and A_j are disjoint when $i \neq j$.

  A **partition** of a set U is a collection of nonempty, pairwise disjoint sets $A_1, \ldots, A_k$ whose union is U.

- A **function** f from A to B is a unique assignment of an element of B for each element a of A. We write $f: A \rightarrow B$. A is called the **domain** of f, and B is the **codomain.**

- The set of all subsets of U is called the **power set** of U and denoted $P(U)$.

- The **greatest integer function** $[x]$ is the integer n satisfying the inequalities $n \leq x < n + 1$.

- The function **n mod m** is the remainder when n is divided by m.

- **Collatz's function** is the function $C(n)$ defined to be $n/2$ when n is even, and $3n + 1$ when n is odd.

- The **Cartesian product** $A \times B$ is the set of ordered couples (a, b) where $a \in A$ and $b \in B$.

- An **operation** on a set A is a function $f: A \times A \rightarrow A$.

- A **sequence** of elements of A is a function $f: \mathbb{N} \rightarrow A$.

- A **finite sequence** of elements of A is a function $f: \mathbb{I}_k \rightarrow A$.

- A **matrix** A is a function A from $\mathbb{I}_m \times \mathbb{I}_n \rightarrow \mathbb{R}$. The value $A(i, j)$ is also written a_{ij}.

- The **characteristic function** of A, denoted f_A, is a function $f_A: U \rightarrow \mathbb{Z}_2$ such that $f(x) = 1$ if $x \in A$, and $f(x) = 0$ when $x \notin A$. U is the universe.

- The **binary logical connectives** are $\vee, \wedge, \rightarrow$, and $\leftrightarrow$, and are used to represent *or, and, if . . . then . . .* , and *if and only if*, respectively.

- The **unary logical connective** is $\neg$ and is used to represent *not*.

- **Priorities**
  $\neg$ has first priority. It applies to as little as possible.
  $\wedge$ has next priority after $\neg$.
  $\vee$ has the next priority after $\neg$ and $\wedge$.

- **Atomic statements** in the propositional calculus are variables $p, q, r, \ldots$, representing statements that can be true or false.

- **Compound statements** are statements built from atomic sentences using the logical connectives.

- **Truth tables** are tables used to indicate whether a compound statement is true (T) or false (F). The truth value of the compound statement is given for all possible combinations of truth and falsity for the atomic statements making up the compound statement.

- A **tautology** is a statement $P = P(p, q, \ldots)$ that is true for all possible truth values of the atomic statements $p, q, \ldots$.

- Some **important tautologies**

Law of the Excluded Middle	$p \lor \neg p$
Law of Contradiction	$\neg(p \land \neg p)$
De Morgan's Laws	$\neg(p \lor q) \leftrightarrow \neg p \land \neg q$
	$\neg(p \land q) \leftrightarrow \neg p \lor \neg q$
Contrapositive Law	$(p \to q) \leftrightarrow (\neg q \to \neg p)$
Reductio ad absurdum	$(\neg p \to p) \to p$
Law of the Double Negative	$\neg\neg p \leftrightarrow p$

- The **contrapositive** of $p \to q$ is $\neg q \to \neg p$. The two statements are equivalent.

- The **converse** of $p \to q$ is $q \to p$. These two statements are *not* equivalent.

- Some methods of proof in the propositional calculus are

  1. modus ponens
  2. use of the contrapositive
  3. case analysis
  4. contradiction
  5. the Deduction Theorem

- **Modus ponens:** $P, P \to Q$ logically imply Q.

- **The Deduction Theorem:** If $h_1, h_2, \ldots, h_n$ are a set of hypotheses that, together with P, logically imply Q, then the hypotheses $h_1, \ldots, h_n$ alone logically imply the statement $P \to Q$.

- **Theorem** Proof by Contradiction
  Suppose $h_1, h_2, \ldots, h_n$ are a set of hypotheses that, together with $\neg P$, logically imply P. Then the hypotheses $h_1, \ldots, h_n$ logically imply P.

- Statements P and Q are **logically equivalent** (written $P \equiv Q$) if the truth value of $P(p, q, \ldots)$ is the same as the truth value of $Q(p, q, \ldots)$ for all possible truth values of the atomic statements $p, q, \ldots$ that appear in them.

- **Theorem**
  $P \equiv Q$ if and only if $P \leftrightarrow Q$ is a tautology.

- Statements $P_1, P_2, \ldots, P_n$ **logically imply** the statement P, provided that P is true whenever $P_1, \ldots, P_n$ are all true. In this case, we say that P is a **valid conclusion** from the hypotheses $P_1, \ldots, P_n$.

- **Theorem**
  The statements $P_1, \ldots, P_n$ logically imply P if and only if $(P_1 \wedge \cdots \wedge P_n) \rightarrow P$ is a tautology.

- A **rule of inference** is a way of proceeding from several statements P_1, $P_2, \ldots, P_k$ to another, P. It is required that these statements P_i logically imply P. We write

$$\frac{P_1, \ldots, P_k}{P}$$

- A **deduction** (in the propositional calculus) from the hypotheses $P_1, \ldots,$ P_n is a sequence of statements $A_1, \ldots, A_N$ where each of the A_i either is one of the P_k, is a tautology, or follows from some of the previous A_j by a valid rule of inference.

- A statement P is a **consequence** of the statements $P_1, \ldots, P_n$ if it is the last statement of a deduction from the hypotheses P_i. We also say that P **is deduced** from the hypotheses P_i.

- **Theorem**
  If it is possible to deduce P from the hypotheses $P_1, \ldots, P_n$, then $P_1,$ $\ldots, P_n$ logically imply P.

- The **universe** in first-order logic is the set of possible values for the variables $x, y, \ldots$.

- There are two kinds of **quantifiers:** the **existential** quantifier $\exists$ and the **universal** quantifier $\forall$. The expression $\exists z$ is interpreted as "there exists a z". The expression $\forall z$ is interpreted as "for any z".

- A **predicate** is a statement involving variables (which are understood to be in the universe). It can take on the two truth values T and F, but these values depend on the value of the variables of the predicate.

- Predicates P and Q are **equivalent** (written $P \equiv Q$) if they have the same truth values for all choices of the variables appearing in these predicates.

- An **atomic predicate** is one that can't be constructed in terms of simpler ones.

- A **free variable** x is not quantified by $\forall x$ or $\exists x$. *Example:* $\forall y(x < y)$. A **bound variable** x is quantified by $\forall x$ or $\exists x$. *Example:* $\exists x(x < y)$.

- A **statement** is a predicate with no free variables.

- **A term** is function of variables in the universe, whose value is also in the universe.

- Rules for the **negation of quantifiers:**
$$\neg \forall x P(x) \equiv \exists x \neg P(x)$$
$$\neg \exists x P(x) \equiv \forall x \neg P(x)$$

CHAPTER 2

Foundations

We tend to take facts about the natural numbers for granted. Most likely this is so because we have been indoctrinated with facts about whole numbers since we were little children. In this chapter we look into some of these principles in some detail. In particular, the important notions of induction and recursion will be introduced, and some very basic ideas on counting and finite sets will be analyzed. The idea of a relation is introduced, and we will also consider the question of how and why numbers are represented in different bases.

2.1 Relations

In this section, we study and illustrate an important class of predicates—the predicates $P(x, y)$ involving two variables.

DEFINITION 2.1.1

A **relation** $R(x, y)$ on a set U, called the universe, is a predicate in two variables x and y. For each x and y in U, $R(x, y)$ is either true or false.

It will often be convenient to use infix notation. This means that we will write xRy rather than $R(x, y)$. For example, we usually write $x < y$ rather than $<(x, y)$ or $\mathrm{LT}(x, y)$.

EXAMPLE 2.1.2 *Some Examples of Relations*

a. $x < y$: universe $U = \mathbb{Z}$.
b. $x \leq y$: $U = \mathbb{R}$.
c. $x = y$: U = any set A.
d. x divides y or $x \mid y$: $U = \mathbb{N}^+$.
e. x PARENT y or x is a parent of y: U = people in the city.
f. $P \equiv Q$ (P is logically equivalent to Q) or $P \leftrightarrow Q$ is a tautology: U = statements of propositional calculus.
g. $x \equiv y$ mod 7 (read "x is congruent to y modulo 7"): $U = \mathbb{Z}$. This is true if and only if $x - y$ is divisible by 7. In general, we write $x \equiv y$ mod n if $n \mid (x - y)$.

In part (d), for example, we have $3 \mid 12$ and $3 \nmid 7$, respectively. ($a \nmid b$ is an abbreviation for $\neg(a \mid b)$.)

In part (f), we always have $P \equiv P$, since $P \leftrightarrow P$ is a tautology for any statement P.

In part (g), $2 \equiv 16$ mod 7, since $2 - 16 = -14$ is divisible by 7.

Certain properties of relations occur so frequently that they are given special names.

DEFINITION 2.1.3

A relation xRy is called **reflexive** if

$$xRx \text{ for all } x \text{ in } U$$

This can be expressed in predicate logic as

$$(\forall x)(xRx)$$

DEFINITION 2.1.4

A relation xRy is called **symmetric** if whenever xRy is true, then yRx is also true:

$$(\forall x)(\forall y)(xRy \rightarrow yRx)$$

DEFINITION 2.1.5

A relation xRy is called **transitive** if whenever xRy and yRz are true, then xRz is also true:

$$(\forall x)(\forall y)(\forall z)(xRy \wedge yRz \rightarrow xRz)$$

EXAMPLE 2.1.6   Reflexive, Symmetric, and Transitive Properties

In previous examples, the relation $x < y$ is transitive. It is not reflexive, because $x < x$ is false, and it is not symmetric because $x < y$ does not imply $y < x$.

The relation $\leq$ may be seen to be transitive and reflexive. It is not symmetric because if $x \leq y$ we need not have $y \leq x$.

Equality is reflexive, symmetric, and transitive.

The relation $x \mid y$ (x divides y) is reflexive because every positive integer divides itself. It is transitive because if $x \mid y$ and $y \mid z$, then $x \mid z$. (For example, $2 \mid 4$ and $4 \mid 36$, so $2 \mid 36$.)

x PARENT y is neither reflexive, symmetric, nor transitive.

$P \equiv Q$ is reflexive because $P \leftrightarrow P$ is a tautology. It is also transitive: If $P \leftrightarrow Q$ and $Q \leftrightarrow R$ are tautologies, then $P \leftrightarrow R$ is a tautology. It is also symmetric: If $P \leftrightarrow Q$ is a tautology, then $Q \leftrightarrow P$ is a tautology.

Remark   *If R is transitive*, it is customary to write a string of relations as $xRyRzRw$ to mean that xRy, yRz, and zRw. Because of transitivity, we can see that any element is related to another element later in the series. Thus, from $xRyRzRw$ we have yRw by inspection because yRz and zRw. This notation is common in the study of inequalities, where we often see inequalities such as

$$0 < x < y < 5$$

However, this convention is *not* used for nontransitive relations. A statement like $x \neq y \neq 0$ is not used because it does not connect x and 0. Nor is the student favorite $0 < x > y$ used, for the same reason.

DEFINITION 2.1.7

A relation R is called an **equivalence relation** if it is reflexive, symmetric, and transitive.

In the foregoing examples, equality, congruence modulo 7, and logical equivalence of statements are equivalence relations. It is no coincidence that the symbol $\equiv$ was used for congruence and logical equivalence. Symbols such as $\equiv$ and $\approx$ are often used for equivalence relations because of the similarity with the properties of $=$. However, the equal sign is always reserved for equality: $x = y$ means that x and y are the same. We study the theory of equivalence relations later in this section.

Relations can be *graphed*. We assume that the universe U is finite and proceed as follows. Place vertices in the plane, one for each element of the universe. These vertices represent the universe. If two vertices represent elements x and y that are related (xRy is true), then draw a line or curve from the vertices representing x to y. (We usually label the vertices themselves as x and y, and we take the liberty of saying "join x to y if xRy." These lines then represent the relation. The lines or curves are called **edges.** We also need a direction on this line because we need to distinguish xRy from yRx. Therefore, if xRy, put an arrow pointing from x to y on the line joining these two vertices. We call an edge with an arrow attached a **directed edge,** though we often simply call it an edge. The resulting figure is called the **graph** of the relation. To see whether xRy, simply determine whether x is joined to y by a directed edge.

Since we shall be working with a finite universe, our examples are usually given with the universe $\mathbf{I}_n$, consisting of the integers from 1 through n.

EXAMPLE 2.1.8

Draw the graph for the relation $x < y$, where the universe is $\mathbf{I}_4$.

Method   See Figure 2.1.1. Other possible graphs are Figures 2.1.2 and 2.1.3. Many drawings are possible, depending on where the vertices are placed and what curves are used to connect related vertices.

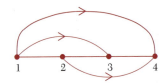

Figure 2.1.1

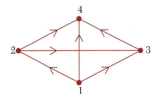

Figure 2.1.2

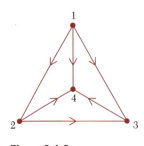

Figure 2.1.3

We make the vertices more visible by exhibiting them as little circles. This makes it easy to see what the universe is and forces us to ignore lines or curves that intersect, as in Figure 2.1.2. Any of these figures represents a **directed graph** or **digraph**. Chapter 5 considers the properties of digraphs in more detail.

EXAMPLE 2.1.9

Draw a digraph for the relation $x \mid y$, where the universe is $\mathbb{I}_6$.

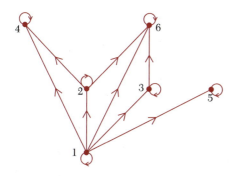

Figure 2.1.4

Method   See Figure 2.1.4. Since $x \mid x$ for every x, we must connect each x to itself. Therefore, it is necessary to introduce the indicated loops about each vertex. This is true of any reflexive relation.

Notice that these digraphs, even for relatively small universes, can be complex. Later, we discuss simplified procedures to cut down "unnecessary" edges.

Conversely, as in the following example, any digraph can be interpreted as the digraph of some relation.

EXAMPLE 2.1.10

The digraph of Figure 2.1.5 is the digraph of a relation R over a universe. Describe the relation and the universe.

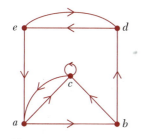

Figure 2.1.5

Method The universe is represented by the vertices. Therefore, take the universe as the set consisting of *a, b, c, d,* and *e*. Each directed edge corresponds to a relation. Thus, the edge from *a* to *c* shows that we have *aRc*. Similarly, define *xRy* if and only if there is an edge from *x* to *y*. This is the required relation.

The relation can be tabulated as in Table 2.1.1.

The body of the table is a square (5 by 5) **matrix.** We use 0 to mean *xRy* is false, and 1 to mean *xRy* is true. Thus, since *aRc* is true, the entry in the first row (opposite *a*) and third column (under *c*) is 1, so $a_{13} = 1$. The matrix gives complete information about the relation, because it

Table 2.1.1

R	a	b	c	d	e
a	0	1	1	0	0
b	0	0	1	1	0
c	1	0	1	0	0
d	0	0	0	0	1
e	1	0	0	1	0

tabulates whether *xRy* is true or false for all choices of *x* and *y* in the universe. It is therefore called the **matrix of the relation.** The matrix representation of a relation is important because it numerically designates a relation and so is useful for computational purposes. A computer looks at a relation through its matrix. Since we used 1 for true and 0 for false, the matrix of a relation is a square (0, 1) matrix; that is, an *n* by *n* matrix in which each entry is 0 or 1.

EXAMPLE 2.1.11

Alice, Bill, Carol, Dan, and Eloise are in the same class. Alice and Carol are friends, and so are Bill and Carol, Dan and Eloise, and Carol and Eloise. Discuss the relation *xFy*, meaning *x* is a friend of *y*, on the universe of these five people.

Method Common usage implies that if *x* and *y* are friends, then *y* and *x* are friends—*F* is a symmetric relation. We assume this. However, we do not expect transitivity, since in this case, a friend of a friend is not necessarily a friend. (Alice is a friend of Carol, who is a friend of Eloise, but Alice and Eloise are not friends.) Nor do we consider *F* to be reflexive, though one might argue that everyone is his or her own friend. The relation is thus symmetric but not reflexive or transitive. A digraph is given in Figure 2.1.6.

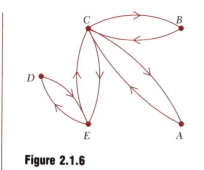

Figure 2.1.6

Note the symmetry in Figure 2.1.6. For each edge joining vertex X to vertex Y, there is an edge joining Y to X. Thus, Figure 2.1.6 can be simplified to Figure 2.1.7. The convention is that *any time vertex X is joined to vertex Y and vertex Y is joined to vertex X, they are joined by one line without a direction.* A graph whose edges are not directed is simply called a graph. Thus, although we can find a digraph for a symmetric relation, it is easier to use a (nondirected) graph in this case. The nondirected graph is called the **symmetric skeleton** for the relation.

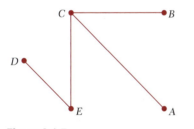

Figure 2.1.7

The matrix of this relation is given in Table 2.1.2. We have abbreviated each name.

We now consider simplification procedures for graphing *transitive* relations. Let us construct the $<$ relation on the universe $\mathbb{I}_{10}$. As we saw in Figures 2.1.1 to 2.1.3, the digraph was relatively complicated for four integers. We can expect a much higher degree of complication for ten integers.

Table 2.1.2

	A	B	C	D	E
A	0	0	1	0	0
B	0	0	1	0	0
C	1	1	0	0	1
D	0	0	0	0	1
E	0	0	1	1	0

EXAMPLE 2.1.12 A Simplified Graph

A commonsense approach for the graph of $<$ on $\mathbb{I}_{10}$ is to write $1 < 2 < \cdots < 10$ and draw the simple digraph of Figure 2.1.8. This digraph is *not* the digraph of the $<$ relation. For example, $4 < 7$ is not shown.

1   2   3   4   5   6   7   8   9   10

Figure 2.1.8

However, we can use Figure 2.1.8 to see that $4 < 7$. We see that $4 < 5$, $5 < 6$, and $6 < 7$. In general, we can say, using the digraph of Figure 2.1.8, that $x < y$ if there is a **path** in the digraph joining x to y. Each edge of the path must have the same direction as in the figure. Think of the edges as one-way streets. Then $x < y$ if we can go from x to y without going against the one-way sign.

Figure 2.1.8 *is* the digraph of a certain relation. If we define xPy (x precedes y) to mean that $y = x + 1$, then Figure 2.1.8 gives the digraph of P. When the graphic discussion in the preceding paragraph is translated into the relations P and $<$, we have

x < y if and only if there is a finite sequence of numbers x_0, x_1, ..., x_n such that $x_0 P x_1$, $x_1 P x_2$, ..., $x_{n-1} P x_n$, with $x = x_0$ and $y = x_n$.

We say that $<$ is the **transitive closure** of the relation P. The analysis can be generalized.

DEFINITION 2.1.13

Let R be any relation on a universe U. Define the relation R' on R as follows:

$xR'y$ if and only if there is a sequence x_0, x_1, ..., x_n such that $x_0 = x$, $x_n = y$, and $x_i R x_{i+1}$ for $i = 0$, ..., $n - 1$.

The relation R' is called the **transitive closure** of R.

Paraphrasing, $xR'y$ is true if and only if there is a path on the digraph for R joining x to y whose edges are all edges in the digraph of R. In the preceding example, $<$ was the transitive closure of P.

DEFINITION 2.1.14

If R' is the transitive closure of R, we say that the digraph of R is a **transitive skeleton** of the digraph of R'.

Thus, the digraph of Figure 2.1.8 is a transitive skeleton of the relation $<$.

Transitive closure may be thought of in this way: If a relation R is transitive, what facts can be used, along with transitivity, to force all the true statements xRy? Such a set of facts is a relation P. The given relation R is the transitive closure P'. Thus, the $<$ relation, known to be transitive, is "forced" by Figure 2.1.8. This makes $<$ the transitive closure of the P relation graphed in Figure 2.1.8 and gives Figure 2.1.8 as its transitive skeleton. The reader may verify that the transitive closure R' of any relation R is a transitive relation.

EXAMPLE 2.1.15

Given the relation

$$xRy: (x \mid y) \wedge (x \neq y)$$

on $\mathbb{I}_{12}$, find a digraph that is its transitive skeleton. Describe the relation of its skeleton.

Method   We know that $x \mid y$ is transitive. Thus, once we know that $1 \mid 2$ and $2 \mid 4$, it follows that $1 \mid 4$. We do not need this latter fact. It is forced by transitivity from the preceding two. Figure 2.1.9 graphs a minimal relation P. There are no loops about any vertex x, since that would imply xRx. The graph can be defined numerically by the condition that xPy if and only if y/x is a prime number.

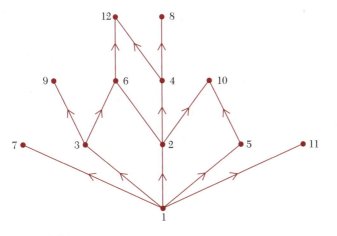

Figure 2.1.9

We can similarly simplify the graphing of reflexive relations. Graphically, once we know that a relation is reflexive, we can omit the loops about each vertex because these loops are all automatic. We then label the graph as the *reflexive skeleton* of the relation. This means we must fill in, mentally, the omitted loops. For example, Figure 2.1.8 is the reflexive skeleton of the relation $(x = y) \lor (x + 1 = y)$. It is the **reflexive-transitive** skeleton of the relation $x \le y$. Similarly, Figure 2.1.9 is the reflexive-transitive skeleton of the relation $x \mid y$. We can summarize with the following definition.

DEFINITION 2.1.16

Let R be a relation on U. Define the relation S by the condition xSy if and only if $xRy \lor x = y$. Then S is called the **reflexive closure** of R. If, in addition, $\neg xRx$ for every x, then the graph of R is called the **reflexive skeleton** of the graph of S.

Equivalence Relations

We now study equivalence relations. We start with an example.

EXAMPLE 2.1.17

Suppose that grades are assigned to the students in a class so that at least one person gets an A, and similarly for the grades B, C, D, and F. Define the relation xRy on the students to mean that x and y have the same grade. The relation R is seen to be an equivalence relation.

The class is partitioned into five groups: the students with grade A, those with grade B, and so on. Let these groups be called A_1 for the A students, $A_2, \ldots, A_5$. Then two students x and y are in the same group if and only if they have the same grade, and any student x determines the entire group, namely all of the other students with the same grade. These are the students y such that yRx.

We now generalize this example to arbitrary equivalence relations.

DEFINITION 2.1.18

Let R be an equivalence relation on the set U. For each x in U, the **equivalence class** $C(x)$ is the set of all y in U such that yRx: $C(x) = \{y \mid yRx\}$.

In Example 2.1.2g ($x \equiv y \bmod 7$), the equivalence class $C(1)$ contains 1, 8, 15, -6, -13, among others. It contains all numbers of the form $7n + 1$. The equivalence class $C(8)$ contains 8, 15, 22, 1, -6, among others. It turns out that the class $C(1)$ is equal to the class $C(8)$; that is, they have the same elements. The congruence $1 \equiv 8 \bmod 7$ leads to equality of sets: $C(1) = C(8)$; the sets contain exactly the same elements.

In Example 2.1.17, suppose Tim had grade B. Then the equivalence class $C(\text{Tim})$ consists of all students with the same grade as Tim, namely the set A_2. If Shirley had grade B, she would determine the same equivalence class: $C(\text{Tim})$ and $C(\text{Shirley})$ would be the same class. Shirley's equivalence class and Tim's are the same. We now proceed in full generality.

THEOREM 2.1.19   Equivalence Classes

Let R be an equivalence relation, and let $C(x)$ denote the equivalence class of x. Then

a. $x \in C(x)$.
b. If xRy, then $C(x) = C(y)$.
c. If $C(x) = C(y)$, then xRy.
d. If $C(x)$ and $C(y)$ have at least one element in common, then $C(x) = C(y)$.

Proof   We constantly use the definition of $C(x)$, so it is worth restating: $y \in C(x)$ if and only if yRx.

a. xRx by reflexivity. Therefore $x \in C(x)$.
b. Suppose xRy. Now let z be an arbitrary element of $C(x)$. Therefore zRx. Therefore we have zRy by transitivity. Therefore $z \in C(y)$, and every element z of $C(x)$ has been shown to be an element of $C(y)$. We may similarly show that every element of $C(y)$ is an element of $C(x)$, and so $C(x) = C(y)$.
c. Suppose $C(x) = C(y)$. Then, since $x \in C(x)$ by part (a), we have $x \in C(y)$. But this means xRy.
d. If z is a common element of $C(x)$ and $C(y)$, we have zRx and zRy. Therefore xRz by symmetry, and hence xRy by transitivity. Therefore $C(x) = C(y)$ by part (b). ■

We recall from Definition 1.1.17 that a set U is partitioned into sets A_i if these sets are nonempty, pairwise disjoint, and have union U.

COROLLARY 2.1.20

The equivalence classes of a relation R over the set U partition U.

Proof   By Theorem 2.1.19a, $x \in C(x)$. Therefore, every equivalence class is nonempty, and every element x of U is in one of the equivalence classes, namely $C(x)$. By Theorem 2.1.19d, no two (different) equivalence classes have an element in common. ■

Conversely, any partition of U leads to a unique equivalence relation in which the sets of the partition are the equivalence classes.

THEOREM 2.1.21

Let $A_1, A_2, \ldots$ be a partition of U. Define the relation xRy to mean that x and y are in the same set A_i. Then xRy is an equivalence relation and the sets A_i are the equivalence classes for R.

Proof Left as an exercise. ∎

EXAMPLE 2.1.22   Partitioning and Equivalence Relations

Going back to Example 2.1.17, we defined the equivalence relation xRy to mean that x and y have the same grade. This is a way of indirectly classifying students; using R, we can say when they are put in the same category. Once we partition into the sets A_1, and so on, we retrieve the equivalence relation by stipulating that xRy if x and y are in the same equivalence class.

Figure 2.1.10

Figure 2.1.10 gives a partition of $\mathbb{I}_6$. The associated equivalence relation is graphed in Figure 2.1.11. We have used the reflexive-symmetric skeleton for the digraph of this relation. Thus, no loops are shown, and a "two-way street" is unmarked.

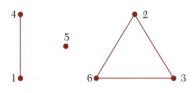

Figure 2.1.11

EXAMPLE 2.1.23

Exhibit the digraphs of *all* equivalence relations R on $\mathbb{I}_6$ in which $1R2$, $2R3$, and $4R5$.

Method We need to partition the universe in such a way that 1, 2, and 3 are in the same set, and so are 4 and 5. All of the digraphs are given in Figure 2.1.12. Actually, we exhibit the reflexive-symmetric-transitive skeletons. The full digraph of skeleton B is indicated on the right.

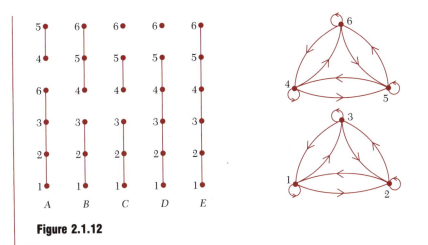

Figure 2.1.12

Partially Ordered Sets

If R is a relation that is reflexive and transitive, it is natural to think of R as an ordering. The division relation of Example 2.1.15 is such a relation, as is $\leq$. However, suppose we define a transitive relation R on the set of integers 1, 2, 3 such that $1R2$, $2R3$, and $3R1$, and we include the reflexive conditions $1R1$, $2R2$, and $3R3$. A transitive skeleton S is given in Figure 2.1.13.

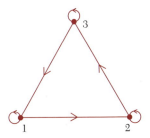

Figure 2.1.13

The **cycle** $1 \to 2 \to 3 \to 1$ in this graph goes counter to what we want an ordering to be. We can avoid these cycles if we stipulate the following condition on a relation.

DEFINITION 2.1.24

A relation R is **antisymmetric** if, for all x and y, whenever xRy and yRx, we have $x = y$:

$$(\forall x)(\forall y)(xRy \land yRx \to x = y)$$

This condition is valid for the division relation as well as the $\leq$ relation on the integers. It is not valid for the transitive closure R of the relation S of Figure 2.1.13, where $1S'3$ and $3S'1$, yet $1 \neq 3$.

DEFINITION 2.1.25

A **partial ordering** on a set P is a relation R that is reflexive, transitive, and antisymmetric. A set P with a partial ordering R defined on it is called a **partially ordered set**, or **poset** for short. To make the ordering relation explicit, we refer to the poset (P, R).

Remark   The usual ordering of the number system has the property that, for any x and any y, either $x \leq y$ or $y \leq x$. A partial ordering with this property is called **linear.** A partial ordering is so-called because it does not necessarily have this property.

Because it is suggestive, we use the symbol $\leqslant$ for an arbitrary partial ordering on a set. We can now show that cycles such as in Figure 2.1.13 cannot appear in the digraph of a partial ordering.

THEOREM 2.1.26

No cycles can appear in any poset. That is, if

$$x \leqslant x_1 \leqslant x_2 \leqslant \cdots \leqslant x_n \leqslant x$$

then $x = x_1 = x_2 = \cdots = x_n = x$.

Proof   The proof is clear from our discussion of a string of relations in the remark following Example 2.1.6. Thus, for example, we can read $x \leqslant x_2$ and $x_2 \leqslant x$. By antisymmetry, $x = x_2$. In the same way, all terms are equal to x. A formal proof by mathematical induction must wait until Section 2.3. (See Exercise 2.3.37.)   ∎

Example 2.1.15 and Figure 2.1.9 give a relatively simple reflexive-transitive skeleton for the poset $\mathbb{I}_{12}$ for the division relation $x \mid y$. Similarly, Figure 2.1.8 does the same for $\mathbb{I}_{10}$ under the usual inequality $\leq$. In both cases we graphed a much simpler relation whose reflexive-transitive closure was the given partial ordering. We now generalize these examples for any finite poset.

DEFINITION 2.1.27

In any poset $(P, \leqslant)$, the relation $x < y$ is defined to mean $x \leqslant y$ and $x \neq y$.

We leave it to the reader to show that $<$ is a transitive relation on the set P.

DEFINITION 2.1.28

In any poset $(P, \leqslant)$, we say that x is **covered** by y if

1. $x < y$.
2. For any z, it is impossible to have $x < z < y$.

When x is covered by y, we write xCy.

To paraphrase condition 2, there is no element strictly between x and y. C is called the **covering relation.**

THEOREM 2.1.29

Let P be a finite poset with relation $\leqslant$, and let C be the covering relation. Then P is the transitive-reflexive closure of C.

Proof   We sketch the proof. A more rigorous proof is given in Example 2.3.16. It is only necessary to show that if $x < y$, then either xCy or there is a sequence $x_1, x_2, \ldots, x_n$ such that

$$xCx_1, x_1Cx_2, \ldots, x_nCy$$

We illustrate this as in Figure 2.1.14. If $\neg xCy$, then we can place an element z strictly between x and y. If $\neg xCz$, we can find another element w between x and z. We now continue this process of saturating the path from x to y. It cannot continue indefinitely because P is finite and no cycles can be introduced. When the process is completed, we will have the required sequence. ■

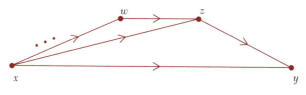

Figure 2.1.14

EXAMPLE 2.1.30

In $\mathbb{I}_{16}$, define $x \leqslant y$ to mean that y/x is a nonnegative power of 2. Show that this is a partial ordering. Find the covering relation C, and graph the skeleton of $\leqslant$.

Method   Since $x/x = 1 = 2^0$, we have $x \leqslant x$.
    To prove transitivity, suppose $x \leqslant y$ and $y \leqslant z$. Then $y/x = 2^m$, so $y = 2^m x$. Similarly, $z = 2^n y$. Therefore $z = 2^{m+n}x$, and so $x \leqslant z$.

To show that $\leqslant$ is antisymmetric, suppose $x \leqslant y$ and $y \leqslant x$. Then $y = 2^m x$ and $x = 2^n y$, where n and m are nonnegative. But these equations give $y = 2^m 2^n y$, or $1 = 2^{m+n}$. Therefore, $m + n = 0$, and since m and n are nonnegative, we have $m = 0$ and $n = 0$. Thus, $y = 2^m x = 2^0 x = x$. This proves antisymmetry.

When is x covered by y? Let us look at $3 \leqslant 12$. Here $12/3 = 4 = 2^2$. We can write $3 \leqslant 6 \leqslant 12$, so 3 is not covered by 12. In general, if $x < y$, we can show that x is covered by y if and only if $y/x = 2$. Thus, xCy if and only if $y = 2x$. The graph of this relation is given in Figure 2.1.15.

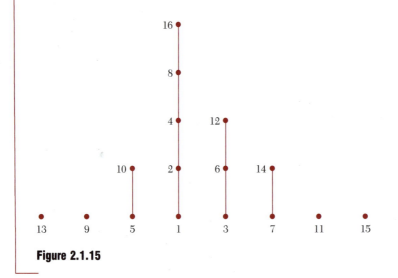

Figure 2.1.15

Exercises

In Exercises 1–6, determine whether

(a) the relation is reflexive.
(b) the relation is symmetric.
(c) the relation is transitive.
(d) the set is a poset with the given relation.
(e) the relation is an equivalence relation.

1. Set: the positive integers; relation $x > y$.
2. Set: all positive divisors of 36; relation $x \mid y$.
3. Set: $\mathbb{I}_{10}$; relation x next to y ($|x - y| = 1$).
4. Set: $\mathbb{Z}$ (the integers); relation $x^2 = y^2$.
5. Set: $\mathbb{N}$ (the natural numbers); relation $x^2 + y^2 + 1 = 0$.
6. Set: $\mathbb{I}_{10}$; relation $|x - y| \leq 2$.

7. Given the relation $|x - y| = 1$ on $\mathbb{I}_{10}$, draw its graph. What is the transitive closure? Prove your result.

*8. Given the relation $|x - y| = 2$ on the set $\mathbb{Z}$ of integers.
   (a) Show that its transitive closure is reflexive.
   (b) Show that its transitive closure is an equivalence relation.
   (c) Find the equivalence classes of the transitive closure of this relation.
*9. Ten people are seated at a round table as in Figure 2.1.16. If the universe is taken as the people around the table, say that xRy if x is seated two seats to the left of y.
   (a) Is R reflexive?
   (b) Draw a graph of the relation R.
   (c) Show that the transitive closure R' of R is an equivalence relation.
   (d) How many equivalence classes are in R'?

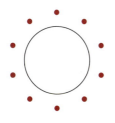

Figure 2.1.16

*10. Repeat Exercise 9, but define xR_3y if x is *three* seats to the left of y.
 11. Discuss the transitive closure of the relation x PARENT y.
 12. In a certain family, Fred and Alice are the children of Gertrude; Harold and Ike are the children of Alice; and Gertrude and Larry are the children of Morris. Draw the graph of the relation PARENT, using this family as the universe. [See Example 2.1.2e.] For which x is the relation Morris PARENT x true? For which x is the relation Morris PARENT' x true? Here, PARENT' is the transitive closure of PARENT.
 13. Prove: If a relation R is symmetric, then its transitive closure R' is symmetric.
*14. What is wrong with the following theorem and proof?
   Theorem(?): A transitive, symmetric relation R is reflexive.
   Proof(?): If xRy, then yRx by symmetry. But by transitivity, once we have xRy and yRx we must have xRx. This shows that R is reflexive.

   In Exercises 15–18 a set U is given. A relation R is defined on U, and all the pairs (x, y) for which xRy is true are given. Draw the full graph of the given relation and give its matrix. If there is a reflexive-symmetric-transitive skeleton for your graph, give it and indicate the type of skeleton you are displaying.

15. U: $\{a, b, c, d, e\}$ (a, a), (b, a), (c, a), (b, c), (b, d), (b, e), (c, d), (d, e), (e, e), (e, d), (c, c)
16. U: $\{X, Y, Z\}$ (X, X), (Y, Y), (Z, Z), (X, Y), (Y, X)
17. U: $\{r, s, t, u\}$ (r, s), (s, r), (r, t), (t, r), (s, t), (t, s), (s, u), (u, s)
18. U: $\mathbb{I}_6$ $(1, 2)$, $(2, 3)$, $(3, 4)$, $(4, 5)$, $(5, 6)$, $(6, 4)$, $(5, 3)$, $(4, 2)$, $(3, 1)$, $(1, 1)$, $(2, 2)$, $(3, 3)$, $(4, 4)$, $(5, 5)$, $(6, 6)$

In Exercises 19–22, the matrix of a relation is given. Draw the graph of this relation, and state whether the relation is reflexive, symmetric, and/or transitive.

19.

	a	b	c
a	1	0	0
b	0	1	1
c	0	1	1

21.

	r	s	t	u
r	1	0	0	1
s	0	0	1	0
t	1	1	0	0
u	0	1	1	1

20.

	1	2	3
1	0	1	1
2	1	0	1
3	1	1	0

22.

	a	b	c	d	e
a	0	1	0	1	1
b	1	0	1	0	0
c	0	1	0	0	1
d	1	0	0	0	0
e	1	0	1	0	0

23. Define a relation R on the positive integers $\mathbb{N}^+$ as follows: xRy if and only if y/x is a nonnegative power of 3. Show that R is a partial ordering of $\mathbb{N}^+$. Let C be the covering relation of R. Explain, in numerical terms, when xCy. Draw the graph of C when R is restricted to $\mathbb{I}_{12}$.
24. Repeat Exercise 23 except let xRy if and only if y/x can be factored into the form $2^m 3^n$, where m and n are nonnegative integers. For example, $4R12$ and $4R8$, but $\neg 2R30$.
25. For the points of the square in Figure 2.1.17, define pRq to mean that you can travel from p to q along the edges of the square by going up and/or right. By convention, we also agree that pRp for all p. Show that R is a partial ordering, and draw its graph.

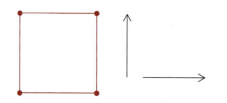

Figure 2.1.17

26. Repeat Exercise 25 for Figure 2.1.18.

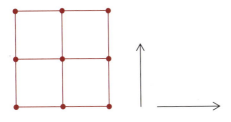

Figure 2.1.18

27. Let the universe be all points (x, y) in the plane, where x and y are positive integers. Define $(x_1, y_1) R (x_2, y_2)$ if and only if $x_1 \leq x_2$ and $y_1 \leq y_2$. Show that R is a partial ordering.

*28. The alphabet $A, B, \ldots, Z$ is ordered in the usual way: $A < B < \cdots < Y < Z$. Define the alphabetic (dictionary) order relation R on strings of length 2 on the alphabet. Is this relation symmetric? Is it transitive? Is it reflexive? Show that this order is a linear ordering. (See the remark following Definition 2.1.25.)

29. Let the universe U consist of all vertices of the cube of Figure 2.1.19. Define the relation R on U by the condition that pRq if and only if it is possible to move from p to q along the edges of the cube using only the indicated direction (right, up, and into the paper). By convention, we also agree that pRp is true for all p. Show that the relation is a partial ordering. (*Hint:* Express the relation algebraically in terms of the coordinates.) Explain what is the covering relation C in this case. Draw the graph of the cover relation C.

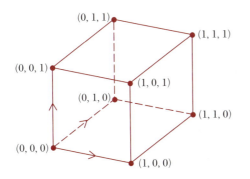

Figure 2.1.19

30. Generalize the·previous exercise by considering all points (x, y, z) of space where x, y, and z are nonnegative integers. Motion is possible only if neither x nor y nor z is decreased.

31. Let the universe consist of the people living in the United States. Define xRy to mean that x and y live in the same state. What are the equivalence classes? How many equivalence classes are there?
*32. Prove Theorem 2.1.21.
*33. We stated after Definition 2.1.27 that the relation $<$ defined on a poset P was transitive. Prove this.

2.2 One-to-One and Onto Functions—Permutations

In this section, we develop some further properties of functions and consider a special class of functions called permutations. We first consider the question of when a function $f: A \rightarrow B$ has an **inverse** $g: B \rightarrow A$. This function reverses the roles of output and input. If b is the output of a by f, then we want a to be the output of b by g, and vice versa. The details follow.

DEFINITION 2.2.1

Let $f: A \rightarrow B$. Then a function f is

1. **onto** if for every element b of B there is an element a of A such that $f(a) = b$. In this case, f is said to *map A onto B*.
2. **one-to-one** (1-1) if $f(a_1) = f(a_2)$ implies $a_1 = a_2$.

In other words, $f: A \rightarrow B$ is onto when the range of f equals the co-domain, and $f: A \rightarrow B$ is 1-1 when each element of the range is the image of exactly one element of A.

Remark   Equivalently, a function is 1-1 if different elements in A have different images. The contrapositive of the statement

$$f(a_1) = f(a_2) \rightarrow a_1 = a_2$$

is

$$a_1 \neq a_2 \rightarrow f(a_1) \neq f(a_2)$$

This latter form is often given as the definition of 1-1. We usually use the former because it is easier to work with equalities than inequalities.

EXAMPLE 2.2.2   1-1 and Onto Functions

The function $f: \mathbb{R} \rightarrow \mathbb{R}$ defined by $f(x) = x^2$ is neither 1-1 nor onto. It is not 1-1 since $f(2) = f(-2) = 4$, yet $2 \neq -2$. It is not onto because $-1 \neq x^2$ for any $x \in \mathbb{R}$. Yet if we use the same formula $g(x) = x^2$ to define

$g: \mathbb{R}^+ \to \mathbb{R}^+$, where $\mathbb{R}^+$ is the set of positive reals, then g will be both 1-1 and onto: Every positive real has a unique positive square root. And if we define $h: \mathbb{N} \to \mathbb{N}$ by the same formula $h(n) = n^2$, then h is 1-1 because every natural number has at most one square root that is a natural number. But h is not onto because the number 2 is not an image: $n^2 = 2$ does not have a solution in $\mathbb{N}$.

Thus, to decide if a function is 1-1 or onto, we must know both its domain and codomain.

We can relate these ideas to counting. What is meant when a set A is counted? For example, what is meant when a poker player says there are 37 chips in the pot? The count goes "1, 2, 3, . . . , 37." For each number counted, the player designates a chip. The honest player makes sure that each chip is counted and that no chip is counted more than once. The player stops when there are no more chips to be counted. This can be succinctly expressed in the language of set theory:

1. Each number from 1 through 37 is assigned a chip. Let C = the set of chips. Then the count is a function $f: \mathbb{I}_{37} \to C$.
2. No chip is counted twice. This means that f is 1-1.
3. Each chip is counted. This means that f is onto.

Thus, when it is stated that the pot has been counted and there are 37 chips in the pot, it means there is a function $f: \mathbb{I}_{37} \to C$ that is onto and 1-1. In general, then, a *count* of a set A is a function $f: \mathbb{I}_n \to A$ that is onto and 1-1. This is the formal meaning of the statement that A has n elements. As we have noted, we write $|A| = n$.

DEFINITION 2.2.3

For any set A, if there is a function $f: \mathbb{I}_n \to A$ that is 1-1 and onto, then we write $|A| = n$. The function f is called a **count** of A, and $|A|$ is its **cardinality.**

Remark   According to this definition, it is conceivable that a set may have 900 elements by one count and 899 by another. This often happens in practice, but it is always due to carelessness. It will be shown in Corollary 2.4.6 that there can only be one value of $|A|$, regardless of the counting function.

We have used $\mathbb{I}_n$ as the prototype of a set with n elements, and we have decreed that any set related to it in a 1-1 and onto manner has the same number n of elements. The general principle is the same for any two sets, finite or infinite: Two sets A and B *have the same number of elements* if there is a function $f: A \to B$ that is 1-1 and onto. Such a function may be thought of as a **pairing** of A with B, and so two sets have the same number if they can be paired.

If $f: A \rightarrow B$ is 1-1 and onto, then there is an *inverse function* f^{-1}: $B \rightarrow A$.† This is the "going-back" function. The function f^{-1} assigns each element b of B to the element a of A if b is the image of a by f.

DEFINITION 2.2.4

Let $f: A \rightarrow B$ be 1-1 and onto. The **inverse function** $f^{-1}: B \rightarrow A$ is a function such that

$$a = f^{-1}(b) \quad \text{if and only if} \quad b = f(a)$$

THEOREM 2.2.5

Let $f: A \rightarrow B$ be 1-1 and onto. Then there is one and only one inverse function $g: B \rightarrow A$. Conversely, if f and g are inverse functions, then each is 1-1 and onto.

Proof   For each b in B, there is an a in A such that $f(a) = b$. (This is the onto hypothesis.) There is at most one such a. (This is the 1-1 hypothesis.) Therefore, we may define $g(b)$ to be the unique a in A such that $f(a) = b$. Thus, $g(b) = a$ if and only if $f(a) = b$. The proof shows that g is the unique inverse.

Now, suppose $f: A \rightarrow B$ and $g: B \rightarrow A$ and that $b = f(a)$ if and only if $a = g(b)$. We shall show that f is necessarily 1-1 and onto. To show that f is onto B, let b be an arbitrary element of B. Define $a = g(b)$. Then, by assumption, $b = f(a)$, so b is the image of a by f. Thus, f is onto. To show that f is 1-1, suppose $f(a_1) = f(a_2) = b$. Then $a_1 = g(b)$ and $a_2 = g(b)$, so $a_1 = a_2$. This shows that f is 1-1. By symmetry, since the roles of f and g are identical, g is also 1-1 and onto. ■

EXAMPLE 2.2.6

Let $f: \mathbb{I}_6 \rightarrow \mathbb{I}_6$ according to the table

x	1	2	3	4	5	6
$f(x)$	3	5	6	4	1	2

Show that f is 1-1 and onto, and find its inverse function f^{-1}.

Method   By observation, we see that the range, which is the set of entries in the last row, consists of all integers of $\mathbb{I}_6$. Thus, f is onto. Further, there are no repetitions in the last row. This means that f is 1-1. The inverse $g = f^{-1}$ is obtained by reversing the roles of the domain and

---

† The notation f^{-1} does not mean $1/f$.

the codomain; we simply interchange the two rows in the definition of f as follows:

x	3	5	6	4	1	2
$g(x)$	1	2	3	4	5	6

As usual, we list the x values in numerical order for ease in evaluating:

x	1	2	3	4	5	6
$g(x)$	5	6	1	4	2	3

This table gives the inverse function $g = f^{-1}$.

We now define composition of functions and the identity function.

DEFINITION 2.2.7

Let $f: A \to B$ and $g: B \to C$. Then the **composition** $g \circ f: A \to C$ is defined by the equation

$$(g \circ f)(x) = g(f(x)) \tag{2.2.1}$$

DEFINITION 2.2.8

Let A be any set. The **identity function**

$$i_A: A \to A$$

is defined by the equation

$$i_A(x) = x \tag{2.2.2}$$

for all x in A.

Informally, the composition of two functions is simply a "function of a function." The output of f is used as the input of g. Note the order: When we write $g \circ f$, we find $f(x)$ *first* and then g, even though g appears first in the notation $g \circ f$. When we write $y = \sin(\log x)$, the function sin is written first but evaluated last.

Remark   Definition 2.2.7 is used when functions are written in prefix form. For postfix form, it is more natural to have the analogue of Equation (2.2.1) read

$$(x)(f \circ g) = ((x)f)g \tag{2.2.3}$$

and so composition is defined in the reverse order. This is consistent with calculator use. In most calculators, sin(30) is calculated by entering 30, then operating with sin. Thus, postfix notation "30 sin" is appropriate. In this case, to calculate sin(log(30)), the entries would be 30 log sin, in that order. Thus, for calculator (postfix) notation, we would want log ∘ sin to represent the composite function of first taking the log and then finding the sin. This is the exact reversal of prefix notation, in which log ∘ sin is thought of as "the log of the sin of." In this text, prefix notation and Equation (2.2.1) are generally used. Occasionally, when postfix notation is used, as for permutations, we revert to Equation (2.2.3). This confusion is, alas, standard.

The following results show that composition is analogous to multiplication, with the identity acting like 1.

THEOREM 2.2.9

Let $f: A \to B$, $g: B \to C$, $h: C \to D$. Then

$$(h \circ g) \circ f = h \circ (g \circ f) \tag{2.2.4}$$

■

THEOREM 2.2.10

Let $f: A \to B$. Then

$$i_B \circ f = f; \qquad f \circ i_A = f \tag{2.2.5}$$

■

The proofs of these results depend on the definition of equality of functions. For example, to prove the first of the formulas (2.2.5), we find

$$(i_B \circ f)(x) = i_B(f(x)) = f(x)$$

The first of these equations is true by the definition of composition, the second by the definition of i_B. Thus, $i_B \circ f = f$ by the definition of equality of functions. We leave the rest of the proof to the reader.

Equation (2.2.4) is called the **associative law** for composition, and Equation (2.2.5) is the **identity law.** Because of the associative law, we usually write $h \circ g \circ f$ for $h \circ (g \circ f)$. Because the location of parentheses doesn't alter the value, they are usually dropped.

THEOREM 2.2.11

Let $f: A \to B$ and $g: B \to A$ be inverses. Then $f \circ g = i_B$ and $g \circ f = i_A$. Thus,

$$f \circ f^{-1} = i_B; \qquad f^{-1} \circ f = i_A \tag{2.2.6}$$

Proof Let b be an arbitrary element of B, and let $a = g(b)$. Then $b = f(a)$ and $f \circ g(b) = f(g(b)) = f(a) = b$. Thus, $f \circ g(b) = i_B(b)$ for each b in B, so $f \circ g = i_B$ by the definition of equality of functions. The other equation follows similarly. ■

The converse is equally true.

THEOREM 2.2.12

If $f: A \rightarrow B$ and $g: B \rightarrow A$, and

$$f \circ g = i_B, \qquad g \circ f = i_A$$

then f and g are each 1-1 and onto and are inverses of one another.

Proof We use Theorem 2.2.5. If $b = f(a)$, then, applying g, we compute $g(b) = g(f(a)) = g \circ f(a) = i_A(a) = a$. Similarly, if $a = g(b)$, we have $f(a) = b$. ■

The following theorem shows how to take the inverse of a composition of functions.

THEOREM 2.2.13

If $f: A \rightarrow B$ and $g: B \rightarrow C$ are 1-1 and onto, then $g \circ f: A \rightarrow C$ is also 1-1 and onto. Further,

$$(g \circ f)^{-1} = f^{-1} \circ g^{-1} \tag{2.2.7}$$

Proof A simple calculation shows that

$$(g \circ f) \circ (f^{-1} \circ g^{-1}) = g \circ f \circ f^{-1} \circ g^{-1} = g \circ i_B \circ g^{-1} = g \circ g^{-1} = i_C$$

Similarly, $(f^{-1} \circ g^{-1}) \circ (g \circ f) = i_A$. By Theorem 2.2.12, it follows that $g \circ f$ and $f^{-1} \circ g^{-1}$ are inverses of each other. ■

Permutations

Imagine a calculator or a computer with memory cells numbered 1, 2, ..., n. Data is stored in each of these cells, perhaps numbers or records. We wish to perform certain data transfers. For example, interchange the contents of cells 1 and 2, or cycle the contents of cells 2, 3, 5, and 7 so that no data is lost. This means send the contents of cell 2 into cell 3, the previous contents of 3 into cell 5, and so on, until the contents of cell 7 are stored in cell 2. Any such transfer of data is given by a function. Namely, let $f: I_n \rightarrow I_n$ be defined as follows: Transfer the data in memory cell k into memory cell $f(k)$. Thus, interchanging memory cells 1 and 2 (called a **swap** of 1 and 2) is given by the function f, where $f(1) = 2$, $f(2) = 1$, $f(k) = k$ for $k > 2$. Any data transfer of this sort is given by a function that has the same domain and codomain and is 1-1 and onto.

In the rest of this section, we shall learn how to compute easily with such functions, and we will see further applications. For example, suppose we cycle the contents of cells 1, 2, 3; cycle the contents of cells 3, 4, 5, 6; and repeat this pair of cyclings three times. We want to get back to our original memory storage. Is there a quick way to proceed? We shall answer this question in Example 2.2.23.

For the remainder of this section, we consider 1-1 and onto functions from a set A onto itself. For simplicity, we take $A = \mathbb{I}_n = \{1, 2, \ldots, n\}$.

DEFINITION 2.2.14

A **permutation** of $\mathbb{I}_n$ is a 1-1, onto function $\sigma: \mathbb{I}_n \to \mathbb{I}_n$. The set of such permutations is denoted by S_n.

It is convenient to adopt *postfix notation* for permutations. The permutations themselves are denoted by Greek letters $\rho, \sigma, \tau, \ldots$. We write the image of k by σ as $(k)\sigma$, or even $k\sigma$. Composition is denoted by juxtaposition: We write $\rho\sigma$ for $\rho \circ \sigma$. We also call $\rho\sigma$ the **product** of ρ and σ. As noted in the remark following Definition 2.2.8, we have $(k)\rho \circ \sigma = ((k)\rho)\sigma = k\rho\sigma$. The function f of Example 2.2.6 was a permutation in S_6. It will be denoted simply as

$$f: \begin{array}{cccccc} 1 & 2 & 3 & 4 & 5 & 6 \\ \downarrow & \downarrow & \downarrow & \downarrow & \downarrow & \downarrow \\ 3 & 5 & 6 & 4 & 1 & 2 \end{array}$$

There is never any question of whether we can compose permutations on $A = \mathbb{I}_n$, since domain and codomain are always the same set A. Also, we need consider only one identity function $i = i_A$. Equations (2.2.4)–(2.2.7) can be summarized for permutations as follows:

$$\rho(\sigma\tau) = (\rho\sigma)\tau \qquad (= \rho\sigma\tau)$$

$$i\rho = \rho i = \rho$$

$$\rho\rho^{-1} = \rho^{-1}\rho = i$$

$$(\sigma\tau)^{-1} = \tau^{-1}\sigma^{-1}$$

We now illustrate how to multiply and take inverses.

EXAMPLE 2.2.15

Let

$$\sigma: \begin{array}{cccccc} 1 & 2 & 3 & 4 & 5 & 6 \\ \downarrow & \downarrow & \downarrow & \downarrow & \downarrow & \downarrow \\ 5 & 2 & 4 & 6 & 1 & 3 \end{array} \qquad \tau: \begin{array}{cccccc} 1 & 2 & 3 & 4 & 5 & 6 \\ \downarrow & \downarrow & \downarrow & \downarrow & \downarrow & \downarrow \\ 2 & 3 & 1 & 5 & 6 & 4 \end{array}$$

Compute $\sigma\tau$, $\tau\sigma$, and σ^{-1}.

Method To compute $\sigma\tau$, we first operate with σ, then with τ, since this is the convention we adopted for permutations. The answer is computed by finding, one at a time, the images of $1, 2, \ldots, 6$. We start by computing $1\sigma\tau$ (the image of 1 by $\sigma\tau$):

$$1\sigma\tau = (1\sigma)\tau = 5\tau = 6$$

This can be done very quickly by following the path of $\sigma\tau$ as indicated:

$$
\begin{array}{c}
\qquad 1 \;\; 2 \;\; 3 \;\; 4 \;\; 5 \;\; 6 \qquad\qquad 1 \;\; 2 \;\; 3 \;\; 4 \;\; 5 \;\; 6 \\
\sigma: \downarrow \qquad\qquad\qquad\qquad \tau: \qquad\qquad\qquad \downarrow \\
\qquad 5 \;\; 2 \;\; 4 \;\; 6 \;\; 1 \;\; 3 \qquad\qquad 2 \;\; 3 \;\; 1 \;\; 5 \;\; 6 \;\; 4
\end{array}
$$

We similarly see what happens to 2: $2 \to 2 \to 3$. Thus $2\sigma\tau = 3$. And 3: $3 \to 4 \to 5$. Therefore, by finding the action of $\sigma\tau$ on each of the elements 1 through 6, we find that

$$
\begin{array}{c}
1 \;\; 2 \;\; 3 \;\; 4 \;\; 5 \;\; 6 \\
\sigma\tau: \downarrow \;\; \downarrow \;\; \downarrow \;\; \downarrow \;\; \downarrow \;\; \downarrow \\
6 \;\; 3 \;\; 5 \;\; 4 \;\; 2 \;\; 1
\end{array}
$$

(The process takes longer to explain than to do.) When carrying this out, one need not actually find the image of the last integer, 6. The composition $\sigma\tau$ is 1-1 and onto, so the image of 6 must be 1, the only missing number. However, it is a good idea to compute the image of 6 as a check on your work.

In the same way, we compute $\tau\sigma$:

$$
\begin{array}{c}
1 \;\; 2 \;\; 3 \;\; 4 \;\; 5 \;\; 6 \\
\tau\sigma: \downarrow \;\; \downarrow \;\; \downarrow \;\; \downarrow \;\; \downarrow \;\; \downarrow \\
2 \;\; 4 \;\; 5 \;\; 1 \;\; 3 \;\; 6
\end{array}
$$

Observe that $\sigma\tau \neq \tau\sigma$. Thus, composition of permutations is not commutative. To find the inverse of σ, reverse arrows:

$$
\begin{array}{c}
\qquad 1 \;\; 2 \;\; 3 \;\; 4 \;\; 5 \;\; 6 \qquad\qquad 1 \;\; 2 \;\; 3 \;\; 4 \;\; 5 \;\; 6 \\
\sigma: \downarrow \;\; \downarrow \;\; \downarrow \;\; \downarrow \;\; \downarrow \;\; \downarrow \qquad \sigma^{-1}: \uparrow \;\; \uparrow \;\; \uparrow \;\; \uparrow \;\; \uparrow \;\; \uparrow \\
\qquad 5 \;\; 2 \;\; 4 \;\; 6 \;\; 1 \;\; 3 \qquad\qquad 5 \;\; 2 \;\; 4 \;\; 6 \;\; 1 \;\; 3
\end{array}
$$

We put σ^{-1} into standard form by turning it upside down and rearranging the top row numerically. Thus,

$$
\begin{array}{c}
1 \;\; 2 \;\; 3 \;\; 4 \;\; 5 \;\; 6 \\
\sigma^{-1}: \downarrow \;\; \downarrow \;\; \downarrow \;\; \downarrow \;\; \downarrow \;\; \downarrow \\
5 \;\; 2 \;\; 6 \;\; 3 \;\; 1 \;\; 4
\end{array}
$$

Multiplication and the taking of inverses are directly related to the memory manipulation example given earlier. The permutation $\rho\sigma$ is the result of performing ρ followed by σ. According to our memory cell description of permutations, ρ^{-1} puts the memory back to what it was before ρ was performed.

Cycle Notation

There is a very compact notation for permutations that is typographically simple and easy to compute with. It is based on a simple graphic idea. Since a permutation σ sends a set A *into itself*, we can represent the elements of A by points, or vertices, in a diagram, and indicate the action of σ by drawing an arrow from a vertex to its image. Let us consider, for example, the previous permutations. The appropriate diagram is indicated in Figure 2.2.1. We see that σ "decomposes" into three "cycles," of lengths 1, 2, and 3. Similarly, τ decomposes into two cycles, each of length 3.

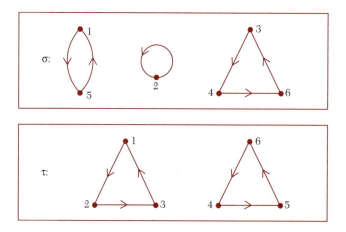

Figure 2.2.1

The general situation is similar. Since each of the vertices corresponds to an element of A, we have exactly one arrow from that vertex to another. Because a permutation is 1-1 and onto, exactly one arrow enters a vertex from some other vertex. If we now follow the arrows around, we must arrive at a cycle. If the cycle does not include all of the vertices, follow the arrows around from another vertex. In this way, we get a picture of the action of a permutation as decomposing into distinct cycles that have no vertex in common. (These are called *disjoint cycles*.)

The picture can be described linearly. We designate the cycle $3 \rightarrow 4 \rightarrow 6 \rightarrow 3$ as (346), closing the parentheses to show that 6 goes back into 3. We can also write this as (463) and (634). Similarly, the cycle $1 \rightarrow 5 \rightarrow 1$ becomes (15) or (51), and $2 \rightarrow 2$ becomes (2). Thus, we can represent σ by $(346)(15)(2)$, and we write $\sigma = (346)(15)(2)$. We usually omit cycles of length 1, so we write

$$\sigma = (346)(15)$$

Similarly, as in Figure 2.2.1,

$$\tau = (123)(456)$$

This representation of σ compactly gives all of the values of $(k)\sigma$. In an instant, we see that $4\sigma = 6$ and $5\tau = 6$. The fact that 2 doesn't show up in the representation of σ simply implies that it is *fixed:* $2 \rightarrow 2$, or $2\sigma = 2$.

Note that a cycle such as (123) may be regarded as a permutation of $\mathbb{I}_6$ in its own right. Here the action of 1, 2, and 3 is indicated by the cycle, whereas 4, 5, and 6 map into themselves. We now summarize this discussion.

DEFINITION 2.2.16

Let $a_1, a_2, \ldots, a_k$ be **distinct** elements of $\mathbb{I}_n$. The **cycle** $\rho = (a_1 a_2 \cdots a_k)$ is defined as follows:

$(a_1)\rho = a_2$
$(a_2)\rho = a_3$ and, in general,
$(a_i)\rho = a_{i+1}$ for $i = 1, \ldots, k-1$
$(a_k)\rho = a_1$ and
$(x)\rho = x$ if $x \in \mathbb{I}_n - \{a_1, a_2, \ldots, a_k\}$

The set $\{a_1, a_2, \ldots, a_k\}$ is called the **orbit** of the cycle ρ. In this case, the cycle ρ is said to have **length** k.

Remark Clearly $(a_1 a_2 \cdots a_k) = (a_2 a_3 \cdots a_k a_1)$.

DEFINITION 2.2.17

Cycles $\rho_1, \rho_2, \ldots$ are said to be **disjoint** if they have disjoint orbits.

Paraphrasing, the elements appearing in the cycles are all distinct. Thus, for example, (123), (685), and (49) are disjoint cycles.

We say that two permutations σ and τ **commute** if $\sigma\tau = \tau\sigma$.

THEOREM 2.2.18

Disjoint cycles commute. Thus, if ρ and σ are disjoint cycles, $\rho\sigma = \sigma\rho$. ∎

Let us illustrate this theorem with an example. Suppose we consider permutations on S_{10}. Why, for example, is $(1345)(629) = (629)(1345)$? What this means, by the definition of equality of functions, is that each of these permutations gives the same result for every number from 1 through 10. A case analysis shows this to be true. For $k \in \{1, 3, 4, 5\}$, both $(1345)(629)$ and $(629)(1345)$ act on k according to the cycle (1345); the cycle (629) has no effect. (Here we use the fact that the cycles (629) and (1345) are disjoint.) Similarly, for $k \in \{6, 2, 9\}$, the cycle (629) gives all the action, and (1345) has no effect. Finally, for $k \in \{7, 8, 10\}$ (the numbers in neither orbit), both cycles keep k fixed. Thus, in every case, the two cycles $(1345)(629)$ and $(629)(1345)$ yield the same result.

The general proof is quite similar and is omitted.

Remark   An example such as this is often used as a substitute for a proof, provided there is nothing special about the example that doesn't apply to the general case. A "general proof" can be given following the exact pattern of this illustration, but it sheds little extra light.

We can now summarize the discussion accompanying Figure 2.2.1.

THEOREM 2.2.19

Any permutation of S_n can be written as the product of disjoint cycles of length 2 or larger. Except for the order of these cycles, this decomposition is unique. This product is called the **cycle decomposition of the permutation.**   ■

We may regard the identity i as the product of zero cycles of length 2 or larger, in order for this result to hold universally. We omit the proof of Theorem 2.2.19. When we write a permutation as a product of disjoint cycles, we call the result the cycle form or the cycle decomposition of the permutation. Permutations are most conveniently manipulated when they are in cycle form.

EXAMPLE 2.2.20

Inverses of permutations are relatively easy to compute in the cycle notation. We first note that **inverses** of cycles are simply obtained by **reversing the cycle.** Thus,

$$(14568)^{-1} = (86541)$$

This follows from the definition of a cycle. We also have $(\sigma\tau)^{-1} = \sigma^{-1}\tau^{-1}$ if σ and τ are disjoint. This is true because disjoint cycles commute: $(\sigma\tau)^{-1} = (\tau\sigma)^{-1} = \sigma^{-1}\tau^{-1}$. Thus, for example, we can compute

$$[(1234)(578)]^{-1} = (4321)(875)$$

We also note that if σ is a cycle of length 2, then $\sigma^{-1} = \sigma$. For example,

$$(12)^{-1} = (21) = (12)$$

EXAMPLE 2.2.21

Let

$$\sigma: \begin{array}{cccccccc} 1 & 2 & 3 & 4 & 5 & 6 & 7 & 8 \\ \downarrow & \downarrow & \downarrow & \downarrow & \downarrow & \downarrow & \downarrow & \downarrow \\ 5 & 1 & 7 & 8 & 6 & 3 & 2 & 4 \end{array} \qquad \tau: \begin{array}{cccccccc} 1 & 2 & 3 & 4 & 5 & 6 & 7 & 8 \\ \downarrow & \downarrow & \downarrow & \downarrow & \downarrow & \downarrow & \downarrow & \downarrow \\ 2 & 1 & 5 & 3 & 8 & 6 & 4 & 7 \end{array}$$

Write σ and τ in cycle form and compute $\sigma\tau$, σ^2, and σ^3. (As usual, $\sigma^2 = \sigma\sigma$, $\sigma^3 = \sigma\sigma\sigma$, and so on.)

Method   To write σ in cycle form, start with 1 and follow its path via σ: $1 \to 5 \to 6 \to 3 \to 7 \to 2 \to 1$. Thus, we have, so far, $\sigma = (156372) \ldots$. Now start with any number not in the orbit, so far. We usually take the first missing number, here 4, but any one will do. We have $4 \to 8 \to 4$, a cycle of length 2. Therefore, we have, so far, $\sigma = (156372)(48) \ldots$. There are no missing elements, so the process is completed and we finally have $\sigma = (156372)(48)$. The process for τ is similar. We have, by inspection, $\tau = (12)(35874)$. [We do not write the cycle (6) of length 1.] Thus,

$$\sigma = (156372)(48); \qquad \tau = (12)(35874) \tag{2.2.8}$$

To compute $\sigma\tau = (156372)(48)(12)(35874)$, we proceed as in Example 2.2.15, except that we compute the answer as a cycle. Thus, following the path of 1, from left to right, we find

$$\sigma\tau = (1\underline{5}6372)(48)(12)(3\underline{5}874)$$

Thus, as indicated, $1 \to 5 \to 8$, so the product starts with the cycle $(18 \ldots)$. Now follow 8. We have, as indicated, $8 \to 4 \to 3$, and

$$\sigma\tau = (156372)(\underline{48})(12)(3587\underline{4})$$

so the cycle for $\sigma\tau$ continues $(183 \ldots)$. Proceeding with 3, we find $3 \to 7 \to 4$; then $4 \to 8 \to 7$; then $7 \to 2 \to 1$, closing the cycle. Thus, we have, so far, $\sigma\tau = (18347) \ldots$. Now follow the action of 2 under $\sigma\tau$. We have $2 \to 1 \to 2$, so 2 doesn't change. Next follow 5, the first number not yet in any orbit. We find $5 \to 6$; then $6 \to 3 \to 5$. This yields the cycle (56), so we obtain the computation $\sigma\tau = (18347)(56)$. Once again, the computation is much simpler than the explanation.

Finding powers of permutations written in cycle form is relatively straightforward. First, note that if α and β are disjoint cycles, then $(\alpha\beta)^2 = \alpha\beta\alpha\beta = \alpha\alpha\beta\beta$, since α and β commute. Therefore, $(\alpha\beta) = \alpha^2\beta^2$

for disjoint cycles α and β, and similar reasoning applies to higher powers. Further, finding the power of a cycle can be done almost by inspection. For example, $(1234567)^3$ is computed by noting that under this transformation, each element jumps three over to find its image. Thus, $1 \rightarrow 4$, $4 \rightarrow 7$, and so on, so that $(1234567)^3 = (1473625)$, and, similarly, $(1234)^2 = (13)(24)$. In this example, we found $\sigma = (156372)(48)$. Therefore we have, directly,

$$\sigma^2 = (167)(532); \qquad \sigma^3 = (13)(57)(62)(48)$$

Earlier, we introduced the idea of a *swap* of positions. This is simply realized as a cycle (ij) of length 2. It is usually called a *transposition*.

DEFINITION 2.2.22

A **transposition** is a cycle of length 2.

A calculation shows that if σ is a transposition, then $\sigma^2 = i$. More generally, if σ is a cycle of length r, then $\sigma^r = i$.

At the beginning of our discussion of permutations, we gave some examples of data transfers among memory cells of a computer or calculator. For n cells, the permutation f transferred the data in cell k into cell $f(k)$. In our notation, the swap that interchanges the contents of cells 1 and 2 is simply given by (12). The cycling of the contents of cells 2, 3, and 4 is given by (234). Following one transfer by another is given by the product (composition) of these permutations. For example, if we swap the contents of cells 1 and 2, and then swap the contents of cells 2 and 3, we can compute the effect as $(12)(23) = (132)$. This is a cycling of cells 1, 3, and 2.

EXAMPLE 2.2.23

Earlier, preceding Definition 2.2.14, we cycled memory cells 1, 2, 3 and then 3, 4, 5, 6, and we did this three times. The question proposed was, how can we retrieve the initial memory storage?

Method   We can now compute

$$(123)(3456) = (124563)$$

Therefore $[(123)(3456)]^3 = (124563)^3 = (15)(26)(43)$. The result can be obtained by three disjoint transpositions (swaps). To retrieve the original memory storage, simply compute the inverse: $[(15)(26)(43)]^{-1} = (15)(26)(43)$. Thus, the answer to the proposed question is to swap the contents of 1 and 5, of 2 and 6, and of 3 and 4.

Suppose we have cards numbered from 1 to n and want to rearrange them in ascending order. We can do this by a sequence of transpositions, as follows. Find the card numbered 1, and, if it is not already in the first position, swap it with the card in the first position. Now do the same for 2, and continue until all cards are in natural order. The whole operation can be performed with at most $n - 1$ swaps. Thus, any permutation of I_n is a product of at most $n - 1$ (not necessarily disjoint) transpositions.

Rubik's Cube

Rubik's cube (Figure 2.2.2) is a complicated puzzle. In its original form, it consists of a cube visualized as 27 minicubes, as indicated in the figure. Its clever construction allows each face of the cube (consisting of nine minicubes) to be rotated 90°, and any sequence of rotations is possible. Since the cube initially comes with the six faces colored differently, the puzzle is to mix it up with a few twists and then to try to retrieve the initial configuration. It is quite difficult for the uninitiated.

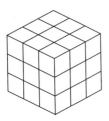

Figure 2.2.2

Any operation on the cube simply moves the minicubes to other positions; it leads to a permutation of the minicubes. The only minicubes that move are the 8 corners, numbered 1 through 8, and the 12 edge cubes, lettered, x, y, z, and so on, some of which can't be seen in the diagram. Let us first concentrate on the corner minicubes for simplicity.

EXAMPLE 2.2.24   Computing in Rubik's Cube

In Figure 2.2.3, the corner 8 is hidden and will be unaffected by our permutations. Some basic permutations allowed for the cube manipulation are F, U, and R (for front, up, and right). These are all 90° rotations,

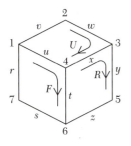

Figure 2.2.3

and thus their action on the corner cubes can be given as cycles:

$$U = (1234); \qquad R = (3564); \qquad F = (1467)$$

It is now possible to see what happens if any of these motions are used consecutively. For example,

$$U^2R^2 = (1234)^2(3564)^2 = (13)(24)(36)(54) = (163)(254)$$

a result not easily foreseen. Cubing, we find

$$(U^2R^2)^3 = [(163)(254)]^3 = I \quad \text{(the identity)}$$

since the cube of any cycle of length 3 is clearly the identity. Thus, $(U^2R^2)^3$ leaves all of the corners in the same position. What does it do to the edges? Let's compute.

Working only with the edge pieces, we see in Figure 2.2.3 that $U = (uvwx)$ and $R = (xyzt)$. Then $U^2 = (uw)(vx)$, and $R^2 = (xz)(yt)$. Therefore,

$$U^2R^2 = (uw)(vx)(xz)(yt) = (uw)(vzx)(yt)$$

and so

$$(U^2R^2)^3 = [(uw)(vzx)(yt)]^3 = (uw)^3(vzx)^3(yt)^3 = (uw)(yt)$$

Thus, we have found that $(U^2R^2)^3$ has the effect of interchanging edge cubes u and w and edge cubes y and t, but keeps all other edges and corner pieces fixed. The reader with a cube is invited to try out this sequence of moves.

The full description of U is $(1234)(uvwx)$. However, since the edge cubes and the corners never mix (they stay disjoint for all operations on the cube), we can always work them separately.

Our analysis is a simplified one. When one plays with a Rubik's cube, one discovers that it is possible for a corner cube to be fixed by a series of moves, but *rotated* 120°, or 240°, or not at all. Similarly, a series of moves might take an edge, move it to its original position, but rotated 180°. This can be observed by noting that the colors of the edge cube have switched positions. A fuller analysis has to take this into consideration.

Most solutions of the cube work with a few little algorithms, such as $(U^2R^2)^3$, that perform specific tasks. It is clear that our algebraic notation, such as $(U^2R^2)^3$, is superior to a verbal description, and that our cycle notation makes computations relatively easy. The algebra described in this section is the framework for any reasonable analysis of the cube.

Conjugation

If σ and τ are permutations, then $\tau^{-1}\sigma\tau$ is said to be the **conjugate of σ by τ.** Since σ and τ need not commute, the conjugate of σ need not be σ. For example, if

$$\sigma = (12345) \quad \text{and} \quad \tau = (12)(345)$$

then we have

$$\tau^{-1}\sigma\tau = (12)(543)(12345)(12)(345) = (14532) \qquad \textbf{(2.2.9)}$$

EXAMPLE 2.2.25   Relabeling $\mathbb{I}_n$

We now give an interpretation of $\tau^{-1}\sigma\tau$ that explains, among other things, why we must expect the conjugate of σ to be a cycle of length 5 also, since σ is. Briefly, we think of τ as giving a **code name,** or a new label, for the integers of $\mathbb{I}_n$. Here, $\tau = (12)(345)$, so 1 is given the code name 2, 2 the code name 1, 3 the code name 4, and so on. Decoding is clearly given by τ^{-1}. Now σ is a permutation of $\mathbb{I}_n$ (here, n may be taken as 5). We do not think of σ as a coding device.

   We now ask the question, "What is the action of σ, using the code names of elements rather than the elements themselves?" (See Figure 2.2.4.) For example, $2\sigma = 3$, so σ sends 2 into 3. Using code names, we see that the $1\sigma' = 4$. Here, 2 and 3 were replaced by code names 1 and 4, and we use σ' as the resulting permutation on code names. The general situation is to replace the equation $a\sigma = b$ by the equation in which a and b are replaced by their code names $a\tau$ and $b\tau$, and σ by σ': $a\tau\sigma' = b\tau$. Using $b = a\sigma$, we get the equation

$$a\tau\sigma' = a\sigma\tau$$

Since this equation holds for every a, we have

$$\tau\sigma' = \sigma\tau$$

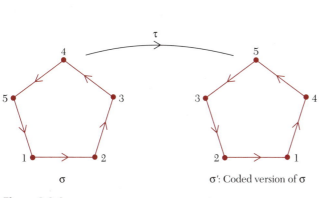

σ σ': Coded version of σ

Figure 2.2.4

Solving for σ', we have

$$\sigma' = \tau^{-1}\sigma\tau \qquad (2.2.10)$$

This interpretation of $\sigma' = \tau^{-1}\sigma\tau$ makes its evaluation simpler than directly evaluating this product. *We evaluate $\sigma' = \tau^{-1}\sigma\tau$ by replacing each element k in the cycle description of σ by its code number $k\tau$.* The reader should check the computation of Equation (2.2.9) to see how this can be easily done. Here, from $\sigma = (12345)$, we simply replace the entries k by their code $k\tau$. Thus, $\tau^{-1}\sigma\tau = (21453)$, which is another way of writing the answer given in Equation (2.2.9).

EXAMPLE 2.2.26

If $\sigma = (1357)(246)$ and $\tau = (4567)$, find $\tau^{-1}\sigma\tau$.

Method   Use τ to replace the numbers appearing in the cycle decomposition of σ. Thus, replace 4 by 5, 5 by 6, 6 by 7, and 7 by 4. The result is $\tau^{-1}\sigma\tau = (1364)(257)$.

Another way of looking at Equation (2.2.10) is as follows: To see how σ looks in coded form, first decode an element (τ^{-1}), then operate with σ, and then code the result (τ).

A simple example involving Rubik's cube (Figure 2.2.3) will illustrate this result. What effect does $R^{-1}FR$ have on the vertices of the cube? We have $F = (1476)$ and $R = (4356)$. We can compute directly, but the previous technique shows that $R^{-1}FR = (1374)$. Similarly, we found a move $\alpha = (uw)(yt)$. Then $F^{-1}\alpha F = (tw)(sy)$. Thus, by using conjugation on known moves, we can obtain other moves in which the roles of the edges, or vertices, are appropriately changed. Many published solutions of the cube use this technique.

Exercises

1. Let $f: \mathbb{Z} \to \mathbb{Z}$ according to the formula $f(n) = n + 1$. Is f a function? Is f onto ? Is f 1-1? Give brief explanations.
2. Repeat Exercise 1, but $f: \mathbb{N} \to \mathbb{N}$ and $f(n) = n + 1$.
3. Repeat Exercise 1, except $f: \mathbb{N} \to \mathbb{N}^+$ and $f(n) = n + 1$.
4. Repeat Exercise 1, except that $f: \mathbb{Z} \to \mathbb{N}$ and $f(n) = n + 1$.
5. Let $g: \mathbb{N} \to \mathbb{N}$ with $g(n) = 2n$, and let $h: \mathbb{N} \to \mathbb{N}$ with $h(n) = [n/2]$. Discuss whether g and h are onto and 1-1. Is $g(h(n)) = n$? Is $h(g(n)) = n$? Explain. Are g and h inverses? Explain.

6. Let $f: \mathbb{Z}_7 \to \mathbb{Z}_7$ according to the formula $f(n) = (4n + 2) \bmod 7$. In a table list all values of n in $\mathbb{Z}_7$ together with corresponding values of $f(n)$. Is f a permutation on $\mathbb{Z}_7$? If so, write f in cycle form.

7. Let $h: \mathbb{Z}_7 \to \mathbb{Z}_7$ with $h(n) = (n^3 + 2) \bmod 7$. Discuss whether $h(n)$ is onto and/or 1-1.

*8. Suppose $f: A \to B$ and $g: B \to A$, and that $f(g(x)) = x$ for all $x \in B$. Prove that f is onto and that g is 1-1.

*9. Suppose $f: A \to B$ is onto. Show that there is a function $g: B \to A$ such that $f(g(x)) = x$ for all $x \in B$.

*10. Suppose $g: B \to A$ is 1-1. Show that there is a function $f: A \to B$ such that $f(g(x)) = x$ for all $x \in B$.

11. Suppose A and B are finite sets and $f: A \to B$ is an onto function that is not 1-1. What can you say about the relative sizes of $|A|$ and $|B|$? Suppose f is 1-1 but not onto. What can you then say about their relative sizes?

12. (a) Prove: If $f: A \to B$ and $g: B \to C$ are each onto, then $g \circ f$ is an onto function from A into C.
    (b) Repeat part (a) but assume f and g are 1-1. Prove that $g \circ f$ is also 1-1.

*13. If $f: A \to B$ and $g: B \to C$, and $g \circ f$ is onto, what can you say about f? about g? Explain.

*14. If $f: A \to B$ and $g: B \to C$, and $g \circ f$ is 1-1, what can you say about f? about g? Explain.

15. Let $\sigma = (1247)(56)$ and $\tau = (3579)(68)$. Find $\sigma\tau$ and $\tau\sigma^{-1}$. Using Equation (2.2.10), find $\tau^{-1}\sigma\tau$ and $\sigma\tau\sigma^{-1}$.

16. If $\sigma = (123)(47)(589)$ and $\tau = (14726)$, find $\tau^{-1}\sigma\tau$ and $\sigma^{-1}\tau\sigma$.

17. For $\sigma = (3471)(45)$ and $\tau = (237)$, find $\tau\sigma\tau^{-1}$.

18. Let $\sigma = (1645)$ acting on $\mathbb{I}_6$. Let $f: \mathbb{I}_6 \to \{a, b, c, d, e, f\}$ be defined by $f(1) = a, f(2) = b, \ldots, f(6) = f$. Express the permutation $f^{-1}\sigma f$ in cycle form. (Note: The notation is understood to be postfix.)

19. A file contains 100 records numbered 1 through 100. It is desired to change the location of some of these records according to the permutation $(2 \quad 43 \quad 48 \quad 34 \quad 51)(4 \quad 87 \quad 31)$. Your assistant has reversed the order of these records, which you think is a good idea, except that the preceding reordering should have been done first. Instead of reversing back, changing the records as indicated, and then reversing again, give a better way of proceeding.

20. Let $\sigma = (123456)$. Compute $\sigma, \sigma^2, \sigma^3, \ldots \sigma^6$ and express your answer in each case as a product of disjoint cycles.

21. Solve for the permutation σ:

$$(123)\sigma(5432) = (123456)$$

*22. Let $\sigma = (123)(5689)$ and $\rho = (352)(8614)$. Find two solutions τ of the equation $\tau\sigma = \rho\tau$. (*Hint:* This equation is equivalent to $\sigma = \tau^{-1}\rho\tau$.)

23. The following sequence of steps leads from an arrangement of letters to the same arrangement. Express this fact in terms of multiplication

of permutations.

$$
\begin{array}{cccccc}
A & B & C & D & E & F \\
\end{array}
$$

1↓

$$
\begin{array}{cccccc}
E & C & A & B & D & F \\
\end{array}
$$

2↓

$$
\begin{array}{cccccc}
A & B & F & C & E & D \\
\end{array}
$$

3↓

$$
\begin{array}{cccccc}
A & B & C & D & E & F \\
\end{array}
$$

[*Hint:* Think of the six columns as six memory cells. Thus, permutation 1 moves the contents (A) of cell 1 to cell 3 (so $1 \to 3$), the contents of cell 2 to cell 4 ($2 \to 4$), and so on.]

24. As in Exercise 23, the transformation of the arrangement *ABCDEFG* into *BDCFAGE* can be expressed by a permutation σ. Compute σ. (Be careful, here. Do not say that "the first letter *A* changes into the second letter *B*." Instead, say that the content of the first position (A) moves into the fifth position, and so on.)

25. Express the permutation (145867)(345) as a product of transpositions. Hence, show how the sequence of rearrangements

$$ABCDEFGH \to GBCADHFE \to GBDCAHFE$$

can be obtained by a sequence of swaps.

*26. A cycle of length 3 is called a triple. Express the permutation (12)(34567)(89) as a product of triples. (*Hint:* Think of this permutation as a rearrangement. Then see how this rearrangement can be formed using triples. Thus, transform *ABCDEFGHI* into *BAGCDEFIH* by means of triples.)

27. Suppose ten objects are listed in some order. Each of the following operations can be given by a permutation. Write the permutation as a product of cycles. In all cases where an object is moved, close up the space it took.
    (a) reversing the order of the objects.
    (b) moving the fifth object to the beginning of the list.
    (c) moving the seventh object to the end of the list.
    (d) moving the first four objects to the end of the list.
    (e) moving the last object to the third, pushing up all objects from the third on.

*28. (Don't stay up late working on this one!) Six cards are placed in a row. The cards are to be rearranged, but the only operations allowed are

σ: (123) [Cyclically change the first three cards.]

τ: (3456) [Cyclically change the last four cards.]

Using these operations, can you change

 ?

*29. Suppose in Rubik's cube we have found an operation α that inter-changes corners 1 and 3, leaving all other corners fixed. (See Figure 2.2.3.) We now want to find an operation that interchanges corners 3 and 4, leaving the others unchanged. Describe how this can be done. (*Hint:* Think of conjugates.)

2.3 Induction and Recursion

Mathematical induction is a powerful tool used to prove results about natural numbers. It is based on a simple idea: Reduce the proof to a known result, called the **basis** of the induction. In an inductive proof, a technique is typically set up to prove a result from the knowledge of a previous result. This technique is powerful enough that it can be used to prove the previous result from the knowledge of the one previous to it, and so on, until the basis is finally used.

We begin with a simple example, intuitively done, which we shall soon do formally by induction.

EXAMPLE 2.3.1

Show that 2^{4n+1} is an integer that ends with the digit 2 for $n = 0, 1, \ldots$.

Method These are the numbers $2, 2^5, 2^9, 2^{13}, \ldots$. Note that each number is 2^4 times the previous one. Since the first number in this sequence ends in 2 (the basis!), and each number is $16 (= 2^4)$ times the previous one, we see that the last digit 2 transmits through the sequence, since $2 \cdot 16 = 32$, which also ends in 2. Thus, $2^5 = 2 \cdot 16 (= 32)$ ends in 2. Thus, $2^9 = 2^5 \cdot 2^4 (= 32 \cdot 16)$ ends in 2, and this method continues indefinitely.

This example starts with a known case to get the next case, which is then used as the next known case to get the next case, and so on. The principle of mathematical induction formalizes this procedure.

The Principle of Mathematical Induction (first formulation)

Suppose $P(n)$ is a statement for each $n \in \mathbb{N}$. Suppose further that

1. $P(0)$ is true.
2. If $P(n)$ is true, then $P(n + 1)$ is also true.

Then, under these circumstances, $P(n)$ is true for all $n \in \mathbb{N}$.

Item 1 is called the **basis** of the induction.

Item 2 is called the **inductive step.** In showing that $P(n + 1)$ is true, we may assume that $P(n)$ is true. This latter assumption is called the **inductive hypothesis** or the **induction hypothesis**.

The truth of this principle will be taken for granted by us. In a foundational treatment of the integers, it is taken as an axiom. The idea behind it is very much like the previous illustration. The *basis* of the induction is the known case. The *inductive step* allows us to get any case, once we know the previous one. Since we know that the case $n = 0$ is true, it follows that the case $n = 1$ is true; hence, the case $n = 2$ is true, and so on. The inductive step allows us to continue from any integer n to the next integer $n + 1$. The basis permits us to start the trip.

We can restate the principle in the language of predicate logic. If the universe is $\mathbb{N}$ and $P(n)$ is any predicate, then the principle of mathematical induction is

$$P(0) \wedge (\forall n)(P(n) \to P(n + 1)) \to (\forall n)P(n)$$

Remark Sometimes the basis is chosen to be $P(1)$. In that case, the proof by induction proves the result $P(n)$ for all $n \geq 1$. Similarly, if the basis $P(n_0)$ is chosen, an inductive proof would prove $P(n)$ for all $n \geq n_0$.

EXAMPLE 2.3.2   Example 2.3.1 Revisited

Prove by induction that 2^{4n+1} ends with the digit 2 for $n = 0, 1, 2, \ldots$.

Method We let $P(n)$ be the statement 2^{4n+1} ends with the digit 2. For $n = 0$, $2^{4n+1} = 2^1 = 2$, which ends with the digit 2. Thus $P(0)$ is true. (This is the basis of the induction.)

Now assume that $P(n)$ is true: 2^{4n+1} ends with the digit 2. (This is the induction hypothesis.) We shall show that $P(n + 1)$ is true. We let $(\ldots 2)$ denote a number ending with 2. Then

$$2^{4(n+1)+1} = 2^{4n+5}$$
$$= 2^{4n+1} \cdot 2^4$$
$$= 2^{4n+1} \cdot 16$$
$$= (\ldots 2) \cdot 16$$
$$= (\ldots 2), \quad \text{a number ending in 2.}$$

Remark When first learning induction, it will sometimes seem unreasonable to simply *assume* that $P(n)$ is true and to call it the induction hypothesis. For this is apparently what we are trying to prove! The difference is this. We are trying to prove that $P(n)$ is true for *every* n. In the induction hypothesis, we are assuming that n is a number *for which $P(n)$ happens to be true*, and then we are trying to show that in that case, $P(n + 1)$ is also true. When we say in an inductive proof, "suppose $P(n)$ is true," this is a shorthand way of saying, "suppose we have found an n for which $P(n)$ is true."

Using the methods of Section 1.4, to prove

$$P(n) \to P(n + 1)$$

we may assume $P(n)$, deduce $P(n + 1)$, and apply the Deduction Theorem. Here, the assumption $P(n)$ is called the induction hypothesis.

EXAMPLE 2.3.3

Prove by induction that $n < 2^n$ for $n \in \mathbb{N}$.

Method   It is convenient to start the induction with $n = 1$. The case $n = 1$ is trivial: $1 < 2 < 2^1$.

Now assume that $n < 2^n$. (This is the induction hypothesis.) To prove the case $n + 1$ (which is the statement $n + 1 < 2^{n+1}$), it seems reasonable to multiply the inequality $n < 2^n$ by 2. Thus, we have $2n < 2^{n+1}$. It is now enough to show that $n + 1 \le 2n$, since then we would have $n + 1 \le 2n < 2^{n+1}$. But $n + 1 \le 2n$ is equivalent to $1 \le n$, which is our assumption. With this analysis, we can proceed: If $n < 2^n$ (the induction hypothesis), then

$$
\begin{aligned}
n + 1 &\le 2n && \text{(since } 1 \le n) \\
&< 2 \cdot 2^n && \text{(induction hyp.)} \\
&= 2^{n+1} && \text{(algebra)}
\end{aligned}
$$

This proves the result by induction for $n \ge 1$. The case $n = 0$ is separately verified to be true: $0 < 1 = 2^0$.

Induction is used to prove results about natural numbers. **Recursion,** which is related, is used to define functions. We can **recursively define** a sequence a_n (or, equivalently, a function $f: \mathbb{N} \to X$, where X is any set). Working with sequences a_n, the method is to define a_0 arbitrarily and then define a_{n+1} as a function F of n and a_n. We first illustrate with an example.

EXAMPLE 2.3.4

Given the equations

$$a_0 = 1 \tag{2.3.1}$$

$$a_{n+1} = a_n^2 + n, \, n \ge 0 \tag{2.3.2}$$

Find a_5.

Method   We use Equation (2.3.2) repeatedly, starting with $n = 0$. Equation (2.3.1) is used as a starting point.

$$n = 0: \qquad a_1 = a_0^2 + 0 = 1^2 + 0 = 1$$

$$n = 1: \qquad a_2 = a_1^2 + 1 = 1^2 + 1 = 2$$

$$n = 2: \qquad a_3 = a_2^2 + 2 = 2^2 + 2 = 6$$

$$n = 3: \qquad a_4 = a_3^2 + 3 = 6^2 + 3 = 39$$

$$n = 4: \qquad a_5 = a_4^2 + 4 = 39^2 + 4 = 1525$$

It is clear that this method will yield a_n for any given value of n by repeated use of Equation (2.3.2).

Recursive Definitions

We now generalize Example 2.3.4 to show how a function $f: \mathbb{N} \to X$ may be recursively defined.

THEOREM 2.3.5

Let c be a fixed element in some set X, and let $F: \mathbb{N} \times X \to X$. Then the equations

$$f(0) = c \qquad\qquad\qquad (2.3.3)$$

$$f(n + 1) = F(n, f(n)), \, n \geq 0 \qquad\qquad\qquad (2.3.4)$$

are satisfied by a unique function $f: \mathbb{N} \to X$. The function f is said to be recursively defined by Equations (2.3.3) and (2.3.4).   ∎

Equation (2.3.3) is called the **basis** of the definition. Equation (2.3.4) is the **recursion.**

In sequence notation a_n (where $a_n \in X$), Equations (2.3.3) and (2.3.4) become

$$a_0 = c \qquad\qquad\qquad (2.3.5)$$

$$a_{n+1} = F(n, a_n), \, n \geq 0 \qquad\qquad\qquad (2.3.6)$$

Remark   Thus, a_n is defined in terms of its previous value, and a_0 is defined independently. This is analogous to an inductive proof, where a statement is proved, assuming that statement is true for the previous value, and the statement for $n = 0$ is proved independently.

If we choose $F(n, x) = x^2 + n$, then Equation (2.3.6) becomes $a_{n+1} = a_n^2 + n$, which is exactly Equation (2.3.2). (Here, X may be taken as the set of positive integers, so $F : \mathbb{N} \times X \to X$.) Example 2.3.4 shows why definition by recursion works and how to use this technique to compute. The next example shows how definitions by recursion may sometimes *hide* what's going on.

EXAMPLE 2.3.6

Suppose $a_0 = 3$ and $a_{n+1} = 2a_n - 1$. Find a_{99}.

Method   Starting the computation, we find $a_1 = 5$, $a_2 = 9$, $a_3 = 17$, $a_4 = 33$. The numbers $3, 5, 9, 17, 33$ look like powers of 2 (plus 1). Since $a_4 = 2^5 + 1$, we *guess* that $a_n = 2^{n+1} + 1$. Checking further, we find that $a_5 = 2{\cdot}33 - 1 = 65$, which is $2^6 + 1$, so we should be convinced! But being convinced is not a proof. However, a proof (by induction) is available.

To prove that $a_n = 2^{n+1} + 1$, we first note that this is true for $n = 0$ ($a_0 = 2^1 + 1$). Now, use the induction hypothesis ($a_n = 2^{n+1} + 1$) to find

$$
\begin{aligned}
a_{n+1} &= 2a_n - 1 && \text{(the recursion)} \\
&= 2(2^{n+1} + 1) - 1 && \text{(induction hyp.)} \\
&= 2^{n+2} + 1 && \text{(the result for } n + 1)
\end{aligned}
$$

This proves the result. We can now say

$$a_{99} = 2^{100} + 1$$

Remark   This example illustrates several useful ideas. First, it is foolish to carry out the computation for a_{99} blindly. Check a few cases and make a good guess. Second, a proof of the guess by induction was very natural because the function was defined recursively. Many results in mathematics are often established simply by observing, guessing, and finally proving. If the result is about nonnegative integers, a proof by induction is often the correct approach.

The following is an example of how we may use recursion to find the cardinality of a set.

EXAMPLE 2.3.7

Prove that the number of strings of bits of length n is 2^n.

Method   Let L_n be the required number. We have $L_0 = 1$, because there is only one string of length 0—the empty string. If we wish, we may take $n = 1$ for the basis. There are two strings of length 1: "0" and "1". Both cases agree with the formula $L_n = 2^n$. To prove the general result, assume

$L_n = 2^n$ as the induction hypothesis. We now count the strings of length $n + 1$. There are two cases, according to whether the string starts with "0" or "1". The number of strings of length $n + 1$ that start with "0" is L_n, since "0" must be augmented by an arbitrary string of length n in order to form a string of length $n + 1$, and similarly for strings that start with "1". Therefore, $L_{n+1} = L_n + L_n = 2L_n = 2 \cdot 2^n = 2^{n+1}$, which is the result for $n + 1$. Thus, the result is proved by induction.

The example may be restated as, "a set with n elements has 2^n subsets." (Compare with Example 1.1.36 and the definition following it, where strings of bits of length n were interpreted as characteristic functions of subsets of a set with n elements.)

EXAMPLE 2.3.8   String Manipulation

Suppose strings are formed using the alphabet $\Sigma = \{A, B, C, X, Y\}$. We are permitted to alter (transform) strings in one of four ways:

T1. Replace any occurrence of "A" with "BAB".
T2. Replace any occurrence of "B" with "XCBCY".
T3. Replace any occurrence of "BCY" with "Y".
T4. Replace any occurrence of "YY" with "Y".

For example, we can transform

"AAC" → "ABABC" → "ABAXCBCYC" → "ABAXCYC"

Prove that no matter how we transform the initial string "ABC", the final result will have an equal number of X's and Y's and exactly one occurrence of "A".

Method   Common sense seems to indicate that the result is true because none of the allowed transformations change the number of A's, and, if we ignore T4, they introduce as many X's as Y's. It looks like T4 will never be used, because "Y" is introduced only with an "XCBC" preceding it, so we will never find "YY" and therefore we can never expect to use T4. This analysis suggests, after some attempts at a proof, the following formulation of $P(n)$.

$P(n)$: After n steps, there are an equal number of X's and Y's and exactly one "A", each occurrence of "Y" is preceded by "C", and each occurrence of "BCY" is preceded by "XC".

To prove $P(n)$, we note that the case $n = 0$ is true, because at $n = 0$ the string is "ABC". (The statements about "Y" are vacuously true.) Let λ_n be the string at step n. Now assume that $P(n)$ is true. (This is the inductive hypothesis.) We apply one of the transformations to arrive at the string λ_{n+1}. The number of A's doesn't change with any of the transformations, so there continues to be exactly one "A" in λ_{n+1}. Since each occurrence of "Y" is preceded by "C", "YY" cannot occur, so we cannot use T4 at the nth step. Therefore the numbers of X's and Y's remain equal. (T1 and T3 kept them the same, whereas T2 increases both by 1.)

To see that each occurrence of "Y" is preceded by "C" and each occurrence of "BCY" is preceded by "XC", we have to consider the effects of each type of transformation on λ_n. T1 doesn't introduce anything before "Y", since it involves replacing "A", which can't precede "Y" by induction. T2 introduces "Y", but introduces "$XCBC$" before it. Since it replaced "B", there was no "Y" after it (by induction). Finally, if we use T3, the occurrence of "BCY" was preceded with "XC", so "$XCBCY$" was replaced by "XCY". So here, too, each occurrence of "Y" is preceded by "C", and no new occurrences of "BCY" were introduced. This completes the proof.

Remark Note that we introduced the unasked-for fact about how "Y" is preceded. It looks like we're trying to prove more, which in fact we are. But it was introduced because, by strengthening the theorem, we are able to use more in the inductive proof—a remarkable process wherein we can prove a result more easily if we try to prove more.

This example is an artificial exercise, but similar to many real-life problems (in form and messiness) that occur in the theory of artificial languages.

An important application of a definition by recursion is the Σ (sigma) notation. If a_n is any sequence of real numbers ($n = 1, 2, 3, \ldots$), we can define

$$S_n = \sum_{k=1}^{n} a_k$$

recursively:

DEFINITION 2.3.9 Definition of $\displaystyle\sum_{k=1}^{n} a_k$

The sum $S_n = \displaystyle\sum_{k=1}^{n} a_k$ is defined recursively by the formulas

$$S_0 = 0 \tag{2.3.7}$$

$$S_{n+1} = S_n + a_{n+1}, \, n \geq 0 \tag{2.3.8}$$

The sum should be read "sigma a sub k from $k = 1$ to $k = n$," or simply "the sum a sub k from 1 to n."

Note that by actually defining the sum from $n = 1$ to $n = 0$ to be $S_0 = 0$ in Equation (2.3.7), we have chosen an especially simple initialization for the sum S_n. Equation (2.3.8) adds the term a_{n+1} to the "running sum" S_n to get the "current sum" S_{n+1}. This is exactly how we manually add a column of numbers. Often, a sum will start from $k = 0$ to n. The definition is the same as before, except that the initial value S_0 is taken to be a_0. A similar definition applies to other starting values.

Because finite sums are defined recursively, many "induction problems" appearing in texts are evaluations of sums. Let us offer a few of them.

EXAMPLE 2.3.10

Find the value of

$$1 + 3 + 5 + \cdots + (2n - 1)$$

and prove your result.

Method   A little computation shows that the sum

$$S_n = \sum_{k=1}^{n} (2k - 1)$$

has values 0, 1, 4, 9 for $n = 0, 1, 2, 3$. A reasonable guess is certainly $S_n = n^2$. We now prove this by induction. By definition, $S_0 = 0$, so $S_0 = 0^2$ is true. Now, assuming $S_n = n^2$ (the inductive step), we have

$$
\begin{aligned}
S_{n+1} &= S_n + (2(n + 1) - 1) && \text{(definition of a summation)} \\
&= S_n + 2n + 1 && \text{(algebra)} \\
&= n^2 + 2n + 1 && \text{(induction hyp.)} \\
&= (n + 1)^2 && \text{(algebra)}
\end{aligned}
$$

This is the case $n + 1$; hence, the result is proved by induction.

THEOREM 2.3.11   The Geometric Series

For n in $\mathbb{N}$ and $x \neq 1$,

$$\sum_{k=0}^{n} x^k = \frac{1 - x^{n+1}}{1 - x}$$

Proof   Let S_n be the required sum. First note that the summing starts with $k = 0$. Thus, $S_0 = x^0 = 1$, which agrees with the formula for $n =$

0. Now, assuming the result for n, we obtain

$$S_{n+1} = S_n + x^{n+1} = \frac{1 - x^{n+1}}{1 - x} + x^{n+1}$$

$$= \frac{1 - x^{n+1} + x^{n+1} - x^{n+2}}{1 - x}$$

$$= \frac{1 - x^{n+2}}{1 - x}$$

This is the result for $n + 1$. ■

Remark Note that the result is even true for $x = 0$. However, this requires $0^0 = 1$, which is a commonly used convention and one that we use throughout the text.

There is another, more powerful, form of induction called **complete induction** or **course-of-values induction.** We state it first, then illustrate with some examples.

The Principle of Complete Mathematical Induction

Let $P(n)$ be a statement for each $n \in \mathbb{N}$, and suppose that

1. $P(0)$ is true.
2. If $P(m)$ is true for all $m < n$, then $P(n)$ is also true.

Then, under these circumstances, $P(n)$ is true for all n.

As in the first formulation of mathematical induction, condition 1 is called the **basis** of the induction.

Condition 2 is called the **inductive step.** We must show that $P(n)$ is true, assuming that $P(m)$ is true for all values of $m < n$. This latter assumption is called the **inductive or induction hypothesis.**

The intuitive idea behind this formulation is as follows. The basis, $P(0)$, is true; $P(m)$ is true for all $m < 1$ (namely $m = 0$). Hence, by condition 2, $P(1)$ is true. Then, since $P(m)$ is true for all $m < 2$ (namely $m = 0$ and $m = 1$), $P(2)$ is true. Thus the cases $n = 1$ and 2 are true. But by the inductive step, this further implies that the case $n = 3$ is true, and so on, for all n. Complete induction is often more powerful than ordinary induction because it allows us to use a more powerful induction hypothesis. Ordinary induction allows us to assume only the *one* case (n) preceding $n + 1$ in order to prove the case $n + 1$, but complete induction allows us to assume *all* cases ($m < n$) preceding n in order to prove the case n.

As with the principle of mathematical induction, the principle of complete induction may be stated in the language of predicate logic. For any predicate P on the natural numbers, the following is true:

$$P(0) \wedge (\forall n)[(\forall m)(m < n \rightarrow P(m)) \rightarrow P(n)] \rightarrow (\forall n)P(n)$$

The inductive step is the portion of the displayed formula in the brackets. The induction hypothesis is the clause $(\forall m)(m < n \rightarrow P(m))$.

There is an analogous result, which is a stronger version of the recursive definition of function. In this version, a function is recursively defined if $f(0)$ is given, and if $f(n)$ can be defined in terms of n and *all* the values $f(m)$ where $m < n$. We illustrate its use in Examples 2.3.17 through 2.3.20.

EXAMPLE 2.3.12

A hat contains a (finite) number of paper slips. John removes one or more of these slips and gives the hat to Mary, who does likewise. They keep alternating until there are no more slips of paper. Show that this process must stop.

Method   Although the result is intuitively clear, let us give a proof by (complete) induction. We'll show that if the hat contains n slips, the process must stop. This is true for $n = 0$, because that's the way the game is played. Now suppose we know that the game stops if there are fewer than n slips in the hat. If there are n slips in the hat, there will be fewer than n after John moves, since he has to take at least one. Therefore, after he moves, it is Mary's turn with fewer than n slips of paper in the hat, and therefore, the game must stop, by our induction hypothesis.

We can restate this example in more mathematical terms: It is not possible to have an infinite, descending sequence of natural numbers. This is a special property of $\mathbb{N}$. It is not true for the positive real numbers or the integers. For example, the sequence 1, 0.1, 0.01, 0.001, ... is a descending sequence of positive reals, and the sequence 0, -1, -2, ... is such a sequence of integers. The following is a more complicated variation of this example.

EXAMPLE 2.3.13

John and Mary are moving a checker on the checkerboard in Figure 2.3.1. The board is understood to extend infinitely to the right and up. The rules are as follows: each may move the checker straight *down*. Or, they can move it *up*, any distance, provided they also move it to the left. Also they must move, if possible. Show that this game must end.

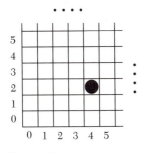

Figure 2.3.1

Method If the checker is in column 0, moving to the left is out of the question, so it must be moved down. This game must end by Example 2.3.12. We now show that the game must end if the checker is in column n. We just showed that the result is true for $n = 0$. (This is the basis for this result.) Now assume the result is true for all $m < n$, and suppose the checker is in column n. We cannot stay in column n indefinitely, since this is only possible by moving down, and this must stop by Example 2.3.12. Therefore, at some time, we must move to column m, where $m < n$. Now the induction hypothesis takes over, and the game must end.

Remark This result is more subtle than Example 2.3.12. In that example we could predict in advance how many moves, at most, it would take for the game to end. In this one, no such prediction is possible. One can prolong the agony, yet we have a guarantee that the game must end.

In Example 2.3.12, each person was changing the "state of the system," which was the number n of slips of paper in the hat. The natural numbers $\mathbb{N}$ are ordered by the relation $<$, so the change always decreases the state. The conclusion was that this had to be a finite process. In Example 2.3.13, there was also a state of the system, given by the location (n, m) of the checker. Here we can also order according to the rule

$$(a, b) < (c, d) \quad \text{iff}† \quad a < c, \quad \text{or} \quad a = c \text{ and } b < d$$

Again, the rules of the game showed that any position moves into one smaller than it. We showed, by induction, that there cannot be an infinite decreasing sequence of positions. Both of these examples illustrate the notion of a well-ordered set.

---

† Here, "iff" is shorthand for "if and only if".

DEFINITION 2.3.14

Let $(A, \leqslant)$ be a linearly ordered poset. This means that A is a poset with the property that for any x and y in A, either $x \leqslant y$ or $y \leqslant x$. Then A is said to be **well-ordered** if there is no infinite decreasing sequence of distinct elements of A.

We can say that the reason the games in Examples 2.3.12 and 2.3.13 ended was that the players were creating a decreasing sequence of elements in a well-ordered set.

The next example is from the theory of numbers. (This topic is considered in Chapter 8.) We first recall that a prime number is an integer greater than 1 whose only divisors are itself and 1. (For example, 3, 23, and 17 are prime numbers, but 6, 35, and 91 are not primes because they are divisible by 3, 5, and 7, respectively.)

THEOREM 2.3.15

Any integer $n > 1$ is either a prime number or a product of prime numbers.

Proof   We can paraphrase the result: "For all $n \in \mathbb{N}$, either $n \leq 1$, or n is a prime, or n is a product of prime numbers." Thus, the result is true for $n = 0$. Now, assume the result for all integers less than a given n. To prove the result for n, we have two cases, depending on whether n is a prime number. (We may assume $n \geq 2$, since otherwise the result is true.) If n is a prime, the result is true for n. If n is not a prime, it can be written $n = rs$, where r and s are both less than n and greater than 1. Hence, by induction, both r and s are primes or a product of primes. Thus, $n = rs$ is a product of primes. This concludes the proof. ■

Remark   Alternatively, the induction may be started with the case $n = 2$.

In Section 2.1, we indicated a proof that any finite poset is the transitive closure of its covering relation. We now restate this and give its proof by induction.

EXAMPLE 2.3.16

Let U be a finite partially ordered set with order relation $\leqslant$. Define $x < y$ in U to mean that $x \leqslant y$ and $x \neq y$. We define xCy (x is covered by y) to mean $x < y$, but there is no z satisfying $x < z$ and $z < y$. Prove that if $x < y$, then there is a sequence of elements starting with x and ending with y such that each element of the sequence is covered by the one after it. Thus, there is a sequence $x_1, x_2, \ldots, x_n$ such that $x = x_1$, $y = x_n$, and $x_1Cx_2, x_2Cx_3, \ldots, x_{n-1}Cx_n$.

Method   The original idea of the proof (Theorem 2.1.29) was to take more and more intermediate points until we ran out of points. This type of argument is a natural for a proof by induction. We prove it *by induction on the number of elements of U.*

The result is vacuously true when U has one element, and it is true when U has two elements, since in this case xCy and $x < y$ are the same. This is our basis.

Now suppose U has n elements and that the result is true for all posets with fewer than n elements. We are given $x < y$. If x is covered by y, we have the result for U. If not, there is a z in U such that $x < z < y$. Consider the set U^+ of all elements t of U satisfying $z \leqslant t$, and the set U^- of elements t of U satisfying $t \leqslant z$. These sets are indicated in Figure 2.3.2. We can verify that U^+ and U^- are posets and that a covering in the poset U^+ is a covering in U, and similarly for U^-. Now we use our induction hypothesis. Both U^+ and U^- have fewer elements than U because $x \notin U^+$ and $y \notin U^-$. Therefore, we can find a sequence joining x to z in U^-, where each term of the sequence is covered by the next, and we can similarly find such a sequence in U^+ joining z to y. Joining these sequences, we obtain the required sequence in U.

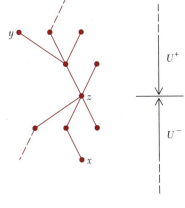

Figure 2.3.2

As we indicated preceding Example 2.3.12, it is possible to recursively define a sequence so that the term a_n is defined as a function of n and of *all* values a_m for $m < n$. We first give a more modest example.

EXAMPLE 2.3.17

The **Fibonacci sequence** f_n has each term defined in terms of the two preceding values. For initialization, both f_0 and f_1 must be separately defined:

$$f_0 = 0$$

$$f_1 = 1$$

$$f_{n+2} = f_n + f_{n+1} \qquad (n = 0, 1, 2, \ldots)$$

The first 12 terms of this sequence are

$$0, 1, 1, 2, 3, 5, 8, 13, 21, 34, 55, 89 \quad (= f_{11})$$

There is an extensive literature on this sequence, including an entire journal devoted to it. We give a few examples involving the Fibonacci sequence.

EXAMPLE 2.3.18

Let Ψ be the positive solution of the equation $x^2 - x - 1 = 0$. ($\Psi = (1 + \sqrt{5})/2 = 1.618033\ldots$) Then

$$f_n \leq \Psi^{n-1}$$

Method   This is true for $n = 0$ and $n = 1$:

$$0 = f_0 \leq \Psi^{-1}$$

$$1 = f_1 \leq 1 = \Psi^0$$

Assuming the result for n *and* $n + 1$, we find

$$
\begin{aligned}
f_{n+2} &= f_{n+1} + f_n \\
&\leq \Psi^n + \Psi^{n-1} \quad \text{(ind. hyp.)} \\
&= \Psi^{n-1}(\Psi + 1) \\
&= \Psi^{n-1}(\Psi^2) \quad \text{(since } \Psi \text{ is a solution of } x^2 - x - 1 = 0) \\
&= \Psi^{n+1}
\end{aligned}
$$

which is the result for $n + 2$.

EXAMPLE 2.3.19

A luncheon counter has n seats. How many seating plans are there in which no one sits next to anyone else? In other words, how many subsets of $\mathbb{I}_n$ are there with no two numbers adjacent? (Recall that $\mathbb{I}_n = \{1, 2, \ldots, n\}$.)

For example, if the counter had four seats, the subsets would be Ø, {1}, {2}, {3}, {4}, {1, 3}, {1, 4}, and {2, 4}. Thus, there are eight such arrangements for $n = 4$.

Method   The number depends on n. Let the required number be b_n. (We just found $b_4 = 8$.) We take a case analysis, according to whether the last seat (or the number n) is used.

Case 1 (n is not in the subset)   In this case, we have the same problem only with l_{n-1} (or $n - 1$ seats). There are just b_{n-1} choices here.

Case 2 (n is in the subset)   In this case, we have to fill the remaining $n - 1$ seats, but seat $n - 1$ cannot be filled because it is adjacent to n. Thus, we have the case $n - 2$, and there are b_{n-2} choices here.

So in all, we have $b_{n-1} + b_{n-2}$ possibilities:

$$b_n = b_{n-1} + b_{n-2} \qquad (2.3.9)$$

But this is precisely the Fibonacci recursion! Now let's check the initial conditions. Clearly $b_0 = 1$ (since the null set $\varnothing$ works for $l_0 = \varnothing$). Also, $b_1 = 2$, since $\varnothing$ and $\{1\}$ work for $l_1 = \{1\}$. Thus, the initial conditions start at 1, 2 rather than the 0, 1 of the Fibonacci sequence. Thus, the sequence is 1, 2, 3, 5, 8, ... But this is just the Fibonacci sequence displaced by 2. Since $f_2 = 1$ and $f_3 = 2$, we have $b_0 = f_2$, $b_1 = f_3$. In general, then

$$b_n = f_{n+2}$$

The formal proof is by induction: The bases $n = 0$ and $n = 1$ have been verified. Assuming the result for all values preceding $n + 1$, we have

$$\begin{aligned}
b_{n+1} &= b_{n-1} + b_n \\
&= f_{n+1} + f_{n+2} \qquad \text{(ind. hyp.)} \\
&= f_{n+3} \qquad \text{(definition of Fibonacci sequence)}
\end{aligned}$$

This is the result.

Remark   Our result $b_4 = 8$ reflected $b_4 = f_6 = 8$. Even if we had never studied Fibonacci sequences, the technique of this solution, leading to the recursion (2.3.9), would be the natural way to study this problem. The example illustrates, as does Example 2.3.7, an important method of counting. We didn't find the count b_n explicitly. Instead, we found a **recursion** that related b_n to previous values so that any value of b_n could be computed easily. Consider the problem of finding b_9 in Example 2.3.19. To actually write out the different seating arrangements and count them is almost guaranteed to lead to some miscalculation. We now know, however, that $b_9 = f_{11} = 89$.

We conclude with an interesting and curious sequence, introduced by the mathematician J. W. Conway.

EXAMPLE 2.3.20

Let

$$c_1 = 1$$

$$c_2 = 1$$

$$c_n = c_k + c_{n-k}, \qquad \text{where } k = c_{n-1} \text{ and } n \geq 3$$

Compute the first ten terms. Show that

$$0 \leq c_n - c_{n-1} \leq 1 \tag{2.3.10}$$

Method   For $n = 3$ we have $k = c_2 = 1$. Thus,

$$c_3 = c_1 + c_2 = 1 + 1 = 2$$

For $n = 4$, $k = c_3 = 2$. Thus,

$$c_4 = c_2 + c_2 = 2$$

Similarly,

$$c_5 = c_2 + c_3 = 3$$

Continuing in this way, we arrive at the first ten terms of the Conway sequence:

$$1, 1, 2, 2, 3, 4, 4, 4, 5, 6, \ldots$$

We now prove the inequality (2.3.10) by induction. By observation, it is clearly true for $n = 2$ through 10, so we have the basis result. Now assume that the inequality (2.3.10) is true for all $m < n$. Let $k = c_{n-1}$ so that $c_n = c_k + c_{n-k}$. Similarly, if $j = c_{n-2}$, we have $c_{n-1} = c_j + c_{n-j-1}$. Now j, being the term preceding $k = c_{n-1}$, is, by induction, either k or $k - 1$.

In the first case $(j = k)$, $c_{n-1} = c_k + c_{n-k-1}$, so

$$c_n - c_{n-1} = c_k + c_{n-k} - (c_k + c_{n-k-1}) = c_{n-k} - c_{n-k-1}$$

Thus, inequality (2.3.10) is true because it is true for $n - k$ by induction.

In the second case $(j = k - 1)$, we have $c_{n-1} = c_{k-1} + c_{n-k}$, and a similar computation shows that

$$c_n - c_{n-1} = c_k - c_{k-1}$$

and the result is again true by induction. (See Exercise 2.3.36 for further elaboration.)

Remark   The computation of any term of this sequence for large n is not easy to do on a computer because its definition involves earlier terms, which have to be remembered. Unlike the Fibonacci sequence, in which we need only keep track of the last two terms, we need to keep track of much of the history of Conway's sequence in order to calculate any specific term. Conway proved that $c_n/n \rightarrow 1/2$ as $n \rightarrow \infty$. The convergence is quite slow. The reader should refer to the *New York Times* of August 30, 1988, for interesting details.

Exercises

1. Evaluate the sum

$$\sum_{k=1}^{n} \frac{1}{k(k + 1)}$$

   by finding its value for a few values of n, making a guess, and proving your answer.

*2. Repeat Exercise 1 for the sum

$$\sum_{k=1}^{n} (-1)^k \frac{4(k + 1)}{(2k + 1)(2k + 3)}$$

3. Prove by induction:

$$\sum_{k=1}^{n} k = \frac{n(n + 1)}{2}$$

4. Prove by induction:

$$\sum_{k=1}^{n} k^2 = \frac{n(n + 1)(2n + 1)}{6}$$

5. Prove by induction:

$$\sum_{k=1}^{n} k(k + 1) = \frac{n(n + 1)(n + 2)}{3}$$

6. Prove by induction:

$$\sum_{k=1}^{n} k^3 = \frac{n^2(n + 1)^2}{4}$$

7. Prove that the product of two consecutive numbers is even (divisible by 2). Use induction.

*8. Prove that the product of any three consecutive integers is divisible by 6. You may use the result of the previous exercise.

*9. (*Generalization of Exercises 7 and 8*) Prove that the product of r consecutive integers is divisible by r! (This is "r factorial," the product of the integers from 1 through r.)

*10. A hat contains finitely many slips of paper, each with a natural number written on it. (The same number may be repeated on more than one slip of paper.) Mary picks one slip of paper, observes the number, and does the following: (a) She destroys the slip she chose, and (b) if desired or if possible, she adds as many slips of paper (finite in number) as she wishes to the hat with (natural) numbers *smaller* than the one she observed. The hat is passed over to John, who does the same thing. Play continues indefinitely or until nothing is left in the hat. Must the game end? Prove your answer.

11. Let $-1 < c$. Prove

$$(1 + c)^n \geq 1 + nc \qquad (n \in \mathbb{N})$$

12. Let $0 < c < 1$. Prove

$$(1 - c)^n \geq 1 - nc \qquad (n \in \mathbb{N})$$

*13. Let $0 < c < 1$. Prove

$$(1 - c)^n \leq 1 - nc + n(n-1)c^2/2 \qquad (n \in \mathbb{N})$$

14. Suppose $a_{n+1} = 2a_n$ and $a_0 = 3$. Find a formula for a_n and prove it by induction.

15. Suppose $a_{n+1} = a_n + 1$ and $a_0 = 1$. Find a formula for a_n and prove it by induction.

16. Suppose $a_{n+1} = 3a_n - 1$ and $a_0 = 1$. Find a formula for a_n and prove it by induction. (*Hint:* Look at powers of 3.)

17. Repeat Exercise 16 for $a_0 = 2$.

18. Let E_n be the number of subsets of $\mathbb{I}_n$ that have an even number of elements, and let O_n be the number of subsets with an odd number of elements. Prove that $E_n = O_n$ for $n \geq 1$.

*19. Let F_n be the number of subsets of $\mathbb{I}_n$ with an even number of elements no two of which are consecutive, and let P_n be the number of such subsets with an odd number of elements.
   (a) Write a suitable recursion formula for F_n and P_n together with initial values.
   (b) Explain why $F_n + P_n = f_{n+2}$. (See Example 2.3.19.)
   (c) Compute a few values of F_n and P_n and guess some rule for computing F_n and P_n.
   (d) Prove your guess in part (c).

20. Let d_n be the number of subsets of $\mathbb{I}_n$ such that any element in the subset is two or more away from any other element. (For example, for $n = 15$ the set $\{1, 5, 8, 14\}$ is one such subset.) Compute d_{15}.

21. Prove by induction that $10n < 2^n$ for all n that are sufficiently large.

*22. Prove that $n^2 < 2^n$ for $n \geq 5$.

*23. Prove that any positive integer can be written as a sum of distinct Fibonacci numbers f_n, where $n \geq 2$, no two of which are consecutive Fibonacci numbers. Thus, using the numbers 1, 2, 3, 5, 8, 13, . . . , we have $4 = 1 + 3$, $10 = 2 + 8$, $13 = 13$, $20 = 2 + 5 + 13$, $55 = 55$, and so on.

*24. Prove by induction:

$$f_{n+2}f_n - f_{n+1}^2 = (-1)^{n+1}$$

Here, f_n is the nth Fibonacci number.

25. Let Ψ be the positive root of the equation $x^2 - x - 1 = 0$, as in Example 2.3.18. Prove that if $n \geq 2$, $f_n \geq \Psi^{n-2}$.

26. Let $a_0 = 1$, and $a_{n+1} = \sqrt{1 + a_n}$.
    (a) Prove $a_n < 2$ for all $n \in \mathbb{N}$.
    (b) Prove $a_n < a_{n+1}$ for all $n \in \mathbb{N}$.

27. The factorial function $f(n) = n!$ may be defined recursively. ($n!$ is defined as the product of all the integers between 1 and n inclusive.) Give a recursive definition of $n!$ and verify that $4! = 24$, using your definition.

28. If $f\colon A \to A$ is any transformation on A, recursively define the nth iterate $f^n\colon A \to A$. (This is $f \circ f \circ \cdots \circ f$ n times.) (*Hint:* Initialize with $f^0 = i$, the identity function.)

*29. Using your definition in Exercise 28, prove that $f^m \circ f^n = f^{m+n}$.

30. Let $S\colon \mathbb{N} \to \mathbb{N}$ according to the formula $S(n) = n + 1$. Using the definition in Exercise 28, find a formula for $S^k(n)$. Prove it.

*31. Using the definition in Exercise 28, find a formula for $SQ^k(n)$, where $SQ\colon \mathbb{N} \to \mathbb{N}$ is defined by $SQ(n) = n^2$. (*Hint:* Try a few low values of k. Then guess an answer and give an inductive proof.)

32. Suppose that a statement $P(n)$ satisfies the following properties:
    (a) $P(0)$ and $P(1)$ are true.
    (b) If $P(n)$ is true, then $P(2n)$ is true.
    Does this imply that $P(n)$ is true for all n? Explain or prove. If not true, give an example of $P(n)$ that is not true yet satisfies properties (a) and (b).

33. Suppose that the statement $P(n)$ in Exercise 32 also satisfies
    (c) If $P(n)$ is true ($n > 1$), then $P(n - 1)$ is true.
    Does this additional condition imply that $P(n)$ is true for all n? Explain or prove. If not true, give an example.

34. Let $a_0 = 1$, and $a_{n+1} = n + a_{\lceil n/2 \rceil} + 2a_{\lceil n/3 \rceil}$ for $n \geq 0$. Explain why this defines a_n. Find the first ten terms.

*35. Using an alphabet $\{p, C, N, L, R\}$, we may transform any string in one of the following ways:

    T1: Replace any occurrence of "p" with "$LpCpR$".
    T2: Replace any occurrence of "p" with "$LNppR$".

    Starting with the initial string "p" and using at least one of these transformations, we end with a string α. Prove that
    (1) α starts with "L" and ends with "R".
    (2) α has an equal number of L's and R's.
    (3) If there are x occurrences of "p", y occurrences of "C", and z occurrences of "N", then $x - y - z = 1$.

*36. In the definition of the Conway sequence in Example 2.3.20, the recursion formula assumed that $k = c_{n-1}$ satisfied the inequality $1 \le k \le n - 1$ when $n \ge 2$. Prove this.

*37. Theorem 2.1.26 stated that no cycle can appear in a poset. That is, if $x \le x_1 \le x_2 \le \cdots \le x_n \le x$ in a poset, then $x = x_1 = x_2 = \cdots = x_n = x$. Prove this by induction.

Exercises 38–41 involve the theory of well-ordered sets, as defined in Definition 2.3.14.

*38. Show that if a set $(A, <)$ is well-ordered, then every nonempty subset of A has a least element.

*39. Show that if every nonempty subset of A has a least element, then A is well-ordered.

*40. Let A and B be disjoint sets, and let $(A, <)$ and $(B, <)$ be well-ordered. Define $(A \cup B, <)$ by defining the relation $<$ on $A \cup B$ by the condition $x < y$ if both x and y are in A and $x < y$, or both x and y are in B and $x < y$, or $x \in A$ and $y \in B$. (Think of putting A before B.) Show that $(A \cup B, <)$ is well-ordered.

*41. Let A and B be disjoint sets and let $(A, <)$ and $(B, <)$ be well-ordered. Define $(A \times B, <)$ by defining the relation $<$ on $A \times B$ by the condition

$$(a_1, b_1) < (a_2, b_2) \quad \text{iff} \quad a_1 < a_2, \quad \text{or } a_1 = a_2 \text{ and } b_1 < b_2$$

(Think of the checkerboard in Example 2.3.13.) Show that $(A \times B, <)$ is well-ordered.

Some Programming Exercises

42. Verify your answer in Exercise 16 by directly programming the sequence a_n and comparing it with the formula you obtained in that exercise. Take about 15 terms.

43. Using the recursion you found in Exercise 19(a), write a program to list the first 20 values of F_n and P_n.

44. Verify the results of Exercise 26 by writing a program and computing the first 20 terms.

45. Write a program to list all the Fibonacci numbers from f_0 through f_{30}.

46. Write a program to compute the quotient f_{n+1}/f_n of successive Fibonacci numbers. List these for $n = 1$ to 10, and make an educated guess as to the exact value of the limit.

47. Write a program to compute the terms through a_{30} of the sequence defined in Exercise 34.

2.4 Finite Sets and the Box Principle

The Box Principle, also known as Dirichlet's Principle, or the Pigeonhole Principle, is a fundamental property of finite sets. We give a more precise description of this principle at the end of this section. For now we illustrate its power with a few examples. Intuitively, it expresses the idea that if many people are in a small room, then there will be crowding somewhere in the room. Similarly, it implies that if a flock of pigeons are occupying fewer pigeonholes than there are pigeons, then one of the pigeonholes will be occupied by more than one pigeon. This accounts for the name Pigeonhole Principle.

Suppose we have 51 boxes and that 52 objects are put into them in any manner. Then we can say that one of the boxes must have at least two objects in it. This commonsense idea is basic to the idea of finite sets and turns out to be a useful technique in many applications. The first version we use is as follows.

The Box Principle

If $n + 1$ or more objects are placed into n boxes, then one box will contain two or more of these objects.

EXAMPLE 2.4.1

Five points are located in a unit square, as in Figure 2.4.1. Prove that at least one pair of these points is within $\sqrt{2}/2$ of each other. (A unit square is a square whose dimensions are 1 by 1.)

Figure 2.4.1   **Figure 2.4.2**

Method See Figure 2.4.2. By the Box Principle, five points in four squares imply that two of these points will be in the same square. But each square has diagonal $\sqrt{2}/2$, which is the largest distance attainable in that square.

Remark Of course it may be possible to get closer than $\sqrt{2}/2$ in particular situations. (This is so in Figure 2.4.1.) But the solution shows that for *any* five points in the unit square, the minimum distance between two of the points is at most $\sqrt{2}/2$.

This problem is a max-min problem. For five points, we find the minimum distance between them. We claim that this is at most $\sqrt{2}/2$; in other words, the maximum possible value of this minimum distance is $\leq\sqrt{2}/2$. In fact, Figure 2.4.3 shows that we have an equality here. We are minimizing over the possible distances in a configuration of five points. We then maximize this minimum over all possible configurations.

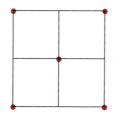

Figure 2.4.3

The same problem with 11 points cannot be as easily handled by breaking up the square into ten smaller figures. An estimate is possible (see the exercises), but we shouldn't expect the Box Principle to give the best possible estimate.

EXAMPLE 2.4.2

Thirty-one different integers are chosen between 1 and 60 inclusive. Show that at least two of them are consecutive.

Method Let the "boxes" be the 30 pairs $(1, 2), (3, 4), \ldots, (59, 60)$. Then if we put each of the 31 integers into the box in which it appears, by the Box Principle one of the boxes must contain two of the given numbers. This is the result. We've actually shown more: there is a consecutive pair of which the first is odd.

There is another way to do this example, based on a slightly different version of the Box Principle.

The Box Principle (alternative version)

Suppose that r red chips and g green chips are put into n boxes and that no box contains two chips of the same color. Suppose also that $r + g > n$. Then one box is occupied by different colored chips.

To see this, note that we have more chips than boxes, so one box is multiply occupied. Since the chips can't be of the same color, we have the result.

Here is a proof of Example 2.4.2 using this alternative version. Let the 31 integers be $a_1, a_2, \ldots, a_{31}$, in increasing order. Call these *green*. Now take the integers $b_1, b_2, \ldots, b_{31}$, where

$$b_i = a_i + 1 \qquad (i = 1, \ldots, 31)$$

Call the b's *red*. Note that the b's are all between 2 and 61, since the a's are between 1 and 60. Thus both the a's and the b's are between 1 and 61, inclusive. We think of the numbers 1 through 61 as the boxes, and each of the a's and b's as chips, and we put each chip into the box with its number. Since $61 < 31 + 31$, it follows that a red and a green (i.e., one of the a's and one of the b's) are in the same box. But this means that they are equal. Thus,

$$a_i = b_j \qquad \text{for some } i \text{ and } j$$

or

$$a_i = a_j + 1 \qquad \text{for some } i \text{ and } j$$

This is the result.

The following example is a simpler version of Example 2.4.1, since the points are on a line. It again illustrates the idea of "crowding."

EXAMPLE 2.4.3

If 1001 positive real numbers are chosen strictly between 0 and 1, show that two of them are less than 0.001 apart.

Method   Divide the interval between 0 and 1 into 1000 parts, using the points of division $0, 1/1000, 2/1000, \ldots, k/1000, \ldots, 1 = 1000/1000$. The original interval is divided into 1000 subintervals: $I_1, I_2, \ldots, I_{1000}$. (See Figure 2.4.4.) Here, I_k is the set of x such that $(k-1)/1000 < x \le k/1000$. By the Box Principle, two of the given points are in the same interval. But the length of any of the intervals is 0.001, and since the left end point is not in any of the intervals, the two points in the same interval are strictly less than 0.001 apart.

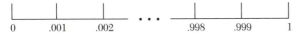

$0 \qquad .001 \qquad .002 \qquad\qquad .998 \qquad .999 \qquad 1$

Figure 2.4.4

Our final illustration involves graphs and uses induction as its principal tool. The Box Principle plays a crucial role in the proof. It shows that if we have points in the plane and enough of them are connected by line segments, then they are bound to form a triangle (crowding again, in a subtle form).

EXAMPLE 2.4.4

Suppose that $2n$ points in the plane are connected by $n^2 + 1$ segments. Prove that a triangle is formed. (Assume that no three of the points are on a line.)

Method   We start the induction with $n = 2$. If four points are connected by $5 = 2^2 + 1$ segments, we have a figure such as Figure 2.4.5. Only one pair is not connected, and we have two triangles formed.

Figure 2.4.5

Now assume the result is true for n (the induction hypothesis), and suppose we have $2(n + 1) = 2n + 2$ points in the plane and that they are connected by $(n + 1)^2 + 1 = n^2 + 2n + 2$ segments. We single out two of the points that are connected, call them A and B, and leave the $2n$ remaining points in a set Σ. (See Figure 2.4.6.)

A ●————————● B

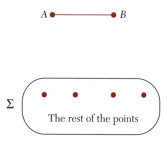
Σ
The rest of the points

Figure 2.4.6

Since A and B are already connected, there are $n^2 + 2n + 1$ remaining connections. We now separate two cases.

Case A   Suppose $n^2 + 1$ or more of the remaining segments connect points of Σ only. In this case, we are reduced to the inductive hypotheses for the segments in Σ, and we have the result.

Case B   Suppose that *fewer* than $n^2 + 1$ of the remaining segments connect points of Σ. Since there are $n^2 + 2n + 1$ of these segments in all, *more* than $2n$ of the remaining segments join either A or B. Of these, call the segments joining A the *azure* segments, and call those joining B *blue*, and regard a point of Σ as a box. Thus each of the segments joining A or B is placed into a box, the end point of that segment in Σ. Since we have more than $2n$ segments and only $2n$ "boxes," it follows that two differently colored segments go into the same box. But this forms the triangle, as in Figure 2.4.7.

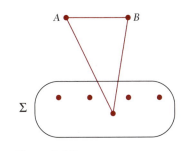

Figure 2.4.7

We have seen that the Box Principle, as mundane as it appears to be, can be used in fairly sophisticated ways. The principle has been stated in terms of boxes, pigeonholes, chips, and other items. Let us now analyze this tool and see what it says in a more mathematical language.

How can we interpret "placing objects in a box"? Let the set of objects be called A, and the set of boxes B. Then the placing of the objects into the boxes is merely an assignment of values: each $a \in A$ is assigned a value $b \in B$. This is simply a function from A to B.

What does it mean that "two different objects are in the same box"? In terms of our previous definition, this means that the function has the same value for two different elements—that is, it is not 1-1. Thus, we have the following formulation of the Box Principle:

The Box Principle (abstract version)

Suppose A and B are finite sets, with $|A| > |B|$ and $f: A \to B$. Then f is not 1-1. Alternatively: If $f: A \to B$ is 1-1, then $|A| \leq |B|$.

This latter statement is the contrapositive of the former statement and has exactly the same meaning.

We shall *prove* a version of the Box Principle for the prototypes of finite sets, namely the sets $\mathbb{I}_n = \{1, 2, \ldots, n\}$.

THEOREM 2.4.5

Let $n > m > 0$ be natural numbers, and let $f: \mathbb{I}_n \to \mathbb{I}_m$. Then f is not 1-1.

Proof The proof is by induction on n. It is clearly true for $n = 2$ because in that case $m = 1$, so f can't be 1-1 because $f(1) = f(2) = 1$.

We now assume the result for n, and we consider any map $f: \mathbb{I}_{n+1} \to \mathbb{I}_{m+1}$, where again $m < n$. It is convenient to speak of $\mathbb{I}_{n+1}$ as the objects and $\mathbb{I}_{m+1}$ as the boxes. Thus, f merely places objects into boxes. We now look at the object $n + 1$ in $\mathbb{I}_{n+1}$ and separate cases according to whether $n + 1$ is placed in $m + 1$.

Case A   If $n + 1$ is placed into the box $m + 1$, the rest of $\mathbb{I}_{n+1}$ is $\mathbb{I}_n$. If $\mathbb{I}_n$ is mapped into $\mathbb{I}_m$, then f is not 1-1 by the induction hypothesis. If one of the elements x of $\mathbb{I}_n$ is sent into an element not in $\mathbb{I}_m$, the only choice is $m + 1$. But then the function f would not be 1-1 because both x and $n + 1$ would be placed into $m + 1$. So f cannot be 1-1 in either case.

Case B   If $n + 1$ is placed into box $k \in \mathbb{I}_m$, we switch k and $m + 1$ by applying the transposition $(k, m + 1)$. The function $g = (k, m + 1) \circ f$ sends $n + 1$ into $m + 1$. [The transposition $(k, m + 1)$ is simply the swap of k and $m + 1$, as explained in Section 1.2.] This means that g is covered by Case A, so g is not 1-1. Therefore, there are two elements $a \neq b$ such that $g(a) = g(b)$. Hence, $(k, m + 1) \circ f(a) = (k, m + 1) \circ f(b)$. Applying $(k, m + 1)$, we find $f(a) = f(b)$. This shows that f is not 1-1. ∎

COROLLARY 2.4.6

Any finite set A has a unique value $|A|$.

Proof   Suppose $|A|$ had two values m and n, where $m < n$. Then, by Definition 2.2.1 there are 1-1 and onto functions f and g such that f: $\mathbb{I}_m \to A$, and g: $\mathbb{I}_n \to A$. But this implies that $f^{-1} \circ g$, which sends $\mathbb{I}_n$ into $\mathbb{I}_m$, is also 1-1 and onto. (See Theorems 2.2.11 and 2.2.12.) However, this is impossible by Theorem 2.4.5. ∎

Thus, the Box Principle is closely related to the result that a finite set has a unique number of elements. The abstract version of the Box Principle is a simple corollary of Theorem 2.4.5.

COROLLARY 2.4.7   The Box Principle

Suppose A and B are finite sets, with $|A| > |B|$. Let $f: A \to B$. Then f is not 1-1.

Proof   Let $n = |A|$ and $m = |B|$. Let $f_A: \mathbb{I}_n \to A$ and $f_B: \mathbb{I}_m \to B$ be 1-1 and onto. We prove this result by contradiction. Suppose f were 1-1. Then $f_B^{-1} \circ f \circ f_A$ would be 1-1 from $\mathbb{I}_n$ to $\mathbb{I}_m$ because the composition of 1-1 functions is 1-1. But since $m < n$, this is impossible by Theorem 2.4.5. ∎

The following corollary is a variant of the Box Principle. It is stated in terms of onto functions.

COROLLARY 2.4.8

Suppose A and B are finite sets, with $|A| < |B|$. Let $f: A \to B$. Then f is not onto.

Proof   Suppose f were onto. Then for every element b of B, we can find at least one preimage a in A, namely an element such that $f(a) = b$. Call any one of these elements $g(b)$. Then $g: B \to A$ and $f(g(b)) = b$, by construction. But this implies that g is 1-1. For if $g(b_1) = g(b_2)$, then, applying f, we get $b_1 = b_2$. But g can't be 1-1 by Corollary 2.4.7, where the roles of A and B are reversed. This proves the result. ■

THEOREM 2.4.9

Let A be a finite set and let $f: A \to A$. Then

(a) If f is onto, it is also 1-1.
(b) If f is 1-1, it is also onto.

Proof   (a) Suppose f is onto. If it is not 1-1, there are two elements a and b of A that have the same image. If we look at f on the set $A - \{a\}$, it remains onto. Thus, we have an onto function from $A - \{a\}$ onto A. This is impossible by Corollary 2.4.8.

(b) Suppose f is 1-1. If it is not onto, then the range is not all of A, so we have a 1-1 function from A into the range, a set that has fewer elements. This is impossible by Corollary 2.4.7. ■

Remark   For example, suppose we list n integers, all between 1 and n, inclusive. Then if they are different, they must include all the integers between 1 and n, inclusive. Conversely, if they include all the integers between 1 and n, inclusive, then they are all different. This is true because such a list is a function from I_n to I_n.

This obvious fact is not true of infinite sets. The function $f: \mathbb{N} \to \mathbb{N}$ defined by $f(n) = n + 1$ is 1-1 but not onto. Similarly, the function $g: \mathbb{N} \to \mathbb{N}$ defined by $g(n) = [n/2]$ (the greatest integer in $n/2$) is onto but not 1-1.

The difference between finite and infinite sets can be illustrated by means of the "Hilbert hotel." A hotel has infinitely many rooms numbered, naturally, 0, 1, 2, ... Suppose all of its rooms are occupied, and an important traveler arrives and wishes to be accommodated without having the current occupants double up. The owner has a simple solution: Tell the occupant of room n to move to $n + 1$ immediately, thereby freeing room 0 for the new arrival. The owner understood that the function $f(n) = n + 1$ was 1-1, so no doubling up occurred, but $f(n)$ was not onto, so a room was now available.

Exercises

1. (a) Generalize the Box Principle to placing $2n + 1$ objects into n boxes.
   (b) Repeat part (a) but place $rn + 1$ objects into n boxes.
   (c) Repeat part (a) but place infinitely many objects into n boxes.
2. Prove your generalizations of Exercise 1(a), (b), and (c).

3. Generalize the alternative version of the Box Principle for placing balls of three different colors into n boxes.

4. Suppose $n^2 + 1$ points are in the unit square. Show that two of them are within $\sqrt{2}/n$ of each other.

5. Suppose that 21 points are in the unit square. You want to guarantee that at least two of them are within x of each other. What x do you choose? Make x as small as you can.

*6. Infinitely many points are in the unit square. Show that two of them are within 0.0001 of each other. (You can use Exercise 4.)

*7. If three points are in the unit square, at least two of them must be within y of each other. Find the smallest possible y that makes this true. (*Note:* There's no guarantee that the Box Principle will help here!)

8. Four points are in an equilateral triangle of side 2. Show that a certain pair of these points are within 1.16 of each other. Show that this result is not true if 1.16 is replaced by 1.15.

9. Seven points are in the unit square. Prove that two of these points are within 0.601 of each other.

*10. Nine points are placed in a unit square. Prove that two of these points are within 0.518 of each other. (*Hint:* See Exercise 7.)

*11. Seventeen different integers are chosen between 1 and 30, inclusive. Show that some two of them differ by 3. Can you show this for 16 integers? Explain. Can you show this for 15 integers? Explain.

*12. If 15 different integers are chosen between 1 and 45, show that some two of them differ by 1, 3, or 4. Is the result true for 14 integers? Explain.

13. A lattice point in the plane is a point with integer coordinates. Prove that if five lattice points are given in the plane, then there are two of them whose midpoint is also a lattice point. [The midpoint of (a, b) and (c, d) is the point $((a + c)/2, (b + d)/2)$.]

14. Generalize the previous exercise to lattice points in 3-space.

*15. Fifty-one integers are chosen between 1 and 100, inclusive. Prove that some one of these divides another. Generalize to $n + 1$ integers chosen from 1 through $2n$.

16. Given any 25 integers, show that some two of them differ by a multiple of 24. (*Hint:* Look at the remainders when each is divided by 24. Put the remainders in a box.)

17. In Example 2.4.4, $n^2 + 1$ segments connected $2n$ points and a triangle was formed. Suppose only n^2 segments are formed. Is a triangle necessarily formed?

*18. Suppose the letters A through J are arranged in one string. Prove that some four of these letters must be arranged in alphabetic order, reading either from left to right or from right to left. For example, "*DBICAJFEHG*" has the letters "*BCEG*" in alphabetic order, and "*HIJEFGBCDA*" has the letters "*HEBA*" in antialphabetic order. [*Hint:* For any of the letters appearing in the string, find the size m of the largest alphabetic substring starting with it and the size n of

the largest antialphabetic substring starting with it. Put the letter
into the "box" (m, n).]

*19. Generalize the previous exercise to a string of $n^2 + 1$ numbers.

20. The proof of Corollary 2.4.8 constructed a function $g: B \rightarrow A$ that
looked a little like an inverse of $f: A \rightarrow B$. Here, f was onto but not
necessarily 1-1. Is g an inverse? Explain. Is $f \circ g = i_B$? Is $g \circ f = i_A$?

21. Give an example of a 1-1 function $f: \mathbb{N} \rightarrow \mathbb{N}$ such that there are
infinitely many points of $\mathbb{N}$ not in the range of f.

*22. Give an example of an onto function $g: \mathbb{N} \rightarrow \mathbb{N}$ such that each point
of $\mathbb{N}$ is an image of infinitely many points of $\mathbb{N}$.

2.5 Number Bases

Numbers are usually written with the ten digits $0, 1, \ldots, 9$. This familiar
system of notation is called decimal notation. It was long understood by
mathematicians that 10 can be replaced by another integer $b > 1$, leading
to different systems of notation that are equally feasible. (The ten digits
were probably originally used because we have ten fingers, hence the
term "digits.") However, it was only with the advent of computers that
different number systems were taken seriously in practice as well as in
theory. Let us begin by showing how a different number system can be
used in a practical situation.

EXAMPLE 2.5.1 *Base 2 Representation*

We considered bytes in Section 1.1. Bytes are strings of length 8 consisting of 0's and 1's. The first popular personal computers had memory
cells that stored only 1 byte. Each memory location may be regarded as
eight switches that can be on or off (coded as 1 and 0, respectively.)
Figure 2.5.1 represents an 8-bit memory cell. A string is stored in this
cell by turning each of these eight switches on or off. Figure 2.5.2 is an
example of what can be in a memory cell.

Figure 2.5.1

| 0 | 1 | 1 | 0 | 0 | 1 | 1 | 1 |

Figure 2.5.2

It is sometimes not convenient to say that this memory cell "contains
01100111." When dealing with many strings, 0's and 1's can be misread,

which can lead to human error. We now show how to code such a sequence into a more manageable number. The method is to label the cells from right to left with the powers of 2: 1, 2, 4, 8, 16, 32, 64, and 128, as in Figure 2.5.3.

n:	7	6	5	4	3	2	1	0
2^n:	128	64	32	16	8	4	2	1

0	1	1	0	0	1	1	1

Figure 2.5.3

Now add the powers of 2 that are labeling the cells storing a 1. (These are colored in the figure.) This number is $1 + 2 + 4 + 32 + 64 = 103$. The sequence 01100111 is then coded as the number 103, and we write

$$(01100111)_2 = 103$$

The subscript 2 is used to indicate that the columns are coded by powers of 2. The number 2 is called the **base**, and we say that the number 103 is written as the string 01100111 using the base 2. Figure 2.5.3 shows how the number 103 is stored in memory. Clearly, the largest number that can be stored has the sequence 11111111, which represents 255, and the lowest number represented is 0. As in Example 2.3.7, there are $2^8 = 256$ possible bytes. This method gives a convenient numbering of these bytes.

In Examples 1.1.3 and 1.1.7, we said that the ASCII characters are a standard set of 128 characters that can be transmitted to the printer. All of these are coded with strings starting with "0". (The first bit in a string is called the high bit, because it corresponds to the highest power of 2. Thus, the standard ASCII characters are coded with 0 as the high bit.) The letter R, for example, has ASCII number 82 and is stored as the string "01010010": $(01010010)_2 = 2^6 + 2^4 + 2^1 = 64 + 16 + 2 = 82$. The strings with high bit 1 correspond to numbers from 128 through 255 and, as we have noted, are usually assigned special printing tasks, depending on the printer involved.

In what follows, we generalize the method of Example 2.5.1 by replacing 2 with an arbitrary base b and using the digits $0, 1, \ldots, b - 1$. (The usual decimal notation has $b = 10$.) We begin with the familiar statement of arithmetic that it is possible to divide one number by another to get a quotient and a remainder. More precisely, we have the following result.

THEOREM 2.5.2 The Division Algorithm

If n and b are integers and b is positive, then there are two integers Q and R, called the quotient and the remainder, such that

$$n = bQ + R \qquad (2.5.1)$$

and

$$R = 0, 1, \ldots, \text{ or } b - 1 \qquad (2.5.2)$$

The integers Q and R are uniquely determined by n and b. ■

The process of finding Q and R is called long division; the algorithm is simply the method given in grade school to find the quotient and the remainder. A proof of this result is indicated in the exercises.

Condition (2.5.2) may be stated $0 \le R \le b - 1$, or simply $R \in \mathbb{Z}_b$. The numbers of $\mathbb{Z}_b$ are called the **base b digits**. There is a difference between an integer of $\mathbb{Z}_{10}$ and the symbol used to denote that integer. (See Example 1.1.6.) Strictly, a digit is a symbol with no numerical properties, and an element of $\mathbb{Z}_{10}$ is an integer with various arithmetic properties. Despite this observation, we continue to identify a base b digit with an integer of $\mathbb{Z}_b$ until such time as this identification leads to confusion.

The numbers of $\mathbb{Z}_{10}$ are called the **decimal digits,** and the numbers of $\mathbb{Z}_2$ are called the **binary digits,** or **bits.** Similarly, the **octal digits** are the digits of $\mathbb{Z}_8$. Equation (2.5.1) can be generalized in a very powerful way that directly generalizes the base 2 representation of Example 2.5.1, as well as the ordinary way we represent decimal integers.

THEOREM 2.5.3

Let b and n be nonnegative integers with $b \ge 2$. Then for some integers k and d_i $(i = 0, 1, \ldots, k)$,

$$n = d_k b^k + \cdots + d_2 b^2 + d_1 b + d_0 \qquad (2.5.3)$$

with $d_i \in \mathbb{Z}_b$.

Proof The proof is by induction on n; b is kept fixed for the duration of the proof. The result is true for $n = 0$ because we can choose $k = 0$ and $d_0 = 0$.

Now assume the result is true for all integers less than n. Using the division algorithm, write

$$n = bQ + R \qquad (2.5.4)$$

Since $Q < n$, we have by induction

$$Q = d_k b^{k-1} + \cdots + d_2 b + d_1 \qquad (2.5.5)$$

Now define $d_0 = R$. Then by (2.5.4) and (2.5.5), we have

$$n = b(d_k b^{k-1} + \cdots + d_2 b + d_1) + d_0$$
$$= d_k b^k + \cdots + d_2 b^2 + d_1 b + d_0$$

∎

This inductive proof actually shows how to find the base b digits d_0, $d_1, \ldots$. We shall show this and illustrate with a numerical example. But first we introduce the standard notation, which helps improve the readability of the formulas.

Base b Notation for Integers

If

$$n = d_k b^k + \cdots + d_2 b^2 + d_1 b + d_0$$

we write

$$n = (d_k \cdots d_2 d_1 d_0)_b$$

(Compare with Example 2.5.1.) The notation is useful because it keeps track of the coefficients of the various powers of b, without actually writing these powers. For example,

$$(321)_8 = 3 \cdot 64 + 2 \cdot 8 + 1 = 209$$

Our usual decimal notation is an example. The notation $n = 345$ is shorthand for $n = 3 \cdot 10^2 + 4 \cdot 10 + 5$, and 10 is itself a representation of $1 \cdot b + 0$, the base b being the number ten, but which is called 10 without further ado, because the decimal notation is very much part of our language.

If we follow the argument in Theorem 2.5.3, we see that the last digit of a number n (base b) is the **remainder** when n is divided by b. The digits before it are the digits of the quotient.

We can see this directly from Equation (2.5.3). Since

$$n = d_k b^k + \cdots + d_2 b^2 + d_1 b + d_0$$

we have

$$n = b(d_k b^{k-1} + \cdots + d_2 b + d_1) + d_0$$

where we can see that the remainder is d_0 and the quotient has the representation $(d_k \cdots d_2 d_1)_b$. We can then repeat this process. The last digit of the quotient is the remainder when *it* is divided by b, and the digits before it are the digits of its quotient when divided by b. We then continue this process until the quotient is less than b, in which case it is its own base b representation. An example will illustrate the procedure.

EXAMPLE 2.5.4

Write 327 in base 8 and in base 4.

Method   Divide by 8 to obtain the quotient and the remainder, and then continue this process on the new quotient in the same way. Continue until the quotient is 0.

$$327 = 8 \cdot 40 + 7$$
$$40 = 8 \cdot 5 \ + 0$$
$$5 = 8 \cdot 0 \ + 5$$

Reading *up* the column of remainders, we have

$$327 = (507)_8$$

The work is entirely similar for base 4:

$$327 = 4 \cdot 81 + 3$$
$$81 = 4 \cdot 20 + 1$$
$$20 = 4 \cdot 5 \ + 0$$
$$5 = 4 \cdot 1 \ + 1$$
$$1 = 4 \cdot 0 \ + 1$$

Thus,

$$327 = (11013)_4$$

This procedure is an example of an **algorithm.** By this, we mean a mechanical, unambiguous procedure to arrive at a result. The procedure illustrates the algorithm for finding the base b representation of a given number n. The base b was fixed. In Chapter 3, we consider algorithms in greater detail, and we present this one in Examples 3.3.1 and 3.3.3 in a somewhat more sophisticated setting.

To convert base b to decimal notation, it is only necessary to use the definition. Thus, $(231)_4 = 2 \cdot 4^2 + 3 \cdot 4 + 1 = 45$.

For large numbers, a shortcut, called **Horner's method,** is often useful. The evaluation of a number given in base b notation can be routinely done by using this method by a straightforward series of additions and multiplications by b. No powers of b are directly computed. For example, $(56326)_8$ can be computed as follows. Find the first digit 5, multiply by the base 8, and add the second digit. This yields $5 \cdot 8 + 6$. Now multiply the answer by the base 8 and add the next digit. This yields $5 \cdot 8^2 + 6 \cdot 8 + 3$. Doing this again, we obtain $5 \cdot 8^3 + 6 \cdot 8^2 + 3 \cdot 8 + 2$, and so on, until we finally arrive at the result. We illustrate the technique.

EXAMPLE 2.5.5   Horner's Method

Find the value of $(3406)_8$.

Method   We systemize the previous procedure by placing the digits along the top row and by placing the base on the side. The first digit is brought down, multiplied by 8 (the base), and brought up to the next column. Then add, multiply this sum by the base, and bring the answer up to the next column. Continue to the end, to get the final answer 1798.

$$
\begin{array}{rrrr|l}
3 & 4 & 0 & 6 & 8 \\
  & 24 & 224 & 1792 & \\
\hline
3 & 28 & 224 & 1798 &
\end{array}
$$

This procedure is perhaps the most efficient way of calculating a conversion to decimal notation, and is very easily done on a hand calculator.

Binary (base 2) Usage

In binary notation, $b = 2$, and the binary digits are 0 and 1. We have already seen some examples of strings of bits. Example 2.5.1 showed how binary notation is used to describe what is stored in a memory cell of a computer. In Section 2.3, we showed how to code subsets of an n-element set as a string of bits of length n and, thus (regarding this string as the binary representation of a number), as a number. We now illustrate another use of binary notation.

EXAMPLE 2.5.6   Dot Matrix Printer Graphics

Dot matrix printers print letters, symbols, and diagrams by printing dots on a blank piece of paper. (This is also true of laser printers.) Depending on the type of printer and the quality of the printing, a letter or a symbol is defined by the dots in an m by n rectangle (m rows of dots by n columns). We illustrate with a 7 by 11 matrix. The printer will print a column of 7 dots at a time, and a letter or symbol will be 11 columns across (including spaces needed to separate one letter from another).

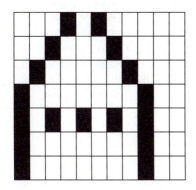

Figure 2.5.4

Figure 2.5.4 illustrates how the letter *A* might be defined by such a 7 by 11 matrix. When letters, symbols, or graphics are specially designed, it is necessary to tell the printer which dots to print—a tedious task. The method suggested by the printer manufacturers is to code each column in binary notation. Table 2.5.1 shows how column 1 of Figure 2.5.4 can be represented ("dot" = 1; "no dot" = 0). Since $(1111000)_2 = 120$, we can say that the number 120 is stored in column 1.

Table 2.5.1

1	0
2	0
4	0
8	1
16	1
32	1
64	1

In this way, the dots in Figure 2.5.4 would be transmitted to the printer by transmitting 11 numbers, one for each column. The reader may verify that this sequence is 120, 4, 18, 1, 16, 1, 18, 4, 120, 0, 0. The alternative would be to send a 7 by 11 matrix of 0's and 1's to the printer, which would be more open to human error. (*Note:* Each printer has its own way of coding graphics and symbols. The foregoing is an illustration only. You can be virtually certain that binary notation is used to print graphics on a dot matrix printer.)

Octal Notation

Although the binary system is the most natural for computers, it has an inherent disadvantage for humans: it is very difficult to read. (Consider the number 103 compared with 01100111.) Relatively small numbers have many binary digits, and it is very easy for a person to make a mistake in transcribing binary numbers. What is needed is a notation that is (a) easily converted to and from binary and is (b) easy for humans to read. The octal (base 8) and the hexadecimal (base 16) systems do this.

Table 2.5.2 contains some common conversions that are worth memorizing.

Table 2.5.2   Octal Conversion Chart

Base 2	Decimal	Octal
000	0	0
001	1	1
010	2	2
011	3	3
100	4	4
101	5	5
110	6	6
111	7	7
1000	8	10

It is an easy matter to convert from binary to octal. The trick is to exploit the fact that 8 is a power of 2 ($8 = 2^3$). We first illustrate the method and then show why it works.

EXAMPLE 2.5.7

Convert $(1011001101100111)_2$ to octal.

Method   First group the digits into groups of three (starting from right to left):

$$1\quad 011\quad 001\quad 101\quad 100\quad 111$$

Now convert each triple to octal, using Table 2.5.2:

$$1\quad 3\quad 1\quad 5\quad 4\quad 7$$

The answer is $(131547)_8$.

Why does this work? Any time we have a power of 2 that is 3 or higher, we can factor out 2^3, or 8. Thus, the underlined triple in the given binary string 1 011 001 <u>101</u> 100 111 corresponds to the sum $1{\cdot}2^8 + 1{\cdot}2^6 = (1{\cdot}2^2 + 1){\cdot}2^6$. The first factor is simply 5, the value of $(101)_2$; the second is $2^6 = 8^2$, since $2^3 = 8$. In short, the underlined sequence corresponds to the coefficient of the power of 8^2 in the octal representation of the number. The same argument works for any of the triples.

This method works backwards just as well. To convert $(7132)_8$ into binary, we write

$$
\begin{array}{cccc}
7 & 1 & 3 & 2 \\
111 & 001 & 011 & 010
\end{array}
$$

and the answer is $(111001011010)_2$. Note that it was never necessary to find the *decimal* representation of any of the numbers with this system.

The method works for any base, even decimal. For example, the number 123,456 (decimal) becomes 12 34 56 in the base 100 (10^2).

Hexadecimal Notation

The base $b = 16$, called hexadecimal or hex notation, is often used. However, one problem arises. The digits for the hex system of notation represent the numbers of $\mathbb{Z}_{16}$: the numbers 0 through 15. Now a "digit" such as 15 might also be interpreted in hex as $1 \cdot 16 + 5$ or 21. Here it is wise to introduce special symbols for the decimal numbers 10 through 15. The special hex digits A through F are used. Table 2.5.3, similar to the one introduced for octal, is useful.

Table 2.5.3   Hexadecimal Conversion Chart

Base 2	Decimal	Hexadecimal	Base 2	Decimal	Hexadecimal
0000	0	0	1000	8	8
0001	1	1	1001	9	9
0010	2	2	1010	10	A
0011	3	3	1011	11	B
0100	4	4	1100	12	C
0101	5	5	1101	13	D
0110	6	6	1110	14	E
0111	7	7	1111	15	F

The method used to convert binary to octal applies for conversion from binary to hex. In this case, since $16 = 2^4$, we group by 4's.

EXAMPLE 2.5.8

Convert $(1110110101)_2$ to hex.

Method   As in Example 2.5.7, except that we group by 4's:

$$
\begin{array}{ccc}
11 & 1011 & 0101 \\
3 & B & 5
\end{array}
$$

The answer is $(3B5)_{16}$. Often, 3B5H rather than the subscript notation is used. Sometimes 3B5h or $3B5 is used.

One reason for using hex notation is to simplify the representation of bytes. Since 8 bits separate neatly into two 4-bit groups, a byte is given by a two (hex) digit number. Thus, the binary number $(01100111)_2$ in Example 2.5.1 is, at a glance, 67H. For this reason, hex notation is commonplace when working with computers.

On a certain personal computer with 8-bit memory cells, the following output shows the status of the memory:

$$
\begin{array}{llllllll}
1000- & 30 & 28 & \text{F8} & 71 & 14 & \text{0D} & 60 \\
1008- & \text{ED} & 56 & \text{0A} & 54 & 20 & 00 & 00 \\
1010- & 21 & 45 & 85 & & & &
\end{array}
$$

The meaning is as follows: Memory location 1000H has 30H stored in it, memory location 1001H has 28H stored in it, and so on. The memory locations and the numbers stored in them are written in hex. If we convert to decimal, the value ED stored at 1008 means that the number 237, or $(11101101)_2$, is stored in memory location 4104. Usually, decimal conversions are not needed at this level. It is reasonable to say that many computer technicians are just as familiar with hex as with decimal notation.

Base *b* Arithmetic

Arithmetic (addition and multiplication) of numbers expressed in base b is quite similar to arithmetic using decimal notation. We first note that $b \cdot b^k = b^{k+1}$. In base b notation, this means that multiplying by b [i.e., by $(10)_b$] is equivalent to shifting one space to the left. For example, $(1234)_b \cdot (10)_b = (12340)_b$. This is quite familiar in decimal notation. In fact, all of the usual laws of addition and multiplication (as well as subtraction and long division) apply in base b notation with the appropriate addition and multiplication tables. Table 2.5.4 gives the arithmetic tables for binary notation.

These are familiar enough except for the entry $1 + 1 = 10$. However, using this table and the usual rules for addition and multiplication, we can add numbers written in binary:

$$
\begin{array}{r}
1001101 \\
+\ 1011010 \\
\hline
10100111
\end{array}
\qquad
\begin{array}{r}
100110 \\
\times\quad 110 \\
\hline
1001100 \\
100110\quad \\
\hline
11100100
\end{array}
$$

Since computers do their work in binary, addition and multiplication routines for the machine are programmed in binary using Table 2.5.4. It is often useful to know that multiplication by a power of 2 in binary amounts to adding 0's to the end of the number. Thus, to find 40 in binary, we compute $40 = 8 \cdot 5 = (1000)_2(101)_2 = (101000)_2$.

Table 2.5.4 Addition and Multiplication Tables (binary notation)

+	0	1		×	0	1
0	0	1		0	0	0
1	1	10		1	0	1

The following results are familiar in decimal notation. We carry them through for arbitrary bases.

THEOREM 2.5.9

$$(b - 1 \quad b - 1 \quad \cdots \quad b - 1)_b = b^n - 1 \qquad (2.5.6)$$

Proof Here, we have n of the digits $b - 1$ in a row. This is a version of the geometric series (Theorem 2.3.11):

$$(b - 1) \sum_{k=0}^{n-1} b^k = b^n - 1$$

As a familiar illustration, $9999 = 10^4 - 1$. For other bases, $(1111111)_2 = 2^7 - 1 = 127$, and $(777)_8 = 8^3 - 1 = 511$. ∎

THEOREM 2.5.10

Let

$$n = (a_k \quad a_{k-1} \quad \cdots \quad a_1 \quad a_0)_b$$

$$m = (b_k \quad b_{k-1} \quad \cdots \quad b_1 \quad b_0)_b$$

be two integers and suppose that $a_k < b_k$. Then $n < m$.

Proof By definition, $n = \sum_{j=0}^{k} a_j b^j$. Thus, increasing or decreasing any of the digits of an integer will increase or decrease its value correspondingly. Since $a_k + 1 \le b_k$, we have

$$m = (b_k \quad b_{k-1} \quad \cdots \quad b_1 \quad b_0)_b$$
$$\ge (a_k + 1 \quad 0 \quad 0 \quad \cdots \quad 0 \quad 0)_b$$
$$= (a_k \quad 0 \quad 0 \quad \cdots \quad 0 \quad 0)_b + (1 \quad 0 \quad 0 \quad \cdots \quad 0 \quad 0)_b$$
$$> (a_k \quad 0 \quad 0 \quad \cdots \quad 0 \quad 0)_b + (0 \quad b - 1 \quad b - 1 \quad \cdots \quad b - 1)_b$$
$$= (a_k \quad b - 1 \quad b - 1 \quad \cdots \quad b - 1)_b$$
$$\ge (a_k \quad a_{k-1} \quad \cdots \quad a_1 \quad a_0)_b$$
$$= n$$

The crucial inequality was $(1 \quad 0 \quad 0 \quad \cdots \quad 0)_b > (0 \quad b-1 \quad b-1 \quad \cdots \quad b-1)_b$, which follows from Theorem 2.5.9. ■

The result is familiar. Thus, $682 < 723$ since $6 < 7$.

This theorem can be strengthened. We can see at a glance that one number is smaller than another if it is **lexicographically** smaller. The representations must both have the same length, so leading 0's may have to be introduced in order to use this result.

DEFINITION 2.5.11

Let

$$\tau = ``a_k \quad a_{k-1} \quad \cdots \quad a_1 \quad a_0"$$

$$\sigma = ``b_k \quad b_{k-1} \quad \cdots \quad b_1 \quad b_0"$$

be strings of length $k + 1$ on the alphabet $\mathbb{Z}_b$. We say that τ **precedes** σ **lexicographically** if, starting from the leftmost digit, $a_j < b_j$ in the first place j in which they differ. Otherwise put, for some j we must have $a_j < b_j$, and $a_r = b_r$ for $r > j$.

COROLLARY 2.5.12

Let

$$t = (a_k \quad a_{k-1} \quad \cdots \quad a_1 \quad a_0)_b$$

$$s = (b_k \quad b_{k-1} \quad \cdots \quad b_1 \quad b_0)_b$$

be two integers, and suppose that the string for t precedes the string for s lexicographically. Then $t < s$.

Proof If j is the first place that s and t differ, simply subtract $(a_k \quad \cdots \quad a_{j+1} \quad 0 \quad \cdots \quad 0)_b$ from s and t. The resulting numbers are

$$\bar{s} = (0 \quad \cdots \quad 0 \quad a_j \quad \cdots \quad a_0)_b \quad \text{and} \quad \bar{t} = (0 \quad \cdots \quad 0 \quad b_j \quad \cdots \quad b_0)_b$$

Dropping the leading 0's, we have $\bar{s} < \bar{t}$ by Theorem 2.5.10. Therefore $s < t$. ■

This is, of course, how we compare numbers in decimal notation. We can see at a glance that $8841 < 8857$.

Real numbers can also be expressed in base b. As in the usual decimal representation of reals, we use negative powers of b and separate the negative and nonnegative powers by a decimal point. The number $(1.1010)_2$ can be calculated (in decimal form) as $1 + 2^{-1} + 2^{-3}$, or 1.625.

A simpler way might be $(1101)_2/(1000)_2 = \frac{13}{8}$. Infinite decimals are also possible. As for infinite repeating decimals, the governing equation of algebra is the formula for an infinite geometric series whose first term is a and with ratio r:

$$\sum_{k=0}^{\infty} a_0 r^k = \frac{a_0}{1 - r} \quad \text{if } |r| < 1$$

This follows from Theorem 2.3.11 (the sum of a geometric series) by letting the number of summands approach ∞. For example, $(1.01010101\ldots)_2$ is an infinite geometric series with ratio $r = (0.01)_2 = \frac{1}{4}$, and first term 1. Thus its sum (value) is $1/(1 - \frac{1}{4}) = \frac{4}{3}$. In binary, this is $1/(1 - (0.01)_2) = 1/(0.11)_2 = (100)_2/(11)_2$.

Note that $(0.111\ldots)_2 = 1$. This is analogous to the decimal formula $0.999\ldots = 1$. Similarly, in base b, $(.b-1\ b-1\ \cdots)_b = 1$.

A **finite** binary expansion of a number x is possible only if x is a rational number whose denominator is a power of 2. For in this case, x is a finite sum of numbers of the form 2^{-i}. Thus, a "simple" decimal number like $x = 1.6$ will have an **infinite** binary expansion. This is significant because humans usually use decimal notation, whereas computers use binary notation. Thus, unless special techniques are used, the number $x = 1.6$ will be stored to a finite number of binary places, and thus will not be exactly stored. Don't be surprised, then, if (in certain calculators or computers) we multiply $x = 1.6$ by 1,000,000,000 and get an answer like 1,599,999,998. Good design will circumvent this problem in many cases.

We have seen that base 2 integers need more digits than do decimal integers. How much more? We now consider the general question of how many digits are needed to represent a positive integer N using base b. We know that a number between 1000 and 9999 needs four decimal digits. Thus, if $10^3 \le N < 10^4$, then N has four decimal digits, and conversely. Our analysis of base b representation shows that the general case is similar:

$$b^{k-1} \le N < b^k \quad \text{iff} \quad N \text{ has } k \text{ base } b \text{ digits}$$

THEOREM 2.5.13

Let $D(N, b)$ be the number of digits N has using base b. Then

$$D(N, b) = 1 + [\log_b N] \tag{2.5.7}$$

where $[x]$ is the greatest integer function.

Proof We noted that $b^{k-1} \le N < b^k$ if N has k base b digits. Thus, setting $k = D(N, b)$, we have

$$b^{D(N,b)-1} \le N < b^{D(N,b)}$$

Now take logarithms to the base b. This yields

$$D(N, b) - 1 \leq \log_b N < D(N, b)$$

Rewriting, we have

$$\log_b N < D(N, b) \leq 1 + \log_b N$$

Since $D(N, b)$ is an integer, this shows that

$$D(N, b) = 1 + [\log_b N]$$

■

EXAMPLE 2.5.14

Formula (2.5.7) gives the simple approximation

$$D(N, b) \approx \log_b N \qquad (2.5.8)$$

valid to within 1 unit. It is a property of logarithms that $\log_b N = \log_{10} N/\log_{10} b$. Therefore, Equations (2.5.7) and (2.5.8) can also be written simply as

$$D(N, b) = 1 + [\log_{10} N/\log_{10} b] \qquad (2.5.9)$$

EXAMPLE 2.5.15   Number of Binary Digits of an Integer

Using Equation (2.5.9), we can now approximately express the number of binary digits of an integer N in terms of the number of its decimal digits. We have $D(N, 2) \approx \log_{10} N/\log_{10} 2$. Since $\log_{10} 2 \approx 0.301$, this gives $D(N, 2) \approx 10(\log_{10} N)/3$.

Since $D(N, 10) \approx \log_{10} N$, *a reasonable rule of thumb to find the number of binary digits of N is*

$$D(N, 2) \approx 10D(N, 10)/3 \qquad (2.5.10)$$

For example, the number of binary digits needed for 50,352 is approximately 50/3 or about 17. By Equation (2.5.9), the accurate answer is as follows:

$$D_2(50,352) = 1 + [\log_{10} 50,352/\log_{10} 2] = 16$$

A simple computation gives $2^{15} = 32,768$ and $2^{16} = 65,536$. This shows that $\log_2 50,352$ is between 15 and 16, which shows directly, without the explicit use of logarithms, that the number of binary digits of 50,352 is exactly 16.

Exercises

1. Write the number 8352 in octal; in base 5; in hex.
2. Write $(7413)_8$ as a decimal number. Use Horner's method.
3. Write $(10011110110011011)_2$ in octal; in hex; in base 4. Do not convert any of these to decimal.
4. Write $(37156)_8$ in hex. It is not necessary to find decimal representations.
5. Write 4EAH as a decimal number. Use Horner's method.
6. Write $(12011202)_3$ as a number in base 9.
7. Convert $(3452)_7$ to a number in base 9.
8. (a) Approximately how many digits are in the binary expansion of 67,812,345? (b) Approximately how many decimal digits are in the expansion of 10011011000111?
9. (a) How many decimal digits are in the expansion of $(10011001)_4$? Give an exact answer. (b) Exactly how many binary digits are in the expansion of 157,654?
10. A balance scale has the weights 1, 2, 4, 8, 16, and 32 grams. Explain how and why any integer number of grams up to 63 grams may be weighed on this scale.
*11. Repeat Exercise 10, but the weights are 1, 3, 9, 27, and 81 grams. What weights can you measure exactly? Explain. (*Hint:* You can put weights on both sides of the scale.)
*12. Theorem 2.5.3 is clearly false for $b = 1$. Explain where the proof fails in this case.
*13. Find a formula for $(1010 \ldots 10)_2$, where this number has k consecutive pairs (10). (*Hint:* Multiply by $3 = (11)_2$ and use binary arithmetic. Alternatively, use the formula for the sum of a geometric series.)
14. Give addition and multiplication tables for the digits in base 4 arithmetic.
15. (*Programming exercise*) Write a program (in any suitable computer language) that converts a nonnegative integer into its octal representation. Your output should be a string of digits in $\mathbb{Z}_8$.
16. (*Programming exercise*) Repeat Exercise 15, but convert from octal to decimal. See Example 2.5.5 (Horner's method) for a possible approach.
17. Using the method of Example 2.5.6, give 11 code numbers to draw the letter E, the letter b, and the Greek letter α. (For mechanical reasons, it is often necessary to avoid printing two adjacent dots on the same horizontal level. For example, a sequence including 18 and 16 on adjacent columns should not be used.)
18. By the method of Example 2.5.6, the sequence 4, 0, 124, 0, 4, 0, 124, 0, 4, 0, 0 is fed to the printer to describe a certain character. Name that character.
19. Evaluate $(0.1111\ldots)_b$ and $(0.121212\ldots)_3$.
20. Evaluate $(0.\underline{110}110110\ldots)_2$. The underlined pattern is repeated indefinitely.

21. Repeat Exercise 20 for $(0.\underline{012012}\ldots)_3$.
*22. Find the infinite binary expansion for 0.6. (*Hint:* Writing

$$0.6 = 0.a_1a_2a_3\ldots$$

we can move the decimal point one place to the right by multiplying by 2 ($= 10_2$). Thus, multiplying by 2, we obtain

$$1.2 = a_1.a_2a_3\ldots$$

This gives $a_1 = 1$, and

$$0.2 = 0.a_2a_3a_4\ldots$$

Now multiply by 2 again to obtain

$$0.4 = a_2.a_3a_4\ldots$$

This yields $a_2 = 0$. So far, then, $0.6 = 0.10.\ldots$ Now continue this process until a pattern emerges.)
*23. Find the binary expansion for 0.7.
24. How many *binary* places (digits after the decimal place) are required if a number is to be accurately given to two *decimal* places (that is, to within 0.01)?
25. Suppose that a positive real number is stored in a computer with 32 binary digits after the decimal place. How accurately is the real number given?
*26. *Prove* by induction on n: If $b \geq 1$ and $n \in \mathbb{N}$, then there are integers Q and R such that $n = bQ + R$ and $0 \leq R \leq b - 1$. (*Hint:* Separate cases according to whether $R = b - 1$ or $R < b - 1$.)
*27. (Uniqueness in the Division Algorithm) Show that Q and R are uniquely determined in the Division Algorithm (Theorem 2.5.2). (*Hint:* Assume $N = bQ_1 + R_1 = bQ_2 + R_2$. Now show that $R_1 - R_2$ is divisible by b. By estimating the size of $R_1 - R_2$, show that $R_1 - R_2 = 0$.)

Summary Outline

● A **relation** $R(x, y)$ on a set U, called the universe, is a predicate in two variables x and y. For each x and y in U, $R(x, y)$ is either true or false.

● A relation R is

reflexive if xRx is true for all x in U.

symmetric if for all x and y in U, xRy implies yRx.

transitive if for all x, y, and z in U, xRy and yRz imply xRz.

antisymmetric if xRy and yRx imply $x = y$.

an **equivalence relation** if it is reflexive, symmetric, and transitive.

a **partial ordering** if it is reflexive, transitive, and antisymmetric.

a **linear order** if it is a partial ordering and if, for any x and y, either xRy or yRx or $x = y$.

● The **digraph** of a relation R has elements of U as vertices, and a directed edge joins x to y if and only if xRy.

● The **matrix** of a relation is an n by n array of 0's and 1's. Each row and each column represent an element of the universe. The number 1 is in the row representing x and the column representing y if and only if xRy.

● The **symmetric skeleton** of a symmetric relation R is a graph (undirected) in which x and y are joined by an edge if and only if xRy.

● The **transitive closure** R' of a relation R is defined by the condition that $xR'y$ if and only if there is a finite sequence $x_0, x_1, \ldots, x_n$ such that $x_0 = x$, $x_n = y$, and x_iRx_{i+1} for $i = 0, \ldots, n - 1$.

● If R' is the transitive closure of R, we say that the digraph of R is a **transitive skeleton** of the digraph of R'.

● If R is a relation on U, the **reflexive closure** S of R is the relation defined by the condition xSy if and only if $xRy \lor x = y$.

● The **reflexive skeleton** of a reflexive relation is the graph of that relation, except that all loops about the vertices are omitted.

● If R is an equivalence relation, the **equivalence class** $C(x)$ of an element x is the set of all y such that yRx.

● **Theorem**   The equivalence classes of an equivalence relation partition the universe U.

● **Theorem**   Let $A_1, A_2, \ldots$ be a partition of U. Define the relation xRy to mean that x and y are in the same set A_i. Then xRy is an equivalence relation, and the sets A_i are the equivalence classes for R.

● A **partially ordered set** or **poset** is a set P with a partial ordering R on it. It is denoted (P, R).

● **Theorem**   No cycles can appear in a poset.

● $x < y$ in a poset $(P, \leqslant)$ if $x \leqslant y$ and $x \neq y$.

● x **is covered** by y (xCy) in a poset $(P, \leqslant)$ if $x < y$ and it is impossible to have $x < z < y$.

● **Theorem**   For a finite poset P with relation $\leqslant$ and covering relation C, P is the transitive-reflexive closure of the covering relation C.

● For $f: A \to B$, f is an **onto** function if for every element b of B there is an element a of A such that $f(a) = b$. The function f is **one-to-one** (1-

1) if whenever $f(a_1) = f(a_2)$ then $a_1 = a_2$. If f is 1-1 and onto, the **inverse function** $f^{-1}: B \to A$ is a function such that $a = f^{-1}(b)$ if and only if $b = f(a)$.

- If $f: A \to B$ and $g: B \to C$, the **composition** $g \circ f: A \to C$ is defined as $(g \circ f)(x) = g(f(x))$.

- The **identity function** $i_A: A \to A$ satisfies $i_A(x) = x$.

- A **permutation** of $\mathbb{I}_n$ is a 1-1, onto function $\sigma: \mathbb{I}_n \to \mathbb{I}_n$. **Postfix notation** is used composing permutations: $(k)\rho \circ \sigma = ((k)\rho)\sigma$.

- The **conjugate of σ by τ** is the permutation $\tau^{-1}\sigma\tau$.

- **The Principle of Mathematical Induction** Suppose $P(n)$ is a statement for each $n \in \mathbb{N}$. Suppose further that

  1. $P(0)$ is true. (This is called the **basis** of the induction.)
  2. If $P(n)$ is true, then $P(n + 1)$ is also true. (This is called the **inductive step.** $P(n)$ is called the **inductive hypothesis.**)

  Then, under these circumstances, $P(n)$ is true for all $n \in \mathbb{N}$.

- **The Principle of Complete Mathematical Induction** Suppose $P(n)$ is some statement for each $n \in \mathbb{N}$. Suppose further that

  1. $P(0)$ is true (the basis).
  2. If $P(m)$ is true for all values of $m < n$, then $P(n)$ is also true (the inductive step).

  Then, under these circumstances, $P(n)$ is true for all n.

- **Recursive definitions** Let c be a fixed element in some set X, and let $F: \mathbb{N} \times X \to X$. Then the equations

$$f(0) = c \qquad \text{(the basis)}$$

$$f(n + 1) = F(n, f(n)) \qquad \text{(the recursion)}$$

are satisfied by a unique function $f: \mathbb{N} \to X$ defined by recursion using these equations.

- The **geometric series** If $x \neq 1$,

$$\sum_{k=0}^{n} x^k = \frac{1 - x^{n+1}}{1 - x}$$

- The **harmonic series**

$$S_n = \sum_{k=1}^{n} \frac{1}{k}$$

- The **Fibonacci sequence** f_n is defined recursively by the equations $f_0 = 0$, $f_1 = 1$, and $f_{n+2} = f_n + f_{n+1}$ ($n = 0, 1, 2, \ldots$). The first few terms of this sequence are 0, 1, 1, 2, 3, 5, 8, 13, 21, 34, 55, and 89.

- The **Conway sequence** is defined recursively by the equations $c_1 = 1$, $c_2 = 1$, and $c_n = c_k + c_{n-k}$, where $k = c_{n-1}$ and $n \geq 3$.

- **The Box Principle**   If $n + 1$ or more objects are placed into n boxes, then one of these boxes will contain two or more of these objects.

- **The Box Principle** (alternative version)   Suppose that r red chips and g green chips are put into n boxes and that no box contains two chips of the same color. Suppose also that $r + g > n$. Then one box is occupied by different colored chips.

- **The Box Principle** (abstract version)   Suppose A and B are finite sets, with $|A| > |B|$ and $f: A \to B$. Then f is not 1-1.

- **The Euclidean Algorithm**   If n and b are integers, $b > 0$, then there are integers Q and R, called the quotient and remainder, such that $n = bQ + R$ and R in $\mathbb{Z}_b$. Integers Q and R are uniquely determined by n and b.

- **Base b representation of integers**   Let b and n be nonnegative integers with $b \geq 2$. Then for some integers k and d_i,

$$n = d_0 + d_1 b + d_2 b^2 + \cdots + d_k b^k$$

with $d_i \in \mathbb{Z}_b$, and n is written $n = (d_k \cdots d_2 d_1 d_0)_b$.

- **Infinite geometric series**

$$\sum_{k=0}^{\infty} a_0 r^k = \frac{a_0}{1 - r} \qquad \text{if } |r| < 1$$

- **The number of binary digits** of $N \approx 10n/3$, where n is the number of decimal digits.

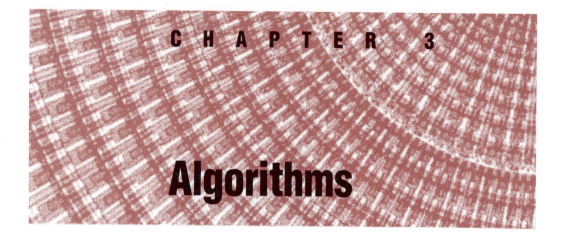

CHAPTER 3

Algorithms

An algorithm tells us how to perform a task. In this chapter we learn how to do this for various situations by using an appropriate language, called a pseudocode. This language is halfway between ordinary English and a full-fledged programming language. Its advantage is that it is not as strict with its rules as a computer language, but is not hard to translate into one. Important aspects of any algorithm are how efficient it is, and how long it takes to use. In what follows, we study these aspects.

3.1 What Is an Algorithm?

An **algorithm** is a precise, mechanical way of performing a task. The task is broken into smaller, elementary subtasks called instructions or commands, each of which is simple to perform. In this way, a complicated task is accomplished. The term "algorithm" was used long before high-speed computers existed, and was applied to simple arithmetic computations. For example, Theorem 2.5.2 is called the **Division Algorithm,** and this is the technique learned in grade school to perform a long division in order to find the quotient and remainder. Much of the arithmetic and some of the mathematics we learn are strongly algorithmic in nature. Here is an example from algebra.

EXAMPLE 3.1.1

How do you solve a linear equation such as $ax + b = cx + d$, where a, b, c, and d are given numbers?

Method

1. Move all the x terms to one side and all the constant terms to the other, using the allowed rules of algebra. [This yields the equation $ax - cx = d - b$.]
2. Now, using the rules of algebra, combine all x terms to obtain a single x term on one side. [This yields $(a - c)x = d - b$.]
3. Combine all constant terms.
4. If the coefficient of x is 0, and the constant term is 0, then every x is a solution. But if the coefficient of x is 0, and the constant term is not 0, then there is no solution to the equation. Otherwise (if the coefficient of x is not 0), divide by the coefficient of x to solve for x. This gives the required solution. [$x = (d - b)/(a - c)$.]

Remark   This algorithm is precise enough for many who have seen this done before. However, the words "rules of algebra" are used, and the presumption is that the user of this algorithm knows these rules of algebra. Several of these rules were used, with results illustrated in the statements in brackets. The bracketed statements are regarded not as part of the algorithm but as helpful explanations. The statement "combine all terms" only makes sense to someone who has had this explained to him or her.

The instructions in an algorithm are simple steps that, presumably, the user of the algorithm knows how to do. Even a step such as "let $x = 1 + 1$," or "$x := 1 + 1$" presumes that the user (human or machine) will be able to perform this computation.

Example 3.1.1 is, approximately, the algorithm learned by many in algebra. It includes an "IF . . . THEN . . . OTHERWISE . . ." clause in step 4 to cover the case that we end with a linear equation with no x term. The example has an important characteristic of an algorithm. Namely, it starts with an input of a certain kind, and it ends with an output. Here the input was a linear equation. The output was either a number or a statement that there is no solution. All algorithms are assumed to have inputs and outputs, and these are specified in advance.

The next example is, in principle, even more elementary. How do you add two two-digit numbers? The presumption is that the user of this algorithm knows the "addition tables" for single digits, for example $3 + 5 = 8$ and $6 + 8 = 14$. The algorithm will then add two numbers given in decimal representation.

EXAMPLE 3.1.2 Adding Two Numbers

Find an algorithm for adding two two-digit numbers.

Method The algorithm is surprisingly complicated. We are explaining how to do the addition

$$a_1 a_0$$
$$+ \; b_1 b_0$$
$$\overline{\text{Output}}$$

For example,

$$57$$
$$+ \; 74$$
$$\overline{131}$$

In this case, all comments will be given after the algorithm is given, so as not to interfere with its instructions. The algorithm is familiar: add the unit's column, and if the sum is greater than 9 use the last digit and carry the first. Then proceed to the next column to the left and do the same.

If x is written in decimal notation and has two digits, then we call the unit's digit its last digit, and the tens' digit its first. In the same way, if a number is written $d_2 d_1 d_0$, then d_2, d_1, and d_0 will be called its first, second, and last digits, respectively. The algorithm is as follows.

Input: a, written $a_1 a_0$ in decimal notation, and b, written $b_1 b_0$ in decimal notation.
Output: $a + b$, written in decimal notation.
1. $L := a_0 + b_0$[1]
2. $c_0 :=$ the last digit of L[2]
3. $M := a_1 + b_1$
4. $c_1 :=$ the last digit of M[3]
5. IF L has one digit THEN[4]
    1. IF M has one digit THEN[5]
        1. The output is the two-digit number $c_1 c_0$
        2. End algorithm

---

[1] We start by adding the last digits. The symbol $:=$ is the assignment symbol. The variable L is assigned the value $a_0 + b_0$.

[2] c_0 is the last digit of the output.

[3] c_1 may be the next to last digit of the output, depending on whether we "carry."

[4] In this case (the whole of step 5) there is no carrying from the unit's (last) column.

[5] Subcases 1 and 2 decide whether the output has two or three digits, depending on whether M has one or two digits.

    2. IF M has two digits THEN
        1. The output is the three-digit number $1c_1c_0$
        2. End algorithm
6. IF L has two digits THEN[6]
    1. IF $c_1 = 9$ THEN
        1. Output is the three-digit number $10c_0$
        2. End algorithm
    2. IF $c_1 \neq 9$ THEN
        1. $c_1 := 1 + c_1$[7]
        2. IF M has one digit THEN
            1. Output is the two-digit number c_1c_0
            2. End[8]
        OTHERWISE[9]
            3. Output is the three-digit number $1c_1c_0$[10]

Remark This algorithm is not particularly elegant. It is filled with enough complications (IFs within IFs) to make us fully appreciate the grade school child who has trouble with "carrying." Not only is this person perhaps a bit shaky on his or her addition table, but the algorithm is really not that simple. Try telling a Martian how to walk gracefully (or designing a robot to walk), and you will find that what seems rather simple can be quite complicated when a detailed analysis is required.

One of the first things you might notice about this algorithm is the indenting of various steps. Some are indented twice and even three times. This is not done solely to help understand the logical flow. If you follow the algorithm down, you will see that there are six main instructions, labeled 1 through 6. Some of these instructions, such as 5, have subinstructions. The labeling of the subinstructions also starts with 1. This is actually a shorthand. Subinstruction 1 of instruction 5 can be labeled 5.1. In fact, this is its official name. We avoid writing it out fully by indenting. In the algorithm, 5.1.1 is the subinstruction 1 of the subinstruction 1 of instruction 5. It reads "1. The output is the two-digit number c_1c_0." The reason we indent (or use subinstructions) is that in an IF . . . THEN . . . instruction there may be several instructions to be

---

[6] This is the alternative to step 5. There is to be carrying into the tens' place.

[7] We carry the 1 by adding it. c_1 is the tens' place digit. The assignment $c_1 := 1 + c_1$ means to add 1 to the known value of c_1 and assign the answer to c_1.

[8] This is the same as the "End algorithm" instruction.

[9] This is the alternative to the previous instruction starting with "IF M has one digit THEN." We could have written "IF M has two digits THEN" instead.

[10] We don't need an End algorithm statement here. When there is nowhere to go, the algorithm automatically ends.

performed following the THEN. By calling and labeling them all sub-instructions, we understand that all of them are to be performed if the IF part of the main instruction is true. Otherwise, all the subinstructions are to be ignored. If one of the steps to be performed is another IF . . . THEN . . . instruction, then it will have its own subinstructions.

We call an IF . . . THEN . . . instruction a **conditional** instruction. The statement following the IF is called the **condition.** It must be true if the indented instructions following the THEN are to be executed.

One of the advantages of indenting is that it allows the user of the algorithm to ignore all the subinstructions of a conditional instruction, if the condition following the IF is not true. We illustrate Example 3.1.2 with specific numbers.

EXAMPLE 3.1.3

Show how the algorithm of Example 3.1.2 works for inputs $a = 41$ and $b = 85$.

Method   We have inputs $a_1 = 4$, $a_0 = 1$, $b_1 = 8$, $b_0 = 5$.

1. "$L := a_0 + b_0$" assigns 6 to L.
2. "$c_0 :=$ the last digit of L" assigns 6 to c_0.
3. "$M := a_1 + b_1$" assigns 12 to M.
4. "$c_1 :=$ the last digit of M" assigns 2 to c_1.

Step 5 is a conditional instruction. We check whether L has a single digit. It does. Therefore we go into the subinstructions of 5. Step 5.1 is a condition that begins "IF M has one digit THEN." Since $M = 12$, the condition is false. Therefore we ignore all of its subinstructions. We go on to step 5.2. Since the condition "M has two digits" is satisfied (i.e., true), we have the output $1c_1c_0$, which in our case is 126. This is not unexpected!

This entire procedure is a **test** or **trace** of the algorithm. As certain as you may be of the logic of an algorithm, it is usually a good idea to try some special cases and to carefully do exactly what the algorithm tells you to do—not your preconceived idea of what you think the algorithm wants to do. The special cases should be critical ones. If we are at all unsure of our work, we might try adding 57 and 77, since this involves a carry from the unit's to the tens' place.

EXAMPLE 3.1.4

Find an algorithm to determine, for a given relation R on I_3, whether it is reflexive.

Method   We have to check whether xRx is true for $x = 1, 2$, and 3. If it is, then it is reflexive. If it fails for one of these x's, it is not reflexive. One algorithm is as follows.

Input: A relation R on $\mathbb{I}_3$.
Output: One of the statements "reflexive" or "not reflexive."
1. IF $1R1$ is false THEN
    1. Output is "not reflexive"
    2. End
2. IF $2R2$ is false THEN
    1. Output is "not reflexive"
    2. End
3. IF $3R3$ is false THEN
    1. Output is "not reflexive"
    2. End
4. Output is "reflexive" [if we get to step 4, xRx was true for $x = 1, 2$, and 3]

The solution is valid, but it is not very pretty. A similar solution would work for a relation on $\mathbb{I}_{10}$, but we would not be too happy repeating instruction 1 and its subinstructions 10 times with a slight change each time. We reconsider this example in Example 3.1.14, where we use the appropriate FOR instruction to perform such repetitive work.

We have unofficially used algorithms throughout the text, and we shall continue to do so. For example, in Section 1.2, we learned how to find the truth table of a statement in the propositional calculus or to determine whether it is a tautology. The method was quite algorithmic in nature, but was not stated in extreme detail. If we had done this, we might have started as follows: "Scan the formula to find which variables are used in it. Now make a table and put all possible truth values for all combinations of these formula variables . . . ." Each of these statements may be further defined by its own algorithm.

Our algorithms will be expressed as **instructions** in an informal language called a **pseudocode.** These will also be called **commands.** Though precise, it is not nearly as fussy as a programming language such as BASIC or Pascal. But it will be precise enough for us to give algorithms and analyze them carefully by stepping through them. The pseudocode we are using is based on an informal language called INFL. It was introduced by G. Stewart, who encouraged flexibility in its use. We now review some of the instructions we have used thus far.

DEFINITION 3.1.5

The instruction $x := y$ may be read "x is assigned the value of y." The previous value of x is lost.

EXAMPLE 3.1.6

Some examples of assignments are

1. $x := 21$
2. $x :=$ "ABCDEF"
3. $x := y^2 - 1$
4. $x := x + 1$

In 1, x is assigned the number 21. From this point on, x is equal to 21. It will change only if it is reassigned. In 2, x is assigned a string "ABCDEF". In 3, x is assigned the value $y^2 - 1$. If y has not been assigned a value, then this instruction will be ignored. If y has been assigned the value 5, then the instruction 3 will assign x the value 24. Finally, 4 computes the value of $x + 1$ and assigns it to x. Thus, if x was 17, then, after this command is executed, x will be 18. If x was not previously assigned a value, it continues to be unassigned.

EXAMPLE 3.1.7

Consider the following algorithm.

1. $x := 1$
2. $y := 5$
3. $z := x + y$
4. $x := y$
5. $y := z$

What are the final values of x and y?

Method   Going through these commands one by one, we find that initially (steps 1 and 2) x and y are assigned the values 1 and 5, respectively. Step 3 assigns their sum 6 to z. Then x is assigned the value 5, and y is then assigned the value 6. The effect of the sequence of statements 3 through 5 is to replace $(x, y) = (1, 5)$ by $(5, 6)$.

We called the instruction $x := y$ an assignment. On the other hand, "$x = y$" is a **statement** or a **condition**, and (assuming that x and y have been assigned values) it is either true or false. Statements usually occur as comparisons or as comparisons connected by logical connectives. Some examples are $x < y$, and ($x \geq 4$ or $y > 0$). They are used in conditional commands. An instruction such as

$$\text{IF } x = 5 \text{ THEN } y := 7$$

is an example. Note that what follows the IF is a **statement** ($x = 5$), but what follows the THEN is a **command** ($y := 7$). Similarly, we can have an instruction such as

$$\text{IF } x = 5 \text{ and } y \leq 1 \text{ THEN } z := x + y$$

Here the statement following the IF is a compound statement of the form $p \land q$, and a command follows **THEN**.

EXAMPLE 3.1.8 *Transposing Two Values*

Find an algorithm that has two values x and y as inputs and as outputs. The input values of x and y are interchanged to obtain the output.

Method If we tried

1. $x := y$
2. $y := x$

we would not obtain the desired output. Step 1 correctly changed x, but the original value of x was lost. Step 2 has no effect on the value of y. To obtain the desired results, we must save the original value of x. This is done by assigning the value of x to a new variable called Save:

1. Save $:= x$ [storing the original x]
2. $x := y$ [part of the output]
3. $y :=$ Save   [the original x value]

EXAMPLE 3.1.9 *Comparison of an Assignment and an Equality*

Suppose x has been assigned the value 3, and y the value 4. Then the command

$$x := x + y$$

will simply reassign to x the value 7. The statement

$$x = x + y$$

is simply false in this case. By algebra, of course, it is only true when $y = 0$. Similarly, the command

$$x := x + 1$$

increases the value of x by 1. The statement

$$x = x + 1$$

is always false.

We now consider the general form of a conditional (IF . . . THEN . . .) instruction.

DEFINITION 3.1.10

A **conditional instruction** is one of the form "IF *statement* THEN *command*". The instruction first computes the truth value of *statement*. If it is true, then *command* is executed. If it is false, then *command* is ignored. If several commands are to be executed if *statement* is true, we use the form

$$IF\ statement\ THEN$$
$$command\ 1$$
$$command\ 2$$
$$. . .$$
$$command\ k$$

If *statement* is true, then each command following it, and indented once, is executed. If it is false, each command is ignored. Each subcommand can itself be a conditional instruction. A sequence of such commands is called a **command list.** Thus, we can abbreviate a conditional command as

$$IF\ statement\ THEN$$
$$command\ list$$

Note that the IF . . . THEN . . . formulation is not the one considered in propositional calculus. In this case, what follows the IF is a statement, but what follows THEN here is an instruction or a list of instructions. This is true of programming languages, too.

A variation of the IF . . . THEN . . . command is of the form

$$IF\ statement\ THEN$$
$$command\ list\ 1$$
$$OTHERWISE$$
$$command\ list\ 2$$

In this case, the instructions of *command list* 1 are executed if *statement* is true, but the instructions of *command list* 2 are executed if *statement* is false. Step 6.2.2 of Example 3.1.2 was of this form.

Remark Many computer languages indicate the beginning and end of a command list with words such as "begin" and "end." Our approach doesn't do this because the indenting shows the statements in the command list. When translating these algorithms into such a language, we must follow the rules of that language.

EXAMPLE 3.1.11   *Tracing an Algorithm*

Determine the values of the variables x and y after the following algorithm is executed:

1. $x := 5$
2. $y := 7$
3. IF $x = 5$ THEN $y := 8$
   OTHERWISE $y := 0$
4. IF $y = 7$ THEN
     1. $x := 6$
   OTHERWISE
     2. $x := 3$
     3. IF $x = y$ THEN $y := 0$

Method   It is convenient to follow the progress of such a sequence of instructions in a table. In Table 3.1.1, the values given are the ones *after* the instruction has been carried out.

Table 3.1.1   Tracing an Algorithm

Step Number	x	y
1	5	—
2	5	7
3	5	8
4.1	5	8
4.2	3	8
4.3	3	8

After following this algorithm step by step, Table 3.1.1 can be constructed, yielding the result $x = 3$ and $y = 8$. As in Example 3.1.3, this is called tracing the algorithm.

DEFINITION 3.1.12   *Inputs and Outputs*

Inputs are values that are taken as given. **Outputs** are values that will be found.

In a real-life situation, inputs are either keyed into the computer or taken as the output from another algorithm. An output might be printed or displayed on the screen, or transmitted to another algorithm that uses the output as *its* input.

In addition to the previous commands, **comments** may be placed in a program, enclosed in brackets. They are meant to clarify the meaning or the intention of a command.

Another type of instruction needed is FOR. The FOR instruction repeats a command a specified number of times. An example is

$$\text{FOR } i := 1 \text{ to } n$$
$$a_i := 1$$

The effect of this command is to substitute $i = 1$, then $i = 2$, and so on, through $i = n$ in the command $a_i := 1$. In effect, this gives the commands $a_1 := 1, a_2 := 1, \ldots, a_n := 1$. The general definition is as follows.

DEFINITION 3.1.13

A **FOR instruction** is one of the form

$$\text{FOR } i := 1 \text{ to } n$$
$$command$$

The instruction first assigns i the value 1 and executes *command*. It then assigns the value 2 to i and executes *command*. The process is repeated after i is assigned all values from 1 to n inclusive. More generally, a **FOR** instruction is of the form

$$\text{FOR } i := j \text{ to } k$$
$$command \text{ } list$$

where we assume $j \le k$, or else the command is ignored. In this case, each instruction of the command list will be executed for $i = j, j + 1, \ldots, k$. A FOR command is also called a **FOR loop.**

A useful variation of the FOR command is one of the form

$$\text{FOR } i := k \text{ down to } j$$
$$command \text{ } list$$

Here $k \ge j$, and the command list is executed for $i = k$, then $i = k - 1$, and so on, through $i = j$.

EXAMPLE 3.1.14

Find an algorithm to determine whether the relation xRy is reflexive. The input is a relation R on I_{10}. The output is a string X, which can have the value "reflexive" or "not reflexive."

Method (Compare with Example 3.1.4) The algorithm is as follows.

1. FOR $i := 1$ to 10
    1. IF $\neg iRi$ THEN
        1. $X :=$ "not reflexive"
        2. End algorithm
2. $X :=$ "reflexive"

EXAMPLE 3.1.15

A matrix a_{ij} is called a $(0, 1)$ matrix if $a_{ij} = 0$ or 1 for all possible i and j. Find an algorithm to determine whether a matrix is a $(0, 1)$ matrix. The input is a square (n by n) matrix a_{ij}; the output is yes or no.

Method For each row i we check whether each element in the row is 0 or 1. If this happens for each row, the answer is yes. If we find a nonzero element, the answer is no. An informal algorithm is as follows.

1. FOR $i := 1$ to n
    Check each a_{ij} for all values of j. If we ever find that $a_{ij} \neq 0$ and $a_{ij} \neq 1$, then the answer is "no."
2. If it checks out, the answer is "yes."

Refining this, we obtain

1. FOR $i := 1$ to n
    1. FOR $j := 1$ to n
        1. IF $\neg(a_{ij} = 0$ or $a_{ij} = 1)$ THEN
            1. Output is "no"
            2. End
2. Output is "yes"

As in Example 3.1.14, we had the negative output as soon as we determined that a_{ij} was not a $(0, 1)$ matrix. Once we went through all the checks (both FOR loops), we arrived at the output "yes." We never arrive at 2 if the output was "no" because instruction 1.1.1.2 ended the algorithm once we determined that the output was "no."

FOR commands are a natural way to compute functions that are given recursively.

EXAMPLE 3.1.16

Consider the function $f: \mathbb{N} \to \mathbb{N}$ defined by the equations

$$a_0 = 1 \tag{3.1.1}$$

$$a_{n+1} = a_n^2 + n; \; n \geq 0 \tag{3.1.2}$$

(These are the same equations as in Example 2.3.4.) Construct an algorithm to compute a_n. The input is $n \geq 0$; output is a_n. (The reader should note that (3.1.2) holds for all $n \geq 0$. We are going to compute a_n for a specific value of n.)

Method   As in Example 2.3.4, we use the recursion (3.1.2) over and over, starting with (3.1.1). We put $i = 0, 1, \ldots, n$ in (3.1.2). The required algorithm is as follows.

Algorithm A
1. $a_0 := 1$
2. IF $n = 0$ THEN End
3. FOR $i := 1$ to n
    1. $a_i := a_{i-1}^2 + i - 1$

Instruction 2 can be omitted because if $n = 0$ our convention is to ignore the "FOR $i := 1$ to 0" instruction.

It is a waste of memory to use instruction 3.1 because it is repeated n times, so we would have to store all n values $a_1, \ldots a_n$. This can be avoided by letting one variable x represent the value of a_i at any stage and using a similar algorithm.

Algorithm B
1. $x := 1$ [initializing a_n for $n = 0$]
2. For $i := 1$ to n
    1. $x := x^2 + i - 1$
3. $a_n := x$

Remark   The distinction is not a trivial one. Any machine computation takes *time* and uses *memory*. Neither is infinitely available. Both algorithms use about the same time, since this would be roughly proportional to the number of assignments made. But Algorithm B uses one memory location for x and perhaps one for i and for n. Algorithm A uses n memory locations for the values a_i, which will be significant when n is large.

In a similar manner, we can compute sums.

EXAMPLE 3.1.17

Find an algorithm to compute the sum of the harmonic series

$$\sum_{i=1}^{n} \frac{1}{i}$$

Method   The input is n, which is assumed ≥ 0. The output is the sum x. Using the recursive definition of the sum (Definition 2.3.9), we initialize x as 0, and we keep adding the ith term. The algorithm is as follows.

1. $x := 0$
2. FOR $i := 1$ to n
    1. $x := x + 1/i$

We conclude with two additional examples.

EXAMPLE 3.1.18

Devise an algorithm to find the largest number in a sequence and its location.

Method   The input is the sequence $\{a_i\}$ and its size n. The output is M, the maximum value of the a_i, and Loc, the first value of i for which $M = a_i$. We shall go through the list, one by one, keeping a temporary maximum M and location Loc. Whenever a larger term is found, we replace M with the larger element and Loc accordingly. We initialize M and Loc by choosing the first element in the sequence. The algorithm is as follows.

1. $M := a_1$
2. Loc $:= 1$
3. FOR $i := 2$ to n
    1. IF $a_i > M$ THEN
        1. $M := a_i$
        2. Loc $:= i$

EXAMPLE 3.1.19

Devise an algorithm to check whether a relation R on $\mathbb{I}_n$ is symmetric. The input is the matrix a_{ij} of the relation, and the number n. The output is the message X, which can be either yes or no.

Method   We have to check whether $a_{ij} = a_{ji}$ for all i and j. Thus the following algorithm works. (Compare with Example 3.1.4.)

Algorithm A
1. $X :=$ "yes" [assume the relation is symmetric until we find otherwise]
2. FOR $i := 1$ to n [for each row]
    1. FOR $j := 1$ to n [and each entry in that row]
        1. IF $a_{ij} \neq a_{ji}$ THEN
            1. $X :=$ "no"
        2. End

This algorithm checks each i and j, and if any $a_{ij} \neq a_{ji}$ its output message is changed to "no." If it always finds that $a_{ij} = a_{ji}$, the message stays "yes."

However, the algorithm, although it works, wastes time. Once we have found, for example, that $a_{24} = a_{42}$, it is unnecessary to check whether $a_{42} = a_{24}$. The first condition is found in the two FOR loops corresponding to $i = 2$ and $j = 4$, the second for $i = 4$ and $j = 2$. It makes no sense to check twice. Also, it approximately doubles the time it takes for the algorithm to work. In addition, we never have to check that $a_{ij} = a_{ji}$ when $i = j$. It always is. Thus, a more efficient algorithm would do all the checking of the condition $a_{ij} = a_{ji}$ only for $i < j$. By changing statement 2.1 to start with the clause "FOR $j := i + 1$ to n" this will be achieved. Further, the case $i = n$ is never checked—there is no case $j > n$. The complete, more efficient algorithm is as follows.

Algorithm B
1. $X := $ "yes"
2. FOR $i := 1$ to $n - 1$
   1. FOR $j := i + 1$ to n $[j > i]$
      1. IF $a_{ij} \neq a_{ji}$ THEN
         1. $X := $ "no"
      2. End

We now compare the times used for these two algorithms. For simplicity let us count the number of times step 2.1.1 is executed. This will be called the number of checks. Assume the "worst case"—that a_{ij} is symmetric, so all (i, j) must be checked. In Algorithm A, n^2 checks are made, once for each (i, j), where $1 \leq i, j \leq n$. However Algorithm B checks (i, j) for $i < j$. When $i = 1$, there are $(n - 1)$ j's. Similarly there are $(n - 2)$ j's when $i = 2$, and so on. The number of instructions is therefore

$$(n - 1) + (n - 2) + \cdots + 2 + 1$$

This is $n(n - 1)/2$. If T_A is the number of checks Algorithm A takes and T_B is the number for Algorithm B, we have

$$T_B/T_A = (n - 1)/(2n) < 1/2$$

Thus, in the worst case, Algorithm A takes more than twice as long as Algorithm B.

Certainly, an impetus for the study of algorithms was the advancement of computing technology. It was realized that computers were able to perform algorithms, but only if the individual steps were exceedingly simple (so simple even a machine could understand them). The study of algorithms concerns itself with how algorithms can be constructed, how to find optimal algorithms, and to investigate how long we may expect a long algorithmic computation to take. But we should also note that impetus has also come from other sources. Gödel, in 1930, gave his famous "incompleteness theorem" of logic, and had to develop some theoretical understanding of what an algorithm, or a computation, was. The study of Turing machines, covered in Chapter 10, was also developed for theoretical needs.

Exercises

1. Discuss the effect of the following algorithm on x, y, and z.

    1. $x := y$
    2. $y := z$
    3. $z := x$

2. Find an algorithm whose input is two numbers x and y and whose output is x and y, but in increasing order.
3. Find an algorithm whose input is three numbers x, y, and z and whose output is x, y, and z, but in increasing order.
4. Find an algorithm that cycles the values of x, y, and z. For example, input $(x, y, z) = (1, 2, 5)$ will have output $(x, y, z) = (2, 5, 1)$. The input is x, y, and z. The output is the new values of x, y, and z.
5. Consider the following algorithm:

    1. $x := 3$
    2. $y := 1$
    3. IF $(x + y) = 5$ THEN
        1. $x := x + 1$
        2. IF $x = 4$ THEN
            1. $x := 0$
            2. $y := 0$
    4. $x := x + y$

    What are the final values of x and y?
6. Consider the following algorithm:

    1. $x := 1$
    2. $y := x + 1$
    3. IF $x < y$ THEN
        1. $x := x + 1$
    4. IF $x < y$ THEN
        1. $x := x + 1$
        OTHERWISE
        2. $y := x + 1$
        3. $x := y + 1$
        4. End
    5. $x := x + y$

    What are the final values of x and y?

7. Consider the following algorithm:

    1. FOR $i := 1$ to n
        1. $a_i := i^2$
    2. FOR $i := 1$ to n
        1. IF $a_i > n$ THEN $a_i := n$

    What is the output $\{a_i\}$ when n is 5? when n is 10? In each case, your answer should be a sequence $\{a_i\}$, $1 \le i \le n$.

8. Consider the following algorithm:

    1. $a_0 := 1$
    2. FOR $i := 1$ to n
        1. $a_i := 2a_{i-1}$
        2. $a_i := a_i - 1$

    What is the value of a_n?

9. Consider the following algorithm:

    1. FOR $i := 1$ to n
        1. IF $a_i < 0$ THEN $a_i := 0$
        2. IF $a_i > 2$ THEN $a_i := 2$

    Describe, in words, what task this algorithm performs on a given sequence a_i $(1 \le i \le n)$.

10. Consider the following algorithm:

    1. FOR $i := 1$ to n
        1. FOR $j := 2$ to n
            1. $a_{ij} := a_{1j}$

    Explain, in words, what task this algorithm performs on a given matrix a_{ij} $(1 \le i, j \le n)$.

11. Consider the following algorithm:

    1. FOR $i := 1$ to n
        1. FOR $j := 1$ to n
            1. IF $a_{ij} > a_{ji}$ THEN $a_{ij} := a_{ji}$

    Does this algorithm accomplish the same task as the following algorithm? Explain.

    1. FOR $i := 1$ to $n - 1$
        1. FOR $j := i + 1$ to n
            1. IF $a_{ij} > a_{ji}$ THEN $a_{ij} := a_{ji}$

*12. Find an algorithm to compute $x + 1$, where x is a two-digit integer whose digits are d_1 and d_2. Assume a knowledge of the addition table. Be sure to state, in advance, your input and output.

*13. Let $\{a_i\}$ be a finite sequence of length n. Find an algorithm that finds the first index i where a_i is 0. If i is found, the algorithm assigns Loc the value i. If a_i is never 0, Loc is set equal to 0. The input is a sequence $\{a_i\}$ of numbers, and the length n of the sequence.

14. Find an algorithm that swaps the first and last places of a given sequence $\{a_i\}$ ($i \in \mathbb{I}_n$). Thus, for the sequence 3, 5, 7, 2, 5, the output sequence will be 5, 5, 7, 2, 3.

15. Find an algorithm that swaps the ith and jth places of a sequence. The integers i and j are inputs with $1 \leq i, j \leq n$.

16. Find an algorithm that replaces all negative terms of a sequence with 0. Thus, the sequence 3, 4, -2, 5, -4 is replaced by 3, 4, 0, 5, 0.

17. Find an algorithm that reverses the order of the sequence. Thus, the sequence 2, 3, 5, 6, 9, 10 changes to 10, 9, 6, 5, 3, 2.

*18. Find an algorithm that takes the ith term of the sequence, moves it to the first position, and pushes up the remaining terms. The integer i is also given as an input. For example, if $i = 3$, the sequence 2, 3, 6, 4, 8 is transformed to 6, 2, 3, 4, 8.

*19. Construct an algorithm that finds the smallest term in the sequence, moves it to the first position, and pushes up the rest. Thus, the sequence 5, 4, 8, 3, 10 changes to 3, 5, 4, 8, 10.

20. Construct an algorithm that finds the minimum value m occurring in a matrix a_{ij} and the location (i, j) where this minimum occurs. Here, $1 \leq i, j \leq n$.

*21. Let the input be a positive integer n and an n by n matrix a_{ij}. Find an algorithm that decides whether the given matrix is the matrix of a reflexive relation.

In Exercises 22–26 the input is understood to be a positive integer n and an n by n matrix a_{ij}. The output is the matrix a_{ij} suitably altered.

22. Find an algorithm to interchange rows and columns of the matrix. Thus, the matrix

$$\begin{bmatrix} 3 & 5 & 2 \\ 3 & 0 & 1 \\ 0 & -7 & 5 \end{bmatrix} \rightarrow \begin{bmatrix} 3 & 3 & 0 \\ 5 & 0 & -7 \\ 2 & 1 & 5 \end{bmatrix}$$

23. Find an algorithm to interchange rows i and j of the matrix. Here i and j are additional inputs with $1 \leq i, j \leq n$.

24. Find an algorithm that interchanges rows i and j of the matrix and then interchanges columns i and j. Here i and j are additional inputs with $1 \leq i, j \leq n$.

25. Find an algorithm that adds row i of the matrix to row j. Here, $1 \leq i, j \leq n$. For example, with $i = 1$ and $j = 3$, we have

$$\begin{bmatrix} 3 & 5 & 2 \\ 3 & 0 & 1 \\ 0 & -7 & 5 \end{bmatrix} \rightarrow \begin{bmatrix} 3 & 5 & 2 \\ 3 & 0 & 1 \\ 3 & -2 & 7 \end{bmatrix}$$

26. Find an algorithm that replaces each positive entry of the matrix with 1 and all other entries with 0. For example, we have

$$\begin{bmatrix} 3 & 5 & 2 \\ 3 & 0 & 1 \\ 0 & -7 & 5 \end{bmatrix} \rightarrow \begin{bmatrix} 1 & 1 & 1 \\ 1 & 0 & 1 \\ 0 & 0 & 1 \end{bmatrix}$$

3.2 Time Estimates and Orders of Magnitude

In Example 3.1.19, while devising an algorithm to decide whether a relation was symmetric, we found a way of cutting down the time of the algorithm by eliminating unnecessary comparisons. We now consider the question of how much time an algorithm takes. Since we can't really determine the speed at which an individual person or computer will perform the steps of an algorithm, we discuss the number of steps executed in an algorithm.

EXAMPLE 3.2.1

How many steps are executed in the algorithm of Example 3.1.18 to find the maximum value in a list?

Method The algorithm was as follows.

1. $M := a_1$
2. Loc $:= 1$
3. FOR $i := 2$ to n
    1. IF $a_i > M$ THEN
        1. $M := a_i$
        2. Loc $:= i$

Step 3.1 was repeated $n - 1$ times, and, together with assignments 1 and 2, we had a total of $n + 1$ steps. In this computation, we ignored steps 3.1.1 and 3.1.2, regarding them as part of step 3. In fact, we are not sure that they ever get executed, because we need the condition $a_i > M$ before this happens.

EXAMPLE 3.2.2

How long does the algorithm of Example 3.2.1 take to execute?

Method How this algorithm translates into time depends on several factors. For simplicity, we assume that the time it takes to compare two values (for example, to decide whether $a_i > M$ is true or false) is approximately the same as the time it takes to assign a value (for example, $M := a_i$). The loop itself takes time, since the loop variable i is assigned values along the way, from 2 to n. We can see that *at most n* assignments of M and Loc are possible. This will happen if the sequence is given in increasing order, so we have to reassign M at each step 3.1. This is the **worst case** from the point of view of time; instructions 3.1.1 and 3.1.2 make an assignment each time 3.1 is executed. The **best case** occurs when there is only one assignment of M. This happens if a_1 was the largest element in the sequence. The **average case** is somewhere between the worst and best cases. For this case, we assume that the order of the given sequence is randomly given.

We usually consider the worst case. In any case, if t is the number of M and Loc assignments within the FOR loop we have $0 \le t \le n - 1$. There are always $n - 1$ comparisons to be made, and the FOR loop itself makes $n - 1$ assignments of i. If we take all of this into consideration, the worst-case time is $T(n) = 2 + 4(n - 1) = 4n - 2$. The best case is $2 + 2(n - 1) = 2n$. Thus, if $T(n)$ is the time of the algorithm in Example 3.2.1, we have

$$2n \le T(n) \le 4n - 2$$

If n is very large in Example 3.2.1, we know that $T(n)$ will be large too. The time taken, in any case, is roughly proportional to the length n of the input sequence.

DEFINITION 3.2.3

The **size** of an input is a measure of how complicated the input is. Usually, if an input is a sequence of length n, we take n as its size. If an input is an integer n, we could take $|n|$ as its size, or perhaps the number of digits or binary digits of n.

A basic consideration in the analysis of an algorithm is how the time of execution of an algorithm is affected as the input data gets more complex. The complexity of data is measured by its size. The **complexity of an algorithm** is measured by the time it takes and by other considerations, such as the memory it uses. In this text, we usually use the worst time, counting assignments and comparisons, as the basic measure of the complexity of an algorithm.† Since time is measured by a function $f(n)$, where n is the size of the input, we need to understand the behavior of functions $f(n)$ as $n \to \infty$ and to *compare* functions when $n \to \infty$.

Big Oh and Little Oh Analysis

EXAMPLE 3.2.4

The functions $f(n) = 20n$ and $g(n) = n^2$ both approach ∞ with n, but n^2 is much larger than $20n$ when n is very large. If we had a choice of

---

† However, see Example 3.1.16, where memory was a factor, and Example 3.3.9, where assignments are ignored.

algorithms to perform a certain task and their time functions were

$$T_1(n) \approx 20n \qquad \text{and} \qquad T_2(n) \approx n^2$$

we would no doubt use the algorithm with time function T_1, provided we anticipated inputs with large n. For example, when n is 100, $T_1(n)$ = 2000, and $T_2(n)$ = 10,000. If we knew that n would be about 20 in our use of the algorithms, we would then look for other reasons for using one algorithm over the other. ($20n = n^2$ when $n = 20$.) Although T_1 and T_2 are both large when n is, T_1 is insignificant when compared with T_2. For example, if $n = 10^6$, then $T_1(n) = 2 \cdot 10^7$, and $T_2(n) = 10^{12}$, 50,000 times as much as T_1. In this case, if the algorithm with time T_1 took 5 seconds to do, the one with time T_2 would take about 250,000 seconds, or slightly under three days. Figure 3.2.1 compares these functions graphically.

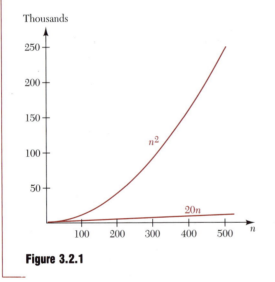

Thousands

Figure 3.2.1

The notation (3.2.1) should remind us that f is small, or insignificant, when compared with g. The equal sign in (3.2.1) is used loosely. For example, we will *not* say that $o(g(n)) = f(n)$. If $g(n) \neq 0$, condition (3.2.2) can be restated using limits:

$$\lim_{n \to \infty} \frac{f(n)}{g(n)} = 0$$

The clause "provided n is sufficiently large" in (3.2.2) is important. Thus, we do not say that

$$20n \leq 0.005n^2$$

for all n. This inequality is true only for all n large enough (in this case, for $n \geq 4000$). The statement $f(n) = o(1)$ is equivalent to the condition $f(n) \to 0$ as $n \to \infty$.

DEFINITION 3.2.6   Big Oh Notation

Let $f(n)$ and $g(n)$ be nonnegative functions of n. We say that

$$f(n) = O(g(n)) \tag{3.2.3}$$

if

$$f(n) \leq Kg(n) \tag{3.2.4}$$

for some fixed number K and n sufficiently large. We read (3.2.3) as "$f(n)$ is big Oh of $g(n)$."

Thus, the statement $f(n) = O(1)$ is equivalent to the statement that $f(n)$ is bounded: $f(n) \leq K$ for some value of K.

Remark   The notation $f(n) = o(g(n))$ or $O(g(n))$ is a rare time in mathematical usage that the equal sign is not used to show that both sides are the same. One could circumvent this usage by defining $O(g(n))$ as the **set of functions** $f(n)$ satisfying (3.2.4) for some fixed number K and n sufficiently large. We could then say $f(n) \in O(g(n))$, a more standard mathematical usage. However, the notation (3.2.1) and (3.2.3) is useful and entrenched. A common usage, which we occasionally use, is to say that $f(n)$ is $O(g(n))$, and similarly for $o(g(n))$.

Using sequence notation for functions, we can also say that $a_n = O(b_n)$ or $o(b_n)$.

Definitions 3.2.5 and 3.2.6 are often stated for functions that might be negative. We do not do this here because we do not need this generality and it introduces additional technicalities.

Intuitively, when we have $f(n) = O(g(n))$, it means that $f(n)$ is *at most the same order of magnitude* as $g(n)$. The function $f(n)$ might be at most five times as much as g (or 10, and so on), but the ratio of f to g never gets too large. Thus, we make the following definition.

DEFINITION 3.2.7

We say that $f(n)$ and $g(n)$ have the **same order of magnitude** if $f(n) = O(g(n))$ and $g(n) = O(f(n))$.

Equivalently, f and g have the same order of magnitude if

$$kf(n) \leq g(n) \leq Kf(n)$$

for positive constants k and K and for n sufficiently large.

It is an immediate consequence of Definitions 3.2.5 and 3.2.6 that a function $f(n)$ that is little Oh of $g(n)$ is big Oh, too.

THEOREM 3.2.8

If $f(n) = o(g(n))$, then $f(n) = O(g(n))$. ■

EXAMPLE 3.2.9

We can put much of the discussion of Example 3.2.4 succinctly as $20n = o(n^2)$.

The sequences $a_n = 2^n$ and $b_n = n!$ are large when n is, but which is larger? We can test for large n. For example, when $n = 20$, $2^{20} \approx 1.048 \cdot 10^6$, and $20! \approx 2.433 \cdot 10^{18}$. Thus, 20! is over 2 *billion* times as large as 2^{20}. Trying $n = 50$, we find $2^{50} \approx 1.126 \cdot 10^{15}$, and $50! \approx 3.041 \cdot 10^{64}$. It is not too hard to guess that 2^n is much smaller than $n!$ when n is large: $2^n = o(n!)$. Table 3.2.1 gives some functions evaluated for relatively large values of n. The value is given to one significant figure.

Table 3.2.1   Orders of Magnitude for Some Common Functions

a_n	$n = 10$	$n = 100$	$n = 1000$
$\log_{10} n$	$1 \cdot 10^0$	$2 \cdot 10^0$	$3 \cdot 10^0$
$\sqrt{n}$	$3 \cdot 10^0$	$1 \cdot 10^1$	$3 \cdot 10^1$
n	$1 \cdot 10^1$	$1 \cdot 10^2$	$1 \cdot 10^3$
n^3	$1 \cdot 10^3$	$1 \cdot 10^6$	$1 \cdot 10^9$
$n^{\sqrt{n}}$	$1 \cdot 10^3$	$1 \cdot 10^{20}$	$7 \cdot 10^{94}$
2^n	$1 \cdot 10^3$	$1 \cdot 10^{30}$	$1 \cdot 10^{301}$
$n!$	$3 \cdot 10^6$	$1 \cdot 10^{158}$	$4 \cdot 10^{2567}$
n^n	$1 \cdot 10^{10}$	$1 \cdot 10^{200}$	$1 \cdot 10^{3000}$

EXAMPLE 3.2.10   Orders of Magnitude for Various Functions

Table 3.2.1 clearly indicates that the functions of n given by the formulas $\log_{10} n$, $\sqrt{n}$, n, n^3, $n^{\sqrt{n}}$, 2^n, $n!$, n^n are in increasing order for large n.

From the calculations in Table 3.2.1, we expect $\log_{10} n = o(\sqrt{n})$, $n = o(2^n)$, and so on. Of course, 2^n is not always larger than n^3. For example, $2^5 = 32$, and $5^3 = 125$. Yet, for large n, $2^n > n^3$. In fact, $n^3 < \epsilon 2^n$ for any positive ϵ for large enough n.

Figure 3.2.2 compares the functions 2^n and n^4. The dominance of the larger function is readily apparent.

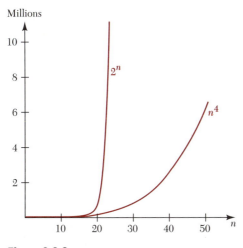

Millions

Figure 3.2.2

The following results give some techniques for manipulating functions using the big Oh and little Oh notation.

THEOREM 3.2.11

If $f(n) = O(g(n))$ and $g(n) = O(h(n))$, then $f(n) = O(h(n))$. ■

THEOREM 3.2.12

If $f(n) = o(g(n))$ and $g(n) = O(h(n))$, then $f(n) = o(h(n))$. If $f(n) = O(g(n))$ and $g(n) = o(h(n))$, then $f(n) = o(h(n))$. ■

The proofs of Theorems 3.2.11 and 3.2.12 are left to the exercises.

THEOREM 3.2.13

If $f(n) = O(h(n))$ and $g(n) = O(h(n))$, then for any nonnegative constants c and d, $cf(n) + dg(n) = O(h(n))$.

Proof   From $f \leq K_1 h$, and $g \leq K_2 h$, we get

$$cf + dg \leq (cK_1 + dK_2)h$$

This gives the result, using $cK_1 + dK_2$ as the constant K in (3.2.4).   ■

We have a similar theorem for little Oh.

THEOREM 3.2.14

If $f(n) = o(h(n))$ and $g(n) = o(h(n))$, then for any constants c and d, $cf(n) + dg(n) = o(h(n))$.   ■

The proof is omitted.

EXAMPLE 3.2.15

Show that

$$6n^2 + 5n + 1 = O(n^2) \tag{3.2.5}$$

and

$$6n^2 + 5n + 1 = o(n^3) \tag{3.2.6}$$

Method   By Theorem 3.2.13, we can prove (3.2.5) by showing that n^2, n, and 1 are each $O(n^2)$. But $n^2 \leq n^2$, so $n^2 = O(n^2)$. Similarly, $n \leq n^2$ and $1 \leq n^2$ for $n \geq 1$. Thus $n = O(n^2)$ and $1 = O(n^2)$. Applying Theorem 3.2.13, we have (3.2.5).

To show (3.2.6), first show that $n^2 = o(n^3)$. If ϵ is an arbitrarily small positive integer, the equation $n^2 \leq \epsilon n^3$ is true if $1 \leq \epsilon n$ or $n \geq 1/\epsilon$. Thus it is true if n is large enough, so $n^2 = o(n^3)$ by definition. Thus,

$$6n^2 + 5n + 1 = O(n^2); \qquad n^2 = o(n^3)$$

and $6n^2 + 5n + 1 = o(n^3)$ by Theorem 3.2.12.

This example generalizes to the following theorem.

THEOREM 3.2.16

Let $f(n) = a_k n^k + a_{k-1} n^{k-1} + \cdots + a_0$ be a polynomial of degree k with $a_k > 0$. Then $f(n) = O(n^k)$ and $f(n) = o(n^{k+1})$.   ■

A similar argument works when f is the sum of possibly fractional powers of n. Thus, $\sqrt{n} + n = O(n)$; $\sqrt{n} + n^2 = o(n^{5/2})$.

THEOREM 3.2.17

The following algebraic results are direct consequences of the definition.

1. If $f = O(g)$ and $k > 0$, then $f^k = O(g^k)$.
2. If $f_1 = O(g_1)$ and $f_2 = O(g_2)$, then $f_1 f_2 = O(g_1 g_2)$.

Similar results are true for little Oh. ■

EXAMPLE 3.2.18

The big and little Oh notations are very helpful in eliminating clutter and keeping the essential features of a function as $n \to \infty$. For example, the sequence

$$a_n = n^5(n + 1/2)^3 + n^3 \qquad (3.2.7)$$

can be simplified to

$$a_n = O(n^8)$$

or, if we want more precision,

$$a_n = n^8 + o(n^8)$$

By taking more terms, we can express the behavior of the function more and more precisely. Thus, using (3.2.7), we can expand Equation (3.2.7) to obtain

$$a_n = n^8 + 3n^7/2 + O(n^6)$$

It is true that we can fully evaluate the expression (3.2.7), but often we only want some estimate for how large this sequence gets for large n. Frequently, $a_n = O(n^8)$ is good enough. For $n = 10^5$, we know that a_n has order of magnitude 10^{40}. The use of the exact formula (3.2.7) can tend to obscure this information.

We observed in Example 3.2.10 that an exponential function (e.g., 2^n) has a higher order of magnitude than any fixed power of n. Thus, $n^k = o(a^n)$ when $a > 1$. Similarly, $a^n = o(n!)$. We state this as a theorem.

THEOREM 3.2.19

If $a > 1$ and k is any integer,

$$n^k = o(a^n) \qquad (3.2.8)$$

$$a^n = o(n!) \qquad (3.2.9)$$

$$\log_a n = o(n) \qquad (3.2.10)$$

■

The proof is left to the exercises. Statement (3.2.10) is a logarithmic way of writing (3.2.8) when $k = 1$. To see this, set $a^n = m$ so that $n = \log_a m$. Then $n < \epsilon a^n$ is equivalent to $\log_a m < \epsilon m$.

EXAMPLE 3.2.20   Logarithms

Logarithms can be computed to any base, but three bases are in common use: 2, e,† and 10. Most calculators have $\log_{10}$ and $\log_e$ as built-in functions. The function $\log_e x$ is called the natural logarithm and is sometimes written $\ln x$. The function $\log_{10} x$ is called the common logarithm. The function $\log_2 x$ will occur naturally in many of our algorithms. There is a simple formula to convert a logarithm from one base to another:

$$\log_b n = \log_a n / \log_a b \qquad (3.2.11)$$

Thus,

$$\log_2 n = \log_{10} n / \log_{10} 2 \approx 3.222 \log_{10} n$$

This shows that $\log_2 n = O(\log_{10} n)$. In fact, (3.2.11) implies that $\log_a n$ and $\log_b n$ have the same order of magnitude. In this text, we use the notations

$$\lg n \quad \text{for} \quad \log_2 n$$

$$\log n \quad \text{for} \quad \log_e n$$

Search Algorithms

EXAMPLE 3.2.21   A Linear Search

Let a_i ($i = 1, \ldots, n$) be a sequence of words. We want to search this list to see if a given word x is there. If it is, and $x = a_i$, then the output Loc is i, the location of the word. If it is not, then the output is 0. Find an algorithm to do this. Estimate the worst-case time $T(n)$.

Method   We check the words one at a time. If we find it, then $x = a_i$ and Loc := i. If not, Loc := 0. One algorithm is as follows:

1. Loc := 0 [assume the worst until the word is found]
2. FOR $i :=$ 1 to n
     1. IF $a_i = x$ THEN [success]
         1. Loc := i
         2. End

---

† The number $e = 2.71828\ldots$ is a constant that is essential in calculus for the study of exponents and logarithms.

The algorithm is simply a methodical series of comparisons. At most n repeats are made in loop 2, but this might be cut short by the End instruction of 2.1.2. Thus, we have at most n comparisons in step 2.1. At most two assignments of Loc are made: one in step 1 and one in step 2.1.1. As in Example 3.2.2, the FOR loop contributes n to the count, since i is reassigned n times. The time consists of at most

2 assignments for Loc

n assignments for the FOR loop

n comparisons

for worst time $T(n) = 2 + 2n$. In big Oh notation, $T(n)$ is $O(n)$.

An algorithm such as this would be used in a search for a word in a text or for a specific name in a randomly assorted pile of records.

We reconsider this example for an alphabetized list. (See Example 3.2.27.) As might be expected, we can find an algorithm more efficient than checking the list from beginning to end, one at a time. Before doing this, however, we introduce a useful instruction into the pseudocode that was discussed in the last section. It is, like the FOR instruction, a convenient way of repeating a process. Unlike the FOR statement, the maximum number of loops executed is not prescribed in advance. Thus it has more flexibility.

DEFINITION 3.2.22

A **WHILE instruction** has the form

WHILE *statement*
command

The instruction first checks whether *statement* is true. If it is not true, the WHILE instruction is ignored. If it is true, it executes *command*. It then rechecks *statement* (which might have a different truth value after *command* is executed), and the process is repeated until *statement* is false. More generally, a WHILE instruction is of the form

WHILE *statement*
command list

In this case, the full *command list* is executed if *statement* is true; it continues to be executed until *statement* is false. A WHILE command is also called a **loop.**

EXAMPLE 3.2.23   An Infinite Loop

Consider the following instruction. The input and the output are the integer n.

1. WHILE $n < 100$
    1. $n := 2n$

The intent is to keep doubling n until $n \geq 100$ when the process stops. The problem with this instruction is that if n is initially 0 or negative, the instruction $n := 2n$ gets repeated indefinitely, since after doubling, n will still be ≤ 0. In this case, we will be in "an infinite loop."

    The moral: (1) Make sure that the condition of the WHILE will not be true all the time, unless you are certain that the algorithm will end within the loop. (2) A WHILE instruction, though powerful, doesn't have a predetermined upper limit on the number of times a loop will be executed.

EXAMPLE 3.2.24   The Division Algorithm (but see also Example 3.2.25)

This was stated as Theorem 2.5.2. How do you do a long division of n by b? One way is to subtract b from a repeatedly, keeping track of how many subtractions are made, until you get a remainder $R < b$. The quotient Q is the number of subtractions. The assumption is that the user of the algorithm is able to subtract. For this algorithm, the input is the natural numbers n and b, with $n > 0$ and $b > 0$. The output is the numbers Q and R with

$$n = bQ + R \qquad 0 \leq R < b$$

The repetitive subtraction of b from n suggests a WIIILE instruction. We initialize $Q := 0$ and $R := n$. Every time we subtract b from R, we increase Q by 1 and decrease R by b. This comes from the formula

$$bQ + R = b(Q + 1) + (R - b)$$

We do this until the remainder is smaller than b. The algorithm is

1. $Q := 0$
2. $R := n$
3. WHILE $R \geq b$
    1. $R := R - b$
    2. $Q := Q + 1$

    We can estimate the time involved. Suppose we let n be the size of the input. In this algorithm, the WHILE loop is tested exactly $Q + 1$ times and executed Q times. In each execution, two assignments are made. Therefore, along with the initializations, $2 + (Q + 1) + 2Q = 3Q + 3$ assignments or comparisons are made. Since $Q = [n/b]$, we have time $T(n) = 3[n/b] + 3$, so $T(n)$ has order of magnitude n.

EXAMPLE 3.2.25   *The Division Algorithm, Revisited*

The algorithm in Example 3.2.24, although mathematically accurate, is *hopelessly inefficient*. Who divides 3,345,565 by 7 by repeatedly subtracting 7's and keeping track of how many subtractions were made? This technique might be necessary if this number were given as a huge pile of pebbles. But it isn't. It is represented by a seven-digit number. Our experience is that a long division is not very different for 7,986,432 than it is for 3,345,565. We certainly don't expect it to take more than twice the time. (It would, approximately, if these numbers did represent pebbles and we were confronted with the task of pulling away seven at a time.)

A more reasonable selection for the size of the input n is the *number of digits d* in n. In this case, we expect a 14-digit number to take approximately twice as long to divide as a 7-digit number. The relation between the number of digits $d(n)$ of an integer n, and n itself, is logarithmic:

$$d(n) = 1 + [\log_{10} n]$$

by Theorem 2.5.13. Thus, $d(n)$ has the same order of magnitude as $\log n$.

We shall not go into a detailed description of the algorithm for long division taught in grade school. But we can make a good case that the time needed to find the quotient and remainder when n is divided by b is $O(\log n)$. To see this, recall that the usual long division starts with the first digit and does a division. There will possibly be carrying. In any event, this process will take $O(1)$ steps. [Let's overestimate and say that not more than 2000 operations are involved when working with base 10. This is $O(1)$.] This operation is repeated $d(n)$ times, once for each digit, so the number of steps is $d(n) \cdot O(1) = O(d(n)) = O(\log n)$ times.

Since $\log n = o(n)$, the grade school algorithm is better by far than the algorithm of Example 3.2.24. If we let the number of digits d of n represent the size of the input, this analysis shows that the grade school algorithm takes time $O(d)$, whereas the inefficient algorithm of Example 3.2.24 has time of order of magnitude 10^d.

Remark   The algorithm of Example 3.2.24 need not be thrown to the dogs. It does not depend on the decimal representation of n, so it is a viable primitive version of long division. It is the basis for a formal proof of the Division Algorithm using the axioms of number theory.

EXAMPLE 3.2.26   Collatz's Function

Let the input be a positive number n, and the output be M. Can we find the order of magnitude of $M = M(n)$ in the following algorithm?

1. $M := 0$
2. IF $n = 1$ THEN End
3. WHILE $n > 1$
   1. IF n is even THEN $n := n/2$
   2. IF n is odd THEN $n := 3n + 1$
   3. $M := M + 1$

Method   Instructions 3.1 and 3.2 compute the **Collatz function** $C(n)$ and set $n := C(n)$. (See Example 1.1.29 and Exercises 1.1.29 and 1.1.30.) The "output" M is the number of times loop 3 is executed, so it is a good measure of the time this algorithm takes. We don't know whether this algorithm will terminate for all values of n. It appears to, for many values that have been checked, but there is no definite answer at this time (1991). If it did always stop, then $M = M(n)$ would be defined for all n, but we would have no way of finding its order of magnitude without understanding the process involved. Once again, this illustrates that in a WHILE statement care must be taken to have the condition (here $n > 1$) become false at some point.

We now return to the problem of looking through an alphabetized list for a particular word.

EXAMPLE 3.2.27   A Binary Search

Let a_i $(i = 1, \ldots, n)$ be a sequence of words *in alphabetic order*. We want to search this list to see if a given word x is in it. If it is, so that $x = a_k$, then the output is k (the location of the word). If it is not, then the output is 0. Find an algorithm to do this.

Method   A reasonable approach is to look at the word halfway through the list. Then, depending on whether we have overshot or undershot, we look halfway through the list that may contain our word, and continue this search until we hit the word or find that it is not in the list.

The following is an example using an eight-word list. Consider the alphabetized list

ABLE   MANY   MARY   NEXT   NO   VERY   WALK   WARY

Imagine that these words are hidden from us, but we are allowed to look at any of them one at a time. We now want to know if the word PERT is on the list. How do we proceed? We have unknown cards, in alphabetic order. Going halfway, we look at the fourth or fifth card. Let's always take the lesser choice if we have two choices. We read NEXT for the fourth card, so we must look among the cards numbered 5 through 8. Go halfway again, and read VERY for the sixth. Since our word PERT is alphabetically before VERY, we check the fifth card. When we discover NO, we conclude that PERT is not in the list. Each time we cut the range of cards in half or less, this procedure will find the word or discover it is not there in no more than four attempts. That is, we need at most three attempts to narrow down the word and one last look to see if we succeeded. With 16 words, we need at most five attempts. For $n = 2^k$ words in the list, alphabetically arranged, we need at most $k + 1 = O(\lg n)$ trials. Now let's write the algorithm.

The input is a positive integer n, a sequence $\{a_i\}$ of words $(i = 1, \ldots, n)$ and a word x that is to be sought in this list. The sequence is arranged alphabetically. Write $u < v$ to indicate that the string u is alphabetically before the string v. The output is the number Loc, which is the location of x if it is found $(x = a_{\text{Loc}})$ or Loc $= 0$ if it is not in the list. Using the halfway approach, we need to restrict the search range from a low of L to a high of H. We take $K = \text{MID}(L, H) = [(L + H)/2]$ as the integer midpoint. The algorithm follows.

1. $L := 1$
2. $H := n$ [initially we take the full range]
3. Loc $:= 0$ [when and if the word is found, this changes to the location of the word]
4. WHILE $H - L > 0$ [not narrowed down to one word]
    1. $K := \text{MID}(L, H)$
    2. IF $a_K = x$ THEN [success]
        1. Loc $:= K$
        2. End
    OTHERWISE [no success yet]
        3. IF $x < a_K$ THEN [K is too high]
            1. $H := K - 1$ [use $K - 1$ for the new high]
            2. IF $H < L$ THEN $H := L$ [This happens if the previous high was one more than the previous low.]
        4. IF $a_K < x$ THEN [K is too low]
            1. $L := K + 1$
5. IF $a_H = x$ THEN Loc $:= H$ [$H = L$ here; we are out of the loop]

We have discussed why this algorithm works. We will now do a time analysis.

EXAMPLE 3.2.28

In the algorithm of Example 3.2.27, find the order of magnitude for the worst-case time $T(n)$.

Method We count the number $T(n)$ of assignments or comparisons made. There were $O(1)$ assignments or comparisons made in steps 1, 2, 3, and 5. We must now find how many times the WHILE instruction is repeated.

In Example 3.2.27, we saw that each time we went through the loop the number of words a_i that must still be searched was at most half of what it was before the loop. We call this number the search range S. For a given L and H, we have $S = H - L + 1$. Steps 4.2.3 and 4.2.4 show how the new values of H and L are found.

We now estimate S recursively. Let S_k be the search range after k loops. Thus,

$$S_0 = n$$

$$S_{k+1} \leq S_k/2$$

A short induction shows that

$$S_k \leq n/2^k$$

The WHILE instruction 4 is exited when $S_k = 1$. This certainly happens by the time $n/2^k \leq 1$. But this is equivalent to

$$n \leq 2^k$$

$$\lg n \leq k$$

Thus, we are guaranteed an exit, if $k = [\lg n] + 1 = O(\lg n)$. Therefore, since $O(1)$ assignments or comparisons are executed during each run of the loop, we have total time $T(n) = O(1) + O(1) \cdot O(\lg n)$. Since $\lg n = O(\log n)$, we see that the worst-time function $T(n)$ satisfies the condition $T(n) = O(\log n)$.

We can be more precise. To find the worst case, we assume that x is not on the list. Therefore, there are

$$1 + [\lg n] = \lg n + O(1)$$

loops. Two assignments are made within each loop (K and either H or L) and two comparisons. In addition, the loop itself needs a comparison ($H - L > 0$). Therefore, we have five steps per repetition of the WHILE loop and a few extra assignments at the beginning. Thus, the worst-time analysis yields $T(n) = 5 \lg n + O(1)$.

Remark   This is quite fast. For a million words, the number of loops in the worst case $\approx \lg 1{,}000{,}000 \approx 20$.

EXAMPLE 3.2.29   How Big Is Big?

We have seen that if we confine ourselves to functions $f(n) \to \infty$ as $n \to \infty$, there can be a substantial difference in how fast these functions go to infinity. Example 3.2.10 and Table 3.2.1 show some of these differences. We generally take $f(n) = n$ as the standard with which other functions are to be compared. From this point of view, $\log n$ goes quite slowly to infinity; $\log(\log n)$ is practically bounded! If a time analysis of an algorithm gives $T(n) = O(n^k)$ for some fixed value of k, then the algorithm is considered manageable for relatively large n. In this case, the algorithm is said to have **polynomial time.** Any algorithm with larger than polynomial time (for example, if the time has order of magnitude 2^n) is considered unmanageable, or **intractable.** See Table 3.2.1 for appropriate examples.

Some care must be taken in using and applying the big and little Oh notations. For example, if we had two algorithms performing the same task and they had worst-time cases $O(n)$ and $O(n \log n)$, we would not necessarily want to choose the former as guaranteeing more time. For example, the functions $10n$ and $n \lg n$ are equal when $n = 2^{10} \approx 1000$. If we knew that the size of our applications was always smaller than, say, 500, we would be better off with time $n \lg n$. Similarly,

$$n^5 = o(n^{\log(\log n)})$$

But for $n = 1,000,000$, $\log(\log n) \approx 2.626$, and a comparison of n^5 with $n^{\log(\log n)}$ shows that n^5 is much *larger*. Equality of these functions occurs only when n is of the order of magnitude of 10^{64}. This does not negate the big and little Oh notations. But it does indicate that we should take seriously the clause "for n sufficiently large" appearing in Definitions 3.2.5 and 3.2.6.

Table 3.2.2 lists some of the standard functions and their names in increasing order of magnitude.

Table 3.2.2

Function	Name	Comments
$\log(\log n)$	log log	Very slow approach to ∞
$\log n$	log	Slow approach to ∞
$\sqrt{n}$	Square root	Slower than linear
$n/\log n$		Slightly slower than linear
n	Linear	Standard
$n \log n$		Slightly faster than linear
n^2	Quadratic	Polynomial
n^3	Cubic	Polynomial
2^n	Exponential	Very fast approach to ∞
$n!$	Factorial	Much faster approach to ∞

Exercises

1. Show that if $f(n) = o(n)$ and $g(n) = O(n)$, then $f(n) + g(n) = O(n)$.
2. Prove that if $f(n) = O(n)$ and $g(n) = O(n)$, then $f(n)g(n) = O(n^2)$.

In Exercises 3–8, find an upper bound, using the big Oh notation for the number of assignments made, as a function of n.

3. 1. FOR $i := 1$ to n
    1. $x := x + 1$
    2. IF $y < x$ THEN $y := y + 1$

4. 1. FOR $i := 1$ to n
    1. FOR $j := 1$ to n
       1. IF $x < ij$ THEN $x := x + 1$

5. 1. $i := 0$
    2. WHILE $i^2 < n$
       1. $i := i + 1$

6. 1. $x := 0$
    2. FOR $i := 1$ to n
       1. IF $i \le x$ THEN $x := x + 1$

7. 1. $k := n$
    2. WHILE $k > 1$
       1. $k = [k/3]$

8. 1. FOR $i := 1$ to n
    1. FOR $j := 1$ to n
       1. IF $j^2 < n$ THEN $x := x + 1$

In Exercises 9–15, $f \le g$ means that $f = O(g)$, and $f < g$ means that $f = o(g)$. Define $f \approx g$ to mean that $f \le g$ and $g \le f$. The relations $\le$ and $<$ are defined on the positive functions of the natural numbers.

9. Show that $\le$ is transitive and reflexive. (This is Theorem 2.3.11.)

*10. Prove the following:
    (a) If $f \le g$ and $g < h$, then $f < h$.
    (b) If $f < g$ and $g \le h$, then $f < h$.
    (This is Theorem 2.3.12.)

*11. Show that $<$ is transitive, but that if $f < g$, then $\neg(g < f)$.

*12. Show that if $f < g$ then $\neg(g \approx f)$.

13. Show that f and g have the same order of magnitude if and only if $f \approx g$. Prove that $\approx$ is an equivalence relation.

14. If $f \le g$ but $\neg(g \le f)$, show that the functions f, $\sqrt{fg}$, and g satisfy $f \le \sqrt{fg} \le g$. Show that no two of these functions have the same order of magnitude. (This shows that there is no such thing as the "next higher order of magnitude.")

*15. If $f \le g$ but $\neg(g \le f)$, can we say that $f < g$? Use the definitions given for $\le$ and $<$.

In Exercises 16–25, find a simple function, such as one of the functions in Table 2.3.2, that has the same order of magnitude as the given function.

16. $n(n - 1)(n - 2)/6$
17. $n!/(n - 5)!$
18. 2^{4n+5}
19. $\log(2n + 15)$
20. $[n^2/8]$
21. $(n^2 + 3n + 1)/(3n^2 + n + 12)$
22. $\log(n^2 + 3)$
23. $\log(2^n + n)$
24. $(n^2 + 6n + 4)^{1/2}$
25. $\log(n \log n)$

In Exercises 26–29, find an upper bound of a sum by replacing each term with the largest of the terms.

26. Prove that $1^2 + 2^2 + \cdots + n^2 = O(n^3)$.
27. Prove that $\log(n!) = O(n \log n)$. (*Hint:* Use the fact that the logarithm of a product is the sum of the logarithms.)
28. Prove that $[\sqrt{1}] + [\sqrt{2}] + \cdots + [\sqrt{n}] = O(n^{3/2})$.
*29. Prove that $1^1 + 2^2 + 3^3 + \cdots + n^n = O(n^n)$.

*30. Show that $n! = o(n^{n-k})$ for any positive integer k.
31. Show that $\log n = o(\sqrt{n})$. In fact, show that $\log n = o(n^p)$ for any positive real number p. You may use Equation (3.2.8) or (3.2.10).
32. Show that $[\sqrt{n}]$ has the same order of magnitude as $\sqrt{n}$.
33. Show that $2^n = o(n!)$. [This is Formula (3.2.9) for $a = 2$.] More generally, show that $a^n = o(n!)$ for any $a > 1$.
*34. (a) Using the result $n < 2^n$ (Example 2.3.3), show that $\log n = O(n)$.
    (b) Now replace n by $[\sqrt{n}] + 1$ to show that

$$\log \sqrt{n} < \log([\sqrt{n}] + 1) = O([\sqrt{n}] + 1) = O(\sqrt{n})$$

    and therefore $\log n = o(n)$.
*35. Let $a > 1$ and let k be arbitrary.
    (a) Using the result $\log n = o(n)$, as in Exercise 34, show that $n^k < a^n$ for n sufficiently large. (Take logs.)
    (b) Show that $n^k = o(a^n)$, using $n^k = o(n^{k+1})$. [This is Formula (3.2.8).]
*36. Show that

$$1 + \frac{1}{2} + \frac{1}{3} + \cdots + \frac{1}{n} = O(\sqrt{n})$$

    (*Hint:* Separate into two sums, one from 1 to $[\sqrt{n}]$ and the other from $1 + [\sqrt{n}]$ to n. Estimate each half separately.)
*37. Prove by induction on k that if $n = 2^k$ then

$$1 + \frac{1}{2} + \frac{1}{3} + \cdots + \frac{1}{n} \geq 1 + \frac{k}{2}$$

*38. Prove by induction on k that if $n = 2^k - 1$ then

$$1 + \frac{1}{2} + \frac{1}{3} + \cdots + \frac{1}{n} \leq k$$

Hence, show that

$$1 + \frac{1}{2} + \frac{1}{3} + \cdots + \frac{1}{n} = O(\log n)$$

39. Using the results of the previous two exercises, show that

$$1 + \frac{1}{2} + \frac{1}{3} + \cdots + \frac{1}{n}$$

has the same order of magnitude as $\log n$. [It is a result in calculus that this sum is actually between $\log n$ and $\log n + 1$, when $n > 1$, where $\log n$ is the natural logarithm.]

*40. It was stated during the discussion of the binary search algorithm of Example 2.3.27 that the search range $S = H - L + 1$ satisfies the inequality $S_{k+1} \leq S_k/2$, where S_k is the search range after k loops. Prove this, using properties of $[(H + L)/2]$ and steps 4.2.3 and 4.2.4 of this algorithm.

41. Let $d(n)$ be the number of binary digits of n. Show that $d(n) = \lg n + O(1)$.

42. Let $n = n(d)$ be a function of d with the property that $n(d)$ has exactly d decimal digits (d and n are positive integers). Prove that $n = O(10^d)$. Show that n has the same order of magnitude as 10^d.

3.3 Further Examples of Algorithms

We now consider the algorithm that we used to convert a number to its representation in base b.

EXAMPLE 3.3.1 *Base b Conversion*

The following algorithm is used to convert a number to its base b representation. It is based on the proof of Theorem 2.5.3. Recall that to convert the number n to base b, so that $n = (d_k \cdots d_0)_b$, we divide n by b, obtaining a quotient Q and the remainder d_0. We then divide Q by b to find its remainder d_1 and a new quotient. We continue this process until we have quotient 0 and remainder d_k. This repetition suggests a WHILE instruction. In what follows, we identify the numerical remainder with its base b digit.

The algorithm takes input n and b, where n and $b > 1$ are natural numbers. The output is a natural number k and a sequence $d_0, d_1, \ldots, d_k$ such that

$$n = (d_k \cdots d_0)_b$$

When n is divided by b with the Division Algorithm, we have

$$n = bQ + R, \qquad 0 \le R < b$$

In this case, we write $R = \text{Rem}(n, b)$ and $Q = \text{Quo}(n, b)$.

1. IF $n = 0$ THEN
    1. $k := 0$
    2. $d_0 := 0$
    3. End
2. $Q := n$ [the initial quotient]
3. $i := 0$ [the initial subscript for d_i]
4. WHILE $Q > 0$
    1. $d_i := \text{Rem}(Q, b)$ [the ith digit from the right]
    2. $Q := \text{Quo}(Q, b)$
    3. $i := i + 1$ [ready to work on the next digit]
5. $k := i - 1$ [we don't need the next digit because $Q = 0$]

EXAMPLE 3.3.2   Time Analysis of the Algorithm in Example 3.3.1

We take n as the size of the input. (We think of b as a fixed integer.) To find an estimate of the time involved in the base b conversion algorithm of Example 3.3.1, we must find out how many repeats of loop 4 are required.

(a) *Loop Analysis*   The WHILE instruction 4 is governed by the condition $Q > 0$. The value of Q is changed only in instruction 4.2, where it is redefined as the quotient when it is divided by b. We let Q_k be the value of Q after the kth loop. Instruction 2 initializes Q, so

$$Q_0 = n \tag{3.3.1}$$

By instruction 4.2, we have

$$Q_{k+1} = \text{Quo}(Q_k, b) \tag{3.3.2}$$

We now want to find when $Q_k = 0$. Since $\text{Quo}(a, b) = [a/b]$, we have, by (3.3.2),

$$Q_{k+1} = [Q_k/b] \le Q_k/b$$

By induction this recursive inequality, together with Equation (3.3.1), yields the inequality

$$Q_k \le n/b^k$$

Therefore (since Q_k is a positive integer), we know that Q_k will be 0 when $n/b^k < 1$ or $b^k > n$. Taking logarithms, we have $k > \log_b n$. Thus, Q_k will be 0 by the time $k = [\log_b n] + 1$. Thus, the number of repetitions of loop 4 is $O(\log n)$.

(b) *Assignment Analysis*   Checking the algorithm, we see that for $n > 0$ each repetition of loop 4 does three assignments and that three extra ones are done outside of the loop. Since there are $O(\log n)$ repetitions, the number of assignments is $3O(\log n) + 3 = O(\log n)$.

From parts (a) and (b), we see that the time for this algorithm is $O(\log n)$.

(c) *Calculations of the Quotient and Remainder*   The preceding analysis took for granted that no time was used in the computation of the Quo and Rem functions. But we have seen in Example 3.2.28 that we can expect the time for this calculation to be $O(\log Q)$. From this point of view, the assignments in the loop are irrelevant; most of the time in the loop is used in calculating the quotient and remainder. We initialized Q_0 as n, and we already know that $Q_k \leq n/b^k$. Thus the time estimate using this consideration is

$$O(\log n + \log n/b + \log n/b^2 + \cdots + \log n/b^k)$$

There are $O(\log n)$ terms. A crude estimate, replacing all terms with $\log n$, is $O(\log n)^2$. It turns out that this crude analysis cannot be improved to give a lower order of magnitude. We leave this to the exercises.

Summarizing, if we regard finding quotients and remainders as built-ins, taking no time, the base b conversion takes $O(\log n)$ time. But if we must do computation for quotients and remainders, the time of this algorithm is $O(\log n)^2$.

If we think of doing a hand computation to convert, say, 13,367,865,345,331,789,401,825 to base 7, we cannot ignore the time spent in performing a long division by 7. This accounts for the extra $\log n$ factor in part (c).

EXAMPLE 3.3.3   *Base b Conversion to a String of Digits*

We noted in Section 2.5 that the decimal representation of an integer is a string of digits. For the hex notation it was crucial that we introduce a special digit for each number from 10 through 15. We now consider this problem in general.

For base b, we assume that we have an alphabet of base b digits D_b. A natural number is to be represented by a string of D_b^*. In addition, we have a coding Symb: $\mathbb{Z}_b \to D_b$. For example, in hex notation, $b = 16$ and we had Symb(11) = "B". We want an algorithm that has n and b as input and the string representation of n as output. The algorithm of Example 3.3.1 almost does it. However, we do not seek the sequence d_i of the remainder. We seek the string representation of n obtained by stringing the symbols for the d_i's in a row, from right to left. A small adjustment in the algorithm is all that is necessary.

The input is a natural number n. The output is the string representation

X of n, using the digits of D_b. We call this algorithm Integer/String(n; X), making its input and output explicit.

Algorithm Integer/String(n; X)
1. IF $n = 0$ THEN
    1. $X := \text{Symb}(0)$
    2. End
2. $Q := n$ [the initial quotient]
3. $X := \epsilon$ [the empty string; we adjoin digits, one by one, to the beginning of X to get the completed output]
4. WHILE $Q > 0$
    1. $d := \text{Rem}(Q, b)$ [the ith digit from the right]
    2. $Q := \text{Quo}(Q, b)$
    3. $X := \text{Symb}(d)*X$ [concatenate digit in front of X]

The time estimate for this algorithm is still $O(\log n)$ or $O((\log n)^2)$ if we take the Division Algorithm time into account.

Sorting

We noted in the last section that, if a sequence of words is alphabetized, we can efficiently locate a given word in the sequence or decide that the word isn't there. If n is the length of the sequence, then this could be done in $O(\log n)$ steps. If the words are not alphabetized, but given in some random order, the number of steps is $O(n)$. Offices and businesses arrange records in numeric or alphabetic order because they occasionally have to find a record. The office manager may not know that $\log n = o(n)$, but he or she certainly knows that for a large number of records, one record can be quickly found if the records are in alphabetic order; otherwise the job may be hopeless. Computers can do this more quickly, but time considerations still apply.

We now come to the question, "How do you alphabetize a sequence of words?" In Figure 3.3.1, a stack of 5000 checks, numbered from 1000 to 99,999, is to be arranged in order. How would you do it?

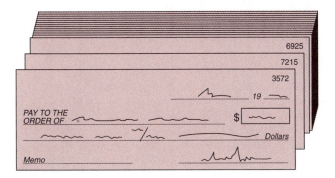

Figure 3.3.1

The problem of arranging numbers in numeric order, or rearranging a sequence of words in alphabetic order, is called a **sorting** problem. In general, suppose we have a finite set A with a transitive relation $<$ on it. We also assume that if $x < y$ then $x \neq y$. In addition, the ordering is assumed to be linear: If $x \neq y$, then either $x < y$ or $y < x$ but not both. The problem of sorting is to rearrange a list $a_1, a_2, \ldots, a_n$ so that $a_i \leq a_j$ whenever $i < j$. (We use $\leq$ because some of the a_i's may be the same. As usual, $x \leq y$ if and only if $x = y$ or $x < y$.) Commercial data base programs attempt to use fast sorting techniques. One of the first tests made when comparing such programs is to compare the times they take in sorting large numbers of records. We usually illustrate sorting techniques by working with a set of integers, using the ordinary inequality $<$. The change to the general setting is not difficult.

The first algorithm illustrated is very inefficient. We look for the largest element in a given sequence of length n, place it at the end of the sequence, do this again for the first $n - 1$ terms, and continue the process until all terms are sorted. In what follows, we always use a_i as the ith term of the sequence and n as the size of the sequence. The sequence itself will be denoted by $\{a_i\}$. The output is again $\{a_i\}$ and is rearranged in ascending order.

EXAMPLE 3.3.4 MaximumLast

We find the maximum value of a_i in steps and bring it to the top. Let M denote the current maximum and Loc its location: $M = a_{\text{Loc}}$. (Compare with Example 3.1.18.) We originally go through the sequence from 1 to n, then from 1 to $n - 1$, and so on, until the sequence is sorted. Let us construct a fragment of this algorithm, called PLACE(j), involving the search from 1 to j (j can be any integer between 2 and n). We want to place the largest term $M = a_i$ $(1 \leq i \leq j)$ at the jth place. This fragment is as follows.

PLACE(j)
1. $M := a_1$
2. Loc $:= 1$ [initializations]
3. FOR $i := 2$ to j
    1. IF $a_i > M$ THEN [new candidate for the maximum]
        1. Loc $:= i$
        2. $M := a_i$
4. $a_{\text{Loc}} := a_j$ [moving a_j to the place where M was found]
5. $a_j := M$ [finishing the swap]

The full sort is given as follows.

MaximumLast
1. FOR $j := n$ down to 2
    1. PLACE (j)

We could copy the full algorithm for PLACE(j), but the present form is suitable because it makes the approach more transparent. Note the convenience of having the FOR ... down to ... variation of the FOR instruction.

The algorithm in Example 3.3.4 breaks up its tasks into smaller, more manageable ones. It made sense to create the algorithm for PLACE(j) before we wrote the full-fledged MaximumLast algorithm.

DEFINITION 3.3.5

An algorithm such as PLACE is called a **routine** or a **procedure.** In programming languages, this can also be called a **subroutine.** The purpose of a procedure is to execute a subtask that will be used in an algorithm.

Typically, a procedure has its own inputs and outputs. Thus, PLACE has input j and the sequence $\{a_i\}$. Its output is the rearranged sequence $\{a_i\}$. We suppressed the sequence $\{a_i\}$ in our notation PLACE(j). Look at the steps of the routine PLACE and notice that the variable M was used to help construct the output. If M were a variable used in an algorithm that used PLACE, then this value of M would be lost. We assume that this does not happen. We can always use special variables not used elsewhere. Our approach is to regard all variables in routines as private to the routine, with the exception of the inputs and outputs that are needed to communicate with the algorithm (or another procedure) that uses it.

EXAMPLE 3.3.6

Estimate the worst-case time for the MaximumLast algorithm of Example 3.3.4.

Method   We first find the time for PLACE(j) as function of j. There are $j - 1$ comparisons, since instruction 3.1 is repeated from $i := 2$ through j. For the full algorithm, PLACE(j) is done for $j = 2$ through $j = n$. The total number of comparisons is

$$(n - 1) + (n - 2) + \cdots + 1 = n(n - 1)/2$$

A fixed bound for the number of comparisons is $n(n - 1)/2$. This is $O(n^2)$. Since each comparison gives at most two assignments (3.1.1–3.1.2), and one assignment for the FOR loop, the entire algorithm has run time $O(n^2)$.

We now consider a somewhat more efficient approach to sorting. How would we sort, by hand, the pile of checks in Figure 3.3.1? One way is to do it one at a time. At each step, we insert a new check into the spot where it belongs. In the usual office procedure, this is the way it would probably be done as the checks came in one at a time.

EXAMPLE 3.3.7   *Insertion Sorting of a Sequence* {a_i}

We wish to sort a sequence {a_i}. The plan is to take the terms a_i one at a time and insert them into an evolving sorted sequence. Once again, we break the task into smaller pieces. Assuming that we have sorted the a_i ($1 \leq i \leq j$), let's find a procedure for inserting the next term a_{j+1}. Then we use this procedure repeatedly to perform the task. We have already done most of the work of inserting a_{j+1} in Example 3.2.27, where we were searching for the word x in a list. Now we search for a_{j+1} in the list to find where it should go. Then we place a_{j+1} there and push the other terms up. Therefore, we work with the following procedures. A sequence {a_i} ($1 \leq i \leq n$) is assumed as input and output.

LOCATE($x, j; k$) is an algorithm that has additional inputs x and j and gives an output k. The sequence {a_i} is assumed to be sorted from $i = 1$ to j: $a_1 \leq a_2 \leq \cdots \leq a_j$. The output k permits x to be inserted at the kth place, moving up the a_i from the kth place on, where k can be any integer from 1 through $j + 1$. The choice of k is as follows. (See Figure 3.3.2.)

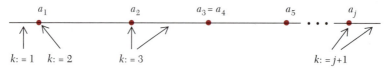

Figure 3.3.2

1. If $x < a_1$, then $k := 1$.
2. If $a_j \leq x$, then $k = j + 1$.
3. If $a_1 \leq x < a_j$, choose k so that $a_{k-1} \leq x < a_k$.

We learned from Example 3.2.27 that if we wanted to save time we shouldn't search for x going through the sequence {a_i} from the first term on, but rather halve the search range and continue to halve it until we find the correct placement. We give details in Example 3.3.11 when we define LOCATE recursively. LOCATE is a variant of the algorithm in Example 3.2.27. We already know, as in Example 3.3.1, that its worst time is $O(\log j)$.

INSERT(j, k) is a routine that moves a_j into the sequence {a_i} ($i \leq 1 \leq n$) at the kth place and pushes up the terms after a_k. We assume $j > k$. For example, if $k = 3$ and $j = 5$ and the sequence {a_i} is

$$1, 5, \underline{13}, 14, \underline{11}, 12, 2, 7$$

then after INSERT(5, 3) is executed, $\{a_i\}$ becomes

$$1, 5, \underline{11}, \underline{13}, 14, 12, 2, 7$$

The routine is as follows (see Figure 3.3.3).

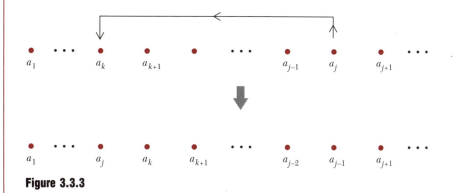

Figure 3.3.3

INSERT(j, k)
1. $s := a_j$ [save a_j; we need it for the new value of a_k]
2. FOR $i := j$ down to $k + 1$
    1. $a_i := a_{i-1}$ [starting with a_j and going down to a_{k+1}, replace each term with the one preceding it]
3. $a_k := s$ [the saved value; a_j has already changed to a_{j-1} in step 2.1, when $i = j$]

If we use assignments as representing time, the time for executing INSERT is $O(j)$. The loop is repeated at most $j - 1$ times.

We now can see how the InsertSort algorithm works. We start with a_1. We find where to place a_2 by using LOCATE(a_2, 1; k). Using this value of k, we place a_2, using INSERT(2, k); then we repeat. To find where to place a_3, we use LOCATE(a_3, 2; k), and then use INSERT(3, k), and so on, until we have placed a_n. The full algorithm, using INSERT and LOCATE, is as follows.

InsertSort Algorithm
1. FOR $j := 2$ to n
    1. LOCATE(a_j, $j - 1$; k)
    2. INSERT(j, k)

EXAMPLE 3.3.8   *Time Analysis for InsertSort*

To make a time analysis, we have to decide whether we want to count assignments and/or conditions checked in a conditional instruction. This will be critical in our analysis of the InsertSort algorithm.

Let us start with a count of conditions checked. This means that we regard assignments and going through loops as taking no time. Under this criterion, INSERT takes no time whatsoever. It consisted of a FOR loop and assignments. As in Example 3.2.28, LOCATE($x, j; k$) has, for the worst case, $O(\log j)$ comparisons—that is, $\leq K \log j$ comparisons for some constant K. The InsertSort algorithm uses PLACE($x, j; k$) for $j = 1, 2, \ldots, n - 1$. Therefore we have at most

$$K \log 2 + K \log 3 + \cdots + K \log(n - 1) \qquad (3.3.3)$$

comparisons. But each term in this sum is $O(\log n)$, and there are $n - 1$ or $O(n)$ terms. Thus, this sum is $O(n \log n)$. The worst-time case takes $O(n \log n)$ time, which is considerably better than the MaximumLast algorithm, which took $O(n^2)$ time.

However, if we count assignments made in the InsertSort Example 3.3.7, the time analysis is $O(n^2)$. The INSERT(i, k) routine has $O(i)$ assignments in the worst case when $k = 1$. Hence, one run through the loop in the InsertSort algorithm takes time $O(i) + O(\log i) = O(i)$. Adding these, we get, in analogy with (3.3.3), a time at most equal to

$$K + 2K + \cdots + (n - 1)K = O(n^2) \qquad (3.3.4)$$

This worst-time case analysis is appropriate for a computer where the assignments in the INSERT routine do take time. This example illustrates, once again, that any time analysis of an algorithm depends on which operations take time and on how much time they take.

EXAMPLE 3.3.9   *Sorting Checks with CheckSort*

How realistic is it to say that INSERT takes little or no time? If we look at the pile of checks in Figure 3.3.1, and we are doing hand sorting of this pile, then it is fairly realistic. Finding where to put a check in a sorted pile of checks of size j may take some time, but inserting it is done in a flash. You never do the assignments in INSERT.

What follows is a rough "algorithm" for hand-sorting a pile of checks, based on the InsertSort algorithm. We first start with a routine Check-Place.

Routine CheckPlace(x, P) [putting check numbered x into a *sorted* pile P of j checks]
1. $P := $ the given pile of checks
2. $L := \varnothing$ [the pile that's too low]
3. $H := \varnothing$ [the pile that's too high]
4. WHILE P has more than one check
    1. Cut the pile in half and see which half x goes into

    2. $P :=$ the half that x goes into
    3. Consolidate the unwanted half into the H or L pile
5. Insert x into pile P, which now has only one check
6. Reconstitute the H and L piles

    A reasonable estimate for the time of CheckPlace is the number of loops entailed in the WHILE statement. This is $O(\log j)$, where j is the number of checks in pile P. The full CheckSort algorithm, in analogy with the InsertSort algorithm, is as follows.

CheckSort Algorithm
  Input: A pile P of checks
Output: The checks sorted in a pile Q
1. $Q :=$ the empty set $\varnothing$ [Q is the growing sorted pile of checks that we add to one by one]
2. WHILE P is nonempty
    1. $x :=$ first check of P
    2. CheckPlace(x, Q)
    3. $P := P - \{x\}$ [physically, this just happens; x is removed from P when it is placed in Q]

    Except for the imprecise instruction 4.1 of CheckPlace, the worst-case analysis for the number of loops is similar to the analysis in Example 3.2.28. The worst time is $O(n \log n)$.

Recursive Algorithms

Many of our algorithms worked by reducing a problem involving a large input to an entirely similar one using a smaller input. This was true, for example, in the CheckPlace routine, where we found where to place to check by looking into a pile half its size and continuing until we had a pile of 1. This is quite reminiscent of the notion of a recursive definition, where a function was defined in terms of its definition at previous values. In terms of algorithms, we can say that an algorithm *uses itself* as a procedure. The only condition is that if this is to continue, we must ultimately come to a version of itself (with small input) in which the algorithm is directly defined. (This is the basis algorithm.) In what follows, we make the input explicit in order to see how an algorithm uses itself, but for a simpler input.

EXAMPLE 3.3.10   A Recursive Function

Consider the recursive function given by

$$a_0 = 1$$

$$a_{n+1} = a_n^2 + n$$

We constructed an algorithm to compute the values of a_n in Example 3.1.16. The method was

1. $x := 1$ [initializing a_n for $n = 0$]
2. FOR $i := 1$ to n
    1. $x := x^2 + i - 1$
3. $a_n := x$

This method starts at $n = 0$ and builds up till we reach n.

We now give a directly recursive method. Let FCN(n; a_n) be the algorithm to find a_n. We list n and a_n, since we want to be explicit as to input and output.

Algorithm for FCN(n; a_n)
1. If $n = 0$ THEN
    1. $a_0 := 1$
    2. End
2. FCN($n - 1$; a_{n-1})
3. $a_n := a_{n-1}^2 + n - 1$

In this illustration, 1 is the base case, 2 uses FCN for the case $n - 1$ and finds a_{n-1}, and 3 finds the required output a_n. How does 2 use FCN with input $n - 1$? By going to this same definition! This algorithm goes *down* from n to the base case 0 rather than building up from the base case 0 to n.

EXAMPLE 3.3.11

Let us consider the algorithm LOCATE(x, j; k) of Example 3.3.7. This algorithm determined the place k where x will be in the increasing sequence $\{a_i\}$ ($1 \le i \le j$). We didn't give this algorithm in Example 3.3.7 because we wanted to use recursion. More generally, we define LOCATE(x, j_1, j_2; k) as the place k where x will be located in the increasing sequence $\{a_i\}$ defined for $j_1 \le i \le j_2$. Our method, as in Example 3.2.27 (the binary search), is to halve the interval and to continue doing this until we find x. The recursive algorithm follows.

LOCATE(x, j_1, j_2; k)
1. IF $j_1 = j_2$ THEN [base case for a one-term sequence]
    1. IF $x < a_{j_1}$ THEN $k := j_1$
        OTHERWISE
    2. $k := j_1 + 1$
2. $m := [(j_1 + j_2)/2]$ [the midpoint; we now find k by working on either the first half or the second]
3. IF $x \le a_m$ THEN [use the first half]
    1. LOCATE(x, j_1, m; k)

OTHERWISE
    2. LOCATE$(x, m + 1, j_2; k)$

This procedure uses itself in steps 3.1 and 3.2. However, it uses a version where $j_2 - j_1$ decreases. Thus, when LOCATE is called once again, this difference will continue to decrease until this difference is 1, when we are covered by step 1. An alternative using a WHILE statement is possible, as in Example 3.2.27.

We end with an example of an efficient sorting technique, called a **merge sort.**

EXAMPLE 3.3.12   MergeSort

If $\{a_i\}$ is a sequence of length n that is to be sorted, the merge sort method uses the idea that if we have two *sorted* sequences of about the same size, it is a simple matter to merge them into one sorted sequence. We do this by building up a new sequence by selecting the smallest term of the two sequences, removing it from the sequence, putting it into the new sequence, and continuing this process until we have the fully merged sequence. How then do we merge sort a sequence $\{a_i\}$? By breaking it into two pieces, merge sorting each piece (the recursion), and merging the sorted pieces. This is an example of a **divide-and-conquer** technique in which a task is done by dividing it into similar tasks, but smaller in scale.

We illustrate on the sequence

$$7 \quad 17 \quad 87 \quad 15 \quad 8 \quad 1 \quad 54 \quad 30$$

We cut in half to form two sequences

$$7 \quad 17 \quad 87 \quad 15 \quad \blacksquare \quad 8 \quad 1 \quad 54 \quad 30$$

and we merge sort each half. We do this by cutting in half and merge sorting each half:

$$7 \quad 17 \quad \blacksquare\blacksquare \quad 87 \quad 15 \quad \blacksquare \quad 8 \quad 1 \quad \blacksquare\blacksquare \quad 54 \quad 30$$

We start reconstituting here. Sort each of the quarter-sequences. We obtain

$$7 \quad 17 \quad \blacksquare\blacksquare \quad 15 \quad 87 \quad \blacksquare \quad 1 \quad 8 \quad \blacksquare\blacksquare \quad 30 \quad 54$$

Now merge the sequences of length 2 (that are sorted) that are separated by $\blacksquare\blacksquare$. This gives

$$7 \quad 15 \quad 17 \quad 87 \quad \blacksquare \quad 1 \quad 8 \quad 30 \quad 54$$

Finally merge the half-sequences (that are now sorted) to obtain the fully sorted sequence

$$1 \quad 7 \quad 8 \quad 15 \quad 17 \quad 30 \quad 54 \quad 87$$

(*Note:* If you do this with the pile of checks of Figure 3.3.1, you'd better have a lot of desk space. This observation indicates that in a computer implementation of a merge-sort technique, we can expect to use lots of memory.)

In order to formalize the algorithm, we see that we need to give a routine DIVIDE($\{a_i\}$, n; $\{b_i\}$, r, $\{c_i\}$, s), for dividing a sequence $\{a_i\}$ of length n into two sequences $\{b_i\}$ and $\{c_i\}$ of lengths r and s, where $0 \le s - r \le 1$, and a routine MERGE($\{b_i\}$, r, $\{c_i\}$, s; $\{a_i\}$, n), for merging two sorted sequences $\{b_i\}$ and $\{c_i\}$ of size r and s, respectively, into one $\{a_i\}$ of size n. Once these are defined, we can give the recursive algorithm for the MergeSort algorithm. Its input is a sequence $\{a_i\}$ of length n, and its output is a sequence $\{a_i\}$ of length n, fully sorted.

MergeSort ($\{a_i\}$, n)
1. If $n = 1$ THEN End [there is nothing to do here]
2. DIVIDE($\{a_i\}$, n; $\{b_i\}$, r, $\{c_i\}$, s)
3. MergeSort($\{b_i\}$, r)
4. MergeSort($\{c_i\}$, s)
5. MERGE($\{b_i\}$, r, $\{c_i\}$, s; $\{a_i\}$, n)

The recursive nature of this algorithm is seen in steps 3 and 4, where it uses a version of itself. The routine for DIVIDE is fairly straightforward.

DIVIDE($\{a_i\}$, n; $\{b_i\}$, r, $\{c_i\}$, s)
1. $r := [n/2]$
2. FOR $i := 1$ to r
    1. $b_i := a_i$
3. $s := n - r$
4. FOR $i := 1$ to s
    1. $c_i := a_{r+i}$

MERGE can itself be defined recursively, and we leave this to the exercises. Once we pull away the smallest element in one of the sequences to be merged, we are left with a "smaller case," and we can use recursion.

EXAMPLE 3.3.13   Time Analysis of MergeSort

Referring to the MergeSort algorithm of the previous example, we can give a big Oh estimate as follows. Since it is defined recursively, it seems reasonable to find a recursive estimate for its worst-time case.

For simplicity, we assume that n is a power of 2. In this case, we merge two sequences of length exactly $n/2$, and the sizes of the divided sequences continue to be powers of 2 until the algorithm finishes. Let $T(n)$ be the worst-case time for a sequence of size n. As a start, we have $T(1) = 1$, since a comparison was used in step 1 of that algorithm.

MERGE can be accomplished with one FOR statement with a few assignments in each loop. Therefore, we take it for granted that MERGE uses $O(n)$ steps, where r and s are $O(n)$. DIVIDE similarly uses $O(n)$ steps when its input is a sequence of length n. Let us suppose, then, that the total number of MERGE and DIVIDE steps is $\leq Kn$. The MergeSort algorithm uses MergeSort itself in steps 3 and 4. However, the inputs r and s are each $n/2$. The algorithm shows that if $n > 1$,

$$T(n) \leq Kn + 2T(n/2) \tag{3.3.5}$$

We now use this for $n/2$ to obtain

$$T(n/2) \leq Kn/2 + 2T(n/4)$$

Multiplying by 2, we obtain

$$2T(n/2) \leq Kn + 4T(n/4)$$

and so, by (3.3.5),

$$T(n) \leq 2Kn + 4T(n/4) \tag{3.3.6}$$

We can now put $n/4$ in (3.3.5) to obtain (after multiplying by 4)

$$4T(n/4) \leq Kn + 8T(n/8)$$

Substituting in (3.3.6) yields

$$T(n) \leq 3Kn + 8T(n/8) \tag{3.3.7}$$

The pattern is clear. If we continue this process, we obtain

$$T(n) \leq rKn + 2^r T(n/2^r) \tag{3.3.8}$$

A formal proof by induction on r, using the preceding technique, proves this inequality. Finally, if $n = 2^r$, we have $r = \lg n$, and by (3.3.8)

$$T(n) \leq rKn + nT(1)$$
$$= Kn \lg n + O(n)$$
$$= O(n \log n)$$

Exercises

*1. We stated in Example 3.3.2 that

$$S(n) = \log n + \log n/b + \log n/b^2 + \cdots + \log n/b^k = O((\log n)^2)$$

and gave a proof using a crude estimate. Prove that this sum has the same order of magnitude as $(\log n)^2$ by proving that

$$\log n + \log n/b + \log n/b^2 + \cdots + \log n/b^k \geq K(\log n)^2$$

for some positive K. [*Hint:* Take only a relatively few terms of this sum through $\log n/b^r$ and get an *underestimate* by taking the least term and multiplying it by the number of terms. Try to arrange for n/b^r to have order of magnitude $\sqrt{n}$.]

*2. Using the technique of Exercise 1, show that the sum $1^5 + 2^5 + \cdots + n^5$ has order of magnitude n^6. [In calculus, this sum is shown to be $n^6/6 + O(n^5)$.]

3. Trace the algorithm of Example 3.3.3 to find a string representation of the number 13 in base 4. Use digits A, B, C, and D to represent the numbers 0, 1, 2, and 3, respectively.

4. Trace the algorithm of Example 3.3.3 to find a string representation of the number 6 in base 2. Use digits O and X to represent 0 and 1, respectively.

5. Trace the algorithm of Example 3.3.1 to find the base 8 digits of the number 53.

6. Find an exact upper bound for the worst time $T(j)$ of the routine PLACE(j) in Example 3.3.4. Count all comparisons, assignments, and loops. Using this result, find an exact upper bound for the MaximumLast sorting algorithm of this example.

Exercises 7–10 develop the BubbleSort technique for sorting a sequence. It is a variation of the MaximumLast algorithm, and is about as efficient. Say that the ith term a_i is out of order if $a_i > a_{i+1}$. If a_i is out of order, then we can interchange the terms a_i and a_{i+1}, and then a_i will not be out of order. Such an interchange will be called a SWITCH of a_i. Now make a pass through the sequence from the first through the $(n-1)$st term, SWITCHing if you come across a term that is out of order. Repeat this as long as necessary to arrive at a sorted sequence.

7. Write a procedure for SWITCH. Write an algorithm for BubbleSort using the SWITCH procedure described.

8. Find an exact worst time $T(n)$ for the BubbleSort algorithm of Exercise 7.

9. Find a big Oh estimate for the BubbleSort algorithm of Exercise 7. Include all assignments, loops, and conditions.

*10. Find an exact worst time $T(n)$ for the MaximumLast algorithm of Example 3.3.4. Include all assignments, loops, and conditions. Compare your result with the previous exercises, and decide whether BubbleSort or MaximumLast has better time.

*11. Write a recursive algorithm for the CheckPlace algorithm of Example 3.3.9.

*12. A sequence a_n is recursively defined by the formulas

$$a_0 = 1$$

$$a_n = na_{[n/2]}(n > 0)$$

Write a recursive algorithm to compute this function. The input is n, and the output is a_n. Also, write an algorithm that starts from $n = 0$ and works up to n. Which algorithm do you think takes less time? Explain.

13. A sequence $\{a_i\}$ is known to be sorted except possibly for one term. Describe an efficient algorithm to sort it, and give a big Oh time estimate for the worst case.

*14. Write a nonrecursive algorithm for the routine MERGE defined in Example 3.3.12.

*15. Write a recursive algorithm for the routine MERGE defined in Example 3.3.12.

16. The integers from 1 to $n + 1$ are arranged in order in a sequence $\{a_i\}$ of length n. It is desired to find the missing integer. Give a full description of an efficient algorithm to do this. Your time should be $O(\log n)$.

17. A number n is known to be divisible by 9 if and only if the sum of its digits is divisible by 9. Using this fact, devise an algorithm to determine whether a number n is divisible by 9. The input is n, represented in decimal notation. The size of the input is the number of digits of n. Take for granted that the digits of n are $d_0, d_1, \ldots, d_k$ and that these numbers, as well as k, are given. [*Hint:* Start by defining the function DigitSum(n).]

18. Trace the algorithm constructed in Exercise 17 for input $n = 568$. Do not compute the function DigitSum algorithmically, but take it as built in.

19. A string on the letters {A, B, . . . , Z} is called **palindromic** if it reads the same backward as forward. (This is true of the string MADAM, for example.) Devise an algorithm in pseudocode to determine whether a string is palindromic. The input is a string α and its length n. The output is "yes" or "no," depending on whether the string is palindromic.

20. Illustrate as in Example 3.3.12 how MergeSort works on the following sequence of 30 numbers:

737	325	349	869	279	362	107	263	925	560
341	523	442	509	9	451	406	84	48	312
420	672	399	712	424	574	966	257	61	221

It is not necessary to do a merge sort on a sequence of length 3 or less. Do this by observation.

21. Illustrate as in Example 3.3.12 how MergeSort works on the following string of 14 numbers:

$$6 \quad 56 \quad 73 \quad 19 \quad 4 \quad 37 \quad 17 \quad 28 \quad 99 \quad 30 \quad 23 \quad 21 \quad 42 \quad 34$$

It is not necessary to do a merge sort on a sequence of length 3 or less. Do this by observation.

22. A specific record must be found among numerous unsorted records. The office manager says to look for this record, one by one. Clyde says that sorting is so efficient that it pays to alphabetize efficiently using MergeSort and then to look for the record with a binary search. Who is right? Explain your answer fully.

In Exercises 23–26, a subset A of I_n is represented by a sequence $\{a_i\}$, $1 \leq i \leq n$, where $a_i = 1$ if $i \in A$ and $a_i = 0$ if $i \notin A$. (See Example 1.1.36 and Definition 1.1.37 of a characteristic function.) Similarly, we let the sequences $\{b_i\}$ and $\{c_i\}$ be the characteristic functions of B and C, respectively.

23. Write an algorithm in pseudocode to determine whether the sets A and B are disjoint.
24. Write an algorithm to compute the set $C = A \cup B$ and $D = A \cap B$.
25. Write an algorithm to determine whether $A \subseteq B$.
*26. Write an algorithm to determine whether the sets $A_1, A_2, \ldots, A_k$ are pairwise disjoint. The sets are given by a k by n matrix a_{ij}, where the jth row $a_{j1}, a_{j2}, \ldots, a_{jn}$ represents the set A_j.

Summary Outline

- An **algorithm** is a precise, mechanical way of performing a task.

- A **divide-and-conquer** technique does a task by dividing it into similar tasks but smaller in scale.

- A **procedure** or a **routine** is an algorithm used within another algorithm.

- A **recursive algorithm** uses a simpler version of itself as a procedure.

- A **test** or **trace** of an algorithm tests it for particular inputs.

- **Pseudocode** is an informal language used to express algorithms.

- **Instructions** or **commands** in pseudocode are the basic steps used to express an algorithm. The types of instructions introduced, together with an example, are as follows.

An **assignment.**   Example: $x := 3$
(Assign x the value 3.)

A conditional.	Example: IF $x = 3$ THEN $y := 0$ (Execute $y := 0$ provided $x = 3$.)
Alternative form:	IF $x = 3$ THEN $y := 0$ OTHERWISE $y := x$
End Algorithm.	Example: End (The algorithm terminates immediately.)
Comments.	Example: [x is greater than y here] (Explanatory note that does not affect the algorithm.)
A FOR loop.	Example: FOR $i := 1$ to n $x_i := i^2$ (Sets $x_i = i^2$ for $i = 1, 2, \ldots, n$)
Alternative form:	FOR $i := n$ down to 1 $x_i := x_i + 1$
A WHILE loop.	Example: WHILE $x > 0$ $\qquad x := x - 1$ $\qquad y := y^2$ (Keeps squaring y and decreasing x by 1 for as long as $x > 0$.)

- **Inputs** are values that the algorithm can use to do its computation. **Outputs** are values that the algorithm finds.

- A **command list** is a sequence of commands, all of which must be performed if part of a conditional, a FOR loop, or a WHILE loop.

- A **condition** or a **statement** is the comparison used following the IF in a conditional statement or the WHILE in a WHILE statement.

- The **time** of an algorithm is the number of operations done by the algorithm. These can be assignments or comparisons. Time depends on the algorithm and the input. The **worst-case** time is the largest time possible over all inputs of a given size. The **best-case** time is the least time possible over all inputs of a given size. The **average-case** time is the average time used over all inputs of a given size.

- The **size** of an input is a measure of how complicated the input is. It is usually given as an integer n. Possibilities are the length of a sequence, the number of digits of an input number, and so on.

- **Little Oh notation** $f(n) = o(g(n))$ [read "$f(n)$ is little Oh of $g(n)$"] means that if ϵ is any positive number, $|f(n)| \leq \epsilon g(n)$ if n is sufficiently large.

- **Big Oh notation** $f(n) = O(g(n))$ [read "$f(n)$ is big Oh of $g(n)$"] means that for some constant K, $|f(n)| \leq Kg(n)$ if n is sufficiently large.

- $f(n)$ and $g(n)$ have the **same order of magnitude** if $f(n) = O(g(n))$ and $g(n) = O(f(n))$.

- In the sequence $\log n$, $\sqrt{n}$, n, $n \log n$, n^3, $n^{\sqrt{n}}$, 2^n, 10^n, $n!$, each one is little Oh of the next.

- If $a > 1$, $n^k = o(a^n)$, $a^n = o(n!)$, and $\log_a n = o(n)$.

- Description and worst-time for
  Division algorithm for dividing n by b:
  $O(n)$ for primitive version.
  $O(\log n)$ for long division method.
  Binary search algorithm for searching an alphabetized list:
  $O(\log n)$.
  Base b conversion algorithm for converting n to base b:
  $O(\log n)$ if division takes no time.
  $O((\log n)^2)$ if long division is not built in.

- Sorting algorithm worst times:
  MaximumLast sort   Methodically pulling the maximum element to the end of the sequence.
  Time: $O(n^2)$, base on about $n^2/2$ comparisons.

- **InsertSort**   Sorting a sequence by methodically inserting new elements into a partially sorted sequence.
  Time: $O(n^2)$, counting assignments and comparisons.
  $O(n \log n)$, counting comparisons only.

- **CheckSort**   Informal way of sorting a sequence based on InsertSort, counting comparisons only.
  Time: $O(n \log n)$

- **MergeSort**   A recursive sorting procedure, halving the sequence, merge-sorting each half, and recombining the halves.
  Time: $O(n \log n)$

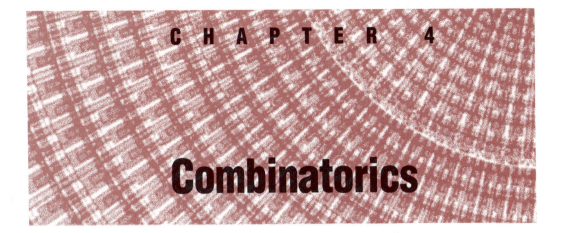

CHAPTER 4

Combinatorics

How does one count the number of elements in a set? One answer is one by one. As we shall see, this direct method is seldom used. Various techniques, including the inclusion-exclusion method, recursion, and generating functions, as well as properties of binomial coefficients, are used. Some of these techniques are quite sophisticated and can solve counting problems that would otherwise appear hopelessly difficult. They will all be introduced and illustrated in this chapter, along with some of the more basic techniques for counting.

4.1 Basic Counting Principles

It is often necessary to count many objects, but it would take too long to count them explicitly. For example, it is doubtful whether anybody actually has counted, one by one, the number of bytes of memory in a personal computer, although this is, in theory, physically possible. We know that a 30-ft by 50-ft area can be tiled with 1500 tiles 1 ft by 1 ft, but nobody actually counts them because a shortcut was learned in grade school. In principle, any finite set A can be counted, one by one, to determine the number $n = |A|$ of elements in A. In practice, however, n may be very large or A may be described in a complicated way. Therefore, shortcuts are essential. The principles introduced in this chapter will go a long way toward helping to find the cardinality of many sets.

EXAMPLE 4.1.1

How many two-letter strings (using the alphabet $A, B, \ldots, Z$) begin with a consonant and end with a vowel? The vowels are A, E, I, O, U. A consonant is a letter that is not a vowel.

Method   There are 21 ($= 26 - 5$) consonants. For any consonant (say B) there are five possible strings, one for each vowel that follows this consonant (for example, BA, BE, BI, BO, BU). Thus, there are $21 \cdot 5 = 105$ strings.

We can think of the solution to this simple problem in the following way. The required string is constructed in two stages. For the first stage, we choose a consonant to start the string (21 possibilities). For the second, we choose a vowel to end it (5 possibilities). We multiply $21 \cdot 5 = 105$ for the total number of strings. This example generalizes as follows.

The Multiplication Principle

Suppose a process is done in two stages. Suppose there are m ways of doing stage 1 and that, for each choice at stage 1, there are n ways of doing stage 2. Then there are a total of mn ways of performing the process.

This principle generalizes to processes that are done in more than two stages. Suppose a process is done in k stages and that, at stage i, there are n_i ways of doing stage i (no matter what was done at the previous stages). Then there are a total of

$$n = n_1 n_2 \cdots n_k$$

ways of performing the process.

Warning   We have assumed that a process can be performed in two or more stages. **We must also assume that it is possible to uniquely determine the process used at stage 1, stage 2, and so on, once the final outcome is given.** Another way of saying this is that there can be only one possible path to the final outcome. We shall show why this warning is necessary after giving a few more examples of how the Multiplication Principle works.

To use the Multiplication Principle, we think of a set as defined in stages.

EXAMPLE 4.1.2

A code number is needed to open a certain file. Unfortunately, this number is lost. However, the user remembers that it was a four-digit decimal number that didn't start with 0 and that all of its digits were different. A search must now be performed to try all possible code numbers. At most, how many code numbers must be tried?

Method   Think of the number being built up one digit at a time. There are nine choices for the first digit (any digit except 0), then nine choices for the second digit, because any of the nine remaining digits can be used once the first digit is chosen. There are then eight choices for the third digit and seven for the fourth. Thus, there are $9 \cdot 9 \cdot 8 \cdot 7 = 4536$ possible code numbers. The extra bit of knowledge that the digits were different cut the possibilities almost in half. (If repeated digits were allowed, there would have been $9 \cdot 10 \cdot 10 \cdot 10 = 9000$ possible numbers.)

Remark   Nothing in this problem indicated stages of a process. It was left to us to take the concept of different digits and to think of building such a four-digit number stage by stage. If we had decided to build the number from the last digit on, we would have been in trouble when we got to the first digit. For then we could not state how many choices we had for the first digit; it would depend on whether 0 was chosen for the other digits. The Multiplication Principle assumes that the number of choices at each stage does not depend on the selections at the previous stages.

EXAMPLE 4.1.3

(a) How many strings of length n are formed using an alphabet of b letters?
(b) How many integers in base b have n digits with no leading 0?
(c) Suppose an integer in some computer language is stored with 16 bits. How many integers can be stored?
(d) Suppose a real number is stored in a computer with 32 bits. How many real numbers can be stored?

Method   (a) The string can be regarded as being formed in n stages, where each letter is chosen one at a time. There are b choices at each stage, so the required answer is b^n. (Compare with Example 2.3.7, where $b = 2$.)

   (b) As in part (a), except that at stage 1 there are only $b - 1$ choices. Thus, the answer is $(b - 1)b^{n-1}$.

   (c) By part (a) there are 2^{16}, or 65,536, possible integers stored. Since negative integers must be stored, such a configuration usually defines an "integer" to be between $-32,767$ and $32,768$. Here, 2 bytes are used to store an integer. The number $2^{16} = 65,536$ is also called 64K. Here, K is not 1000 but $2^{10} = 1024$. Since $64 = 2^6$, $64K = 2^6 \cdot 2^{10} = 2^{16}$.

(d) Thirty-two bits yield 2^{32} ($\approx$4.3 billion) possible reals that can be stored.

In part (d), nothing is stated about *how* the real numbers are stored. The exercise shows how many different real numbers can be stored if 32 bits are allotted for the storage.

We now show why our earlier warning was given. Suppose we want to know how many sets of two elements can be formed with the digits 0 through 9. An incorrect approach is to say that there are ten choices for the first element and nine for the second, yielding a total of 90 choices. The fallacy is that when a set such as {3, 5} is given, we have no way of determining which digit (3 or 5) was chosen first, so there is no unique path to the given set. In the next section we learn how to do this sort of counting. In Section 6.4, the Multiplication Principle will be stated in a more precise form.

The Addition Principle is a restatement of Theorem 1.1.16.

The Addition Principle

If A and B are disjoint finite sets and $C = A \cup B$, then

$$|C| = |A| + |B| \qquad (4.1.1)$$

This is most conveniently restated for several disjoint sets as follows.

THEOREM 4.1.4

If a finite set A is partitioned into the sets $A_1, A_2, \ldots, A_k$, then

$$|A| = |A_1| + |A_2| + \cdots + |A_k| \qquad (4.1.2)$$

∎

Remark   Recall that the sets A_i must be pairwise disjoint in the partition of A.

EXAMPLE 4.1.5

How many integers between 100 and 100,000 inclusive are there that use different odd digits?

Method   This is easier if we know the number of digits in the number. There are either three, four, or five digits (the cases). Thus, we create a **case analysis.**

Case 1   The number has three digits. Building the number up, digit by digit, we have (since there are five odd digits) five choices for the first digit, four for the second, and three for the third. Thus there are $5 \cdot 4 \cdot 3 = 60$ cases.

Case 2   The number has four digits. Hence, we have $5 \cdot 4 \cdot 3 \cdot 2 = 120$ cases.

Case 3   For a five-digit number, we have $5 \cdot 4 \cdot 3 \cdot 2 \cdot 1 = 120$ cases.

Finally, the Addition Principle gives $60 + 120 + 120 = 300$ numbers of the required type.

Remark   The cases three digits, four digits, and five digits are clearly mutually exclusive. Thus, the required set of integers was partitioned into the three-digit, four-digit, and five-digit integers, and Equation (4.1.2) was used.

Sometimes it is much easier to find the number of elements *not* satisfying a condition. For this, we have to use complements of sets.

THEOREM 4.1.6

If $A \subseteq B$ (where B is finite), then

$$|B - A| = |B| - |A| \qquad\qquad (4.1.3)$$

Proof   Set B is the union of $B - A$ and A. (In words, an element is in B if and only if it is in B but not A, or if it is in A.) Therefore, by the Addition Principle,

$$|B| = |B - A| + |A|$$

But this is equivalent to Equation (4.1.3).   ■

As a special case, if $B = U$, the universe, we have

$$|A'| = |U| - |A| \qquad\qquad (4.1.4)$$

Remark   We can dignify this result by calling it the Subtraction Principle. We used it, for example, in Example 4.1.1, when we found the number of consonants. We illustrate with a simple example.

EXAMPLE 4.1.7

Using the alphabet $\{A, B, \ldots, Z\}$, how many strings of length 3 contain at least one vowel?

Method   It is simpler to find how many do not contain any vowel. Here, each letter must be a consonant, so we have 21 choices for each letter.

Therefore there are 21^3 such strings. But there are 26^3 strings in all. Thus, there are $26^3 - 21^3 = 8315$ strings of the required type.

A direct approach is more complicated because it involves the number of vowels used and the position in which they are used. (See Exercise 4.1.20.)

We can now generalize the Addition Principle. In this version, sets A and B are allowed to overlap.

THEOREM 4.1.8

If A and B are finite sets, then

$$|A \cup B| = |A| + |B| - |A \cap B| \tag{4.1.5}$$

Proof The proof uses Equation (4.1.3). We write $A \cup B$ as a disjoint union by noting that an element is in $A \cup B$ if and only if it is in A or in $B - (A \cap B)$. (See Figure 4.1.1.) Thus, we can use the (ordinary) Addition Principle to obtain

$$
\begin{aligned}
|A \cup B| &= |A \cup (B - (A \cap B))| \\
&= |A| + |B - (A \cap B)| \\
&= |A| + |B| - |A \cap B| \quad \blacksquare
\end{aligned}
$$

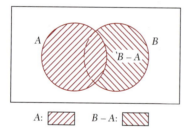

$A:$ $B - A:$

Figure 4.1.1

Equation (4.1.5) is used when a set is defined by cases that are not necessarily mutually exclusive. This equation will be generalized in Theorem 4.1.18 as the Inclusion-Exclusion Formula.

EXAMPLE 4.1.9

How many 8-bit binary numbers begin with 11 or end with 1?

Method We break into cases that are *not* mutually exclusive.

Case 1 The number begins with 11. Here there are six other binary digits, and so there are $2^6 = 64$ numbers.

Case 2 The number ends with 1. Now there are seven digits to choose, so there are $2^7 = 128$ numbers.

Finally, the overlap (intersection) of the cases considers numbers that begin with 11 *and* end with 1. Five digits are still free to be selected, so we have $2^5 = 32$ cases.

Using Theorem 4.1.8, we have $64 + 128 - 32 = 160$ numbers in all.

The following is another illustration of the generalized Addition Principle.

EXAMPLE 4.1.10

In a class of 125 people, 35 are smart and 95 are silly. The class has 10 silly smart people. How many are neither silly nor smart?

Method If SM is the set of smart people in the class and SI is the set of silly people, we have

$$|SM| = 35, \quad |SI| = 95, \quad |SM \cap SI| = 10$$

Therefore, by (4.1.5),

$$|SM \cup SI| = 35 + 95 - 10 = 120$$

This is the number of people who are either silly or smart. Since there are 125 people in the class, this leaves $125 - 120 = 5$ people who are neither silly nor smart.

The Venn diagram of Figure 4.1.2 gives the cardinality of the various sets in this problem. It can be used to do the problem. We can fill in the number for $SM \cap SI$ because it was given. This leads quickly to the count of the sets $SM - SI$ and $SI - SM$. Since we know the universe U has 125 elements, the remaining (required) set can be counted.

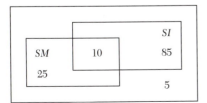

Figure 4.1.2

EXAMPLE 4.1.11

Example 4.1.10 used the fact that the complement of the set of students who were either silly or smart was the set of students who were neither silly nor smart. In the language of sets, $(SI \cup SM)' = SI' \cap SM'$. This is De Morgan's Law for sets (Theorem 1.3.2). The following results show the effects of complementation on various set operations:

$$(A \cup B)' = A' \cap B' \qquad \text{(De Morgan's Law)}$$
$$(A \cap B)' = A' \cup B' \qquad \text{(De Morgan's Law)}$$
$$(A')' = A \qquad \text{(Double Complement Law)}.$$
$$\text{If } A \subseteq B, \text{ then } B' \subseteq A' \qquad \text{(Reversed Inclusion Law)}$$

Each law may be verified by a Venn diagram or by using the various set operations as in Exercises 1.3.7 and 1.3.8. We shall give an *algebraic* verification of these laws, and they will also be discussed from yet another point of view in Section 7.2.

We now interpret these results of set theory with the help of characteristic functions, and we derive the generalized addition formula for more than two sets. Recall that the characteristic function f_A of a set A is a function from U (the universe) into $\mathbb{Z}_2$ such that $f_A(x) = 1$ if and only if $x \in A$ (see Definition 1.1.37). Our plan of action now is to try to replace facts about sets with facts about their characteristic functions. Characteristic functions are numerical, so the analysis is easier because we have arithmetic and algebra at our disposal. We first relate characteristic functions to unions, intersections, inclusion, and complementation.

THEOREM 4.1.12

Let f_A and f_B be the characteristic functions of A and B. Then

$$f_{A \cap B} = f_A f_B \tag{4.1.6}$$
$$f_{A \cup B} = f_A + f_B - f_A f_B \tag{4.1.7}$$
$$f_{A'} = 1 - f_A \tag{4.1.8}$$
$$A \subseteq B \quad \text{iff} \quad f_A \leq f_B \tag{4.1.9}$$

Proof of (4.1.6) If $x \in A \cap B$, then $f_{A \cap B}(x) = 1$. In this case, $x \in A$ and $x \in B$, so $f_A(x)f_B(x) = 1 \cdot 1 = 1$ also. If $x \notin A \cap B$, then either $x \notin A$ or $x \notin B$, so both sides of Equation (4.1.6) are equal to 0 in this case. Thus (4.1.6) is always true.

***Proof* of (4.1.7)**   This is similarly true by a case analysis. If $x \in A$ and $x \in B$, the right-hand side is $1 + 1 - 1 = 1$, and so is the left-hand side. If x is in one set but not the other, both sides are similarly equal to 1. Finally, if x is not in A or B, both sides are 0.

***Proof* of (4.1.8)**   If $x \in A$, then $f_A(x) = 1$ and $f_{A'}(x) = 0$. Thus, $f_{A'}(x) = 1 - f_A(x)$. The proof is similar for $x \notin A$.
  We leave (4.1.9) to the reader.   ∎

EXAMPLE 4.1.13

Using the results in Theorem 4.1.12, we can give an *algebraic* proof of De Morgan's Laws. For example, to find $(A \cup B)'$, we can work with characteristic functions:

$$
\begin{aligned}
f_{(A \cup B)'} &= 1 - f_{A \cup B} \\
&= 1 - (f_A + f_B - f_A f_B) \\
&= 1 - f_A - f_B + f_A f_B \\
&= (1 - f_A)(1 - f_B) \\
&= f_{A'} f_{B'} \\
&= f_{A' \cap B'}
\end{aligned}
$$

This shows that $(A \cup B)' = A' \cap B'$, since they both have the same characteristic function. The other De Morgan Law can be proved similarly.

The number of elements in a set A can also be related to its characteristic function.

DEFINITION 4.1.14

If f is any function on a finite universe U, define

$$
\Sigma f = \sum_{x \in U} f(x) = \sum_{i=1}^{N} f(x_i)
$$

as the sum of all values of $f(x)$ for all x in U. In the latter summation, the elements of U are listed as $x_1, \ldots, x_N$.

We use this definition when $f(x)$ is a characteristic function, but the definition is also valid when f is any real-valued function. The symbol

Σ has some familiar properties of summation:

$\Sigma(f + g) = \Sigma f + \Sigma g.$
$\Sigma cf = c\,\Sigma f,$ if c is constant.
$\Sigma 1 = |U| = N,$ the number of elements in $U.$

The number of elements in A can now be expressed in terms of its characteristic function.

THEOREM 4.1.15

$$|A| = \Sigma f_A \qquad\qquad (4.1.10)$$

Proof   The right side is the sum of 0's and 1's. There are as many 1's as there are elements of A. ∎

Theorems 4.1.12 and 4.1.15 have the power to generate simple, algebraic proofs of some of our previous results.

EXAMPLE 4.1.16

$$|A'| = \Sigma f_{A'}$$
$$= \Sigma(1 - f_A)$$
$$= |U| - |A|$$

This is Equation (4.1.4).

EXAMPLE 4.1.17

Similarly, we have the following, more algebraic, proof of Theorem 4.1.8.

$$|(A \cup B)| = \Sigma f_{A \cup B}$$
$$= \Sigma(f_A + f_B - f_A f_B)$$
$$= \Sigma f_A + \Sigma f_B - \Sigma f_{A \cap B}$$
$$= |A| + |B| - |A \cap B|$$

We now use the same technique for the union of a finite number of sets. This general result is called the **Inclusion-Exclusion Formula.** It is so called because Equation (4.1.5) may be thought of as follows: To find $|A \cup B|$, we proceed to add the number of elements of A to the number of elements of B. But this includes too much, namely the elements common to both. So we exclude the overlap, which has been counted twice, and subtract the extra term. As we shall see, this process continues for three or more sets.

THEOREM 4.1.18 *The Inclusion-Exclusion Formula*

Let $A_1, A_2, \ldots, A_n$ be n sets (subsets of U). Define

$$S_1 = \sum_i |A_i|$$

$$S_2 = \sum_{i<j} |A_i \cap A_j|$$

$$\cdot$$
$$\cdot$$
$$\cdot$$

$$S_n = |A_1 \cap A_2 \cap \cdots \cap A_n|$$

Then

$$|A_1 \cup \cdots \cup A_n| = S_1 - S_2 + S_3 + \cdots \pm S_n \qquad \textbf{(4.1.11)}$$

We now illustrate for $n = 3$. (We will clarify the definition of S_k after the proof for $n = 3$.)

Let f_i be the characteristic function of A_i. Using Formulas (4.1.6) and (4.1.8), we obtain

$$
\begin{aligned}
|A_1' \cap A_2' \cap A_3'| &= \Sigma(1 - f_1)(1 - f_2)(1 - f_3) \\
&= \Sigma(1 - f_1 - f_2 - f_2 + f_1 f_2 \\
&\quad + f_1 f_3 + f_2 f_3 - f_1 f_2 f_3) \\
&= |U| - |A_1| - |A_2| - |A_3| + |A_1 \cap A_2| \\
&\quad + |A_1 \cap A_3| + |A_2 \cap A_3| - |A_1 \cap A_2 \cap A_3| \\
&= |U| - S_1 + S_2 - S_3
\end{aligned}
$$

Using $(A_1 \cup A_2 \cup A_3)' = A_1' \cap A_2' \cap A_3'$, we also have, by (4.1.4),

$$|A_1' \cap A_2' \cap A_3'| = |U| - |A_1 \cup A_2 \cup A_3|$$

Comparing these two expressions for $|A_1' \cap A_2' \cap A_3'|$, we get

$$|A_1 \cup A_2 \cup A_3| = S_1 - S_2 + S_3$$

This is the result for $n = 3$. The general case is quite similar.

We used S_1 as the sum of all the cardinalities $|A_i|$. Similarly, S_2 was, here, $|A_1 \cap A_2| + |A_1 \cap A_3| + |A_2 \cap A_3|$, namely, the sum of the cardinalities $|A_i \cap A_j|$, with $i < j$. In general, S_2 is simply the sum of the cardinalities of all intersections of two sets A_i and A_j with different indices i and j. A similar remark applies to S_3. For example, if $n = 4$, the explicit formulas for S_2, S_3, and S_4 are

$$S_2 = |A_1 \cap A_2| + |A_1 \cap A_3| + |A_1 \cap A_4|$$
$$+ |A_2 \cap A_3| + |A_2 \cap A_4| + |A_3 \cap A_4|$$

and

$$S_3 = |A_1 \cap A_2 \cap A_3| + |A_1 \cap A_2 \cap A_4|$$
$$+ |A_1 \cap A_3 \cap A_4| + |A_2 \cap A_3 \cap A_4|$$

and

$$S_4 = |A_1 \cap A_2 \cap A_3 \cap A_4|$$

As with the generalized addition formula, this result is used when an analysis has overlapping cases. In this situation any number n of cases are possible and any may overlap.

EXAMPLE 4.1.19

The letters A, B, C, D, and E are scrambled to form five-letter strings. How many such strings can be formed? How many can be formed with either an A in the first place, a C in the third, or an E in the fifth?

Method   By the Multiplication Principle, the number of strings is clearly $5! = 5 \cdot 4 \cdot 3 \cdot 2 \cdot 1 = 120$. The condition has many cases that overlap: For example, we can have an A in the first place and C in the third. Therefore, we use the Inclusion-Exclusion Formula. Let A_1 be the set of five-letter strings where A is in the first place. Similarly, let A_3 be the set where C is in the third place, and let A_5 be the set where E is in the fifth place. We are looking for $|A_1 \cup A_3 \cup A_5|$.

We compute $|A_1| = 4! = 24$, with the same answer for $|A_3|$ and $|A_5|$. Therefore,

$$S_1 = 24 + 24 + 24 = 72$$

Computing intersections two at a time, we find that $|A_1 \cap A_3| = 3! = 6$, because after obtaining the string "$AxCxx$", we have three places to fill with the remaining three letters. Similarly, $|A_1 \cap A_5| = |A_3 \cap A_5| = 6$. Therefore,

$$S_2 = 6 + 6 + 6 = 18$$

Finally, $|A_1 \cap A_3 \cap A_5| = 2$, so

$$S_3 = 2$$

Using (4.1.11), the required result is

$$|A_1 \cup A_2 \cup A_3| = 72 - 18 + 2 = 56$$

We now consider an application to number theory. We say that two integers r and s are **relatively prime** if their only common positive factor is 1. Do not confuse relatively prime numbers with a **prime number** n, whose only positive divisors are 1 and n. For example, 15 and 28 are relatively prime, though neither is a prime number. We give an easy way to compute the number of positive integers between 1 and n, inclusive, that are relatively prime to n. For example, with $n = 10$, only 1, 3, 7, and 9 are relatively prime to 10, so there are only four such integers. What is the answer for $n = 480$? A direct count is possible, but it takes a long time. The required method is given in Theorem 4.1.21.

DEFINITION 4.1.20

If n is a positive integer, **the Euler φ function** $\varphi(n)$ is the number of integers x such that x and n are relatively prime and $1 \le x \le n$.

This function is named in honor of the mathematician Leonard Euler, who invented it.

THEOREM 4.1.21

Assume the only prime divisors of n are $p_1, p_2, \ldots, p_k$. Then

$$\varphi(n) = n(1 - 1/p_1)(1 - 1/p_2) \cdots (1 - 1/p_k) \qquad (4.1.12)$$

Proof We find the number of integers x between 1 and n inclusive that are *not* relatively prime to n and subtract these from n to find the number of integers that are relatively prime. We take for granted that a number is not relatively prime to n if and only if there is a prime number dividing both it and n. (Chapter 8 goes into this topic in more detail.) Let

$$A_1 = \text{set of integers } x \text{ divisible by } p_1 \text{ with } 1 \le x \le n$$

$$\vdots$$

$$A_k = \text{set of integers } x \text{ divisible by } p_k \text{ with } 1 \le x \le n$$

Then the integers x with $1 \leq x \leq n$ and that are divisible by at least one of the p_i constitute the set $A_1 \cup \cdots \cup A_k$. Thus, the required count is

$$\varphi(n) = n - |A_1 \cup \cdots \cup A_k| \qquad \text{(4.1.13)}$$

We now use the Inclusion-Exclusion Formula (4.1.11).

We claim that the cardinality of A_i is n/p_i. To see this, note that the integers of A_i are the numbers between a and n, inclusive, that are divisible by p_i. These numbers are all possible multiples yp_i satisfying $1 \leq yp_i \leq n$. But this is equivalent to $1/p_i \leq y \leq n/p_i$ or, since y is an integer, $1 \leq y \leq n/p_i$. There are thus n/p_i such y, so

$$|A_i| = \frac{n}{p_i}$$

(This is a commonsense answer. For example, the number of integers between 1 and 200, inclusive, that are divisible by 5 is $40 = 200/5$. These numbers are $5{\cdot}1$, $5{\cdot}2$, $5{\cdot}3$, . . . , $5{\cdot}40$, 40 in all.)

The same line of reasoning gives $|A_i \cap A_j|$ for $i \neq j$, which is the number of integers between 1 and n, inclusive, that are divisible by p_i and by p_j or, equivalently, by $p_i p_j$. We have

$$|A_i \cap A_j| = \frac{n}{p_i p_j} \qquad (i < j)$$

Similarly,

$$|A_i \cap A_j \cap A_k| = \frac{n}{p_i p_j p_k} \qquad (i < j < k)$$

and so on, for any number of distinct intersections.

We now use the full power of the Inclusion-Exclusion Formula (4.1.11) and Formula (4.1.13) to obtain

$$\varphi(n) = n - \sum_i \frac{n}{p_i} + \sum_{i<j} \frac{n}{p_i p_j} - \sum_{i<j<k} \frac{n}{p_i p_j p_k} + \cdots$$

But, by algebra, this is precisely the expansion of Formula (4.1.12). ∎

EXAMPLE 4.1.22

Evaluate $\varphi(162)$.

Method Since $162 = 2{\cdot}81 = 2{\cdot}3^4$, the only prime divisors of 162 are 2 and 3. Thus, by (4.1.12), $\varphi(162) = 162{\cdot}\frac{1}{2}{\cdot}\frac{2}{3} = 54$.

Exercises

1. The menu in the EatFast restaurant has three choices for soup, four choices of appetizers, five main dishes, and three desserts. How many different meals can Betty eat at this restaurant? (Betty must have a main dish and a dessert, but she can skip the rest.)

2. How many different four-digit (decimal) numbers do not have leading 0's? Of these, how many
   (a) begin with a digit 5 or more?
   (b) begin with a digit 5 and end with an odd digit?
   (c) begin with an even digit and have different digits?
   (d) do not contain the digit 1 or 2?
   *(e) contain the digit 1 or 2 but not both?

3. An apartment has four rooms to paint. Three colors are available. How many different color schemes are there?

4. A computer language allows the name of a variable to be one or two letters long. (The letters are capital letters from A to Z.) How many different variable names are there?

*5. How many different four-letter strings on the letters A through M do not have two consecutive letters the same? How many have some two consecutive letters the same?

6. How many strings on the alphabet $\mathbb{Z}_2$ of length 5 contain the substring "111"? (A **substring** of a string is a string that is part of the string. Thus, "*MEN*" is a substring of "*WOMEN*", and "*AT*" is a substring of "*BRATS*". By definition, the null string ϵ is a substring of any string, and any string is a substring of itself.)

7. Repeat Exercise 6 for strings of $\mathbb{Z}_2$ of length 8, which contain the substring 0110.

8. J. L. Borges, in his short story "The Library of Babel," tells of a library containing all possible books (including nonsense and meaningless ones). Each book has 410 pages; each page has 40 lines; each line has some 80 letters. Further, exactly 25 letters are used: 22 characters, 2 punctuation marks, and a space. Borges notes that the number of different books in his library is finite, though very large. How large?

9. An anagram on the letters *AEBRS* is some rearrangement of these to form a five-letter word.
   (a) A software package considers all rearrangements and checks each such rearrangement in its dictionary to determine whether it is a genuine word in the English language. How many words are checked?
   (b) The software will also find two-, three-, and four-letter words made up of these letters. (No repetitions are allowed.) How many additional words must it check?

10. John classifies movies as dull or interesting, long or short, musical or nonmusical, and as must see, don't see, or see if you have a chance. How many classifications does John have?

11. Mary remembers that a house is located on Elm Street. She remembers that the number of the house has three digits and that it either begins or ends with 4. How many possibilities are there?

*12. Generalize the proof of Formula (4.1.7) to sets A, B, and C to show that

$$f_{A \cup B \cup C} = f_A + f_B + f_C - f_A f_B - f_A f_C - f_B f_C + f_A f_B f_C$$

Explain how this formula gives the Inclusion-Exclusion Formula for $n = 3$.

*13. Give a direct proof of the Inclusion-Exclusion Formula for three sets, using Theorem 4.1.8. Thus, show that

$$|A \cup B \cup C| = |A| + |B| + |C| - |A \cap B| - |A \cap C|$$
$$- |B \cap C| + |A \cap B \cap C|$$

by writing $A \cup B \cup C$ as $(A \cup B) \cup C$ and using Theorem 4.1.8 repeatedly.

14. A class of 100 students has 54 women, 48 computer science majors, and 12 men who are also computer science majors. How many of the women are not computer science majors?

15. Carol's library of 124 books contains math books, great books, and paperbacks. She has 59 math books, 24 great books, 73 paperbacks, 16 math books in paperback, 13 great paperback books, and 8 great math books. How many great math paperbacks does she have?

*16. How many different rearrangements of *ABCDE* are there in which at least one of the letters does not move? (In terms of permutations, how many permutations on $\mathbb{I}_5$ have a fixed point? A fixed point of a permutation $\sigma: A \to A$ is an element of A such that $\sigma(x) = x$.)

*17. How many different rearrangements of the letters *FRUIT* are there in which no letter occupies its original position? (Such an arrangement is called a **derangement** of the original string.)

*18. How many rearrangements of the letters *ABSOLUTE* are possible in which no vowel $\{A, E, O, U\}$ occupies its original position?

19. An auto show displays 69 cars, of which 25 are red cars, 31 are sports cars, and 40 are expensive cars. Of these, 8 are red sports cars, 10 are expensive red cars, and 13 are expensive sports cars. How many expensive red, sports cars are there?

*20. How many four-digit numbers have a "1" in the thousands' place, a "2" in the hundreds' place, a "3" in the tens' place, or a "4" in the unit's place? The number has no leading zeros. For example, 3261 is such a number.

21. Do Example 4.1.7 directly, using the Inclusion-Exclusion Formula.

22. How many strings on the alphabet $\{A, B, C\}$ of length 6 use each letter at least once? (*Hint:* Use complements.) How many of length n, where $n \geq 3$?

23. Redo Exercise 22, but use the letters $\{A, B, C, D\}$. Do for strings of length 6 and for strings of length $n \geq 4$.
24. How many relations are on $\mathbb{I}_5$?
25. How many reflexive relations are on $\mathbb{I}_5$?
26. How many symmetric relations are on $\mathbb{I}_5$?
27. How many equivalence relations R are on $\mathbb{I}_5$ in which $1R2$ and $3R4$?
28. How many functions $f: \mathbb{I}_2 \to \mathbb{I}_5$ are there? How many functions send $\mathbb{I}_5$ into $\mathbb{I}_2$?
29. How many functions $f: \mathbb{I}_5 \to \mathbb{I}_2$ are nondecreasing? [Nondecreasing means that if $a < b$ then $f(a) \leq f(b)$.]

Exercises 30–34 concern the symmetric difference of two sets. Some of these are verified in Section 7.2, where the symmetric difference plays an important role. The **symmetric difference** $A \triangle B$ of A and B is defined as the set of points in A or in B but not in both A and B.

30. Find the characteristic function of $A \triangle B$ in terms of the characteristic functions of A and B.
*31. Show that $\triangle$ is associative [i.e., $(A \triangle B) \triangle C = A \triangle (B \triangle C)$] by showing that the characteristic functions of both sides of this equation are equal.
32. Draw a Venn diagram for $A \triangle B \triangle C$ and give a simple criterion for an element to be a member of this set.
33. Find a formula for $|A \triangle B|$, using characteristic functions.
*34. Generalize the previous exercise to find the number of elements in the symmetric difference of three sets. Guess the answer for n sets.

35. Find $\varphi(360)$ and $\varphi(1792)$.
36. Find $\varphi(540)$ and $\varphi(1232)$.
37. Show that if n is odd, $\varphi(2n) = \varphi(n)$.
38. Express $\varphi(2n)$ in terms of $\varphi(n)$ when n is even.
39. Find all numbers n between 100 and 1000 such that $\varphi(n) = 6n/7$.
40. For what numbers n is $\varphi(n) = n - 1$? Explain.

4.2 Permutations and Combinations

In Definition 2.2.14, permutations were defined as 1-1, onto functions. One application of this definition was that a permutation σ could be regarded as a rearrangement. For example, the permutation $\sigma = (1345)$ rearranges the string $ABCDE$ into $EBACD$. (Because $1 \to 3 \to 4 \to 5 \to 1$, the letter in the first position is moved to the third, the letter in the third position is moved to the fourth, and so on.) Classically, the word "permutation" is defined as an arrangement as well as a function. We now give this definition of permutation. It also generalizes our previous concept, because not all letters need to be used.

DEFINITION 4.2.1

A **permutation** of n objects is a listing of these objects in a definite order. If $r \leq n$, an **r-permutation** of n elements is a listing of r of these objects in a definite order. In each case, the listing must consist of distinct objects.

For example, if $n = 4$ and the objects are A, B, C, and D, then $ACDB$ is a permutation of these elements, and DA is a 2-permutation of these four objects.

More precisely, an r-permutation of the n objects in the set Σ is a 1-1 function from $\mathbb{I}_r$ into Σ. For the rest of the text, unless otherwise specified, we use the term "permutation" as defined in Definition 4.2.1.

DEFINITION 4.2.2

An **r-combination** of n objects is a subset of these objects consisting of r distinct elements.

For example, $\{A, C\}$ is a 2-combination of the four objects A, B, C, D. Because a set depends only on the elements in it and not on the *order* in which these elements are chosen, $\{C, A\}$ is the same combination as $\{A, C\}$, and we say that the *combinations* AC and CA are identical, whereas the *permutations* AC and CA are different.

EXAMPLE 4.2.3

In a baseball league consisting of ten teams, the top three teams are a 3-combination of the set of ten teams. The final standings of the top three (first, second, and third places) are a 3-permutation of the set of ten teams.

EXAMPLE 4.2.4

Find all 2-permutations and all 2-combinations of the set $\mathbb{I}_5$.

Method The objects are 1, 2, 3, 4, 5. The 2-permutations starting with 1 are 12, 13, 14, 15. In the same way, we may find the permutations

Table 4.2.1

—	12	13	14	15
21	—	23	24	25
31	32	—	34	35
41	42	43	—	45
51	52	53	54	—

starting with 2, 3, 4, and 5. All of the 2-permutations are listed in Table 4.2.1. The 2-combinations are also in this table. Because ij and ji represent the same combination, all 2-combinations are in the upper half of this table (i.e., above the diagonal, which is indicated by the dashes). The 2-combinations are therefore 12, 13, 14, 15, 23, 24, 25, 34, 35, and 45.

This example had ten 2-combinations of the five numbers and twenty 2-permutations. We write these as $C(5, 2) = 10$ and $P(5, 2) = 20$, respectively.

DEFINITION 4.2.5

The symbol $P(n, r)$ is the number of **r-permutations** of a set with n elements, and $C(n, r)$ is the number of **r-combinations** of a set with n elements.

Sometimes $P(n, r)$ and $C(n, r)$ are written $_nP_r$ and $_nC_r$, respectively. The notation

$$\binom{n}{r}$$

is also used for $C(n, r)$. The symbol $C(n, r)$ is sometimes called "n choose r," because it represents how many ways r elements can be chosen from among n elements.

The following theorem gives formulas for $P(n, r)$ and $C(n, r)$.

THEOREM 4.2.6

$$P(n, r) = n(n - 1) \cdots (n - r + 1) = \frac{n!}{(n - r)!} \qquad (4.2.1)$$

$$C(n, r) = \frac{P(n, r)}{r!} = \frac{n!}{r!(n - r)!} \qquad (4.2.2)$$

For example, $P(8, 3) = 8 \cdot 7 \cdot 6 = 336$, and $C(8, 3) = \dfrac{8 \cdot 7 \cdot 6}{3 \cdot 2 \cdot 1} = 56$.

Proof The proof of Equation (4.2.1) uses the Multiplication Principle. An r-permutation is formed in r stages, namely the first element in the list, the second, . . . , and the rth. Because the elements must be different, there are n choices for the first element, $n - 1$ for the second, . . . , and $n - r + 1$ for the rth. The Multiplication Principle gives the formula. As a special case of (4.2.1), we have

$$P(n, n) = n! \qquad (4.2.3)$$

Here, $P(n, n)$ is simply the number of permutations of n elements.

The proof of Equation (4.2.2) is indirect. We find $P(n, r)$ in *two* stages, as follows.

First, choose the r elements to be arranged. By definition, there are $C(n, r)$ possibilities. *Next* arrange those r elements. By the formula $P(r, r) = r!$, there are $r!$ possibilities at this stage. Therefore, by the Multiplication Principle,

$$P(n, r) = C(n, r) \cdot r!$$

This gives Equation (4.2.2). ∎

Remark The division in Equation (4.2.2) can also be understood as follows. All of the $C(n, r)$ r-combinations are represented by the $P(n, r)$ r-permutations, but there are duplications. For example, in Example 4.2.4, 12 and 21 were two different permutations but represented the same combination. In this example, the number of combinations was exactly half the number of permutations. In general, there are $r!$ duplications of any combination, so it is necessary to divide by $r!$ to obtain the correct answer.

As a special case, Definition 4.2.5 gives

$$C(n, 0) = C(n, n) = 1 \quad \text{for all } n \geq 0 \qquad \textbf{(4.2.4)}$$

In particular, $C(0, 0) = 1$. Formula (4.2.2) is consistent with Equation (4.2.4), using $0! = 1$.

EXAMPLE 4.2.7

How many 8-bit strings have exactly four 0's and four 1's? Of these, how many begin with 1?

Method There are eight places to be filled with 0's and 1's. It is necessary to choose four places in which to put 0's. (The 1's will go into the other four places.) Therefore, four places must be chosen from among the eight. The order in which the places are chosen is irrelevant. Therefore, there are $C(8, 4)$ choices:

$$C(8, 4) = \frac{8 \cdot 7 \cdot 6 \cdot 5}{4 \cdot 3 \cdot 2 \cdot 1} = 70$$

by Equation (4.2.2).

For the second part of the problem, we choose four 0's from the remaining seven spots. There are $C(7, 4)$ such strings:

$$C(7, 4) = \frac{7 \cdot 6 \cdot 5 \cdot 4}{4 \cdot 3 \cdot 2 \cdot 1} = 35$$

We could have chosen three 1's from the remaining seven spots. This also yields $C(7, 3) = 35$ strings.

In this example, the answers $C(7, 4)$ and $C(7, 3)$ are equal because both counted the same set, although by different methods. This result can be generalized.

THEOREM 4.2.8

$$C(n, r) = C(n, n - r) \qquad (4.2.5)$$

Proof By Equation (4.2.2), we have

$$C(n, n - r) = \frac{n!}{(n - r)!(n - (n - r))!} = \frac{n!}{r!(n - r)!} = C(n, r)$$

A direct proof can be given by using the definition. The number $C(n, r)$ is the number of ways that r objects can be chosen from n objects. To determine which r objects are chosen, we can decide which $n - r$ objects are *not* chosen. There are $C(n, n - r)$ choices here, and they determine the same $C(n, r)$ choices. ■

For example, $C(20, 18) = C(20, 2) = 20 \cdot 19/2 = 190$, which is a far easier way to obtain the answer than the formula, which gives the computation $P(20, 18)/18!$.

EXAMPLE 4.2.9

In an alphabet of 26 letters, how many 5-letter strings consist of different letters in alphabetic order?

Method If it weren't for the alphabetic order, the answer would be simply $P(26, 5)$. Attempting this problem by the Multiplication Principle doesn't work very well. (After the first letter is chosen, the choices for the second letter depend very much on the first.) However, we can solve this problem by combinations. Simply find the number of 5-combinations of the alphabet. Once a combination is chosen, there is a unique way of alphabetizing it. Therefore, the answer is $C(26, 5)$ ($= 65,780$). The method is noteworthy because the problem calls for strings—an ordered listing of letters—yet the answer is found using combinations.

Card Games, Coin Tossing, and Probability

The analysis of card games has historically been a strong motivation for developing counting methods of the type discussed in this section. A card hand usually consists of distinct cards (either ordered or unordered according to the game), all chosen from a deck (set) of 52 cards. There are four **suits:** hearts, diamonds, clubs, and spades. There are 13 **denominations,** called ace (or 1); the numbers 2 through 10; and jack, queen, and king. Each possible suit occurs with each denomination, so there are $4 \cdot 13 = 52$ cards in all. We now consider some aspects of the game of poker.

EXAMPLE 4.2.10

A **poker hand** is an unordered set of 5 cards chosen from an ordinary deck of 52. How many poker hands are possible? How many of these are **full houses**? How many are **two pairs**? How many are **one pair**?

Method   First, some definitions for those unfamiliar with this game. A full house is a hand consisting of three of one denomination and two of another: for example, three 8's and two jacks. Two pairs is a hand consisting of two (only) of one denomination and two (only) of another: for example, two 8's, two aces, and a king. Similarly, a one-pair hand consists of two (only) of any one denomination, and three cards of different denomination from each other: for example, 887K9. We now address the problem.

(a) *Number of Hands*   Since unordered hands are called for, the answer is $C(52, 5) = 2,598,960$ different hands.

(b) *Full Houses*   A full house can be determined in stages.

1. Choose the denomination of the triple.
2. Choose the denomination of the pair.
3. Choose the specific cards in the triple.
4. Choose the specific cards in the pair.

The number of choices for stage 1 is 13. The number of choices for stage 2 (after stage 1 is done) is 12. The number of choices for stage 3 is $C(4, 3) = 4$. Finally, the number for stage 4 is $C(4, 2) = (4 \cdot 3)/2 = 6$.

Thus, by the Multiplication Principle, the number of full houses is $13 \cdot 12 \cdot 4 \cdot 6 = 3744$.

(c) *Two Pairs*   Again, the Multiplication Principle is used. First, decide which two denominations will be used for the two pair. There are $C(13, 2) = 78$ choices. Deciding which two cards are to be used for the lowest pair gives $C(4, 2) = 6$ choices. Similarly, there are six choices for the other pair. Finally, the odd card must be selected from $52 - 8 = 44$ cards. Therefore, there are $78 \cdot 6 \cdot 6 \cdot 44 = 123,552$ possible two-pair hands.

(d) *One Pair*   The stages are (1) the denomination of the pair; (2) the particular cards in the denomination; (3) the three other denominations filling out the hand; and (4) the particular cards representing these denominations. The numbers are (1) 13, (2) $C(4, 2) = 6$, (3) $C(12, 3) = 220$, and (4) $4^3 = 64$. Therefore there are $13 \cdot 6 \cdot 220 \cdot 64 = 1,098,240$ pairs.

Probability

The poker player will not be surprised to learn that there are many more pairs than full houses. Full houses do not occur often. The **probability** that a pair is drawn is defined as the number of possible pairs divided by the number of possible hands: $1,098,240/2,598,960 \approx 0.423$. Thus, a pair will turn up in poker roughly 42 times out of 100. Similarly, the probability of two pair is $123,552/2,598,960 \approx 0.0475$, and the probability of a full house is 3744/2,598,960 ≈ 0.00144, or a little over 1 chance in 1000.

DEFINITION 4.2.11

The **probability** p that an event E occurs is defined by $p = S/N$, where S is the number of ways the event E can occur (the number of possible successes) and N is the total number of possible outcomes. For this definition to be valid, each outcome must be equally likely.

A simple example follows: The probability that a fair coin lands heads when it is tossed is $\frac{1}{2}$. Here, $N = 2$ because there are two possible outcomes (heads and tails), and $S = 1$ because success is possible in only one case. Thus, $p = \frac{1}{2}$. (We assume that heads and tails are equally likely.) It is obviously incorrect to say that the probability of a full house is $\frac{1}{2}$ because there are two possibilities: either a full house or no full house. In the calculation of the poker probabilities, we assumed that each hand was as likely as another. Much of the impetus for discovering the counting techniques we discuss originally came from an analysis of probability with a view toward computing the correct odds in games of chance.

EXAMPLE 4.2.12

Ten coins are tossed. What is the probability that five are heads and five are tails?

Method   If we code heads as 1 and tails as 0, a toss of ten coins may be regarded as a 10-bit string. The number of possibilities is the number of 10-bit strings: $N = 2^{10} = 1024$. The number S of successes is the number of 10-bit strings with five 0's and five 1's. Such a string is determined by choosing five places from among the ten for the heads to occur. Thus, $S = C(10, 5) = 252$, and the required probability is $S/N = 252/1024 ≈ 0.246$, or about 1 chance in 4.

The Binomial Coefficients   Thus far, we have considered two ways of looking at the number $C(n, r)$. One is the definition as the number of ways of choosing r (unordered) objects from n objects. The other is Equation (4.2.2):

$$C(n, r) = \frac{n!}{r!(n - r)!}$$

Two proofs of the equation $C(n, n - r) = C(n, r)$ were given. One used the definition. This is called the **combinatorial**, or counting, approach. The other used the formula, which is the **algebraic** approach. Often, the two methods work together.

EXAMPLE 4.2.13

We can immediately say that $99!/(17!82!)$ is an integer, because it is the formula for $C(99, 17)$, and $C(n, r)$ is clearly an integer by its combinatorial definition. Yet, we can't say the fraction is obviously an integer because it is not clear that the denominator divides the numerator.

We now introduce another way in which the numbers $C(n, r)$ can be interpreted. These numbers are coefficients occurring in the **Binomial Theorem.** For example, in the formula

$$(1 + x)^4 = 1 + 4x + 6x^2 + 4x^3 + x^4$$

the coefficients of the various powers of x are (in ascending powers of x): 1, 4, 6, 4, 1. These are precisely the numbers $C(4, r)$ as r goes from 0 to 4. The general result, which we shall shortly prove, is Theorem 4.2.14.

THEOREM 4.2.14   The Binomial Theorem

$$(1 + x)^n = 1 + nx + C(n, 2)x^2 + \cdots + C(n, r)x^r + \cdots + x^n$$

or

$$(1 + x)^n = \sum_{r=0}^{n} C(n, r)x^r \qquad (4.2.6)$$

■

This theorem will be proved by induction on n. But first, we need a preliminary result.

LEMMA 4.2.15

If $n \geq 1$ and $1 \leq r \leq n$,

$$C(n + 1, r) = C(n, r) + C(n, r - 1) \qquad (4.2.7)$$

Proof   We prove the lemma combinatorially, leaving the algebraic proof to the exercises.

Consider the $n + 1$ numbers of $\mathbb{I}_{n+1}$. The number of ways of choosing r numbers from these is $C(n + 1, r)$. This is the left side of Equation (4.2.7). We now separate cases. When $n + 1$ numbers are chosen from $\mathbb{I}_{n+1}$, the number $n + 1$ can be chosen or not chosen. If chosen, then to choose a total of r numbers we must choose the remaining $r - 1$ from $\mathbb{I}_n$. By definition, there are $C(n, r - 1)$ ways of doing this. However, if the number $n + 1$ is not chosen, we must choose r numbers from $\mathbb{I}_n$, and there are $C(n, r)$ ways of doing that. Hence, there are a total of $C(n, r)$ + $C(n, r - 1)$ ways of choosing r numbers from among the $n + 1$ numbers of $\mathbb{I}_{n+1}$. This gives the result.   ■

We can now prove the Binomial Theorem.

Proof of Theorem 4.2.14 We prove the theorem by induction. The basis $n = 0$ is trivial: $(1 + x)^0 = 1 = C(0, 0)$. Similarly, $(1 + x)^1 = 1 + x$, and the coefficients of 1 and x are $C(1, 0) = C(1, 1) = 1$.

Now assume (4.2.6) to be true for some n. We now use the algebraic identity

$$(1 + x)^{n+1} = (1 + x)^n(1 + x) \qquad (4.2.8)$$

to find the coefficient of x^r in $(1 + x)^{n+1}$. If $r = 0$, this coefficient is clearly $1 = C(n + 1, 0)$. Similarly, if $r = n + 1$, the coefficient of x^{n+1} is $1 = C(n + 1, n + 1)$. Thus, we may assume that $1 \leq r \leq n$. On the right, the term involving x^r is obtained from exactly two terms of $(1 + x)^n$—the term involving x^r (which is multiplied by 1) and the term involving x^{r-1} (which is multiplied by x). These coefficients are $C(n, r)$ and $C(n, r - 1)$, by induction. Therefore, by combining these terms, the coefficient of x^r is $C(n, r) + C(n, r - 1)$. But this is $C(n + 1, r)$ by Equation (4.2.7). This is the result for $n + 1$. ■

Another proof of the Binomial Theorem is given in Section 4.4. Because of the Binomial Theorem, the numbers $C(n, r)$ are also known as the binomial coefficients.

It is convenient to define $C(n, r)$ for $r < 0$ and for $r > n$. The definition is simply

$$C(n, r) = 0 \qquad \text{for } r < 0 \text{ and for } r > n \qquad (4.2.9)$$

These equations make Equation (4.2.7) true for all $r \in \mathbb{Z}, n \in \mathbb{N}$.

Equation (4.2.7) may be used to compute $C(n, r)$ recursively, with Equation (4.2.9) and the equation $C(0, 0) = 1$ as initial values. For example, using these equations, we can construct the entire table of values for $C(n, r)$ (see Table 4.2.2). Equation (4.2.7) means that any entry in

Table 4.2.2 Pascal's Triangle

r	0	1	2	3	4	5	6	7	•
$n = 0$	1								
1	1	1							
2	1	2	1						
3	1	3	3	1					
4	1	4	6	4	1				
5	1	5	10	10	5	1			
6	1	6	15	20	15	6	1		
7	1	7	21	35	35	21	7	1	
•	•	•	•	•	•	•	•	•	•

any row of the table (except the top row) is the sum of the entry directly above it and the one directly above it moving once to the left. Equation (4.2.9) and the formula $C(0, 0) = 1$ give the top row. (The zero entries are replaced by blanks.) Table 4.2.2 is called **Pascal's triangle.**

Ordering, Repetition, and Replacement   It is possible to classify permutations and combinations according to certain criteria that help in the analysis of a counting problem. It is convenient to think of a **population** P of size n and to take a **sample** from this population of size r. The distinction between permutations and combinations depends on whether the sample is **ordered**. For permutations, order counts; for combinations, it doesn't. In each case, however, the sample consisted of different elements.

EXAMPLE 4.2.16

A problem can sometimes be looked at in different ways. Consider counting the number of strings consisting of five 0's and five 1's. This can be regarded as taking the population $\{0, 0, 0, 0, 0, 1, 1, 1, 1, 1\}$ and finding the number of ordered samples of size 10 in it. In this case, the population is said to have **repetitions**. But the method we used was to choose as the population the ten places and to find how many ways we might choose five unordered places from them in which to place the 1's. Thus, the ordered sample 1101000101 from our population would correspond to the unordered sample $\{1, 2, 4, 8, 10\}$ from the population of places.

In general, we can classify a sampling problem as follows.

1. *Order.*   Is the sample ordered?
2. *Replacement.*   Can an element be chosen for the sample more than once?
3. *Repetition.*   Does the population consist of distinct objects, or are some of them regarded as identical?

So far, in all the problems of this section, the point of view was that there were n distinct elements (no repetition), and the elements chosen were all different (no replacement). In the previous section, we consid-

Table 4.2.3

Population Size	Sample Size	Replace?	Ordered?	Number
n	r	No	Yes	$P(n, r)$
n	r	No	No	$C(n, r)$
n	r	Yes	Yes	n^r
n	r	Yes	No	???

ered the question of how many strings of bits have length n. The answer 2^n was based on the Multiplication Principle. Such a string may be regarded as an ordered sample of size n from a population of size 2 (the binary digits), **with replacement.** We can summarize our results so far in Table 4.2.3. The population has no repetition. The last row has not yet been considered in this text.

EXAMPLE 4.2.17

George has 12 coins that are a combination of pennies, nickels, and quarters. How many possible combinations are there?

Method   A sample of size $r = 12$ is to be chosen. The choices are pennies, nickels, and quarters. Thus, the population has size $n = 3$. Certainly replacement is necessary, since duplication of some coin denominations is necessary. Further, an unordered sample is called for, since the order in which these coins are listed is not important for this problem. Therefore, this problem asks for the number of **unordered samples with replacement,** of size 12, from a population of size 3. Having stated the problem, how do we solve it?

If we imagine the pennies, nickels, and quarters arranged in a line (pennies first, then nickels, then quarters), a typical sample might look like PPPNNNNNNNQQ. We need a method to count all such strings. A nice trick does this. Place an x as a separator between different denominations:

$$\text{PPP x NNNNNNN x QQ}$$

This configuration is given more simply by

$$= = = \text{x} = = = = = = = \text{x} = =$$

In the same way, once 2 x's are placed among the 12 ='s, a sample of the required type is formed. But this problem can be done with combinations. There are 14 places $(12 + 2)$ in which to place 12 ='s. This yields a total of $C(14, 12)$ configurations, and hence there are $C(14, 12) = 91$ possibilities.

This method clearly generalizes.

THEOREM 4.2.18

The number of unordered samples, with replacement, of size r chosen from a population of size n is $C(n + r - 1, r)$. ■

We can thus have Table 4.2.4, the complete version of Table 4.2.3.

Table 4.2.4 Population without Repetitions

Population Size	Sample Size	Replace?	Ordered?	Number
n	r	No	Yes	$P(n, r)$
n	r	No	No	$C(n, r)$
n	r	Yes	Yes	n^r
n	r	Yes	No	$C(n + r - 1, r)$

We end this section with a consideration of samples from a population with repetitions.

EXAMPLE 4.2.19

How many different strings can be formed by rearranging the string $AAABBBBCCCCCCDD$?

Method   There are three A's, four B's, six C's, and two D's. The population has 15 elements with some **repetitions**. What is desired is an **ordered** sample of size 15, from this population of size 15, **without** replacement. (If replacement were allowed, then any number of A's, and so on, could be used.) Note that if there were only two letters, this would be a straightforward problem in combinations, because we would ask where the A's are to be placed.

Our solution to the problem is very similar to the proof of Equation (4.2.2). First, imagine the A's are labeled (A_1, A_2, A_3) to distinguish one from another, and similarly for the other letters. In this way, we have 15 distinct letters, but they are grouped into the A's, B's, and so forth.

Let x = the required number of strings. We find x indirectly, by considering the number of permutations on the 15 distinct letters (with subscripts). This can be found directly as 15!. But we can also find it in stages as follows: Determine the pattern of the A's, B's, C's, and D's (x ways); then arrange the A's (3! ways), the B's (4! ways), the C's (6! ways), and the D's (2! ways). Therefore, there are $x(3!4!6!2!)$ ways of arranging the 15 distinct letters. But this is also 15!, so we have $15! = x(3!4!6!2!)$. Thus

$$x = \frac{15!}{3!4!6!2!}$$

This formula generalizes the formula for combinations. In general, if there are r_1 repetitions of a letter, r_2 repetitions of a second letter, . . . , and r_k repetitions of the kth letter, then the number of distinct strings using all of these letters is

$$\frac{(r_1 + r_2 + \cdots + r_k)!}{r_1!r_2! \cdots r_k!}$$

In the language of samples, we may state the theorem.

THEOREM 4.2.20

If a population of size n has repetitions, namely r_1 repetitions of object A_1, r_2 repetitions of object A_2, and so on, then the number of different **ordered samples without replacement** of size n from this population is

$$\frac{n!}{r_1! r_2! \cdots} \quad \blacksquare$$

Exercises

1. Compute $C(9, 3)$ and $P(9, 3)$.
2. Compute $C(25, 22)$ and $P(5, 4)$.
3. Simplify $C(90, 35)/C(89, 34)$. Write your answer as a fraction in lowest terms.
4. Simplify $C(100, 23)/C(99, 23)$. Write your answer as a fraction in lowest terms.
5. (a) Ten people at dinner are toasting Tom and Mary's anniversary. For good luck, everybody decides to click glasses with everybody else. How many clicks are heard?
   (b) Suppose five men and five women are doing the toasting. Each man clicks glasses with each woman. How many clicks are heard?
6. A spelling check program must check all words made from the letters *ARBSED*. The words are to be anywhere from two to six letters long and must not contain any repetitions. How many words must be checked?
*7. As in the previous exercise, except that the letters are *ARRESS* and repetitions are allowed to the extent that the given list contains repetitions. How many different strings must be checked?
8. How many strings on the letters from A through J are there in which each letter appears once, and only once, and in which the word *FACE* appears as a substring? For example, *BGJFACEDHI* is such a string.
9. How many strings of 0's and 1's of length 8 can be formed in which 1011 appears in that order? For example, 00101110 is one such string.
*10. How many strings of 0's and 1's of length 8 can be formed in which 101 appears in that order? For example, 00101010 is one such string. (*Hint:* You may want to use the Inclusion-Exclusion Formula.)
11. In how many ways can all of the integers from 1 to 10, inclusive, be arranged in a finite sequence with no repetitions? In how many ways can they be arranged if the numbers 7 through 10 appear in a block in some order? (For example, 1, 2, 4, 3, 10, 7, 9, 8, 5, 6 is such a sequence.)

Exercises 12–14 concern the game of bridge. A bridge hand consists of 13 (unordered) cards out of the 52.

12. How many bridge hands are there? How many bridge hands are there with five spades, three hearts, three diamonds, and two clubs? How many bridge hands have a 5-3-3-2 distribution? (This means five of one suit, three of another, and so on.) What is the probability of picking up a hand in bridge with a 5-3-3-2 distribution?

*13. Using the results of the previous exercise, determine whether it is more likely to pick up a hand in bridge with a 5-3-3-2 distribution or one with a 4-3-3-3 ("square") distribution? (*Hint:* Find the probability for each case. The one with the larger probability is the more likely occurrence.)

14. What is the probability that a bridge hand contains all 13 hearts?

15. A flush in poker is a hand consisting of all cards of the same suit. If they are consecutive cards, the hand is called a straight flush, which is more valuable than a flush and is *not* considered a flush. What is the probability of drawing a flush in five-card poker?

16. A poker game is played with three cards from a standard deck of 52 cards. Is it more likely to draw a hand with one pair (for example, king, king, 8) or a flush (for example, all hearts)?

17. A roulette wheel (in the United States) has 38 slots, numbered 1 through 36, 0, and 00. Eighteen numbers are red, 18 are black, and the 0 and 00 are green. The roulette wheel is spun, and the ball lands in a hole on one of these numbers. What is the probability that on five consecutive spins only red numbers occur?

18. Figure 4.2.1 represents an area map of a city. In how many different ways can one walk from A to B if one must always walk either north or east? How many walks pass by the famous monument at C?

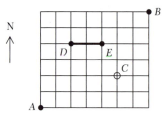

Figure 4.2.1

*19. In Figure 4.2.1, the street marked DE is blocked off due to construction. How many walks are there from A to B (again always north or east) that avoid this street? It is possible to go through the intersections D and E, but not along or through any part of the street DE.

20. How many 4-permutations of $\mathbb{I}_4$ have at least one of the integers i in the ith place? For example, 3241 is one such permutation because 2 is in the second place. [*Hint:* Let A_1 be the permutations in which 1 is in the first place, A_2 those in which 2 is in the second place, and so on. You must find $|A_1 \cup A_2 \cup A_3 \cup A_4|$. Now use the Inclusion-Exclusion Formula (2.1.13).]

21. A **derangement** of 1, 2, ..., n is an n-permutation of $\mathbb{I}_n$ such that i is not in the ith place, for each i. (In other words, all integers move out of place.) The number of derangements of $\mathbb{I}_n$ is called $D(n)$. Using the previous exercise, compute $D(4)$.

*22. Generalize the previous exercise to compute $D(n)$.

23. There are ten points in the plane in general position. (This means that no three of these points are on a line.) How many lines can be formed that are determined by two of these points? How many triangles can be formed, each of whose vertices are one of these points?

*24. Ten integers are to be chosen, one at a time from 1 through 20. Once an integer is chosen, the next one chosen must not be smaller than the previous one, although it can be repeated. How many possible sequences can be chosen by this method?

   Exercises 25–30 concern lotteries. In Exercises 25–28, six different numbers are chosen at random from the numbers 1 through 36. (Different lotteries may use more or fewer numbers and may choose more or fewer than six.) The winning combinations do not depend on the order in which these numbers are drawn.

25. How many different lottery outcomes are possible?

26. A jackpot prize occurs if all six numbers are chosen correctly. What is the probability of winning the jackpot?

27. If you choose five out of six, you share second prize. What is the probability of this happening?

28. If you choose four out of six correctly, you win a free game. What is this probability?

*29. If the lottery consists of choosing 6 out of 40, instead of 6 out of 36, what is the probability of winning the jackpot (choosing all 6 numbers)? Divide your answer in Exercise 26 by your answer here to find out how your chances improve in a 36-number lottery compared with a 40-number lottery.

*30. You are asked to create a lottery consisting of drawing six numbers, as in the previous exercises. The numbers are to be chosen from 1 to N, and the third-prize winner is anyone who correctly chooses exactly four numbers. The advertising agency wants to state that there is better than 1 chance in 50 of being a third-place winner. How large can you choose N?

31. Prove Equation (4.2.7) from Equation (4.2.2).

32. Uncle Al has obtained 13 identical tapes that he wants to distribute to his friends and family. He must send them all to nine people.
   (a) In how many different ways can he do this if each person is to get at least one tape?
   (b) In how many different ways can he send out these tapes if it is possible that some people receive none?

33. Uncle Al has discovered there are 20 people on his mailing list. He has 13 identical tapes to distribute.
    (a) In how many ways can he send out these tapes so that no one gets more than one?
    (b) How many ways are there to send out the tapes, if there are no restrictions on the number of tapes that anyone can receive?

34. How many different solutions are there to the equation

$$x + y + z + w = 12$$

subject to the condition that x, y, z, and w are all nonnegative integers. (*Hint:* Interpret this equation as giving 12 objects to 4 people.)

35. Repeat Exercise 34, except that x, y, z, and w must be positive. (*Hint:* By setting $x = 1 + x'$, and so on, you can reduce the equation to one having nonnegative variables.)

In Exercises 36–40 all numbers (except for 0 itself) are written *without* leading 0's.

36. How many natural numbers less than 1,000,000 are there whose (decimal) digits are increasing? For example, 45,689 is one such number. (*Hint:* See Example 4.2.9.)

*37. How many natural numbers less than 1,000,000 are there whose (decimal) digits never decrease? For example, 344,688 is one such number.

*38. How many natural numbers less than 1,000,000 are there whose (decimal) digits never increase? For example, 84,400 is one such number.

39. How many natural numbers less than 2^{30} are there whose (octal) digits never increase? The number is assumed to be written in octal notation.

40. Repeat Exercise 39 for numbers smaller than 8^n written in the octal notation.

In Exercises 41–43, S_n is the set of all 1-1 maps of $\mathbb{I}_n$ onto itself. These were called permutations in Section 2.2, but we use this word for a slightly different purpose in this chapter. An element of S_n is called a **transformation** in the following exercises.

41. How many different transformations in S_6 map $\mathbb{I}_3$ onto itself?

*42. How many different transformations in S_6 are the product of disjoint cycles of length 2 and 4? [Be careful. Recall that $(1245) = (2451)$, and so on.]

*43. How many different transformations in S_6 are the product of two disjoint cycles of length 3?

4.3 Linear Difference Equations

We consider certain kinds of recursive definitions that generalize the definition of Fibonacci sequences. Techniques are developed to solve them explicitly in many cases, and several applications are given.

DEFINITION 4.3.1

A **second-order, linear difference equation with constant coefficients** is an equation of the form

$$a_n + ra_{n-1} + sa_{n-2} = c_n \quad (n \geq 2) \tag{4.3.1}$$

where r and s are constants ($s \neq 0$) and c_n is a given sequence. The equations

$$a_0 = a, \quad a_1 = b \tag{4.3.2}$$

where a and b are constants, are called **initial conditions** for Equation (4.3.1). Clearly, Equations (4.3.1) and (4.3.2) recursively define the sequence a_n.

The Fibonacci numbers are a special case, with $r = s = -1$, $c_n = 0$, $a = 0$, and $b = 1$.

Equation (4.3.1) is second order because a_n is defined in terms of the two previous terms. Higher-order equations are similarly defined, and their theory is quite similar to the one we develop here. Equation (4.3.1) is linear because the left side is a linear function of a_n, a_{n-1}, and a_{n-2}.

DEFINITION 4.3.2

Equation (4.3.1) is called **homogeneous** if $c_n = 0$.

We first study the homogeneous equation

$$a_n + ra_{n-1} + sa_{n-2} = 0 \quad (n \geq 2) \tag{4.3.3}$$

and then proceed to the more general Equation (4.3.1). Recall that we are requiring $s \neq 0$.

First-order homogeneous equations are especially easy to solve. The equation $a_n - ba_{n-1} = 0$ gives $a_1 = ba_0$, $a_2 = ba_1 = b^2a_0$, and, in general, $a_n = a_0b^n$. We shall see that **powers** are the key to solving Equation (4.3.3). Any solution of (4.3.3) of the form a_0b^n is called a **power solution**.

The solution to the homogeneous Equation (4.3.3) is based on the following two lemmas.

LEMMA 4.3.3

The sequence $a_n = \lambda^n$ ($\lambda \neq 0$) satisfies Equation (4.3.3) if and only if λ is a solution of the quadratic equation

$$x^2 + rx + s = 0 \qquad (4.3.4)$$

■

DEFINITION 4.3.4

Equation (4.3.4) is called the **characteristic equation** of (4.3.1).

LEMMA 4.3.5

If the sequences x_n and y_n satisfy Equation (4.3.3), then so does the sequence $z_n = cx_n + dy_n$, for any constants c and d. (z_n is called a **linear combination** of x_n and y_n.)

Proof of Lemma 4.3.3 The sequence $a_n = \lambda^n$ satisfies Equation (4.3.3) if and only if

$$\lambda^n + r\lambda^{n-1} + s\lambda^{n-2} = 0 \qquad (n \geq 2)$$

Dividing by λ^{n-2}, we obtain

$$\lambda^2 + r\lambda + s = 0$$

Conversely, if this equation is satisfied, we can multiply by λ^{n-2} to see that $a_n = \lambda^n$ satisfies Equation (4.3.3). This proves Lemma 4.3.3. Note that if λ is a solution of (4.3.4), then $\lambda \neq 0$ because we are assuming that $s \neq 0$.

Proof of Lemma 4.3.5 Suppose

$$x_n + rx_{n-1} + sx_{n-2} = 0$$

and

$$y_n + ry_{n-1} + sy_{n-2} = 0$$

Multiplying these equations by c and d and adding gives

$$(cx_n + dy_n) + r(cx_{n-1} + dy_{n-1}) + s(cx_{n-2} + dy_{n-2}) = 0$$

or

$$z_n + rz_{n-1} + s\dot{z}_{n-2} = 0$$

which is the result.   ■

Lemma 4.3.3 is the result of an inspired guess—that a solution might be a power, just as for first-order equations. Lemma 4.3.5 shows how to combine solutions to get new solutions. We now use these two lemmas to solve difference equations.

EXAMPLE 4.3.6

Solve the system

$$a_n = a_{n-1} + 6a_{n-2} \qquad (n \geq 2) \qquad \text{(4.3.5)}$$

$$a_0 = 3, \qquad a_1 = -1$$

Method   We first solve the characteristic equation

$$x^2 - x - 6 = 0$$

as in Lemma 4.3.3. This factors as $(x - 3)(x + 2) = 0$, with solutions $x = 3$ and $x = -2$. By Lemma 4.3.3, two solutions of Equation (4.3.3) are $a_n = 3^n$ and $a_n = (-2)^n$. By Lemma 4.3.5,

$$a_n = c3^n + d(-2)^n \qquad \text{(4.3.6)}$$

is a solution of (4.3.5) for any choice of constants c and d. We now use (4.3.6) and determine the constants c and d so that the initial conditions of (4.3.5) are also satisfied. Substitute $n = 0$ and $n = 1$ into (4.3.6), and use the initial conditions of (4.3.5) to get

$$3 = a_0 = c + d$$

and

$$-1 = a_1 = 3c - 2d$$

Solving these equations, we find $c = 1$ and $d = 2$. Finally, substituting in (4.3.6), we find the required solution

$$a_n = 3^n + 2(-2)^n \qquad \text{(4.3.7)}$$

If we use (4.3.5) to directly compute values of a_n, we find that the first few terms are 3, -1, 17, 11, 113, and 179, so Equation (4.3.7) is not easy to guess.

EXAMPLE 4.3.7   Fibonacci Numbers, Again

If f_n is the nth Fibonacci number, show that f_n is the nearest integer to $\Psi^n/\sqrt{5}$, where $\Psi = (\sqrt{5} + 1)/2$.

Method   As in Example 4.3.6, we can find a formula for f_n. Since the difference equation is $f_n = f_{n-1} + f_{n-2}$, we first solve the characteristic equation $x^2 - x - 1 = 0$ in order to find power solutions. The roots of this equation are, by the quadratic formula, $x = (1 \pm \sqrt{5})/2$. These are $\Psi = (1 + \sqrt{5})/2 = 1.618\ldots$ and $\Psi_1 = (1 - \sqrt{5})/2 = -0.618\ldots$. As in Example 4.3.6, we can find $f_n = c\Psi^n + d\Psi_1^n$ if we can find c and d to make $f_0 = 0$ and $f_1 = 1$. But this leads to the equations

$$0 = f_0 = c + d$$

$$1 = f_1 = c\Psi + d\Psi_1$$

From the first equation, $d = -c$. By the second equation, we obtain $1 = c(\Psi - \Psi_1) = c\sqrt{5}$, by definition of Ψ and Ψ_1. Thus, $c = 1/\sqrt{5}$ and $d = -1/\sqrt{5}$; hence,

$$f_n = \frac{1}{\sqrt{5}}\Psi^n - \frac{1}{\sqrt{5}}\Psi_1^n$$

This is the explicit formula for f_n. We find that $f_n - \Psi^n/\sqrt{5} = -\Psi_1^n/\sqrt{5} = -(-0.618\ldots)^n(0.447\ldots)$. Thus, $\Psi_1^n/\sqrt{5}$ is always less than $1/\sqrt{5} \approx 0.477$ in absolute value. Because f_n is an integer and its difference from $\Psi^n/\sqrt{5}$ is smaller than 0.5, we see that f_n is the nearest integer to $\Psi^n/\sqrt{5}$. Note that this method also shows that $\Psi^n/\sqrt{5}$ alternates above and below f_n.

The preceding technique is a direct generalization of the example at the beginning of this section for first-order equations. The equation $a_n = ba_{n-1}$ has characteristic equation $x - b = 0$, leading to the solution $a_n = cb^n$.

A characteristic equation may have **imaginary** roots. For example, the quadratic equation $x^2 + 1 = 0$ has imaginary roots $x = \pm i$, where $i = \sqrt{-1}$. The methods of this section work for such quadratic equations, but we do not pursue them.

The analogies of Lemmas 4.3.3 and 4.3.5 are valid for difference equations of order 3 or more. However, in this case it is necessary to solve higher order polynomial equations.

The technique used in Examples 4.3.6 and 4.3.7 *fails* when the characteristic equation has *two equal roots*. For example, consider the equations

$$a_n = 4a_{n-1} - 4a_{n-2} \qquad (n \geq 2) \tag{4.3.8}$$

The characteristic equation is $x^2 - 4x + 4 = 0$. But this is $(x - 2)^2 = 0$, which has $x = 2$ as a double root. Therefore, we have $a_n = c2^n$ as a solution. ($d2^n$ is the same solution.) There is only one undetermined value of c here, but each initial condition (4.3.9) gives a different value for c. In this case, however, it turns out that the sequence $n2^n$ is also a solution of (4.3.8). Therefore, by Lemmas 4.3.3 and 4.3.5,

$$a_n = c2^n + dn2^n = (c + dn)2^n$$

is also a solution of (4.3.8). The constants c and d can now be determined by using (4.3.9). Substituting $n = 0$ and $n = 1$, we obtain the equations $-1 = c$ and $0 = 2(c + d)$. Thus, we have $c = -1$ and $d = 1$, and the solution of (4.3.8) and (4.3.9) is $a_n = (n - 1)2^n$. The general situation for repeated roots in a second order, linear difference equation is covered by the following lemma.

LEMMA 4.3.8

If the characteristic equation of Equation (4.3.3) is $(x - \lambda)^2 = 0$, then $a_n = \lambda^n$ and $a_n = n\lambda^n$ are solutions of (4.3.3).

Proof We already have the result for λ^n. Equation (4.3.3) is

$$a_n = 2\lambda a_{n-1} - \lambda^2 a_{n-2}$$

It is now a matter of algebra to show that $a_n = n\lambda^n$ also satisfies this equation. We leave the details to an exercise. ∎

The foregoing results may be summarized as follows.

THEOREM 4.3.9 *Solving Homogeneous Equations*

Let

$$a_n + ra_{n-1} + sa_{n-2} = 0 \qquad (n \geq 2) \tag{4.3.3}$$

be a linear, homogeneous, difference equation with constant coefficients, $s \neq 0$, and let its characteristic equation

$$x^2 + rx + s = 0 \tag{4.3.4}$$

have distinct roots λ_1 and λ_2. Then

$$a_n = c\lambda_1^n + d\lambda_2^n \qquad (c \text{ and } d \text{ constants}) \tag{4.3.10}$$

is a solution of (4.3.3).
    If the characteristic equation has a repeated root λ, then

$$a_n = (c + dn)\lambda^n \qquad (c \text{ and } d \text{ constants}) \tag{4.3.11}$$

is a solution of (4.3.3). Depending on whether (4.3.4) has distinct or equal roots, all of the solutions of (4.3.3) are given either by (4.3.10) or by (4.3.11).

Proof   This follows from Lemmas 4.3.3, 4.3.5, and 4.3.8. To show that all solutions to (4.3.3) are included among the solutions (4.3.10) or (4.3.11), we may use the method as illustrated in previous examples to solve for c and d in order to satisfy the given initial conditions (4.3.2). The proof is left to the exercises. ■

Remark   The solutions (4.3.10) and (4.3.11) are called the **general solutions** of Equation (4.3.3).

We illustrate this result with one more example.

EXAMPLE 4.3.10

How many strings of length 10 on the alphabet $\{A, B, C\}$ have the letter C appearing at least once in any two successive letters?

Method   Call such a string **C-abundant**. Let a_n be the number of such strings of length n. We now find a recursion. If $n \geq 2$, we separate into cases depending on whether the string begins with C.

Case I   If the string begins with C, the next $n - 1$ letters can be any C-abundant string. There are a_{n-1} such strings.

Case II   If the string begins with A or B (two possibilities), the next letter must be C and the remaining $n - 2$ letters can be any C-abundant string. Therefore, there are $2a_{n-2}$ strings in this case.
    Thus,

$$a_n = a_{n-1} + 2a_{n-2} \tag{4.3.12}$$

The initial conditions are

$$a_0 = 1, \qquad a_1 = 3 \tag{4.3.13}$$

corresponding to the empty string for $n = 0$, and to the strings "A", "B", and "C" for $n = 1$. The characteristic equation is

$$x^2 - x - 2 = 0$$

The roots are 2 and -1. Therefore, the general solution of (4.3.4) is $a_n = c2^n + d(-1)^n$. The initial conditions (4.3.5) lead to $c = 4/3$ and $d = -1/3$, so $a_n = 4(2^n)/3 - (-1)^n/3$. Because a_n is an integer, it follows that a_n is the nearest integer to $4(2^n)/3$. For the given problem, we have $n = 10$, so $a_{10} = (4096 - 1)/3 = 1365$.

The Nonhomogeneous Case

The theory for Equation (4.3.1) is based on the theory of the homogeneous Equation (4.3.3).

THEOREM 4.3.11

Suppose that the equation

$$a_n + ra_{n-1} + sa_{n-2} = c_n \qquad (4.3.1)$$

has solution $a_n = p_n$. Let g_n be any solution of the homogeneous Equation (4.3.3). Then $a_n = p_n + g_n$ is a solution of Equation (4.3.1). Further, *any* solution a_n of (4.3.1) is of this form.

Remark   In brief, the general solution of Equation (4.3.1) is a particular solution p_n of (4.3.1) plus the general solution of the homogeneous Equation (4.3.3).

Proof   If p_n satisfies (4.3.1) and g_n satisfies (4.3.3), then

$$p_n + rp_{n-1} + sp_{n-2} = c_n \qquad (4.3.14)$$

and

$$g_n + rg_{n-1} + sg_{n-2} = 0$$

Adding, we obtain

$$(p_n + g_n) + r(p_{n-1} + g_{n-1}) + s(p_{n-2} + g_{n-2}) = c_n$$

showing that $a_n = p_n + g_n$ is a solution of (4.3.1).

Now if a_n is any solution of (4.3.1), then subtracting (4.3.14) from (4.3.1), we have

$$(a_n - p_n) + r(a_{n-1} - p_{n-1}) + s(a_{n-2} - p_{n-2}) = 0$$

This shows that $a_n - p_n$ satisfies the homogeneous Equation (4.3.1), so $a_n - p_n = g_n$, where g_n is some solution of (4.3.1). This gives the result. ■

The theorem and proof are valid for higher- and lower-degree equations. The theorem gives a method for solving (4.3.1) *if* we know one of its solutions: Find a solution p_n, find the general solution of the homogeneous Equation (4.3.3), and then add. Then, fit the initial conditions to the constants in the general solution. Some methods for finding particular solutions are indicated in Example 4.3.13.

EXAMPLE 4.3.12

A string is to be formed from the letters A through F, subject to the following conditions: A, B, and C must be followed by D, unless they are the last letters of the string, and F must be followed by F, unless F is the last letter of the string. How many such strings of length n are there?

Method   We can form a recursion as in Example 4.3.10. Assume $n \geq 2$. Letting a_n be the number of such strings, we separate cases as follows.

Case I   The string starts with A, B, or C. Then the string must be of the form $AD*$, $BD*$, or $CD*$, where $*$ is any string of the required type of length $n - 2$. There are $3a_{n-2}$ such strings.

Case II   The string starts with D or E. It can then be followed by any string of the required form. There are $2a_{n-1}$ such strings.

Case III   The string starts with F. There is only one such string: $FFF \ldots F$. Thus,

$$a_n = 2a_{n-1} + 3a_{n-2} + 1 \quad (n \geq 2) \tag{4.3.15}$$

The initial conditions are

$$a_0 = 1 \quad \text{and} \quad a_1 = 6$$

Equation (4.3.15) is solved by using Theorem 4.3.11. We first solve the homogeneous equation

$$a_n = 2a_{n-1} + 3a_{n-2}$$

The characteristic equation is $x^2 - 2x - 3 = 0$ with roots $x = 3$ and $x = -1$. Thus, the general solution of the homogeneous equation is

$$g_n = c3^n + d(-1)^n$$

To find a particular solution, we *guess* $p_n = A$, a constant. Substituting into (4.3.15), we find that $A = 2A + 3A + 1$, or $A = -\frac{1}{4}$. Therefore, the general solution of (4.3.15) is

$$a_n = -\tfrac{1}{4} + c3^n + d(-1)^n$$

We now find the constants by substituting $n = 0$ and 1 and using the initial conditions:

$$1 = -\tfrac{1}{4} + c + d \quad \text{and} \quad 6 = -\tfrac{1}{4} + 3c - d$$

This gives $c = 15/8$ and $d = -5/8$. Hence,

$$a_n = -\tfrac{1}{4} + [15(3^n) - 5(-1)^n]/8$$

or

$$a_n = [15(3^n) - 5(-1)^n - 2]/8$$

EXAMPLE 4.3.13

How do we guess a "particular" solution of a difference equation? The answer depends on the right side of (4.3.1). If it is constant, we would guess a constant as in Example 4.3.12. If it is a power b^n, we guess a constant times that power. Table 4.3.1 gives some useful guesses and second guesses for a particular solution of (4.3.1), with special right sides c_n. Here, A, B, C, and D are constants.

Table 4.3.1

c_n	Best Guess	If This Guess Doesn't Work
A	B	Bn, then Bn^2, and so on
Ab^n	Bb^n	Bnb^n, then Bn^2b^n, and so on
$An + B$	$Cn + D$	$Cn^2 + Dn$, then $Cn^3 + Dn^2$, and so on

EXAMPLE 4.3.14

A guess for a particular solution of the equation

$$a_n = 3a_{n-1} - 4a_{n-2} + 2^n$$

would be $p_n = A2^n$. This leads to the equation

$$A2^n = 3A2^{n-1} - 4A2^{n-2} + 2^n$$

or (dividing by 2^{n-2}) $4A = 6A - 4A + 4$. Thus, $A = 2$, and a particular solution is $p_n = 2(2^n)$.

The equation

$$a_n = 3a_{n-1} - 2a_{n-2} + 5$$

does not yield a solution for the guess $a_n = A$ because we get the equation $0 = 5$. The next guess $a_n = An$ yields

$$An = 3A(n - 1) - 2A(n - 2) + 5, \quad \text{or} \quad 0 = A + 5$$

Thus, $A = -5$ here, and the particular solution is $p_n = -5n$.

We end this discussion with a financial application.

EXAMPLE 4.3.15

Suppose money earns interest at a rate r in each period in which it is compounded. Starting with an amount A, and depositing an amount P at the end of each period, how much money F (the final amount) will there be after N periods?

Method   Let a_n be the amount after n periods so that

$$a_0 = A$$

The recursion is

$$a_n = (1 + r)a_{n-1} + P \qquad\qquad (4.3.16)$$

This is so because the amount a_{n-1} yields the amount $(1 + r)a_{n-1}$ after one period, and the fixed amount P is then added to it. To solve (4.3.16), let us try a particular solution $p_n = C$, a constant. Substituting in (4.3.16), we have

$$C = (1 + r)C + P$$

so $C = -P/r$, and a particular solution is $p_n = -P/r$. The homogeneous equation is simply $a_n = (1 + r)a_{n-1}$ and has general solution $a_n = c(1 + r)^n$. Thus, the general solution of the given equation is $a_n = c(1 + r)^n - P/r$. Using the initial condition $a_0 = A$, we obtain $A = c - P/r$, so $c = A + P/r$. Thus, the solution is

$$a_n = (A + P/r)(1 + r)^n - P/r = A(1 + r)^n + (P/r)[(1 + r)^n - 1]$$

Since $F = a_N$, we have

$$F = A(1 + r)^N + (P/r)[(1 + r)^N - 1] \qquad\qquad (4.3.17)$$

For example, suppose we start with \$50 in the bank and put in \$1 every month. Assuming that the annual interest rate is 7% compounded monthly, how much money will be in the bank after one year? In this problem, the period is one month, and the interest for this period is $r = 0.07/12$. Here $A = \$50$, $P = \$1$, and $N = 12$. A computation using (4.3.17) gives $F = \$66.01$.

EXAMPLE 4.3.16

Equation (4.3.17) is used to find fixed monthly mortgage payments. Suppose an amount M is borrowed, and there is to be a payment P in each of n periods. We take $A = -M$ and $F = 0$, since there is to be no debt at the end. Therefore,

$$0 = -M(1 + r)^n + (P/r)[(1 + r)^n - 1]$$

This yields the payment formula known to many bankers and some borrowers:

$$P = rM \frac{(1 + r)^n}{(1 + r)^n - 1} \qquad (4.3.18)$$

Linear Inequalities

We have come across linear inequalities in our discussion of time estimates in Section 3.3. (See Examples 3.3.2 and 3.3.13.) In certain cases, the analysis of an inequality can be nicely accomplished by replacing the inequality with an equality.

EXAMPLE 4.3.17

Given the inequality $a_n \geq 2a_{n-1}$ $(n \geq 1)$. If $a_0 \geq 5$, show that $a_n \geq 5 \cdot 2^n$.

Method   We can prove this by induction. It is true for $n = 0$ by the initial inequality. Assuming it is true for n, we have

$$a_n \geq 5 \cdot 2^n$$

Then, by the given inequality, we have

$$a_{n+1} \geq 2a_n$$

Multiplying the first inequality by 2, we get

$$2a_n \geq 5 \cdot 2^{n+1}$$

Combining these last two inequalities, we get

$$a_{n+1} \geq 5 \cdot 2^{n+1}$$

which is the result for $n + 1$.

The method works as well if the inequalities are reversed or replaced by equalities. Since we have some techniques for solving linear equalities, we may ask if some of them can be transferred to the study of linear inequalities. The answer is yes, in some cases. The main algebraic difference in manipulating inequalities is that we can multiply them only by positive quantities in order for the inequality to be preserved. We use this fact in the following theorem, which relates linear inequalities to linear equalities.

THEOREM 4.3.18

Let b_n satisfy the **inequality**

$$b_n \leq cb_{n-1} + db_{n-2} + c_n, \qquad n \geq 2 \qquad (4.3.19)$$

where $c, d \geq 0$. Let a_n satisfy the **equality**

$$a_n = ca_{n-1} + da_{n-2} + c_n, \qquad n \geq 2 \qquad (4.3.20)$$

and let $b_0 \leq a_0$, $b_1 \leq a_1$. Then $b_n \leq a_n$, for $n \geq 0$.

Proof The proof is by induction on n. We have the result for $n = 0$ and 1. Now let $n \geq 2$. Assume the result is true for all values less than n. By the induction hypothesis,

$$b_{n-1} \leq a_{n-1}$$

and

$$b_{n-2} \leq a_{n-2}$$

Multiply the first inequality by c and the second by d (both nonnegative by hypothesis) and add the results:

$$cb_{n-1} + db_{n-2} \leq ca_{n-1} + da_{n-2}$$

Add c_n to both sides of this inequality:

$$cb_{n-1} + db_{n-2} + c_n \leq ca_{n-1} + da_{n-2} + c_n$$

Now, using (4.3.19) and (4.3.20), we obtain

$$b_n \leq a_n$$

This is the result, by induction. ■

Remark The result is true if $\leq$'s are replaced by $\geq$'s. The proof is exactly the same.

EXAMPLE 4.3.19

Let $a_n \geq 2a_{n-1} + 5$ ($n \geq 1$), and suppose $a_0 = 1$. Find an estimate for a_n.

Method Simply solve the **equality**

$$b_n = 2b_{n-1} + 5 \quad (n \geq 1) \quad \text{with } b_0 = 1$$

A particular solution is $p_n = C$, where C must satisfy $C = 2C + 5$. Therefore $C = -5$. The general solution of the homogeneous equation is, by observation, $b_n = c2^n$. Therefore, the solution to the difference equation is $-5 + c2^n$. Matching the initial conditions, we get $-5 + c = 1$, so $c = 6$, and the solution is

$$b_n = 6 \cdot 2^n - 5$$

By Theorem 4.3.18, we have $a_n \geq 6 \cdot 2^n - 5$.

Exercises

Solve the difference equations of Exercises 1–15.

1. $a_n = 5a_{n-1} - 6a_{n-2}$ $(n \geq 2)$; $a_0 = 2$, $a_1 = 5$.
2. $a_n = 3a_{n-1} + 10a_{n-2}$ $(n \geq 2)$; $a_0 = 3$, $a_1 = 1$.
3. $a_n = 2a_{n-1} + 3a_{n-2}$ $(n \geq 2)$; $a_0 = 2$, $a_1 = 0$.
4. $a_n = 6a_{n-1} - 5a_{n-2}$ $(n \geq 2)$; $a_0 = 2$, $a_1 = 3$.
5. $a_n = 3a_{n-1} + 10a_{n-2} + 12$ $(n \geq 2)$; $a_0 = 0$, $a_1 = -1$.
6. $a_n = 7a_{n-1} - 10a_{n-2} + 8$ $(n \geq 2)$; $a_0 = 3$, $a_1 = 1$.
7. $a_n = 3a_{n-1} + 10a_{n-2} + 6 \cdot 2^n$ $(n \geq 2)$; $a_0 = -2$, $a_1 = -3$.
*8. $a_n = 5a_{n-1} - 6a_{n-2} + 24(-1)^n$ $(n \geq 2)$; $a_0 = 1$, $a_1 = -2$.
*9. $a_n = 3a_{n-1} + 10a_{n-2} + 7 \cdot 5^n$ $(n \geq 2)$; $a_0 = 1$, $a_1 = -12$.
*10. $a_n = 3a_{n-1} - 2a_{n-2} + 3 \cdot 2^n$ $(n \geq 2)$; $a_0 = 1$, $a_1 = 6$.
11. $a_n = 4a_{n-2}$ $(n \geq 2)$; $a_0 = 7$, $a_1 = 11$.
*12. $a_n = 6a_{n-1} - 9a_{n-2}$ $(n \geq 2)$; $a_0 = 1$, $a_1 = 2$.
13. $a_n = 4a_{n-1} - 4a_{n-2}$ $(n \geq 2)$; $a_0 = 1$, $a_1 = 3$.
*14. $a_n = 4a_{n-1} - 4a_{n-2} + 3^n$ $(n \geq 2)$; $a_0 = 1$, $a_1 = 2$.
*15. $a_n = 6a_{n-1} - 9a_{n-2} + 2^n$ $(n \geq 2)$; $a_0 = 1$, $a_1 = 5$.
*16. Suppose the Fibonacci recursion $g_n = g_{n-1} + g_{n-2}$ has initial values $g_0 = 1$ and $g_1 = 3$.
   (a) Find a formula for the general solution g_n.
   (b) In analogy to Example 4.3.7, find a simple expression for g_n for large n.
*17. The sequence g_n of the previous exercise can be simply expressed in terms of f_n and f_{n+1}, where f_n is the Fibonacci sequence. Show how this can be done, and prove your result.
18. Suppose a_n is the sequence $0, 1, 3, 10, 33, 109, \ldots$, where each term is 3 times the preceding term plus the term preceding that. We want to find the 1000th term to two significant figures, but this number

is so large that a computation via computer is not routine. Find a_{1000} to two significant figures. (*Hint:* See Example 4.3.6. Take logarithms.)

19. Let $a_n = 2^n - 7(-1)^n$. Write the first few terms of this sequence. Find a difference equation with constant coefficients and initial conditions that this equation satisfies.

20. Repeat Exercise 19 for the sequence $a_n = 3n \cdot 2^n$.

21. Solve $a_n = 2a_{n-1} - a_{n-2} + 1$ $(n \geq 2)$, with initial conditions $a_0 = 1$ and $a_1 = 0$.

22. Solve $a_n = 2a_{n-1} + 5 \cdot 2^n$, with initial condition $a_0 = 1$.

23. Solve $a_n = 3a_{n-1} + 3^n$, with initial condition $a_0 = 1$.

24. Suppose $a_n = a_{n-2}$, $a_0 = 7$, and $a_1 = 11$. Find a_{901} and a_{5040}.

In Exercises 25–27, the sequences a_n, b_n, ... are understood to be defined for *all* $n \in \mathbb{Z}$. In particular, a_{-1}, a_{-2}, ... are defined.

25. If a_n is any sequence, the **difference sequence** Δa_n is defined as the sequence whose nth term is $a_n - a_{n-1}$. Now suppose that $a_n = A + Bn + Cn^2$ for all n. Let $b_n = \Delta a_n$, $c_n = \Delta b_n$, and $d_n = \Delta c_n$. (We write $c_n = \Delta^2 a_n$; $d_n = \Delta^3 a_n$.) Show that $d_n = 0$.

26. Let a_n be a sequence for which $\Delta a_n = 0$. Show that $a_n = A$, a constant. Similarly show that if $\Delta^2 a_n = 0$, then $a_n = A + Bn$.

*27. Show that if a_n is a polynomial of degree k in n, then $\Delta^{k+1} a_n = 0$.

28. Jerry puts $1000 in the bank at 6% interest compounded annually. Every year, after the interest is added to his account, he adds $300 to his account. Determine how much he has after n years.

29. Mr. Warbucks wants to borrow $1,000,000 to help repair equipment at his factory. He wishes to pay a fixed amount annually to pay off part of the loan. The annual interest is 9%. He wants to have a balance of $250,000 after ten years, at which time he expects to pay off the loan in a lump sum. What is his annual payment on this loan?

30. Lemma 4.3.8 stated that the sequence $a_n = n\lambda^n$ satisfied the equation $a_n = 2\lambda a_{n-1} - \lambda^2 a_{n-2}$. Verify this and show your work.

31. The first-order equation $a_n - ra_{n-1} = 0$ $(n \geq 1)$ has the solution $a_n = cr^n$ for $n \geq 0$. Show that this is true for $r = 0$, too. (Recall our convention that $0^0 = 1$.)

*32. It was stated, without proof, in Theorem 4.3.9 that all solutions to Equation (4.3.3) are given by (4.3.10) or (4.3.11). Prove this.

33. How many strings of length 15 on $\{A, B, C, D, E\}$ are there in which a nonterminal A, B, and C must be followed by D?

34. Repeat Exercise 33, except that A, B, and C cannot terminate the string.

35. How many strings of length 15 on $\{A, B, C, D\}$ are there in which a nonterminal A must be followed by either C or D, and a nonterminal B must be followed by C?

36. Repeat Exercise 35, except that A cannot terminate the string.

37. A sequence b_n satisfies the inequality

$$0 \le b_n \le 2b_{n-1}$$

Show that $b_n = O(2^n)$.

*38. A sequence b_n satisfies the inequality

$$0 < 2b_{n-1} \le b_n \le 2b_{n-1} + 5n$$

Show that b_n has the same order of magnitude as 2^n.

39. A sequence a_n satisfies the inequality

$$0 \le a_n \le 2a_{n-1} + 3a_{n-2} + n, \quad n \ge 2$$

Show that $a_n = O(3^n)$.

40. A sequence a_n satisfies the inequality

$$0 \le a_n \le a_{n-1} + 6a_{n-2} + n, \quad n \ge 2$$

Show that $a_n = O(3^n)$.

41. A sequence b_n satisfies the inequality

$$0 \le b_n \le 3b_{n-1} + 10b_{n-2}$$

Show that $b_n = O(a^n)$, for some $a > 1$.

*42. A sequence a_n satisfies the inequality

$$0 \le a_n \le a_{n-1} + a_{n-2} + 3n, \quad n \ge 2$$

Show that $a_n = o(2^n)$.

43. A sequence a_n satisfies the inequality

$$0 \le a_n \le 8a_{n-3}, \quad n \ge 3$$

Show that $a_n = O(2^n)$.

*44. A sequence a_n satisfies the inequality

$$0 \le a_n \le a_{n-1} + 6a_{n-2} + 1/n, \quad n \ge 2$$

Show that $a_n = O(3^n)$. (*Hint:* Don't fuss with $1/n$. It is ≤ 1, for example.)

*45. Let a_n be a sequence defined for $n \ge 1$. Suppose $a_1 = 1$ and $a_n = 2a_{\lfloor n/2 \rfloor} + Kn$ for $n > 1$, where K is a positive constant. Experiment to find a formula for a_n when n is a power of 2. It will be convenient to use $\lg n = \log_2 n$ in your answer.

*46. For the sequence a_n of Exercise 45, show that $a_n < a_{n+1}$. (*Hint:* Prove this by induction.)

*47. Let b_n be a sequence defined for $n \ge 1$. Suppose $b_1 = 1$ and $b_n \le 2b_{\lfloor n/2 \rfloor} + Kn$ for $n > 1$, where K is a positive constant. Prove that $b_n \le a_n$, where a_n satisfies the equation in Exercise 45.

*48. Let b_n satisfy the inequality $0 \le b_n \le 2b_{\lfloor n/2 \rfloor} + Kn$ for $n > 1$, where K is a positive constant, and let $b_1 = 1$. Using the results of Exercises 45–47, show that $b_n = O(n \log n)$. (Compare with Example 3.3.13 on the time analysis of MergeSort.)

4.4 Generating Functions and Binomial Identities

A **generating function** is an algebraic device in which a sequence a_n ($n = 0, 1, \ldots$) is coded as an **infinite series** in a variable x. The algebra of series can then be used to give useful information about sequences.

DEFINITION 4.4.1

The generating function $f(x)$ of a sequence a_n is the infinite series

$$f(x) = \sum_{k=0}^{\infty} a_k x^k = a_0 + a_1 x + a_2 x^2 + a_3 x^3 + \cdots \qquad (4.4.1)$$

a_n is called the **sequence of** $f(x)$. We often abbreviate (4.4.1) as $f(x) = \Sigma a_k x^k$.

Some Algebraic Remarks   Logically, Equation (4.4.1) is simply another way of writing a sequence a_n. Instead of referring to the nth term of a sequence, we consider the coefficient of x^n in its generating function. The sum is taken from 0 to ∞, which means that the sequence is defined for all $n \in \mathbb{N}$. If the value of a_n is 0 from some point on, we replace (4.4.1) by a finite sum. Similarly, we usually omit all 0 terms. For example, the sequence 1, 0, 3, 0, 0, 0, . . . has generating function $1 + 0x + 3x^2 + 0x^3 + 0x^4 + \cdots$, which in turn is identified by the polynomial $1 + 3x^2$. It is only finite series of this sort (the polynomials) that we evaluate for some specific value of x. Thus, no **infinite** arithmetic processes are implied by this notation. The series $f(x)$ is called a **formal power series** to indicate that it is not intended to be evaluated for any specific value of x, nor is there any question of convergence.

We can add and multiply two such series, using the usual algebraic techniques, to find each of the resulting coefficients in a finite number of steps. We take for granted that the usual laws of algebra apply.

EXAMPLE 4.4.2

Suppose $a_n = n$ and $b_n = (-1)^n$. Let these sequences have generating functions $f(x)$ and $g(x)$, respectively. Compute the product $h(x) = f(x)g(x)$, and find some terms of the sequence of $h(x)$.

Method   We have

$$f(x) = \Sigma k x^k = x + 2x^2 + 3x^3 + 4x^4 + \cdots$$

$$g(x) = \Sigma (-1)^k x^k = 1 - x + x^2 - x^3 + \cdots$$

To multiply, we start with the lowest power of x:

$$x + 2x^2 + 3x^3 + 4x^4 + 5x^5 + \cdots$$
$$1 - x + x^2 - x^3 + x^4 + \cdots$$

$$x + 2x^2 + 3x^3 + 4x^4 + 5x^5 + \cdots$$
$$- x^2 - 2x^3 - 3x^4 - 4x^5 - \cdots$$
$$x^3 + 2x^4 + 3x^5 + \cdots$$
$$- x^4 - 2x^5 - \cdots$$
$$+ x^5 + \cdots$$
$$+ \cdots$$

$$x + x^2 + 2x^3 + 2x^4 + 3x^5 + \cdots$$

For $h(x) = f(x)g(x)$, the sequence c_n of $h(x)$ has 0, 1, 1, 2, 2 as its first few terms. A reasonable (and it turns out, accurate) guess for c_n is given by the recursions

$$c_n = 1 + c_{n-2} \quad (n \geq 3); \qquad c_0 = 0, \quad c_1 = c_2 = 1$$

It can be seen that the computation of any coefficient of $h(x)$ involves finitely many operations. For general formal power series $f(x)$ and $g(x)$ with $h(x) = f(x)g(x)$, the computation of the coefficient c_n of $h(x)$ is given by the formula

$$c_n = \sum_{k=0}^{n} a_k b_{n-k} \tag{4.4.2}$$

To see this, note that the term $a_k x^k$ of $f(x)$ multiplies with the term $b_{n-k}x^{n-k}$ to equal $a_k b_{n-k}x^n$. This occurs for $k = 0$ through $k = n$, and these are the only ways to get the term x^n. Combining all these terms, we obtain (4.4.2). (This is actually the *definition* of the product of two formal power series.)

We now show how certain operations on sequences a_n correspond to algebraic operations on its generating function $f(x)$. We adopt the convention that $a_s = 0$ **for all negative values of** s. Thus, $0 = a_{-1} = a_{-2}$ and so forth. This convention simplifies our formulas considerably.

LEMMA 4.4.3

If the sequences a_n and b_n have generating functions $f(x)$ and $g(x)$, respectively, and if c and d are constants, then the sequence $ca_n + db_n$ has generating function $cf(x) + dg(x)$.

Proof This is precisely how $cf(x) + dg(x)$ is computed. ∎

LEMMA 4.4.4

Let the sequence a_n have generating function $f(x)$. Then the sequence $b_n = a_{n-1}$ has generating function $xf(x)$.

Proof Writing

$$f(x) = a_0 + a_1x + a_2x^2 + \cdots$$

we compute

$$xf(x) = 0\ + a_0x + a_1x^2 + a_2x^3 + \cdots$$
$$= b_0 + b_1x + b_2x^2 + b_3x^3 + \cdots$$

This is the result. ∎

Remark Thus, **multiplying** a generating function by x corresponds to a **right shift** of its sequence.

LEMMA 4.4.5

If the sequence a_n has generating function $f(x)$, then the sequence $b_n = a_n - a_{n-1}$ has generating function $(1 - x)f(x)$.

Proof The proof follows from the previous two results. ∎

The sequence $b_n = a_n - a_{n-1}$ is called the **backward difference,** or simply the **difference,** of the sequence a_n, and is often written Δa_n. (The sequence $a_{n+1} - a_n$ is called the **forward difference** of a_n.) The following tabulation illustrates this differencing procedure. Note that $\Delta a_0 = a_0$, since our convention is $a_{-1} = 0$.

$n\!:0$	1	2	3	4	5	6 $\cdots$	
$a_n\!:2$	3	5	11	12	51	23 $\cdots$	has difference
$b_n\!:2$	1	2	6	1	39	$-28\cdots$	which has difference
$c_n\!:2$	-1	1	4	-5	38	$-67\cdots$	and so on.

Here, c_n is called the **second difference** of a_n, and we write $c_n = \Delta^2 a_n$. By applying Lemma 4.4.5 twice, we know that the generating function of c_n is $(1 - x)^2 f(x)$. Here, $f(x) = 2 + 3x + 5x^2 + 11x^3 + \cdots$, and so $(1 - x)^2 f(x) = 2 - x + x^2 + \cdots$.

Differences are used extensively in numerical calculations and have an extensive literature. Many results of calculus were first understood in terms of differences, and differential calculus was initially understood as the continuous analogue of difference calculus.

DEFINITION 4.4.6

The **division of formal power series** is defined as the inverse of multiplication. That is, if $f(x)g(x) = h(x)$ with $g(x) \neq 0$, then we write $f(x) = h(x)/g(x)$.

Before applying series to solve difference equations, let us compile a short list of some simple sequences and their associated generating functions. We prove the formulas by using the preceding lemmas.

THEOREM 4.4.7

$$1 + x + x^2 + \cdots = \sum x^k = \frac{1}{1 - x} \qquad (4.4.3)$$

$$1 + 2x + 3x^2 + \cdots = \sum (k + 1)x^k = \frac{1}{(1 - x)^2} \qquad (4.4.4)$$

$$x + 4x^2 + 9x^3 + \cdots = \sum k^2 x^k = \frac{x + x^2}{(1 - x)^3} \qquad (4.4.5)$$

$$1 + ax + a^2 x^2 + \cdots = \sum a^k x^k = \frac{1}{1 - ax} \qquad (4.4.6)$$

Proof Let $f_0(x) = 1 + x + x^2 + \cdots$. This is the generating function of the sequence $1, 1, 1, \ldots$. The difference sequence is $1, 0, 0, \ldots$, which has generating function $1 + 0x + 0x^2 + \cdots = 1$. Thus, $(1 - x)f_0(x) = 1$. This proves (4.4.3).

Now let $f_1(x) = 1 + 2x + 3x^2 + \cdots$. This is the generating function of $1, 2, 3, 4, \ldots$. The difference sequence is $1, 1, 1, 1, \ldots$, which is the sequence of (4.4.3). Thus, $(1 - x)f_1(x) = 1/(1 - x)$. This yields (4.4.4).

Let $f_2(x) = x + 4x^2 + 9x^3 + \cdots$. Its sequence is $0, 1, 4, 9, \ldots, k^2, \ldots$. The difference sequence is $0, 1, 3, 5, \ldots, k^2 - (k - 1)^2 = 2k - 1, \ldots$. The second difference is $0, 1, 2, 2, 2, \ldots$, where the 2's continue indefinitely. The third difference is $0, 1, 1, 0, 0, \ldots$, whose generating function is $x + x^2$. Thus, $(1 - x)^3 f_2(x) = x + x^2$, and this yields Equation (4.4.5). (Note that the third difference is not identically 0 because of our convention that $a_s = 0$ for $s < 0$. Thus, the formula $a_k = k^2$ gives $a_{-1} = (-1)^2 = 1$, but this is not used because we have agreed that $a_s = 0$ for all negative values of s.)

Finally, if $g(x) = 1 + ax + a^2 x^2 + \cdots$, then

$$axg(x) = ax + a^2 x^2 + \cdots = g(x) - 1$$

Solving for $g(x)$, we obtain (4.4.6). Alternatively, we may substitute ax for x in (4.4.3), a valid operation for infinite series. ∎

We have observed that the difference $\Delta a_n = a_n - a_{n-1}$ corresponds to the multiplication of the generating function by $1 - x$. We now show that division by $1 - x$ corresponds to finding sums.

THEOREM 4.4.8

Let a_n be a sequence with generating function $f(x)$. Let $s_n = \sum_{k=0}^{n} a_k$, and let $s(x)$ be the generating function of s_n. Then

$$s(x) = f(x)/(1 - x) \qquad (4.4.7)$$

Proof We can do this directly by using Equations (4.4.2) and (4.4.3). We have

$$f(x)(1/(1 - x)) = \left(\sum a_n x^n\right) \sum x^n$$
$$= \sum c_n x^n$$

where $c_n = \sum_{k=0}^{n} a_k \cdot 1$, by Equation (4.4.2). But this gives $c_n = s_n$, which is the result.

We can also proceed less directly. The series $s(x)(1 - x)$ has sequence $s_n - s_{n-1} = a_n$, so $s(x)(1 - x) = f(x)$. This also yields (4.4.7). ■

The relationship between linear difference equations and generating functions is given by the following theorem.

THEOREM 4.4.9

Let a_n be defined by the linear recursion

$$a_n + r a_{n-1} + s a_{n-2} = c_n \qquad (n \geq 2) \qquad (4.4.8)$$

Let $f(x)$ be the generating function of a_n, and let $g(x)$ be the generating function of c_n. Then

$$f(x) = \frac{g(x) + A + Bx}{1 + rx + sx^2} \qquad (4.4.9)$$

for certain constants A and B depending on the initial values a_0 and a_1.

Proof We compute $f(x)(1 + rx + sx^2)$. The coefficient of x^n is $a_n + r a_{n-1} + s a_{n-2}$. For $n \geq 2$, this is simply c_n, by (4.4.8). Thus, we have $f(x)(1 + rx + sx^2) - g(x) = A + Bx$ because all powers of x larger than 1 vanish. ■

Remark   To find A and B, we compare coefficients on both sides of the equation

$$f(x)(1 + rx + sx^2) = g(x) + A + Bx$$

Comparing constant coefficients, we obtain

$$a_0 = c_0 + A$$

Comparing the coefficients of x, we have

$$a_1 + ra_0 = B + c_1$$

Thus, we have

$$A = a_0 - c_0; \qquad B = a_1 + ra_0 - c_1 \qquad\qquad (4.4.10)$$

This latter formula is not worth memorizing. It can be worked out by this method in any example.

EXAMPLE 4.4.10

The reader may verify that the generating function of the Fibonacci sequence is $f(x) = x/(1 - x - x^2)$.

Of course, results analogous to Theorem 4.4.9 may be stated for higher- or lower-order difference equations. Note that the denominator in Equation (4.4.9) is not the characteristic equation of the difference equation—the coefficients are reversed.

When applying generating functions in a practical situation, we often need to break up a fraction such as

$$\frac{3 - x}{(1 - 2x)(1 + 3x)}$$

into a sum of simpler fractions (called **partial fractions**).

EXAMPLE 4.4.11

Write $\dfrac{3 - x}{(1 - 2x)(1 + 3x)}$ as a linear combination of $\dfrac{1}{1 - 2x}$ and $\dfrac{1}{1 + 3x}$.

Method   We use "undetermined coefficients" A and B to force the following equality:

$$\frac{3 - x}{(1 - 2x)(1 + 3x)} = \frac{A}{1 - 2x} + \frac{B}{1 + 3x}$$

Clear of fractions to obtain

$$3 - x = A(1 + 3x) + B(1 - 2x)$$

Combining like terms, we get

$$3 - x = (A + B) + (3A - 2B)x$$

Equating coefficients, we obtain

$$3 = A + B$$

$$-1 = 3A - 2B$$

Solving, we find $A = 1$ and $B = 2$. Finally, substituting in the original equation, we get

$$\frac{3 - x}{(1 - 2x)(1 + 3x)} = \frac{1}{1 - 2x} + \frac{2}{1 + 3x}$$

We now illustrate how generating functions can be used as an alternative method for solving difference equations.

EXAMPLE 4.4.12

Solve the difference equation

$$a_n = 3a_{n-1} + 1 \qquad (n \geq 1)$$

$$a_0 = 2$$

$$(4.4.11)$$

Method   Let $f(x)$ be the generating function of a_n. We know that the generating function of the constant sequence $c_n = 1$ is $g(x) = 1/(1 - x)$. Thus, Equation (4.4.11) can be written

$$f(x) = 3xf(x) + 1/(1 - x) + C$$

The constant C is necessary since Equation (4.4.11) fails for $n = 0$. To evaluate C, find the constant term on both sides of this equation. This gives $2 = 0 + 1 + C$, so $C = 1$. Thus

$$f(x) = 3xf(x) + 1/(1 - x) + 1$$

Solving for $f(x)$, we obtain

$$(1 - 3x)f(x) = 1/(1 - x) + 1$$

and so

$$f(x) = \frac{1/(1 - x) + 1}{1 - 3x} = \frac{2 - x}{(1 - x)(1 - 3x)}$$

The reader may now verify that this latter fraction can be decomposed into partial fractions as

$$\frac{2 - x}{(1 - x)(1 - 3x)} = -\frac{1}{2}\frac{1}{1 - x} + \frac{5}{2}\frac{1}{1 - 3x}$$

Thus, by Equations (4.4.3) and (4.4.6), $a_n = (5 \cdot 3^n - 1)/2$.

The Binomial Coefficients

We conclude this section by looking at the binomial coefficients from the point of view of generating functions. For fixed n, the sequence $C(n, 0), C(n, 1), C(n, 2), \ldots$ has $\sum C(n, k)x^k$ as its generating function. This is exactly the formula for $(1 + x)^n$, by the Binomial Theorem. Thus, the sequence

$$C(n, 0), C(n, 1), C(n, 2), \ldots$$

has generating function $(1 + x)^n$.

We can rederive this formula as follows. The key equations are the ones needed to create Pascal's triangle:

$$C(n, r) = C(n - 1, r) + C(n - 1, r - 1) \qquad \text{(the recursion)} \qquad \textbf{(4.4.12)}$$

and

$$C(0, 0) = 1, \qquad \text{with } C(0, r) = 0 \text{ for } r \neq 0 \qquad \textbf{(4.4.13)}$$

[See Equations (4.2.7) and (4.2.9).] Let $P_n(x)$ be the generating function for $C(n, 0), C(n, 1), \ldots$ (for fixed n). Then

$$P_n(x) = \sum_{r=0}^{\infty} C(n, r)x^r$$

By (4.4.13), we have

$$P_0(x) = 1 \qquad \textbf{(4.4.14)}$$

Because the generating function of $C(n - 1, r - 1)$ is $xP_{n-1}(x)$, we have, by (4.4.12),

$$P_n(x) = P_{n-1}(x) + xP_{n-1}(x)$$

or

$$P_n(x) = (1 + x)P_{n-1}(x) \qquad (4.4.15)$$

But (4.4.14) and (4.4.15) show that $P_n(x) = (1 + x)^n$, and this is the Binomial Theorem:

$$(1 + x)^n = \sum_{r=0}^{n} C(n, r)x^r \qquad (4.4.16)$$

The terms stop at $r = n$, since $C(n, r) = 0$ for $r > n$. Since this is a finite sum, we can substitute values for x to obtain a binomial identity.

THEOREM 4.4.13

For $n > 0$,

$$C(n, 0) + C(n, 2) + C(n, 4) + \cdots = C(n, 1) + C(n, 3) + C(n, 5) + \cdots$$

Proof Put $x = -1$ in (4.4.16), to obtain

$$0 = \sum_{r=0}^{n} C(n, r)(-1)^r$$

or

$$0 = C(n, 0) - C(n, 1) + C(n, 2) - \cdots$$

This is the result. ■

Remark Because $C(n, r)$ is the number of sets of size r that are subsets of a given set of size n, this theorem states that there are exactly as many subsets with an odd number of elements as there are with an even number. (Compare with Exercise 2.3.18, where this was to be proved by induction.)

EXAMPLE 4.4.14

Similarly, putting $x = 1$ into (4.4.15) gives

$$2^n = \sum_{r=0}^{n} C(n, r)$$

This too has a combinatorial significance. The left side is the number of subsets of a set with n elements. The right side computes this count by a case analysis according to the number of elements in the subset. Equivalently, the left side is the number of binary strings of length n, and the right side counts these strings by a case analysis according to the number r of 0's in the string.

EXAMPLE 4.4.15

There is an interesting direct combinatorial proof of the Binomial Theorem. We write

$$(1 + x)^n = (1 + x)(1 + x) \cdots (1 + x) \qquad (n \text{ factors})$$

Now think of how this latter product is evaluated. For each factor $1 + x$, we must decide whether to use the term 1 or x, and we must do this for all n of the factors. Further, all possibilities must occur. Now suppose we have n objects $A_1, A_2, \ldots, A_n$. Let us think of the term 1 in $1 + x$ as representing "don't choose the object," and the term x represents "choose the object." The evaluation of the product $(1 + x)^n$ may thus be regarded as a computation of all 2^n subsets of the objects. Every time we choose r objects, and don't choose the remaining $n - r$ objects, a term x^r is created. The coefficient of x^r is therefore the count of how many subsets of size r there are. This is how $C(n, r)$ was originally defined.

We can relate the counting technique of Example 4.4.15 to a previous problem in counting, namely the number of unordered samples of size r from a population of size n, allowing (any number of) replacements.

EXAMPLE 4.4.16

We now reconsider Example 4.2.17. This called for the number of unordered samples of size 12 taken from a population of size 3 (pennies, nickels, and quarters) allowing replacement. We compute

$$(1 + x + x^2 + \cdots)(1 + x + x^2 + \cdots)(1 + x + x^2 + \cdots)$$

(one factor for each element of the population). The terms $x^2, x^3, \ldots$ are used because replacements are allowed. Using x^3 in the first factor corresponds to choosing three copies of the first object. Thus, for example, the underlined entries in

$$(1 + \underline{x} + x^2 + \cdots)(1 + x + \underline{x^2} + \cdots)(1 + x + x^2 + \underline{x^3} + \cdots)$$

correspond to choosing one penny, two nickels, and three quarters, for a total sample of size 6, obtained by multiplying the underlined terms.

This leads to the term x^6. Thus, the number of unordered samples of size 6 is the coefficient of x^6 in the expansion of this product. In general, then, the series $1/(1 - x)^3$ is the generating function for the sequence a_r, where a_r is the number of unordered samples of size r, with replacements, taken from a population of size 3. We already know, from Theorem 4.2.18, that this is $C(r + 2, 2)$. Therefore, we have a **combinatorial** proof of the equation

$$1/(1 - x)^3 = \sum C(r + 2, 2)x^r$$

If we compare with Equations (4.4.4) and (4.4.3), which can be written

$$1/(1 - x)^2 = \sum C(r + 1, 1)x^r$$

and

$$1/(1 - x)^1 = \sum C(r + 0, 0)x^r$$

we are led to the following theorem.

THEOREM 4.4.17

For any positive integer n,

$$\frac{1}{(1 - x)^n} = \sum_{r=0}^{\infty} C(r + n - 1, n - 1)x^r \qquad (4.4.17)$$

Proof One proof is combinatorial. The left side is the product of n identical factors $(1 + x + x^2 + \cdots)$. The resulting coefficient of x^r may be regarded, as we have seen, as the number of unordered samples of size r, with replacement, taken from a sample of size n. By Theorem 4.2.18, this is $C(n + r - 1, r) = C(n + r - 1, n - 1)$.

A direct proof by induction is also possible. We have the result for $n = 1$, as we have seen. Now assume (4.4.17) for a given n and consider the sum

$$f(x) = \sum_{r=0}^{\infty} C(r + n, n)x^r$$

We compute the sequence of $(1 - x)f(x)$ by finding the difference sequence $C(r + n, n) - C(r + n - 1, n)$. (This is Lemma 4.4.5.) But by the basic recursion (4.4.12) for binomial coefficients (replacing n by $r + n$, and r by n), this is just $C(r + n - 1, n - 1)$. Thus, we have

$$(1 - x)f(x) = \sum C(r + n - 1, n - 1)x^r$$

which is $1/(1 - x)^n$ by induction. Thus, $f(x) = 1/(1 - x)^{n+1}$, completing the inductive proof. ■

This latter proof of Equation (4.4.17) gives an alternative combinatorial proof of Theorem 4.2.18 on unordered samples without replacement.

EXAMPLE 4.4.18

How many unordered samples of size 4 can be formed from the alphabet $\{A, B, \ldots, Z\}$, in which each letter may be repeated no more than twice? How many are of size 8?

Method   Each letter may be chosen zero, one, or two times. We thus choose the factor $1 + x + x^2$ for each letter. The product $(1 + x + x^2)^{26}$ is the generating function for the number of such samples. (The coefficient a_n of x^n is the number of such samples of size n.) We have to find the coefficients of x^4 and of x^8. We can proceed algebraically by writing

$$1 + x + x^2 = (1 - x^3)/(1 - x)$$

We thus compute, using the Binomial Theorem and Equation (4.4.17),

$$(1 + x + x^2)^{26} = (1 - x^3)^{26}/(1 - x)^{26}$$
$$= [1 - C(26, 1)x^3 + C(26, 2)x^6 - \cdots]$$
$$\cdot[1 + C(26, 25)x + C(27, 25)x^2 + C(28, 25)x^3 + \cdots]$$

The coefficient a_4 of x^4 is $C(29, 25) - C(26, 1)C(26, 25)$, or 23,075. Similarly, the coefficient a_8 is

$$C(33, 25) - C(26, 1)C(30, 25) + C(26, 2)C(27, 25) = 10{,}293{,}075$$

Remark   We may regard this problem as a sample from a population with repetition. The population is $\{A, A, B, B, \ldots, Z, Z\}$, and we were asking for the number of unordered samples of size 4 and 8, without replacement, from this population.

Exercises

1. Use generating functions to solve $a_n = 2a_{n-1} + 3\,(n \geq 1)$ with initial condition $a_0 = 1$.
2. Using generating functions, solve $a_n = 2a_{n-1} + 3^n\,(n \geq 1)$ with initial condition $a_0 = 1$.
3. Using generating functions, solve the difference equation
$$a_n = a_{n-1} + 2a_{n-2} \quad (n \geq 2); \qquad a_0 = 0, \quad a_1 = 1$$
4. Using generating functions, solve the difference equation
$$a_n - 5a_{n-1} + 6a_{n-2} = 0 \quad (n \geq 2); \qquad a_0 = 1, \quad a_1 = 1$$

*5. Using generating functions, solve the difference equation

$$a_n - 5a_{n-1} + 6a_{n-2} + 1 = 0 \quad (n \geq 2); \quad a_0 = 1, \quad a_1 = 1$$

6. Let $\sum a_n x^n = (7 + 11x)/(1 - x^2)$. Find a_{30} and a_{31}. (*Hint:* Use Equation (4.4.3) with x^2 substituted for x.)

7. Following the introduction of generating functions, the computation $h(x) = f(x)g(x) = \sum c_n x_n$ was carried out through the x^5 term, where $f(x) = \sum kx^k$ and $g(x) = \sum(-1)^k x^k$. Find the generating functions $f(x)$, $g(x)$, and $h(x)$.

*8. In the previous exercise, prove that the sequence c_n of coefficients of $h(x)$ is 0, 1, 1, 2, 2, . . . , where $a_0 = 0$, $a_1 = a_2 = 1$, and $a_n = a_{n-2}$ for $n \geq 3$, as asserted in the text. (*Hint:* One method is to multiply numerator and denominator by $1 + x$. Then use the technique given in the hint of Exercise 6.)

9. Describe the sequence whose generating function is $(1 - 8x + 3x^2)/(1 - x^3)$.

*10. (a) Prove that

$$\sum_{r=0}^{n} C(n, r)2^r = 3^n$$

(b) Give a combinatorial argument to prove this formula. (*Hint:* Count the number of strings of length n on the alphabet $\mathbb{I}_3$. Then make an appropriate case analysis.)

11. Using the identity $(1 + x)^n(1 + x)^2 = (1 + x)^{n+2}$, find a binomial identity equating $C(n + 2, r + 2)$ with terms involving $C(n, k)$. Do this by equating the coefficients of x^{r+2} on both sides of the equation.

*12. Generalize the previous exercise to find a binomial identity equating $C(n + k, r + k)$ with terms involving $C(n, s)$ and $C(k, t)$.

*13. Give a combinatorial argument for the identity of the previous exercise. (*Hint:* Imagine n pennies and r nickels in a pile. In how many ways can $r + k$ coins be chosen from this set? Make a case analysis according to how many pennies are in the chosen set.)

*14. The binomial coefficients for $n = 2$ are 1, 2, 1. We take differences twice to obtain

$$
\begin{array}{cccccc}
1 & 2 & 1 & & & \\
1 & 1 & -1 & -1 & & \\
1 & 0 & -2 & 0 & 1 &
\end{array}
$$

The result is the original sequence, interspersed with 0's, but with alternating signs. Show that this is true for the binomial coefficients $C(n, 0)$, $C(n, 1)$, $C(n, 2)$, . . . , $C(n, n)$, provided n differences are taken. (*Hint:* Use generating functions and Lemma 4.4.5.)

15. Use the identity $(1 + x)^{2n} = (1 + x)^n(1 + x)^n$ to prove

$$C(2n, n) = \sum_{r=0}^{n} C(n, r)^2$$

*16. As in the previous exercise, use the identity

$$(1 + x)^n (1 - x)^n = (1 - x^2)^n$$

to derive a binomial identity. (*Hint:* Compare the coefficients of x^n on both sides. Separate cases according to whether n is odd or even.)

In Exercises 17–22, find the generating function of the indicated sequence. Simplify if possible. (Make a reasonable guess of the intended infinite sequence.)

17. 1, 2, 3, 4, . . .
18. 0, 1, 0, 1, 0, 1, . . .
*19. 1, 0, -2, 0, 4, 0, -8, 0, . . .
20. 41, 30, 1, 1, 1, 1, . . .
*21. 1, -2, 3, -4, . . .
22. 1, 2, 3, 1, 2, 3, 1, 2, 3, . . .

*23. Using generating functions, solve the difference equation

$$a_n = n + 2a_{n-1} - a_{n-2} \ (n \geq 2)$$

with initial conditions $a_0 = a_1 = 1$.

*24. Using the method used to prove Equation (4.4.5), find a formula for

$$\sum_{k=0}^{\infty} k^3 x^k.$$

*25. Repeat Exercise 24 for $\sum k^4 x^k$.
26. Explain the formula

$$(1 + x)(1 + x^2)(1 + x^4)(1 + x^8) = 1 + x + x^2 + x^3 + \cdots + x^{15}$$

combinatorially. Similarly, explain the formula

$$(1 + x)(1 + x^2)(1 + x^4)(1 + x^8) \cdots = 1/(1 - x)$$

*27. Prove the identity

$$(1 + x)(1 + x^2)(1 + x^3) \cdots = \frac{1}{(1 - x)(1 - x^3)(1 - x^5) \cdots}$$

(*Hint:* Use Exercise 26 and substitute x^3 for x, x^5 for x, and so on. Interpret this identity combinatorially. The result is due to Euler.)

*28. How many unordered samples of size 15 can be chosen from the alphabet $\{A, B, C, D, E\}$, where each letter can be repeated at most five times? How many of size 25 can be chosen?

*29. There are no unordered samples of size 26 of the type given in the previous exercise. Use this fact and the appropriate generating function to find a combinatorial binomial identity.

*30. Generalize the previous exercise for samples of an arbitrary size K, where $K \geq 26$.

*31. Suppose we ignore suits in a deck of cards. Thus, we regard the four aces as identical, and similarly for each denomination. How many (unordered) five-card poker hands are there?

Summary Outline

- **The Multiplication Principle**   If a process is done in two stages, and there are m ways of doing stage 1 and for each choice at stage 1 there are n ways of doing stage 2, then there is a total of mn ways of performing the process.

- **The Addition Principle**   If A and B are disjoint finite sets and $C = A \cup B$, then $|C| = |A| + |B|$.

- **The Inclusion-Exclusion Formula** (two sets)   If A and B are finite sets, then $|A \cup B| = |A| + |B| - |A \cap B|$.

- **Properties** of the characteristic function:

$$f_{A \cap B} = f_A f_B; \quad f_{A \cup B} = f_A + f_B - f_A f_B; \quad f_{A'} = 1 - f_A$$

- **The Inclusion-Exclusion Formula** (n sets)

$$|A_1 \cup \cdots \cup A_n| = S_1 - S_2 + S_3 + \cdots \pm S_n,$$

where S_1 is the sum of the $|A_i|$, S_2 is the sum of the $|A_i \cap A_j|$ $(i < j)$, and so on, until $S_n = |A_1 \cap \cdots \cap A_n|$.

- **The Euler φ function** $\varphi(n)$ is defined as the number of integers x such that x and n are relatively prime and $1 \le x \le n$.

- **Theorem**   If the prime divisions of n are $p_1, \ldots, p_k$, then

$$\varphi(n) = n(1 - 1/p_1) \ldots (1 - 1/p_k)$$

- **A permutation** of n objects is a listing of these objects in a definite order. If $r \le n$, an **r-permutation** of n elements is a listing of r of these objects in a definite order. In each case, the listing must consist of distinct objects.

- **An r-combination** of n objects is a subset of these objects consisting of r distinct elements.

- **Notation:** $P(n, r)$ is the number of r-permutations of a set with n elements. $C(n, r)$ is the number of r-combinations of a set with n elements.

- **Theorem**

$$P(n, r) = n(n - 1) \cdots (n - r + 1) = \frac{n!}{(n - r)!}$$

$$C(n, r) = \frac{P(n, r)}{r!} = \frac{n!}{r!(n - r)!}$$

- **Formulas** $C(n, r) = C(n, n - r); \; P(n, n) = n!; \; C(n, 0) = 1$

- **A card hand** usually consists of distinct cards (either ordered or unordered according to the game), all chosen from a deck of 52 cards. There are four **suits:** hearts, diamonds, clubs, and spades. There are 13 **denominations,** called ace (or 1), the numbers from 2 through 10, and jack, queen, and king. Each possible suit occurs with each denomination, so there are 4·13 or 52 cards in all.

- The **probability** p that an event E occurs is defined as $p = S/N$, where S is the number of ways the event E can occur and N is the total number of possible outcomes. It is assumed that each of the individual outcomes is equally likely.

- **The Binomial Theorem**

$$(1 + x)^n = 1 + nx + C(n, 2)x^2 + \cdots + C(n, r)x^r + \cdots + x^n$$

  or

$$(1 + x)^n = \sum_{r=0}^{n} C(n, r)x^r$$

- The number of **unordered samples, with replacement,** of size r chosen from a population of size n is $C(n + r - 1, r)$.

- If a population of size n has **repetitions,** namely r_1 repetitions of object A_1, r_2 repetitions of object $A_2, \ldots$, then the number of different **ordered samples without replacement** of size n from this population is

$$\frac{n!}{r_1!r_2! \cdots}$$

- **Second-order, linear difference equation with constant coefficients**

  Definition An equation of the form $a_n + ra_{n-1} + sa_{n-2} = c_n$ for $n \geq 2$, where r and s are constants $(s \neq 0)$ and c_n is a given sequence.

  The **characteristic equation** of this difference equation is

$$x^2 + rx + s = 0$$

  Initial conditions: Equations $a_0 = a$, $a_1 = b$, where a and b are constants.

  A difference equation is **homogeneous** when the constant term $c_n = 0$.

  A **power solution** of the homogeneous equation is a solution of the form $A\lambda^n$.

  A **linear combination** of solutions x_n and y_n of the homogeneous equation is the sequence $z_n = cx_n + dy_n$, for constants c and d.

- **Theorem** The sequence $a_n = \lambda^n$ satisfies the homogeneous equation if and only if $\lambda^2 + r\lambda + s = 0$ with $\lambda \neq 0$. (Thus, λ satisfies the characteristic equation.)

- **Theorem** A linear combination of solutions of a homogeneous equation is also a solution.

● **Theorem** Any solution of the equation $a_n + ra_{n-1} + ra_{n-2} = c_n$ may be written as $a_n = p_n + g_n$, where p_n is a particular solution of this equation and g_n is some solution of the homogeneous equation obtained by setting $c_n = 0$.

● **Theorem** If b_n satisfies the **inequality**

$$b_n \le cb_{n-1} + db_{n-2} + c_n, \qquad n \ge 2$$

where $c, d \ge 0$, and a_n satisfies the **equality**

$$a_n = ca_{n-1} + da_{n-2} + c_n, \qquad n \ge 2$$

with $b_0 \le a_0$, $b_1 \le a_1$, then $b_n \le a_n$ for $n \ge 0$. A similar theorem is true with $\le$ replaced by $\ge$.

● **The generating function** $f(x)$ of a sequence a_n is the infinite series

$$f(x) = \sum_{k=0}^{\infty} a_k x^k = a_0 + a_1 x + a_2 x^2 + a_3 x^3 + \cdots$$

● The **difference** of the sequence a_n (often written Δa_n) is the sequence $b_n = a_n - a_{n-1}$.

● **Sequence-Generating Function Correspondences**

Sequence Operation	Series Operation
$a_n \to a_{n-1}$	$f(x) \to xf(x)$
$ca_n + db_n$	$cf(x) + dg(x)$
$a_n \to a_n - a_{n-1}$	$f(x) \to (1 - x)f(x)$
$a_n \to s_n$	$f(x) \to f(x)/(1 - x)$

(where s_n is the sum of the a_i from $i = 0$ to $i = n$)

● **The generating function solution** to the difference equation

$$a_n + ra_{n-1} + sa_{n-2} = c_n \qquad (n \ge 2)$$

if $f(x)$ is the generating function of the solution a_n, and $g(x)$ is the generating function of c_n. Then

$$f(x) = \frac{g(x) + A + Bx}{1 + rx + sx^2}$$

for certain constants A and B depending on the initial values a_0 and a_1.

● **Generating function for $C(r + n - 1, n - 1)$**

$$\frac{1}{(1 - x)^n} = \sum_{r=0}^{\infty} C(r + n - 1, n - 1)x^r$$

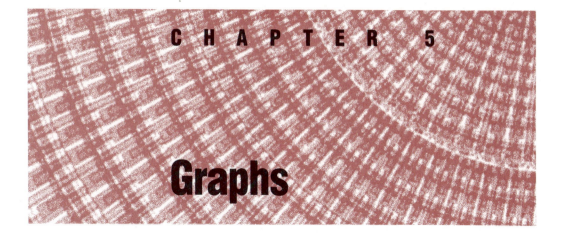

CHAPTER 5

Graphs

A graph is a way of showing connections between things. This humble idea leads to many applications and has also led to a significant body of research. In this chapter, we study some of the theory and applications of graphs. The applications include map coloring, travel optimization, games of strategy, and the use of networks to solve traffic flow problems.

5.1 Graphs

A **graph** is a set of vertices, some pairs of which are connected. In this section, we see how certain data may be represented by graphs that give instant pictorial information about the data. We have already seen its use in Section 2.1 on relations. We start with some simple examples before proceeding to the definition and theory.

EXAMPLE 5.1.1

Abe, Beth, Carol, Dan, Ellen, Frieda, and George work in an office. Abe is a friend of Ellen and Frieda, Beth is a friend of Carol and Dan, Carol is a friend of George and Dan, and Dan and Ellen are friends of Frieda. (a) Using this information, show how Abe can be introduced to Beth via a series of introductions by friends. (b) Abe confided a secret to a friend and later found out that George found out about it through the grapevine. He accused Dan of passing the secret. Was he justified?

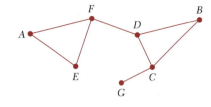

Figure 5.1.1

Method Let the people in the office be represented by vertices, and connect vertices that represent friends. The resulting graph of the friendship relation is given in Figure 5.1.1. We see immediately that Abe can be introduced to Dan via Frieda, and Dan can then introduce him to Beth. It is also immediately clear from the figure that Dan did indeed pass the secret (and that Frieda and Carol did too). The entire question is thus reduced to paths that can be formed on a graph.

Often, in debugging programs, it is necessary to follow the logic of the program to see which steps are executed. Let us do this for an exceedingly simple algorithm, the algorithm given in Example 3.2.26 for a calculation using the Collatz function. The algorithm was

1. $M := 0$
2. IF $n = 1$ THEN End
3. WHILE $n > 1$
    1. IF n is even THEN $n := n/2$
    2. IF n is odd THEN $n := 3n + 1$
    3. $M := M + 1$

EXAMPLE 5.1.2

Draw a graph showing the flow of the preceding algorithm. Using this graph, show why it is possible to go on forever (or "go into an infinite loop"). Once step 3.1 is executed, can step 3 be reached? Can 2?

Method Let each of the instructions be represented by a vertex, as in Figure 5.1.2. We also let the instruction End be represented by a vertex.

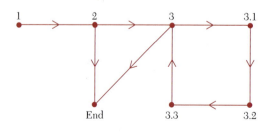

Figure 5.1.2

By the definition of a WHILE instruction, once all subcommands are executed, we return to it to further check the WHILE condition. If the condition is not satisfied, we go to the instruction after the complete WHILE statement, which is an implicit End instruction. This accounts for the edge from 3 to End. From the figure, we can see the possible infinite loop:

$$3 \rightarrow 3.1 \rightarrow 3.2 \rightarrow 3.3 \rightarrow 3 \rightarrow \cdots$$

We can also see the path from 3.1 to 3, and we can also see from the graph that the path from 3.1 to 2 is not possible.

This analysis only gives the possibilities. As noted before, it is unknown whether an infinite loop does occur.

Once again, as in Example 5.1.1, the proposed questions are easily phrased and understood in terms of paths. The path from 3.2 to 2 is impossible because the edges all have a specified direction. This makes sense because the paths have a specified direction. Figure 5.1.2 represents a **directed graph** (see Definition 5.1.4), which is the case with the logical flow of algorithms.

The previous examples illustrate two broad categories of graphs: directed graphs (called **digraphs**) and undirected or simple graphs (called **graphs**). We start by defining a simple graph, and we then show how a small variation defines a directed graph, as well as more general graphs.

DEFINITION 5.1.3   A Graph

A **simple graph** G, or a **graph** G, is a finite set V, called the vertices of G, together with a function $C: V \times V \rightarrow \mathbb{N}$ that satisfies three conditions:

$$C(v, w) = C(w, v) \tag{5.1.1}$$

$$C(v, v) = 0 \tag{5.1.2}$$

and

$$C(v, w) = 0 \text{ or } 1 \tag{5.1.3}$$

If $C(v, w) = 1$, we say that v and w are **adjacent** in G. If $C(v, w) = 0$, then they are not adjacent. If v and w are adjacent, we say that vw is an **edge** of G. The function C is called the **counting function** of G.

The idea behind the definition is that once a set V of vertices is given, we must also be able to determine whether two of the vertices are adjacent. We may think of 1 as "true" and 0 as "false," so the counting function $C(v, w)$ tells whether v and w are adjacent. Then $C(v, w)$ is the **number** of edges joining v and w, hence its name. Equation (5.1.2) makes

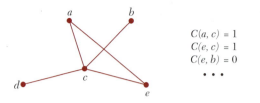

$C(a, c) = 1$
$C(e, c) = 1$
$C(e, b) = 0$

$\bullet \ \bullet \ \bullet$

Figure 5.1.3

it impossible for an edge to connect a vertex to itself. Equation (5.1.1) shows that the relation "v is adjacent to w" is symmetric; vertex v is adjacent to w if and only if w is adjacent to v. Equation (5.1.1) is called the **symmetry condition,** and we say that the counting function is **symmetric.** This equation reflects that the graph is not directed. Any set V and any counting function C satisfying Equations (5.1.1) to (5.1.3) determines a graph. When constructing a diagram of this graph, the value of C determines when to join two vertices with an edge. Figure 5.1.3 illustrates this definition.

For a directed graph, a vertex v may be adjacent to w (pictorially, the arrow goes from v to w), but it is not necessary that w also be adjacent to v. The definition is as follows.

DEFINITION 5.1.4

A **directed graph** (or simply a **digraph**) G is given by a set V of vertices and a counting function $C: V \times V \rightarrow \mathbb{N}$ that satisfies Equations (5.1.2) and (5.1.3).

Thus, C is not necessarily symmetric. A digraph is illustrated in Figure 5.1.4.

As in Definition 5.1.3, we say that vw is an **edge,** or a directed edge, of the digraph G if $C(v, w) = 1$.

Remark   Some confusion is possible. According to Definitions 5.1.3 and 5.1.4, a simple graph can also be regarded as a directed graph,

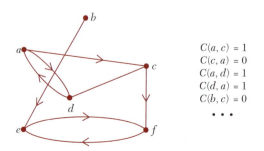

$C(a, c) = 1$
$C(c, a) = 0$
$C(a, d) = 1$
$C(d, a) = 1$
$C(b, c) = 0$

$\bullet \ \bullet \ \bullet$

Figure 5.1.4

since its counting function certainly satisfies Equations (5.1.2) and (5.1.3). If a counting function C is symmetric, the graph can be pictured as a simple graph, with edges v and w joined if and only if $C(v, w) = 1$. However, if the graph is regarded as a digraph by Definition 5.1.4, then, if $C(v, w) = 1 = C(w, v)$, we would envision two directed edges, one from v to w and one from w to v. In Section 2.1, the diagram of the simple graph was called the symmetric skeleton of the corresponding digraph. To avoid a more cumbersome notation, we adopt the following convention:

> A **symmetric** counting function is always understood to represent a simple, nondirected graph, unless otherwise stated. A **nonsymmetric** counting function necessarily represents a directed graph.

By loosening conditions (5.1.1) to (5.1.3), we can consider more general types of graphs. For example, if Equation (5.1.2) is omitted, we allow edges joining a vertex to itself. The resulting graph is then called a **general graph**. And if condition (5.1.3) is omitted, we allow the possibility of several edges joining two vertices. This is called a **multigraph**. In that case, $C(v, w)$ is the number of edges joining v to w, and this number may be greater than 1. In either case, we can have directed or nondirected graphs. (Our convention on whether we have a directed or a nondirected graph still holds and depends on whether the counting function is symmetric.)

Condition (5.1.3) held for relations, but we did not necessarily have (5.1.1) and (5.1.2). Thus, the graph of a relation was, in these terms, a directed general graph. The most general case is a graph whose counting function satisfies *none* of the conditions (5.1.1) to (5.1.3). (However, C must still have value in $\mathbb{N}$.) This is a general directed multigraph. Figure 5.1.5 illustrates some of these possibilities. Unless otherwise noted, either explicitly or by context, the word "graph" represents a *simple graph* as in Definition 5.1.3.

Having given these definitions, we often use a figure to prove or illustrate cases. But one should not regard these definitions as mere formalities. The values $C(i, j)$ are precisely what is needed to put a graph into a computer in order to analyze it. For if we had a digraph with 100 vertices, a picture would appear quite complicated. But by putting the appropriate values $C(i, j)$ as data in a two-dimensional array C, we would "have" the graph in the computer, using 10,000 pieces of information [the values of $C(i, j)$]. A computer analysis of the graph could then be undertaken. The vertices in this case would be the set $\mathbb{I}_{100}$.

Figure 5.1.6 illustrates different drawings of the same graph. Why do we say that they are the same graph? Simply because they have the same vertices (the set $\{a, b, c, d, e\}$) and the same connections. By our definition, a graph is merely a set of vertices and an appropriate counting function. The figure is a way of illustrating the graph, and we have enough leeway in drawing the graph to allow many different-looking diagrams.

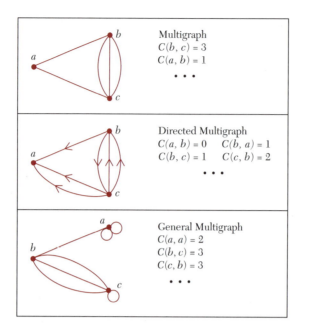

Figure 5.1.5

We occasionally wish to restrict our attention to some of the vertices of a graph and to some of the edges joining them, and to regard those vertices as a graph in themselves. This leads to the following definition.

DEFINITION 5.1.5

Let G be a graph with vertices V and counting function C. A **subgraph** G' of G is a graph whose vertices $V' \subseteq V$ and whose counting function C' satisfies $C' \leq C$ for the vertices of V'. If $C' = C$ for vertices of V', then G' is called the subgraph **spanned** by the vertices V'.

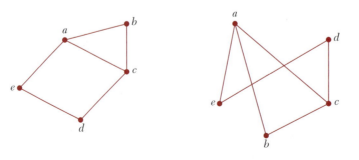

Figure 5.1.6

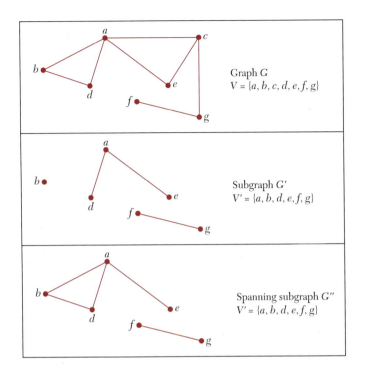

Figure 5.1.7

In short, we choose some of the vertices of G and some or all of the edges connecting them to form a subgraph G'. See Figure 5.1.7 for some examples.

If the vertices V of a graph are labeled $1, 2, \ldots, n$, then the counting function of a graph is a function $C: \mathbb{I}_n \times \mathbb{I}_n \to \mathbb{N}$. The value $C(i, j)$ is written as c_{ij}. These numbers are the entries of a **matrix**, which we also call C. Using Equations (5.1.1) to (5.1.3), we have the following result.

THEOREM 5.1.6

The matrix C of a graph is a symmetric, $(0, 1)$ matrix whose diagonal terms are all 0.† Conversely, any such matrix is the matrix of a graph. ∎

For example, the matrix

$$C = \begin{bmatrix} 0 & 1 & 0 & 0 & 1 \\ 1 & 0 & 1 & 0 & 0 \\ 0 & 1 & 0 & 1 & 1 \\ 0 & 0 & 1 & 0 & 1 \\ 1 & 0 & 1 & 1 & 0 \end{bmatrix}$$

† See the appendix for the definitions.

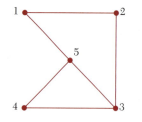

Figure 5.1.8

is one such matrix. It evidently is the matrix of the graph G of Figure 5.1.8. Simply connect vertices i and j if and only if the ith row and jth column are 1 ($c_{ij} = 1$).

Just as symmetric, (0, 1) matrices with 0 diagonal elements represent simple graphs, (0, 1) matrices with 0 diagonals represent digraphs. Similarly, multigraphs, general graphs, directed or not, can be represented by square matrices with nonnegative, integer coefficients. Figure 5.1.9 illustrates a digraph and its matrix. Each can be constructed from the other.

There is a certain arbitrariness in labeling the vertices of a graph and, hence, in finding its matrix. This is seen in Figure 5.1.10, where the simple three-vertex graph has several matrices, depending on the way we number the vertices. A renumbering of the vertices amounts to a rearrangement of the rows and columns of a matrix. Even for moderately sized graphs, this leads to many different matrices for one graph. If there are n vertices, there are $n!$ ways of numbering them from 1 to n, each leading to its own matrix. This problem will be considered in Section 5.4. The matrix representation of a graph can, however, be used to find properties of a graph. For example, see Theorem 5.1.16.

Remark   If a subgraph of G is spanned by the vertices numbered $n_1, n_2, \ldots, n_r$, its matrix can be obtained by taking the full matrix of G and striking out all rows and columns whose vertices are not one of the n_i's.

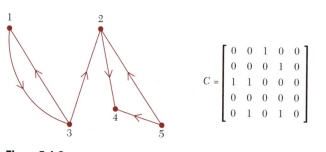

Figure 5.1.9

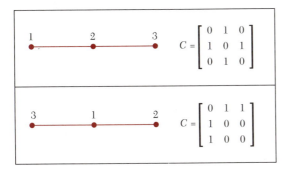

Figure 5.1.10

Paths and Connectedness

In Examples 5.1.1 and 5.1.2, we were led naturally to consider paths along graphs and digraphs. We now define these notions by using our definition of a graph.

DEFINITION 5.1.7

Let P and Q be vertices of a graph G. Then a **path of length k** joining P to Q in the graph is a sequence of vertices

$$P = P_0, P_1, P_2, \ldots, P_{k-1}, P_k = Q$$

where $P_{i-1}P_i$ is an edge of the graph for $i = 1, \ldots, k$; P is called the **initial vertex** of the path, and Q is the **terminal vertex**. If $P = Q$, the path is called a **cycle**. The cycle P, of length 0, is called the **trivial cycle**.

Note that there are k edges and $k + 1$ vertices. The foregoing path is usually denoted $P_0P_1 \cdots P_k$. It may be regarded as a string on the alphabet V of vertices. However, it is not an arbitrary string, because consecutive vertices must be an edge of the graph G: $C(P_{i-1}, P_i) = 1$ for $1 \le i \le k$.

(In Chapter 6, it will be convenient to redefine paths and cycles so that no vertex appears more than once in a path, except possibly for the initial and terminal vertices. These redefinitions will not affect the results on connectivity given in this chapter.)

As in the graphs of Figure 5.1.11, it is possible that no path can be found connecting P and Q. (Example 5.1.1 depended on a path from A to B for its solution. In that example, if they couldn't be connected, other means would have had to be found to introduce Abe to Beth.)

DEFINITION 5.1.8

Define $P \approx Q$ (read P connects to Q) to mean that there is a path with P and Q as initial and terminal vertices.

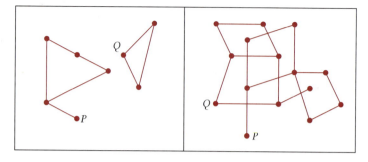

Figure 5.1.11

Remark   The relation $P \approx Q$ is the transitive closure of the relation "P is adjacent to Q."

The same definition applies to digraphs. The connecting path must be consistent with the directions of the edges. For example, we saw in Figure 5.1.2 that vertex 3.1 cannot be connected to vertex 2.

DEFINITION 5.1.9

Let G be a graph or a digraph. Then G is **connected** if for all vertices P and Q of G there is a path joining P to Q.

Equivalently, G is connected if $P \approx Q$ for all vertices P and Q.
We now develop some of the machinery to determine whether a graph is connected.

LEMMA 5.1.10

In any simple, nondirected graph G, the relation $\approx$ is an equivalence relation on the vertices V. Thus,

$P \approx P$.
If $P \approx Q$, then $Q \approx P$.
If $P \approx Q$ and $Q \approx R$, then $P \approx R$.   ■

The proof is left to the exercises.

DEFINITION 5.1.11

Let v be a vertex of the graph G. The **component** $C(v)$ is the subgraph of G spanned by the equivalence class of v under the relation $\approx$. Thus, w is a vertex of the component of v if and only if $w \approx v$.

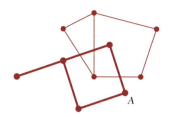

Figure 5.1.12

An example is given in Figure 5.1.12. The edges of the component of *A* are given by the darker edges. The lighter ones are also edges in the full graph.

THEOREM 5.1.12

Let *A* be a fixed vertex in a nondirected graph *G*, and let $A \approx P$ for all vertices *P* in the graph. Then the graph is connected.

Proof   Let *P* and *Q* be arbitrary vertices of *G*. Then by hypothesis, $A \approx P$ and $A \approx Q$. Thus, $P \approx A$ and $A \approx Q$, and finally, $P \approx Q$, by Lemma 5.1.10. This is the result, since *P* and *Q* were arbitrary vertices.   ■

Thus, a graph is connected if and only if it is the component of one of its vertices.

We need the following definition and consequence for several results.

DEFINITION 5.1.13

Let Σ be a set of vertices in a graph *G*. A vertex *v* is **adjacent** to Σ if it is not in Σ but is connected by an edge of the graph to some vertex of Σ.

THEOREM 5.1.14

Let Σ be a nonempty set of vertices in a graph. Let *w* be a vertex not in Σ, but $v_0 \approx w$ for some vertex v_0 in Σ. Then there is a vertex *v* adjacent to Σ.

Proof   (See Figure 5.1.13.) Join v_0 to *w* by some path. Starting from v_0 along the path there is a first vertex *v* of the path that is not in Σ. Then

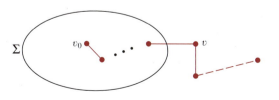

Figure 5.1.13

v is the required vertex because it is not in Σ, but, by definition, the vertex preceding it is and it is joined to v by an edge of the graph. ∎

COROLLARY 5.1.15

If Σ is any nonempty set of vertices in a connected graph G, and if Σ does not consist of all the vertices of G, then there is a vertex v of G that is adjacent to Σ. ∎

It is possible to get precise information about the *number* of paths connecting P to Q by using the matrix representation of the graph. To do this, we must use the algebra of matrices. See the appendix for details.

THEOREM 5.1.16

Let the graph G have matrix C. Let p be a nonnegative integer, and let r_{ij} be the number of paths with p edges (or $p + 1$ vertices) joining vertex i to vertex j. Then r_{ij} is the entry in the ith row and jth column of the matrix C^p.

Proof The theorem is clearly true for $p = 0$ because the number of one-vertex (or zero-edge) paths joining i to j is 0 if $i \neq j$, and is 1 for $i = j$. This is the definition of the identity matrix $I = C^0$. Now assume the result is true for p. Let $R = C^p$, so that r_{ij} is the number of paths with p edges joining i to j. Let us now compute the number of paths with $p + 1$ edges joining i to j. As in Figure 5.1.14, we break this into n cases, according to whether the path starts by joining i to 1, or to 2, . . . , or to n.

If a path starts by joining i to k, then it ends as a path with p edges joining k to j. The number of such paths is, by induction, r_{kj}. The number of paths (with one edge) joining i to k is c_{ik}. This is the definition of the matrix of a graph. Therefore, the total number of paths joining i to j, which start out by going to k, is $c_{ik}r_{kj}$. To find the total number of paths joining i to j, we sum these totals for the different cases $k = 1, 2, \ldots,$ n to find $\sum_{k=1}^{n} c_{ik}r_{kj}$. This is precisely the formula used in the definition of the product CR. But this is $C \cdot C^p = C^{p+1}$. By induction, this is the result. ∎

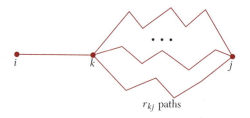

r_{kj} paths

Figure 5.1.14

Remark The proof shows that this result is true for general graphs, digraphs, and multigraphs. The paths counted in this result may have repeated vertices. We can think of C^p as measuring how many walks of length p there are on the graph.

COROLLARY 5.1.17

The number of paths s_{ij} from vertex i to j with no more than p edges is given by the matrix

$$I + C + C^2 + \cdots + C^p \quad \blacksquare$$

EXAMPLE 5.1.18

In the graph of Figure 5.1.8, how many paths of length 4 join vertex 1 to vertex 3?

Method We have

$$C = \begin{bmatrix} 0 & 1 & 0 & 0 & 1 \\ 1 & 0 & 1 & 0 & 0 \\ 0 & 1 & 0 & 1 & 1 \\ 0 & 0 & 1 & 0 & 1 \\ 1 & 0 & 1 & 1 & 0 \end{bmatrix}$$

We now square this matrix. Because it is symmetric, it is convenient to use row-by-row multiplication.

$$C^2 = \begin{bmatrix} 2 & 0 & 2 & 1 & 0 \\ 0 & 2 & 0 & 1 & 2 \\ 2 & 0 & 3 & 1 & 1 \\ 1 & 1 & 1 & 2 & 1 \\ 0 & 2 & 1 & 1 & 3 \end{bmatrix}$$

Note that the square is still symmetric. This may be used as a check of your work. To find C^4, square again:

$$C^4 = \begin{bmatrix} 9 & 1 & 11 & 6 & 3 \\ 1 & 9 & 3 & 6 & 11 \\ 11 & 3 & 15 & 8 & 7 \\ 6 & 6 & 8 & 8 & 8 \\ 3 & 11 & 7 & 8 & 15 \end{bmatrix}$$

By Theorem 5.1.16, there are 11 paths of length 4 joining vertex 1 to vertex 3. We need only read $r_{13} = 11$.

Exercises

Draw the graphs that have the following matrices.

1.
$$\begin{bmatrix} 0 & 1 & 0 & 0 & 0 & 1 \\ 1 & 0 & 1 & 0 & 0 & 0 \\ 0 & 1 & 0 & 1 & 0 & 0 \\ 0 & 0 & 1 & 0 & 1 & 0 \\ 0 & 0 & 0 & 1 & 0 & 1 \\ 1 & 0 & 0 & 0 & 1 & 0 \end{bmatrix}$$

2.
$$\begin{bmatrix} 0 & 1 & 1 & 1 & 1 \\ 1 & 0 & 1 & 1 & 1 \\ 1 & 1 & 0 & 1 & 1 \\ 1 & 1 & 1 & 0 & 1 \\ 1 & 1 & 1 & 1 & 0 \end{bmatrix}$$

Draw the digraphs that have the following matrices.

3.
$$\begin{bmatrix} 0 & 1 & 1 & 0 & 0 \\ 0 & 0 & 1 & 1 & 0 \\ 0 & 1 & 0 & 0 & 1 \\ 1 & 1 & 0 & 0 & 0 \\ 1 & 1 & 0 & 1 & 0 \end{bmatrix}$$

4.
$$\begin{bmatrix} 0 & 1 & 0 & 1 & 0 \\ 0 & 0 & 1 & 0 & 1 \\ 1 & 0 & 0 & 1 & 0 \\ 0 & 1 & 0 & 0 & 1 \\ 1 & 0 & 1 & 0 & 0 \end{bmatrix}$$

For each of the following matrices, identify the type of graph that has it as its matrix and draw the graph.

5.
$$\begin{bmatrix} 0 & 1 & 2 \\ 0 & 0 & 3 \\ 0 & 0 & 0 \end{bmatrix}$$

6.
$$\begin{bmatrix} 0 & 1 & 2 & 3 \\ 1 & 0 & 1 & 0 \\ 2 & 1 & 0 & 1 \\ 3 & 0 & 1 & 0 \end{bmatrix}$$

7.
$$\begin{bmatrix} 1 & 0 & 0 \\ 1 & 0 & 1 \\ 0 & 1 & 1 \end{bmatrix}$$

Find the matrix for each of the following graphs.

8.

9.

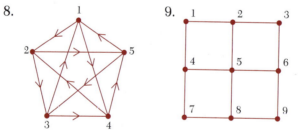

Find the matrix for each of the following graphs.

10.

11.

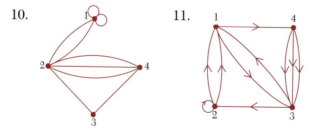

*12. Suppose G is a graph with matrix C. Let G' be the graph with the same vertices. However, vertices in G' are connected if and only if they are not connected in G. If C' is the matrix of G', find $C + C' + I$.

13. Suppose G is a simple graph with $2n$ vertices. At most, how many edges can G have? Suppose n of the vertices of G are colored red, the other n are colored black, and no edge can connect two vertices of the same color. In this case, at most how many edges can G have?

14. Prove Lemma 5.1.10.

*15. Let

$$C = \begin{bmatrix} 0 & 1 & 0 & 0 \\ 0 & 0 & 1 & 0 \\ 0 & 0 & 0 & 1 \\ 1 & 0 & 0 & 0 \end{bmatrix}$$

By interpreting this matrix graphically, compute the matrix

$$C + C^2 + C^3 + C^4$$

What matrix is C^4?

*16. Repeat Exercise 15 for the matrix

$$C = \begin{bmatrix} 0 & 1 & 0 & 1 \\ 1 & 0 & 1 & 0 \\ 0 & 1 & 0 & 1 \\ 1 & 0 & 1 & 0 \end{bmatrix}$$

*17. In Figure 5.1.15, how many paths of length 3 or less are there from vertex P to vertex Q? Use Corollary 5.1.17.

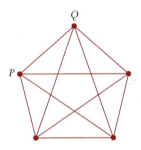

Figure 5.1.15

*18. In the multigraph of Figure 5.1.16, how many cycles of length 4 or less are there starting and ending at P? Use Corollary 5.1.17.

Figure 5.1.16

*19. In the digraph of Figure 5.1.17, how many cycles of length 4 are there starting and ending at P? Use Theorem 5.1.16.

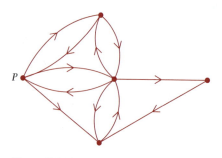

Figure 5.1.17

*20. Show that Theorem 5.1.16 is true for general multigraphs. Using this result, find the number of paths of length 3 or less joining P to Q in Figure 5.1.18.

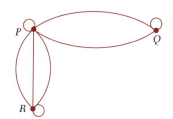

Figure 5.1.18

21. Show that Lemma 5.1.10 is not true for digraphs. Which part of it is true?
22. State and prove an analogue of Theorem 5.1.12 for digraphs.
23. State an analogue of Definition 5.1.13 for digraphs. Using your definition, state and prove an analogue of Theorem 5.1.14 for digraphs.
*24. Prove: If a graph has n vertices, then, if $v \approx w$, there is a path joining v to w that has length $\le n - 1$. (*Hint:* Show this by induction on n. If V is the set of vertices, consider the subgraph spanned by the vertices of $V - \{w\}$.)
*25. Using the result in the previous exercise, show that a graph with n vertices and matrix C is connected if and only if

$$I + C + C^2 + \cdots + C^{n-1}$$

has no zero entries. (*Hint:* Use Corollary 5.1.18.)
*26. Is the result of the previous exercise true for digraphs? Explain.

27. A top-secret organization has seven people working for it. Their code names are A, B, C, D, E, F, and G. They are each fluent in at least one language according to the following table:

Name	English	French	German	Arabic
A		Yes		
B	Yes	Yes		
C	Yes			
D	Yes			Yes
E			Yes	Yes
F		Yes	Yes	
G				Yes

(a) Draw a graph indicating which people can communicate fluently with each other.

(b) A ordinarily communicates with C with the help of B. However, B has gone on an extended tour of duty. Can A communicate with C? How?

(c) A wants to send a top-secret office memo to everybody in the office. The rules are that it is to be read once and transmitted, possibly in translation, to someone who has not read it. Finally, it is to make a full cycle and land back at A's desk, so he or she (even the sexes are top secret) can verify that nothing was lost in translation. Give a possible routing.

(d) With B on extended tour, A has discovered that the routing system in part (c) has failed. To salvage this system, one person must learn to speak French fluently. Which one? Explain by using the graph.

5.2 Euler and Hamiltonian Cycles

The Königsberg Bridge Problem was first stated by the famous Swiss mathematician Leonhard Euler (1707–1783). The problem is to take a walk over *all* of the seven bridges of the city of Königsberg (see Figure 5.2.1) without going over the same bridge twice. Euler showed that such a walk could not be done, and he also gave necessary and sufficient conditions for such a tour to exist on any such configuration. We shall do this in this section. The first step is to make a multigraph out of this system of bridges. Take the land masses as *vertices* A, B, C, D and, naturally, take the bridges as *edges*. We then obtain the multigraph of Figure 5.2.1. The proposed walk is thus a path on this multigraph in which all of the edges are used just once.

There is a simple observation that shows that the walk *cannot* be done. Except for the start and end of the walk, whenever one passes through

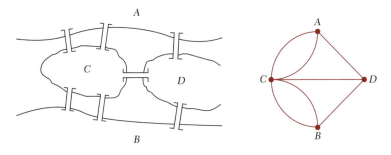

Figure 5.2.1

the vertex along an edge, one passes out of that vertex along another edge. Therefore, the edges about a vertex are used up in pairs. Since each edge is to be used, it follows that, except for the starting or ending point, there must be an even number of edges leaving any vertex. However, we may observe in Figure 5.2.2 that each vertex has an odd number of edges leaving it. This shows that no walk is possible. We shall generalize this proof and show when such a path is possible. We also find an algorithm for generating such a walk.

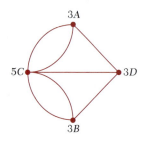

Figure 5.2.2

EXAMPLE 5.2.1 Drawing Puzzles

Puzzles that ask for a drawing of a figure without lifting the pencil and without retracing lines are exactly of this type. In Figure 5.2.3, the figure

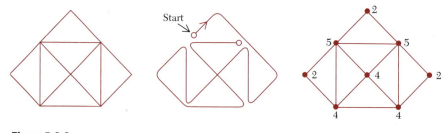

Figure 5.2.3

at the left may be drawn as illustrated. The analysis that this is possible is given at the right. Each vertex is assigned a number, called the *degree* of that vertex, namely, the number of edges leaving it. Two vertices have odd degree. By the foregoing analysis, these vertices have to be the beginning and end points of the path. Start at some other point for this tracing, and you are doomed to failure!

DEFINITION 5.2.2

If G is a graph, or more generally a multigraph, and v is a vertex of G, the **degree** of v, written $\deg(v)$, is the number of edges with end point at v. Equivalently, for a graph G, the degree of v is the number of vertices adjacent to v.

Since $C(v, w)$ is the number of edges leaving v and ending at w, $\deg(v)$ is simply the sum of the $C(v, w)$ over all w:

$$\deg(v) = \sum_{w \in V} C(v, w)$$

This is easily computed from the matrix of the graph. Here, $C(i, j)$ is written c_{ij}, and we have the formula

$$\deg(i) = \sum_{j=1}^{n} c_{ij} \qquad (5.2.1)$$

This is precisely the row sum of the matrix C. Thus, we have the following theorem.

THEOREM 5.2.3

The degree of the vertex i in a graph or multigraph G is the sum of the terms in the ith row of the matrix of G. Since the matrix of C is symmetric, it is also the sum of the ith column ■

For directed multigraphs, we can find two degrees: the number of edges coming into a vertex and the number leaving. These are called the *in-degree* and the *out-degree*, respectively:

DEFINITION 5.2.4

Let G be a directed multigraph with counting function C and matrix c_{ij}. We define the **in-degree** and **out-degree** as follows:

$$\text{in-deg}(v) = \sum_{w \in V} C(w, v)$$

and

$$\text{out-deg}(v) = \sum_{w \in V} C(v, w)$$

As before, these correspond to the column sums and row sums of the matrix of the graph.

THEOREM 5.2.5

If the matrix of a directed multigraph has entries c_{ij}, then

$$\text{out-deg}(i) = \sum_{j=1}^{n} c_{ij}$$

This is the sum of the terms in the ith row of C. Similarly, the in-degree(j) is the sum of the terms in the jth column of C. ■

For a general nondirected graph, in which a vertex is allowed to be connected to itself, it is convenient to count each of the edges that start and end at the same vertex *twice*. In Figure 5.2.4, the vertex v has degree 4, even though only three edges have v as end point. This means that in computing the degree of a vertex from its matrix, we count the diagonal term c_{ii} twice:

$$\deg(i) = c_{ii} + \sum_{j=1}^{n} c_{ij} \tag{5.2.2}$$

Equation (5.2.2) is consistent with Equation (5.2.1) because, for simple graphs, we had $c_{ii} = 0$. All of these definitions are illustrated in Figure 5.2.5.

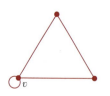

Figure 5.2.4

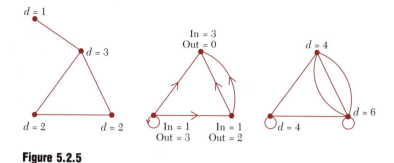

Figure 5.2.5

The following theorems relate the number of edges of a graph to the degrees at each vertex.

THEOREM 5.2.6

Let G be a multigraph with e edges and with vertices $v_1, v_2, \ldots, v_n$. Suppose vertex v_i has degree d_i. Then

$$d_1 + d_2 + \cdots + d_n = 2e \qquad (5.2.3)$$

Proof Each edge has two end points. Thus, there are exactly $2e$ end points of edges in the graph. We now make this same count by looking at the vertices. The number of end points that terminate at vertex v_i is the degree d_i. Therefore, the total number of end points using this method of counting is the sum of the d_i's. This proves (5.2.3). ■

Here is another way of looking at this proof. For every edge, place a little dot at each end point. The total number of dots is $2e$. However, each vertex v has exactly $\deg(v)$ dots. So the total number of dots is also $\Sigma \deg(v)$.

This result is true for general graphs, because an edge connecting v to itself was deemed in this proof to have two end points. An alternative proof using incident matrices is developed in Exercises 5.2.26–32. For digraphs, the situation is similar. We count beginning points, and end points, of edges to obtain the next result.

THEOREM 5.2.7

If a digraph G has e edges and vertices $v_1, v_2, \ldots, v_n$, then

$$\sum_i \text{in-deg}(v_i) = e \qquad (5.2.4)$$

$$\sum_i \text{out-deg}(v_i) = e \qquad (5.2.5)$$

■

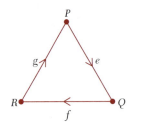

Figure 5.2.6

Suppose we take a directed graph and simply eliminate the arrows on the edges to obtain a graph. Then, because the direction of the edge is no longer a consideration, the degree of any vertex will be the sum of the in-degree and the out-degree:

$$\deg(v) = \text{in-deg}(v) + \text{out-deg}(v)$$

This is consistent with our convention in which any one-edged loop is counted as having two end points. In this case, Equation (5.2.3) is obtained by adding Equations (5.2.4) and (5.2.5).

In what follows, we work with multigraphs G. It is convenient to label each of the edges as well as the vertices. In this way, for example, if there are four edges connecting v to w [$C(v, w) = 4$], we simply have edges $e_1, \ldots, e_4$ connecting them. Then, when we walk over the graph, we indicate which edge we use. In Figure 5.2.6, the indicated path, starting at P, is denoted $PeQfRgP$. We now return to the general Euler problem.

Euler Paths and Cycles

DEFINITION 5.2.8

An **Euler path** on a multigraph G is a path in which each edge of the multigraph is used exactly once. If an Euler path begins and ends at the same vertex, then it is called an **Euler cycle**.

It is more convenient to consider Euler cycles on a multigraph than Euler paths. As we shall see, the analysis for cycles will yield, with little extra effort, the theory for paths. Following the discussion at the beginning of the section, we have the following result.

THEOREM 5.2.9

Suppose there is an Euler cycle on a multigraph G. Then each vertex of G must have an even degree.

Proof   If we follow the path as it passes through any vertex p, we find that it must enter via an edge and exit via another edge. Therefore, the

edges occur two at a time, and the total number of edges is even. The
same applies to the start and end vertex P of the cycle. The cycle begins
by leaving it along an edge and ends by entering it along another edge.
So here, too, the edges occur in paths.   ■

The situation for Euler **paths** is given by the next theorem.

THEOREM 5.2.10

Suppose there is an Euler path on a multigraph G that begins and ends
at different vertices. Then the beginning and ending vertices have odd
degree, and the rest have even degree.

Proof   Simply adjoin an extra edge connecting the beginning and ending
vertices. By using this new edge, we can take the Euler path and complete it to a cycle by continuing the path onto the beginning point. But
by Theorem 5.2.9, each vertex of the new multigraph will have even
degree. But the introduction of the new edge increases the degree by
one at the vertices at the beginning and the end of the cycle, and otherwise leaves the degrees unchanged. Therefore, a vertex of the original
multigraph has odd degree at the beginning and end points of the cycle
and even degree elsewhere.   ■

COROLLARY 5.2.11

For a multigraph to have an Euler path, at most two of its vertices must
have odd degree.   ■

Remark   By Theorem 5.2.9, it is impossible that exactly one of the
vertices of any multigraph has odd degree. Thus, the necessary
condition in Corollary 5.2.11 is that either two or none of the vertices have odd degree.

Theorem 5.2.10 is useful for constructing Euler paths because it gives
information on where the starting point must be. For example, you may
wish to draw an Euler path in Figure 5.2.7, using this result.
The converse of Corollary 5.2.11 concerns the existence of Euler paths.

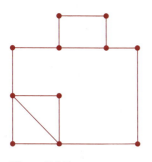

Figure 5.2.7

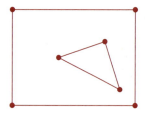

Figure 5.2.8

It turns out that having even degrees at every vertex isn't enough, because we must be able to connect all vertices whose degrees are greater than 0. (See Figure 5.2.8, where this is *not* so.) It is possible to have isolated points because they won't affect the existence of an Euler path, which concerns itself only with edges. We have the following converse of Theorem 5.2.9.

THEOREM 5.2.12

Let G be a **connected** multigraph such that every one of its vertices has even degree. Then there is an Euler cycle on G.

Proof If we attempt to create an Euler cycle starting at a vertex P, we might come back to P with no exits, because all of the edges through P might be used and, possibly, some edges will have been omitted from the attempted path. (See Figure 5.2.9, where the dashed edges have not been used.) Our proof will be given in three steps: (1) Find any potential Euler cycle starting at P. (2) Extend this cycle if there are any free edges leaving any vertex on this cycle. (3) Show that if we can't extend further, we have arrived at the required Euler cycle.

(1) Call a path a **walk** if no edge is repeated. Thus, it is not required that all edges be used, as in an Euler path. Let P be a vertex of G, and e an edge starting at P. Form a walk starting $Pe \ldots$, and continue this path so that no edge is used more than once and so that when we arrive at the end point Z, the walk cannot proceed further, because all edges from Z will have been used in the walk. Now we claim that this occurs

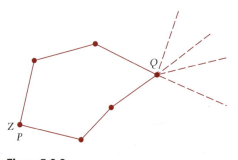

Figure 5.2.9

only when $Z = P$, as in Figure 5.2.9. Here we use the hypothesis that each vertex has even degree. For as we enter a vertex other than P, we use up one edge, so an odd number of edges will be left, hence at least one. Therefore, we can leave that vertex, and, as we leave, the number of unused edges left at the vertex is again even (possibly 0). Thus, we can always continue the partial Euler path until we arrive back at our starting point P. The constructed walk is called a **partial Euler cycle** α at P.

(2) Call an edge **free** in the cycle α if it has not been used in α. Suppose Q is any vertex on the cycle α that has some free edges on it. As shown in part (1) there are an even number of free edges at Q. We now show how to extend α to a walk that includes at least two of those free edges. The method is to build a partial Euler path at Q by *using only free edges*. The argument in part (1) shows that we can continue building such a cycle β at least until we arrive back at Q. We can now augment the cycle α as follows: Go along the cycle α until we hit Q. Then continue along the cycle β until we come back to Q, and, finally, finish out the cycle α. (See Figure 5.2.10.) We have thus extended α. Now continue this process until it is impossible to proceed further. (At each stage we reduce the number of edges available to us.) We finally arrive at a cycle α such that there is no free edge available at any vertex v on the cycle.

(3) We now show that the cycle constructed in part (2) is the required Euler cycle on G. We need to show that every edge e is in the cycle α.

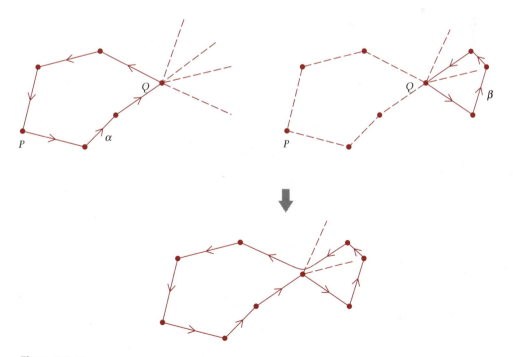

Figure 5.2.10

Let $e = RS$ be any edge of G. If R was in the cycle α, then by the construction in part (2) the edge e would be used in α and the proof would be complete. We now show that R is necessarily on the cycle. Join P, the starting vertex of the cycle α, to R by some path $PP_1P_2 \cdots R$. (Here is where we use the connectivity of the multigraph G.) We claim that every vertex on that path is necessarily in the cycle α. The proof is by induction. It is certainly true for P. Now suppose that P_i is on the cycle. Since α has no free edge available from any vertex on it, the edge P_iP_{i+1} must be an edge in the walk. Therefore, P_{i+1} is also in the cycle. This proves the result. Thus R is on the cycle, and RS is in the walk. ■

Euler **paths** are dealt with as in the proof of Theorem 5.2.10. Simply adjoin an extra edge e connecting the vertices with odd degree. Then find the cycle guaranteed by Theorem 5.2.12 and eliminate e from that cycle. The details are left for the reader. The result is the next theorem.

THEOREM 5.2.13

Let G be a connected multigraph whose vertices all have even degree, with the exception of P and Q, which have odd degree. Then there is an Euler path on G joining P to Q. ■

The situation is similar for **directed** multigraphs. Here, we imagine that the bridges all have one-way signs on them, and the tour is to be done legally, by automobile. The details are left to the reader in the exercises.

EXAMPLE 5.2.14 *An Algorithm To Construct an Euler Cycle*

We can summarize the method used in the proof of Theorem 5.2.12 by the following algorithm. G is assumed to be connected, and all vertices have even degree. We let E denote the set of edges, and V the set of vertices. A **walk**

$$\alpha = P_1e_1P_2e_2 \cdots P_ne_nP_{n+1}$$

is a sequence of vertices P and edges e such that the edges e_i are distinct. An edge e is free at the vertex P of the walk α if e is not one of the edges in the path α, but P is one of the vertices in the path. A partial Euler cycle at P is a walk whose initial and terminal vertices are P but that has no free edges at P. The first part of the proof of Theorem 5.2.12 shows how to construct a partial Euler path at P, which we summarize in the following procedure. The procedure PartialEulerCycle has graph G and vertex P as input, and output PEC = PEC(P), a partial Euler cycle in G at P.

PartialEulerCycle $(G, P; PEC)$ procedure
1. PEC := P [initializing the walk]
2. WHILE there are free edges at P
    1. Q := terminal vertex of PEC
    2. e := edge QR not in PEC
    3. PEC := PEC*eR [adjoin the edge e and its end vertex R to the walk α]

The conclusion of the proof of Theorem 5.2.12 shows how to find the required Euler cycle α. The input is a vertex P. The output is an Euler cycle EC(P).

EulerCycle $(G, P; EC)$
1. α := PEC(G, P)
2. WHILE there is an edge $e = QS$ free for some vertex Q in α
    1. E' := the set of edges used in the path α
    2. F := $E - E'$ [the free edges]
    3. H := subgraph of G with same vertices but that has F as its set of edges
    4. β := PEC(H, Q) [the PEC at Q using only free edges; now insert β into α]
    5. γ := part of path α before Q
    6. δ := part of α after Q [so $\alpha = \gamma Q\delta$]
    7. α := $\gamma\beta\delta$ [inserting β into the cycle α]
3. EC := α [the require Euler cycle]

Hamiltonian Paths and Cycles

In contrast to Euler paths, in which all edges must be used exactly once, a *Hamiltonian path* is one in which all *vertices* are used exactly once. Despite the similarity of definitions, the theory of Hamiltonian paths is significantly more intricate than the theory of Euler paths. In this section and the next, we give some necessary, and some sufficient, conditions for a Hamiltonian path to exist. But the general theory is quite difficult. And in a practical case of a complicated graph, considerable computing time might be required to find a Hamiltonian path or to prove that no such path exists.

DEFINITION 5.2.15

If G is a graph, a **Hamiltonian path** on G is a path in which each vertex of G appears exactly once. A **Hamiltonian cycle** is a cycle in which each vertex of G appears exactly once, except for the beginning and ending vertex, which appears just twice.

Some Hamiltonian paths and cycles are indicated in Figure 5.2.11. A little experimentation shows that the leftmost graph does not contain a

Figure 5.2.11

Hamiltonian cycle, though it does contain a Hamiltonian path. Hamiltonian paths are named after the famous nineteenth-century Irish mathematician Sir William Hamilton, who invented and tried to market a game in which the player was to traverse a route covering all vertices only once and end back at the starting point. (See Exercise 5.2.18.) Unlike Euler paths and cycles, there is no known efficient algorithm† for finding Hamiltonian paths or for showing that they do not exist. Of course, one could list all of the $n!$ arrangements of the n vertices of a graph and check each one to see if it is a Hamiltonian cycle. But the time for this sort of direct approach has order of magnitude at least $n!$ and is not tractable.

Hamiltonian cycles arise in routing problems. For example, Exercise 5.1.27c involved Hamiltonian paths. A salesperson who must visit various cities by train might try to find a Hamiltonian cycle so that all cities are visited once, with home base as the final location.

One way of looking at Hamiltonian cycles on G is that they involve deleting edges from a graph so that the resulting subgraph is connected, has the same vertices as G, and each vertex has degree 2 in the subgraph. The actual cycle looks like a polygon of Euclidean geometry. (See Figure 5.2.12.) Thus, when attempting to form a Hamiltonian cycle on a graph, we might start at a likely vertex, eliminate all edges but two, and proceed

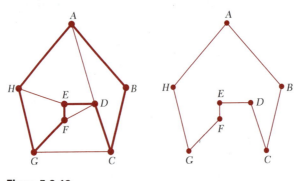

Figure 5.2.12

---

† That is, an algorithm with polynomial time.

by trial and error in this way. In particular, if a vertex has degree 2, both edges on this vertex must be used in any Hamiltonian cycle.

Of course, if *every* vertex is connected to every other vertex, there is no problem in constructing Hamiltonian cycles. Just move freely from vertex to vertex. (This sort of graph is called a **complete graph**.) The following theorem, due to O. Ore, improves this condition by cutting down how many exits there are from any vertex. It gives a sufficient condition for the existence of a Hamiltonian cycle on a graph.

THEOREM 5.2.16   Ore's Theorem

Let a graph have n vertices. Suppose that for any pair of vertices that are not connected by an edge and that have degree d_1 and d_2, we have

$$d_1 + d_2 \geq n$$

Then there is a Hamiltonian cycle on the graph.

Proof   Think of the vertices as *people*, and regard two people as *friends* if they are connected by an edge. We are now going to seat these n people about a round table so that everyone has friends on his or her right and left. The resulting arrangement can then be interpreted as a Hamiltonian cycle! We proceed as follows: We first seat them arbitrarily, but between any two seats we put a marker. We then remove the markers, one by one, and rearrange the seating if necessary in such a way that any two people who do not have a marker between them are friends. When the last marker is removed, we will have the required seating and, hence, the Hamiltonian cycle.

If a marker is between two friends, we simply remove that marker and keep the seating arrangement. Now suppose there is a marker between two people who are not friends. In Figure 5.2.13, suppose these people are A and B. We now show how to remove the marker and rearrange the seating.

It is first necessary to find friends C and D, of A and B respectively, who are sitting next to each other, as in Figure 5.2.13. We shall show how to do this. Once they are found, we use the cycle indicated in Figure

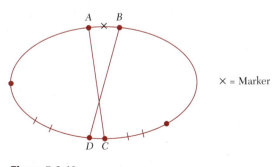

× = Marker

Figure 5.2.13

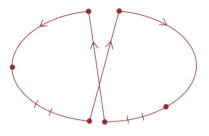

Figure 5.2.14

5.2.14 to change the seating at the table and eliminate the marker between A and B. Thus, follow the arrows in Figure 5.2.14 to find the new
seating arrangement. In this way, A and C will sit near each other, as
will B and D. (If C and D have a marker between them, this can be
discarded, too.)

The only question remaining is to show that friends C and D can be
found as in Figure 5.2.13. This we do as follows. Suppose A has d_1 friends
and B has d_2 friends. Then we know that these friends are among the
$n - 2$ people left at the table. By hypothesis, $d_1 + d_2 \geq n$.

We now proceed by contradiction. Assume that friends C and D *cannot*
be found as in Figure 5.2.14. Mark off the d_1 friends of A. Then the
$d_1 - 1$ chairs to the left of each these friends of A cannot be occupied
by friends of B. (We omit the leftmost friend of A who might be adjacent
to A.) But this then leaves $n - 2 - (d_1 - 1) = n - d_1 - 1$ possible
choices for B's friends. Since there are d_2 such choices, we must have

$$d_2 \leq n - d_1 - 1$$

or

$$d_1 + d_2 \leq n - 1$$

But this contradicts $d_1 + d_2 \geq n$, proving the result. ■

Remark   A less general, but simplified, version is as follows: If each
vertex of a graph with n vertices has degree $\geq n/2$, then there is a
Hamiltonian cycle on that graph. This theorem is due to G. A. Dirac,
and was published in 1952, eight years before Ore generalized it.

We can use Theorem 5.2.16 to give facts about Hamiltonian **paths.**

THEOREM 5.2.17

Let a graph G have n vertices. Suppose for any two vertices having degrees d_1 and d_2, which are not connected by an edge, we have

$$d_1 + d_2 \geq n - 1$$

Then there is a Hamiltonian path on G.

Proof Introduce a new vertex A that is connected to all of the original vertices of G. (A is, artificially, a "friend" of all of the people in G.) Then the new graph G' has $n + 1$ vertices, and the degree of each of the original vertices is one more than its original degree. Thus, if two vertices are not adjacent in G and have degrees d_1 and d_2 in G, they will not be adjacent in G' and have degrees $d_1 + 1$ and $d_2 + 1$ in G'. But

$$(d_1 + 1) + (d_2 + 1) \geq n + 1$$

by hypothesis. This is precisely the condition of the previous theorem (for the graph G') that there is a Hamiltonian cycle on G'. Now let $AP_1P_2 \cdots P_nA$ be a Hamiltonian cycle on G'. Dropping A, we find that $P_1P_2 \cdots P_n$ is the required Hamiltonian path on G. ∎

Remark Once again, we stress that the last two theorems are sufficient conditions that a graph have Hamiltonian paths or cycles. A simple n-gon, which has degree 2 at each vertex, doesn't satisfy these conditions for $n > 5$, yet obviously does have a Hamiltonian cycle.

Exercises

1. A multigraph has 30 vertices, each with degree 5 or more. What can you say about the number of edges?
2. Sixty-one people are at a political reception. During the reception, several people shook hands. Prove that at least one person shook hands an even number of times.
3. A multigraph has 6 vertices and 11 edges. Five of the vertices have degree 4. What is the degree of the other vertex?
4. Let G be a general multigraph. Show that the number of vertices with odd degree is even.
*5. Let A be the matrix of a graph, and let S be the sum of all the entries of A. How is S related to the number of edges and the number of vertices of the graph?
6. Suppose G is a directed graph. An Euler cycle on G is defined as a cycle that uses all of the directed edges of G. Prove that if G has an Euler cycle then for any vertex v of the graph G, the in-degree equals the out-degree. What is the situation for Euler paths on G?
*7. State and prove an analogue of Theorem 5.2.12 for directed multigraphs. (See the previous exercise.)
8. Prove Theorem 5.2.13.
*9. State and prove an analogue of Theorem 5.2.13 for directed multigraphs. You may use the result of Exercise 7.
*10. Prove: A necessary and sufficient condition for the existence of an Euler cycle on a multigraph G is that each vertex have even degree and that the subgraph spanned by the vertices of positive degree is connected.

For each of the following figures, determine whether it has an Euler path or cycle. If it does not have one, state why. If the figure does have an Euler path or cycle, indicate how it can be drawn.

11. 13.

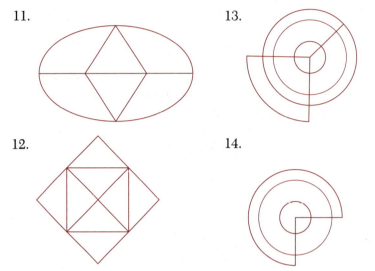

12. 14.

As in the previous exercises, analyze the following digraphs for Euler paths or cycles.

15. 16.

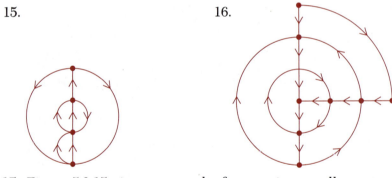

17. Figure 5.2.15 gives a network of streets in a small town, some of

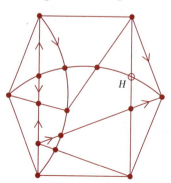

which are one-way as indicated. The rest are two-way streets. Can you arrange a legal drive through all of the streets of the town starting and ending at your hotel H? Explain. (*Hint:* Try partial Euler cycles.)

18. Hamilton's game was played on a dodecahedron, a regular solid with 12 pentagonal faces, 20 vertices, and 30 edges. Such solids are often used as calendar souvenirs, because each face can be used for a month. In Hamilton's game, each vertex was regarded as a famous city, and the game was to make a tour of all the cities along the edges and to end at the city at which the game began. A flattened version of the solid is given in Figure 5.2.16. Can you win the game?

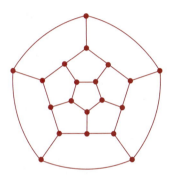

Figure 5.2.16

*19. Try to find a Hamiltonian cycle for the graph (a) of Figure 5.2.17. If one cannot be found, try for a Hamiltonian path. If none can be found, explain.

20. Repeat Exercise 19, except for graph (b) of Figure 5.2.17.

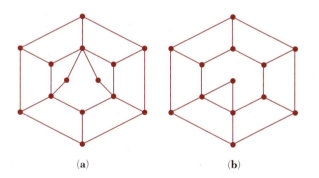

(a)   (b)

Figure 5.2.17

*21. Suppose the vertices of a graph G are divided into two disjoint sets A and B, with $|A| = n$ and $|B| = m$. Suppose further that all edges of the graph necessarily join a vertex of A with a vertex of B. Show that if there is a Hamiltonian cycle on G, then $n = m$. Show that if there is a Hamiltonian path on G, then n and m cannot differ by more than 1.

22. Using the result of Exercise 21, show that the graph of Figure 5.2.18 does not have a Hamiltonian cycle.

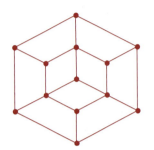

Figure 5.2.18

23. It was stated that there was no Hamiltonian cycle on the left graph of Figure 5.2.11. Prove this.
24. Seating plans must be made for a table of 12. It is known that none of the guests knows more than five of the other guests. The host wants to have no one at the table sitting next to anyone he or she knows. Show that this is possible.
25. Twenty-nine parents of children in the International School want to set up a "telephone tree." This is a procedure, well known to parents, whereby one parent calls just one other parent, who has not been called, to pass on some information. The problem is that not all the parents speak the same language. However, each parent can communicate with at least 14 other parents. The director of the school can speak all the languages. Show that the director can set up a telephone tree so that all parents can be informed of important school matters.

   The following series of exercises relate to the **incidence matrix** D of a graph or multigraph G. Suppose the graph has m vertices $v_1, \ldots, v_m$ and n edges $e_1, \ldots, e_n$. Then D is defined as the m by n $(0, 1)$ matrix in which $d_{ij} = 1$ if the vertex v_i is one of the end vertices of edge e_j. Otherwise, $d_{ij} = 0$.

26. Let D be the incidence matrix of G. Show that each column sum of D is 2.
27. Let D be the incidence matrix of G. Show that the ith row sum of D is $\deg(v_i)$.
*28. Using the results of the previous two exercises, prove Theorem 5.2.6. (*Hint:* How many 1's are in the matrix D?)
29. Suppose D is an m by n $(0, 1)$ matrix in which each column sum is 2. Show that D is the incidence matrix of a multigraph G. How is the matrix C of G determined from D?

30. Draw a diagram of the multigraph G whose incidence matrix is

$$D = \begin{bmatrix} 0 & 1 & 1 & 0 & 0 & 0 & 1 & 0 & 0 \\ 1 & 0 & 0 & 0 & 1 & 1 & 0 & 1 & 0 \\ 0 & 1 & 0 & 1 & 1 & 0 & 1 & 0 & 1 \\ 0 & 0 & 1 & 0 & 0 & 1 & 0 & 1 & 1 \\ 1 & 0 & 0 & 1 & 0 & 0 & 0 & 0 & 0 \end{bmatrix}$$

31. Draw a diagram of the multigraph G whose incidence matrix is

$$D = \begin{bmatrix} 0 & 0 & 1 & 0 & 0 \\ 0 & 1 & 0 & 0 & 1 \\ 1 & 0 & 1 & 1 & 0 \\ 0 & 1 & 0 & 1 & 1 \\ 1 & 0 & 0 & 0 & 0 \end{bmatrix}$$

*32. Generalize the definition of incidence matrices to cover general multigraphs, so that an edge can connect a vertex to itself. Prove Theorem 5.2.6 for general multigraphs, using the appropriate generalizations of the previous exercises.

5.3 Planar Graphs and Coloring Problems

A **coloring** of a graph is an assignment of colors to the vertices of that graph in such a way that adjacent vertices have different colors. Another way of looking at a coloring is that if a graph is colored, then certain edges are forbidden in the construction of the graph, namely, no edge can join vertices of the same color.

EXAMPLE 5.3.1   Marketing

A marketing company is analyzing a population to find whether the people in this population recognize various products that it is marketing. Thus, the graph it is looking for consists of the population and the products as vertices, with an edge joining a person to a product if that person recognizes the product. In this approach, for example, the degree of a product is the number of people recognizing the product. Now it is clear that the graph consists of edges joining people to products, and that an edge joining people or one joining products is not allowed. Imagine labeling the people vertices with one color and the product vertices with another. The result is a coloring of the graph with two colors. The analysis of this graph is considerably simpler than the analysis of a more general one. If, for example, there are 10 products and 25 people, we need only analyze a 10 by 25 matrix (250 entries) for the graph. The more general graph with 35 vertices would involve a 35 by 35 matrix (1225 entries).

EXAMPLE 5.3.2

Another example is the common checkerboard, colored red and black. We can understand why two colors are appropriate, since adjacent squares should be colored differently to visually distinguish squares that border each other. We may take the center of each square as a vertex, and then join vertices that have a common border. The checkerboard will then be colored with the colors R and B. Again, each edge joins vertices of different colors. Notice that we have replaced a **region** by a **vertex** in this analysis.

EXAMPLE 5.3.3   The Four-Color Problem

Similarly, a map of various countries must use different colors for bordering countries so that the countries can be visually distinguished. For example, it is necessary to use four colors for the map of Figure 5.3.1 because of the way the regions border each other. The **Four-Color Problem** is the problem of determining how many colors are needed for all maps in the plane. Mathematicians were long convinced that only four colors were needed for any conceivable map in the plane, and many proofs were attempted. However, the proof of this classical problem was first given in 1976 by Haken and Appel. They reduced the problem to a complicated algorithm and used high-speed computers to solve it. Thus, it is now known as the **Four-Color Theorem.** Later, we give an elementary proof of the much simpler result that it is possible to use only five colors.

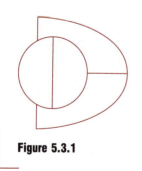

Figure 5.3.1

In the following definition, the numbers $1, 2, \ldots, n$ are used as colors. In our illustrations, we sometimes use R(ed), Y(ellow), and B(lue).

DEFINITION 5.3.4

A graph G is said to be **colored** by the integers $1, 2, \ldots, n$, or simply **n-colored**, if each vertex is assigned one of these integers and if adjacent vertices have different integers assigned to them. The assignment of these integers is a **coloring** of G.

Remark   If fewer than n colors are used, we still have an n-coloring.

DEFINITION 5.3.5

A **bipartite graph** is one that is 2-colorable.

Bipartite graphs are so called because they are graphs whose vertices are partitioned into two sets (vertices colored 1 and vertices colored 2) so that any edge has its end points in each of the sets. The marketing example given earlier is bipartite. Before beginning the general analysis of colorings, we give some additional information about the **components** of a graph. (See Definition 5.1.11.)

THEOREM 5.3.6

The components of a graph G are connected. No vertex is adjacent to a component.

Proof   If a component is $C(v)$, and if w is any vertex in $C(v)$, then by definition there is a path joining v to w. All vertices on this path are also $\approx v$ and are in $C(v)$. Therefore the path joining v to w is in $C(v)$, and $C(v)$ is connected.

To prove the second part of the theorem, suppose w is adjacent to $C(v)$. This means that $w \notin C(v)$, but there is an edge $v'w$ where $v' \in C(v)$. (See Definition 5.1.13.) But this implies that $v \approx v' \approx w$, so $w \in C(v)$, which is a contradiction. ■

The following lemma shows why questions concerning colorings can be restricted to connected graphs.

LEMMA 5.3.7

Suppose each vertex of a graph is assigned a color, and suppose this assignment is a coloring for each component. Then the assignment is a coloring of the original graph. In brief, if we can color each component of G, then the result is a coloring of G.

Proof   Any edge of G joins vertices in the same component, so its end points are colored differently. ■

We now continue our discussion of bipartite graphs.

Suppose we start with a red square of a red and black checkerboard and take a walk, always moving into a neighboring square. Then these squares necessarily alternate colors: $R, B, R, B, \ldots$. Therefore, if we end at the square where we began, we will have taken an even number of steps. This is easily generalized as follows.

THEOREM 5.3.8

All cycles α on a bipartite graph G have an even number of edges.   ■

The conclusion of Theorem 5.3.8 is a statement about a graph that does not involve coloring. Its converse is true, and it gives a condition for a graph to be bipartite.

THEOREM 5.3.9

Suppose all cycles in a graph G have an even number of edges. Then G is bipartite.

Proof   By Lemma 5.3.7, it is enough to color each component of G with two colors. We may therefore suppose that G is connected. We now show how to color G with the colors 0 and 1. Choose any vertex v and color it 0. Now for each other vertex w of G, we proceed as follows. Join v to w by some path $\alpha = \alpha(v, w)$. If the length of α (the number of edges in α) is even, assign w the color 0. If the length is odd, color it 1. In this way, each vertex of G is assigned a color, though the coloring is apparently dependent on the path used to join v to w.

We now show that adjacent vertices cannot have the same color. Suppose w_1 and w_2 are adjacent and have the same color. Let $\alpha(v, w_i)$ have n_i edges ($i = 1$ and 2) so that n_1 and n_2 are both even or both odd. Then $n_1 + n_2$ is even. We now form a cycle starting at v by first using $\alpha(v, w_1)$, moving on the edge from w_1 to w_2, and then going back to v by reversing the direction of $\alpha(v, w_2)$. (See Figure 5.3.2.) The number of edges in that cycle is clearly $n_1 + n_2 + 1$, which is odd. This contradicts the hypothesis that all cycles have an even number of edges, and proves the result.   ■

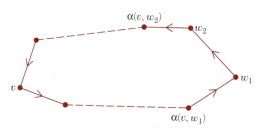

Figure 5.3.2

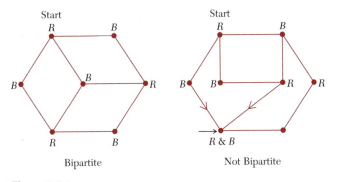

Figure 5.3.3

Remark   Once the vertex v is colored 0, the coloring of the component of v is uniquely determined. Namely, the color of any vertex w is 0 if an even number of edges is needed to join v to w, and the color is 1 if an odd number of edges is needed.

One method of 2-coloring a graph is as follows. Start with one vertex v_1 and color it 0. Let $\Sigma_1 = \{v_1\}$. We adjoin adjacent vertices one at a time to Σ_1 to form a sequence Σ_k of colored vertices. For each adjunction to Σ_k, we must make sure that the vertex adjoined is adjacent to Σ_k and assigned a color different from all of its adjacent vertices in Σ_k. This will be possible only if all such neighbors in Σ_k have the same color. If they don't, the graph is not bipartite. In this way we will color the component of v_1. We can then proceed in the same manner with any vertex not yet colored.

On small graphs, this can be done fairly quickly. See Figure 5.3.3 for one successful and one failed attempt. The next example elaborates this procedure.

EXAMPLE 5.3.10

The algorithm BIPARTITE is based on the method outlined in the previous example. The input is a graph. The output is a (0, 1) coloring of the graph if it is bipartite or else an announcement that the graph is not bipartite. We let the vertices of G be the integers $1, 2, \ldots, n$.

BIPARTITE Algorithm
1. $\Sigma := \{1\}$ [initializing Σ]
2. $f(1) := 0$ [coloring the only vertex in Σ with color 0]
3. WHILE $\Sigma \neq G$ [now pick out an adjacent vertex]
    1. IF there is a vertex adjacent to Σ THEN
        1. $w :=$ the first vertex adjacent to Σ [now check that the adjacent vertices in Σ all have the same color]
        2. $v :=$ the first vertex in Σ adjacent to w
        3. $f(w) := 1 - f(v)$ [changes 1 to 0 and 0 to 1]

4. IF (every u in Σ which is adjacent to w satisfies $f(u) = 1 - f(w)$) THEN
    1. $\Sigma := \Sigma \cup \{w\}$ [w checks out and its color has been properly assigned]
  OTHERWISE
    2. Output "Graph is not bipartite"
    3. End
2. IF there is no vertex adjacent to Σ THEN
    1. $w :=$ any vertex in $G - \Sigma$
    2. $f(w) := 0$
    3. $\Sigma := \Sigma \cup \{w\}$ [start a new component]

One use of bipartite graphs is in the theory of Hamiltonian cycles. Clearly, any **cycle** in a bipartite graph (colored red and blue, for example) must contain the same number of red and blue points. Therefore, if a Hamiltonian cycle exists **in a bipartite graph,** the graph must have the same number of red and blue points. Similarly, if a Hamiltonian **path** exists on a 2-colored graph and it is not a cycle, there is at most one more of one color than the other.

THEOREM 5.3.11

Suppose that a connected graph G is 2-colored with m vertices having color 0 and n having color 1. Then if there is a Hamiltonian cycle on G, $m = n$. Further, if there is a Hamiltonian path on G, then $m = n$ or $m = n \pm 1$. ∎

Remark The condition $m = n$ is a necessary but not sufficient condition for a Hamiltonian cycle to exist on a connected bipartite graph. We need only find a 2-coloring in which $m \neq n$, and we are then guaranteed that no Hamiltonian cycle exists on G. However, if $m = n$, G need not necessarily have such a cycle. Theorem 5.3.11 is a *necessary* condition for the existence of a Hamiltonian graph or cycle. Ore's Theorem (Theorem 5.2.15) gave a sufficient condition.

Figure 5.3.4 illustrates a graph that has no Hamiltonian path or cycle on it. This can be seen by the coloring on the right, in which six vertices are colored B and four are colored A. Without Theorem 5.3.11, and the good fortune that the graph is bipartite, this might be a hard problem.

Map Coloring

We now consider the question of coloring maps. Although Definition 5.3.4 dealt with coloring vertices of a graph, a mapmaker is concerned with coloring countries or regions of a graph. A simple transformation allows one to be converted to the other. If we have any map, it is possible

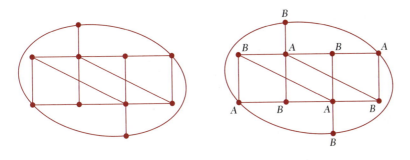

Figure 5.3.4

to make a graph by choosing each **region** as a vertex. Introduce an edge joining two vertices when the two regions they represent have a common border. In this way, as in Figure 5.3.5, we can convert the notion of coloring a map to coloring a graph in the sense of Definition 5.3.1. We used this idea previously on a checkerboard.

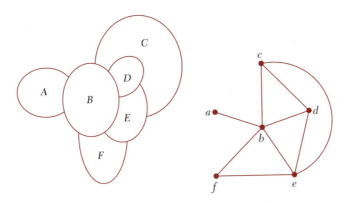

Figure 5.3.5

The graphs we consider are called *planar graphs*. These graphs can be drawn so that no two edges intersect except at vertices. Even very simple graphs can be drawn with intersecting edges. Figure 5.3.6 illus-

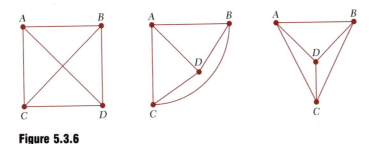

Figure 5.3.6

trates this, but shows how it is sometimes possible to redraw the graph as a planar graph. Once this is done, the regions referred to in the previous discussion become clearly visible and have boundaries consisting of edges. Some of the terms we use, such as *boundary* and *region*, will not be defined, but taken intuitively. A rigorous treatment of the subject (the topology of the plane) is outside the realm of this text.

DEFINITION 5.3.12

A **planar graph** is one that can be drawn in the plane so that edges can meet only at vertices.

DEFINITION 5.3.13

The **dual graph** G' of a planar graph G is the graph whose vertices are the regions of G. If regions R and S share a common edge e, we introduce a corresponding edge e' in G' joining R and S.

We can think of G' as follows. Within each region, choose a point—the capital city of the region/country—representing that region. Now if any two region/countries have a common edge, construct a royal road through that edge connecting those capitals and avoiding all other region/countries and royal roads. The dual graph consists of the capital cities as vertices, and the royal roads as edges. This is illustrated in Figure 5.3.7. Note especially the **infinite region** P of that graph. This is the unique region that has no outside boundary, and it occurs in any finite planar graph. It was omitted in Figure 5.3.5, where the idea of a dual graph was first encountered.

We take for granted that G' is also planar. A little experimentation will convince the reader that the dual of G' is simply G again. The regions

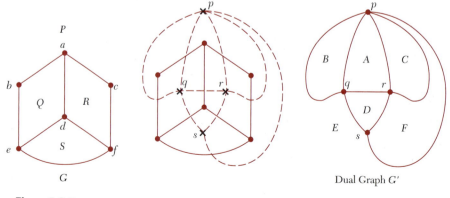

Figure 5.3.7

of G' correspond to the vertices of G, much as the vertices of G' correspond to the regions of G.

These definitions also apply when G is a multigraph. If G is a simple graph, G' might still be a multigraph, as in Figure 5.3.7. This happens when two regions of G have several edges in common. In this case, the dual graph will have several edges joining the vertices that correspond to these regions.

If G is a planar graph, we let $v = v(G) =$ number of its vertices, $e = e(G) =$ number of its edges, and $r = r(G) =$ number of its regions. For G', e remains the same, but r and v are interchanged:

$$e(G) = e(G'), \quad v(G) = r(G'), \quad \text{and} \quad r(G) = v(G')$$

The following remarkable theorem of Euler relates v, e, and r for connected, planar graphs.

THEOREM 5.3.14

Let G be a connected, planar, general multigraph. Then

$$v - e + r = 2 \tag{5.3.1}$$

Proof We take for granted that a connected, planar multigraph can be built up, step by step, from a single vertex u, using any of the following operations:

(a) Adjoin a new vertex and connect it by an edge to an existing vertex without intersecting any existing edge.
(b) Join two existing vertices with an edge.
(c) Join a vertex to itself with a loop. (This is for general graphs.)

See Figure 5.3.8 for an illustration of these operations.

The proof of the theorem is by induction on e, the number of edges in the graph. Note that each of the operations (a), (b), and (c) introduces a new edge. Thus, any graph with e edges ($e > 0$) can be obtained from a graph with $e - 1$ edges by using one of the operations (a), (b), or (c). We start with a single vertex ($e = 0$) and show that for any of the steps (a), (b), or (c), $v - e + r$ is left unchanged.

For $e = 0$, there are no edges, so there is a single vertex u because

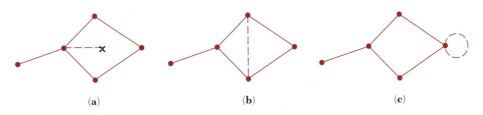

(a)   (b)   (c)

Figure 5.3.8

the graph is connected. Here we have $v = 1$, $e = 0$, $r = 1$, and Equation (5.3.1) is true. This is the basis for the induction.

As in Figure 5.3.8, operation (a) adds an edge and a vertex but leaves r unchanged. Thus, v and e are each increased by 1, and $v - e + r$ is unchanged. Similarly, operation (b) adds an edge and a region but leaves v unchanged, so $v - e + r$ is unchanged. Finally, operation (c) also increases e and r by 1, so Equation (5.3.1) persists. This completes the proof. ∎

Which simple graphs are planar? We shall soon show that the graph in Figure 5.3.9 is not planar. This graph is called K_5. (In general, K_n is the simple graph with n vertices, any two of which are connected by an edge.) Note that K_5 is clearly *not* 4-colorable. The following example gives another nonplanar graph.

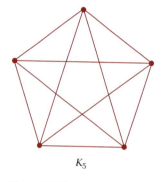

K_5

Figure 5.3.9

EXAMPLE 5.3.15

The graph in Figure 5.3.10 is the basis for the following puzzle. Three houses A, B, and C are to be connected by cables or pipes to the three suppliers of (E)lectricity, (O)il, and (W)ater. The connecting pipes and cables are to be laid at a depth of 10 ft, and are not allowed to intersect. The problem is to show how this can be done, if possible. Since the

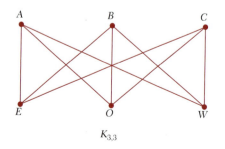

$K_{3,3}$

Figure 5.3.10

pipes and cables are the edges, and they are not supposed to intersect, what is needed is a planar version of this graph. However, it turns out that this graph is not planar. (This graph is called $K_{3,3}$. In general, $K_{m,n}$ is the bipartite graph in which m vertices are colored 0, n are colored 1, and all connections between 0's and 1's are edges.)

We can show that a graph is planar by exhibiting it without having any of the edges intersect. But how can we show that a graph is not planar? The next few theorems give some help in that direction and, in particular, cover the graphs K_5 and $K_{3,3}$.

THEOREM 5.3.16

Let G be a simple, connected, planar graph with v vertices and e edges, where $v \geq 3$. Then

$$e \leq 3v - 6 \tag{5.3.2}$$

Proof If we draw a planar representation of the graph G, the plane will be divided into r regions. First, suppose that each region is bounded by at least three edges. This means that each *vertex* of the dual graph G' has degree 3 or more. Thus, the sum of the degrees of the vertices of G' is $3r$ or more. Since, by Theorem 5.2.6, the sum of the degrees is twice the number of edges, we have

$$3r \leq 2e$$

We now combine this with Euler's formula $v - e + r = 2$. This yields $r = e - v + 2$, so $3r = 3e - 3v + 6$. Since $3r \leq 2e$, we have $3e - 3v + 6 \leq 2e$, or $e \leq 3v - 6$.

If a region is *not* bounded by three edges, then the only possibility for a connected simple graph with three or more vertices is given in Figure 5.3.11, where the region is the infinite region and is bounded by two edges. However, formula (5.3.2) is valid here, too.

These two cases complete the proof. ■

Remark As a consequence of this result, we see that K_5 is not planar, since $e = 10$ and $v = 5$ for this graph, and formula (5.3.2) is not valid for these numbers.

However, Theorem 5.3.16 is not strong enough to show that $K_{3,3}$ (Figure 5.3.10) is not planar, since $e = 9$ and $v = 6$ in this case, and inequality

Figure 5.3.11

(5.3.2) is satisfied. However, for *bipartite* graphs, we have a stronger theorem.

THEOREM 5.3.17

Let G be a simple, connected, planar, bipartite graph with v vertices and e edges, where $v \geq 3$. Then

$$e \leq 2v - 4 \qquad (5.3.3)$$

Proof   In a bipartite graph, the boundary of any region, being a cycle, must have four or more edges. (The exceptions are as in Figure 5.3.11, where the boundary does not "close up.") Thus, in the nonexceptional case, the reasoning of Theorem 5.3.16 shows that $4r \leq 2e$, or $2r \leq e$. Combining this with Euler's formula $r = e - v + 2$, we find $2r = 2e - 2v + 4 \leq e$, so that $e \leq 2v - 4$ in this case. We leave the exceptional cases to the reader. ∎

Since this inequality does not hold for $e = 9$ and $v = 6$, it follows that $K_{3,3}$ is not planar.

Remark   Theorems 5.3.16 and 5.3.17 give necessary conditions for a simple graph to be planar. They are not sufficient.

We now relate coloring and planarity by proving the **Five-Color Theorem** for planar graphs.

THEOREM 5.3.18

Any planar graph G is 5-colorable.

Proof   We prove this by induction on the number of vertices of G. For a basis, note that any graph with four or fewer vertices is clearly 5-colorable. We may assume, by induction, that any planar graph with fewer vertices than G is 5-colorable. Thus, if a vertex is removed from G (together with all edges connected to that vertex), the resulting graph, which we call H, is 5-colorable.

If any vertex of G has degree 4 or less, we can easily show that G is 5-colorable. Simply remove this vertex, color the remaining graph, and finally color this vertex with a color not used on its adjacent vertices. Thus we may assume that all vertices of G have degree 5 or more.

We claim that at least one vertex u must have degree 5, for otherwise, each vertex would have degree 6 or more. Since the sum of the degrees is twice the number of edges, this would imply that $2e \geq 6v$, or $e \geq 3v$, which contradicts Theorem 5.3.16. We therefore have a vertex u of degree 5 in G. Let us remove u from G, along with the five edges at this vertex, and 5-color the remaining graph. If the five adjacent vertices don't use all the five colors, simply color u with an unused color and we have

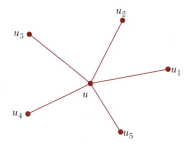

Figure 5.3.12

colored the graph G. If, however, all colors are used, as in Figure 5.3.12, we shall show how to find a way to recolor the remaining graph so that only four colors will be adjacent to u, and so the fifth color can be assigned to u. Therefore, let us assume that we have a situation as in Figure 5.3.12, where the vertices $u_1, \ldots, u_5$ are labeled cyclically around u, and each of the u_i's is colored differently.

We use Lemma 5.3.7 together with the following result: *If we have a coloring of any graph, then if any two colors are interchanged the result is still a coloring.*

Choose the vertices u_1 and u_3, colored 1 and 3. We now consider the subgraph K spanned by the vertices of H that are colored either 1 or 3. There are two cases, according to whether u_1 and u_3 can be joined by a path in K.

Case 1   The vertices u_1 and u_3 cannot be connected in K by using the edges of H. In this case, consider the set C of all vertices of K that *can* be connected to u_1 in this way. Simply interchange colors 1 and 3 in C, and the result will be a coloring of H in which the color 1 adjacent to u has been changed to 3, but the color 3 adjacent to u is unchanged. We can now assign the color 1 to u, and G is 5-colorable.

Case 2   The vertices u_1 and u_3 can be connected in K. Connect them with a path in K (i.e., one that avoids u) as in Figure 5.3.13. If we adjoin

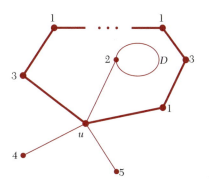

Figure 5.3.13

the vertex u, and the edges uu_1 and uu_3, we have a cycle consisting of vertices colored 1 or 3, except for u. Now look at vertices u_2 and u_4. Any path joining them must meet this cycle. (Here we use an intuitive property of planar graphs.) As before, we now consider the subgraph L of H spanned by the vertices of H that are colored either 2 or 4. Let D be the set of vertices that can be connected to u_2 in L by using the edges of H. Now, D cannot contain u_4, for if it did there would be a path joining u_2 and u_4 consisting of points either colored 2 or 4, and this is impossible because the constructed cycle in Figure 5.3.13 consists of u and points colored either 1 or 3. We now recolor D as in case 1 by interchanging colors 2 and 4. The result is a coloring of H in which the color 2 is not adjacent to u. Therefore, we may assign the color 2 to u, and we have a coloring of G. ■

Exercises

1. Given $2n$ vertices, of which n are colored R and n are colored B. How many graphs on these vertices are possible that make this assignment of colors a coloring? How many graphs on these vertices are there?

2. Explain why any graph with n vertices is n-colorable. Give an example of a graph with n vertices that is not $(n-1)$-colorable. Show that, with the exception of one graph on n vertices, any graph with n vertices is $(n-1)$-colorable.

3. Which graphs are 1-colorable?

4. Show that the graph in Figure 5.3.14 does not have a Hamiltonian path on it.

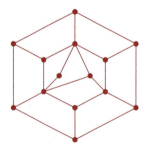

Figure 5.3.14

5. Show that the graph in Figure 5.3.15 does not have a Hamiltonian cycle on it. Does it have a Hamiltonian path?

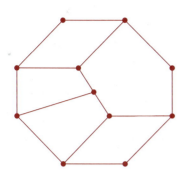

Figure 5.3.15

6. Decide whether the graph of Figure 5.3.16 has a Hamiltonian path on it. If it has one, exhibit it. If not, explain.

Figure 5.3.16

7. Decide whether the graph of Figure 5.3.17 has a Hamiltonian path on it. If it has one, exhibit it. If not, explain.

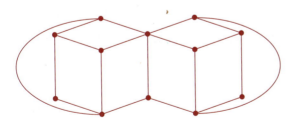

Figure 5.3.17

*8. Give an example of a connected bipartite graph in which the same number of vertices are colored red as are colored black, but which does not have a Hamiltonian path on it.
*9. Suppose a planar graph G is not connected and has k components.

Give a formula for $v - e + r$ in this case and prove it. If operation (b), defined in the proof of Theorem 5.3.14, is used to connect two vertices in different components of a graph, then the number of components of that graph decreases by one. How do you reconcile this fact with your formula and the proof of Theorem 5.3.14, in which it was stated that operation (b) keeps $e - v + r$ the same?

*10. A planar graph is called a **triangulation** if every one of its regions has a triangle for its boundary—that is, a boundary consisting of exactly three edges. Show that if a graph is a triangulation, then $e = 3v - 6$. (*Hint:* See the proof of Theorem 5.13.16.)

*11. As in the previous exercise, suppose that in a planar graph every finite region has a triangle for its boundary and that the infinite region has a boundary consisting of p edges. Find a relationship among e, v, and p.

12. The proof of Theorem 5.3.18 used formula (5.3.2), which is valid for simple graphs. Explain why this theorem is nevertheless true for planar multigraphs.

13. Which theorem is stronger, the Four-Color or the Five-Color? Explain. (Theorem A is said to be stronger than Theorem B if you can prove Theorem B when you know that Theorem A is true.)

14. Give a 2-coloring for the graphs of Figure 5.3.18.

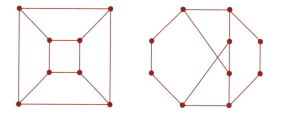

Figure 5.3.18

15. Give a 3-coloring for the graph of Figure 5.3.16. Explain why a 2-coloring is impossible.

16. Give a 3-coloring for the graph of Figure 5.3.19. Explain why a 2-coloring is impossible.

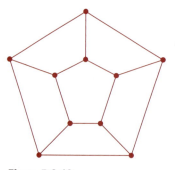

Figure 5.3.19

17. Give a 4-coloring for the graph of Figure 5.3.20. Explain why a 3-coloring is impossible.

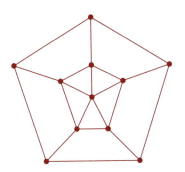

Figure 5.3.20

18. Figure 5.3.21 is an attempt to 4-color a graph, but no color is available at x. Is the graph 4-colorable? Explain.

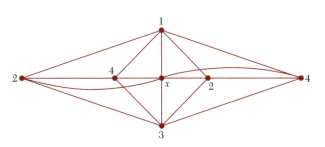

Figure 5.3.21

19. Sketch the dual graph for the graph of Figure 5.3.22.

Figure 5.3.22

20. Sketch the dual graph for the graph of Figure 5.3.23.

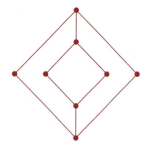

Figure 5.3.23

21. Draw a planar graph with $v = 5$, $e = 8$, and $r = 5$.
22. Show that if v, e, and r are positive integers satisfying

$$v - e + r = 2$$

then there is a planar connected general multigraph with v vertices, e edges, and r regions.

The BIPARTITE Algorithm of Example 5.3.10 had a few loose ends. Exercises 23–26 ask you to fix these. Assume that the set of V vertices is $\mathbb{I}_n$, and that Σ is a set of vertices of V whose characteristic function is $\sigma: V \to \mathbb{Z}_2$. Let c_{ij} be the matrix of the graph.

23. Step 3.1 used the condition "there is a vertex adjacent to Σ." Write a procedure that will determine whether this is true or false.
24. Step 3.1.1 made the assignment "$w :=$ the first vertex adjacent to Σ." Write a procedure that makes this assignment.
25. Step 3.1.4 used the condition "every u in Σ which is adjacent to w satisfies $f(u) = 1 - f(w)$." Write a procedure that will determine whether this is true or false for a given w.
26. Give a big Oh estimate for the worst-time case for the algorithm of Exercise 23. The inputs are taken to be the characteristic function, the number n of vertices, and the counting matrix c_{ij}.

27. Is the graph of Figure 5.3.24 planar? Give reasons.

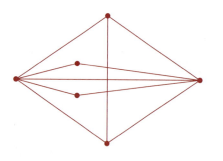

Figure 5.3.24

*28. Is the graph of Figure 5.3.25 planar? Give reasons.

Figure 5.3.25

29. Is the graph of Figure 5.3.26 planar? Give reasons.

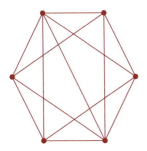

Figure 5.3.26

*30. Is the graph of Figure 5.3.27 planar? Give reasons.

Figure 5.3.27

*31. Give an example of a simple nonplanar graph where $e \leq 3v - 6$.
*32. Give an example of a simple bipartite nonplanar graph where $e \leq 2v - 4$.

5.4 Isomorphisms of Graphs—Some Graph-Related Algorithms

It is possible for two graphs to have different sets of vertices but the same structure. This is the situation in Figure 5.4.1, where the drawings

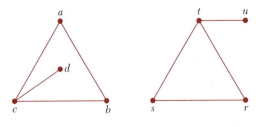

Figure 5.4.1

evidently represent similar graphs. In the first graph, we have

$$C(a, b) = C(b, c) = C(c, d) = 1, \qquad C(a, d) = 0, \ldots$$

In the second, we have

$$C'(r, s) = C'(s, t) = C'(t, u) = 1, \qquad C'(r, u) = 0, \ldots$$

with similar equations for all corresponding pairs of vertices. Here, we have used C' to denote the counting function for the second graph. Thus, if we make the correspondence

$$a \rightarrow r; \quad b \rightarrow s; \quad c \rightarrow t; \quad d \rightarrow u$$

the vertices of the first graph are put into a 1-1 correspondence with those of the second graph in such a way that all incidences are preserved. We may say that except for the labeling of the vertices, the graphs are the same. Such a correspondence is called an *isomorphism*.

DEFINITION 5.4.1

Let G and G' be graphs with counting functions C and C', respectively. Let the vertices of G and G' be denoted V and V', respectively. An **isomorphism** f of the graph G into G' is a 1-1, onto mapping $f\!:\!V \rightarrow V'$ such that

$$C(v, w) = C'(f(v), f(w)) \tag{5.4.1}$$

for all vertices v and w in V.

Remark This definition applies to multigraphs, digraphs, and general graphs. The idea is that, except for the labeling of the vertices, the graphs are identical.

It might be supposed that we can determine whether two graphs are isomorphic by comparing the matrices of the graphs. This is not so, be-

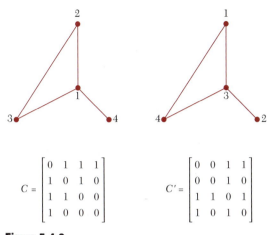

$$C = \begin{bmatrix} 0 & 1 & 1 & 1 \\ 1 & 0 & 1 & 0 \\ 1 & 1 & 0 & 0 \\ 1 & 0 & 0 & 0 \end{bmatrix} \qquad C' = \begin{bmatrix} 0 & 0 & 1 & 1 \\ 0 & 0 & 1 & 0 \\ 1 & 1 & 0 & 1 \\ 1 & 0 & 1 & 0 \end{bmatrix}$$

Figure 5.4.2

cause the matrix of a graph depends on which vertex we label 1, 2, ... Figure 5.4.2 gives two labelings of a very simple graph that has different matrices. Rather than look at matrices, we attempt to find other facts about graphs that may help to determine whether two graphs are isomorphic. The following theorem shows that degrees must correspond under an isomorphism.

THEOREM 5.4.2

If v and v' are vertices that correspond under an isomorphism of G into G', then $\deg(v) = \deg(v')$.

Proof Suppose we write $f(X) = X'$, where X is a vertex of G and f is the isomorphism. By definition of isomorphism, we have $C(X, Y) = C'(X', Y')$. Then

$$\deg(X) = \sum_Y C(X, Y) = \sum_Y C'(X', Y') = \sum_{Y'} C'(X', Y') = \deg(X')$$

Here, as Y runs through the vertices of V, Y' runs through the vertices of G', since f is 1-1 and onto. ∎

This theorem gives a *necessary* condition that two graphs are isomorphic. The vertices that correspond to each other must have the same degrees. If G is any graph, we can list the degrees of the vertices in ascending order, repeating degrees if two vertices have the same degree. We call this listing the **degree list** of a graph. A simple consequence of Theorem 5.4.2 is the next corollary.

COROLLARY 5.4.3

If two graphs are isomorphic, then they have the same degree list. ∎

The converse is not true, as we shall see. The corollary can be used to show that graphs are *not* isomorphic. Merely find their degree lists; if they are different, then the graphs are not isomorphic.

EXAMPLE 5.4.4

Find the degree lists for the graphs of Figure 5.4.3 and show that the graphs are not isomorphic.

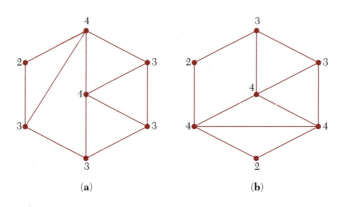

(a)   (b)

Figure 5.4.3

Method   We simply count the degrees:

Graph 5.4.3a has degree list 2, 3, 3, 3, 3, 4, 4.
Graph 5.4.3b has degree list 2, 2, 3, 3, 4, 4, 4.

Therefore the graphs are not isomorphic by Corollary 5.4.3. Other approaches are possible. It is enough to note that graph 5.4.3a has only one vertex of degree 2, and graph 5.4.3b has two.

EXAMPLE 5.4.5

Discuss whether the graphs of Figure 5.4.4 are isomorphic.

Method   The degree lists agree (2, 2, 2, 3, 3, 3, 3), but this does not show that the graphs are isomorphic. In fact, they are nonisomorphic because Figure 5.4.4b contains a triangle and Figure 5.4.4a does not.

Using another approach, notice that in graph 5.4.4a, each vertex of degree 2 is connected to vertices having degree 3, but this is not true in graph 5.4.4b.

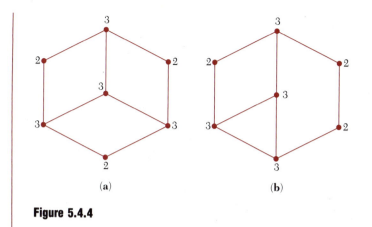

Figure 5.4.4

EXAMPLE 5.4.6

Discuss whether the graphs of Figure 5.4.5 are isomorphic.

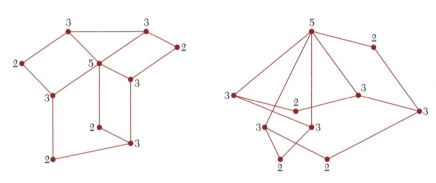

Figure 5.4.5

Method A check of the degree lists show that they agree. If there *were* an isomorphism, the vertices of degree 5 would correspond. But then all of the vertices connecting these vertices would have to correspond, again with degrees corresponding, and so on. If we follow the possibilities, we find there *is* an isomorphism, even though the graphs certainly appeared to be different at first sight. Thus, we can find an isomorphism of the graphs as in Figure 5.4.6. In that diagram, corresponding vertices are labeled with the same letter. The only practical way available for us to show that graphs are isomorphic is to actually find the isomorphism. The degree list helps in this task.

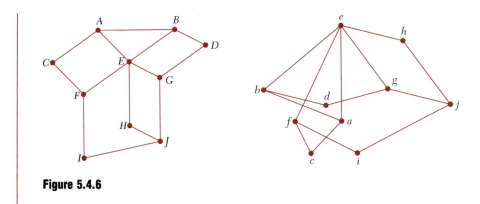

Figure 5.4.6

EXAMPLE 5.4.7

A graph has matrix

$$\begin{bmatrix} 0 & 1 & 1 & 0 & 1 \\ 1 & 0 & 0 & 0 & 1 \\ 1 & 0 & 0 & 1 & 0 \\ 0 & 0 & 1 & 0 & 0 \\ 1 & 1 & 0 & 0 & 0 \end{bmatrix}$$

We decide to rename the vertices of the graph by calling the first vertex 3, and the third vertex 1. What is the matrix of the relabeled graph?

Method Under the relabeling, the only difference is that the first row and the third row are interchanged, and similarly for the first and third columns. The result is thus obtained in two steps as follows:

$$\begin{bmatrix} 0 & 1 & 1 & 0 & 1 \\ 1 & 0 & 0 & 0 & 1 \\ 1 & 0 & 0 & 1 & 0 \\ 0 & 0 & 1 & 0 & 0 \\ 1 & 1 & 0 & 0 & 0 \end{bmatrix} \rightarrow \begin{bmatrix} 1 & 0 & 0 & 1 & 0 \\ 1 & 0 & 0 & 0 & 1 \\ 0 & 1 & 1 & 0 & 1 \\ 0 & 0 & 1 & 0 & 0 \\ 1 & 1 & 0 & 0 & 0 \end{bmatrix} \rightarrow \begin{bmatrix} 0 & 0 & 1 & 1 & 0 \\ 0 & 0 & 1 & 0 & 1 \\ 1 & 1 & 0 & 0 & 1 \\ 1 & 0 & 0 & 0 & 0 \\ 0 & 1 & 1 & 0 & 0 \end{bmatrix}$$

We considered algorithms in Exercises 3.1.22–3.1.26 that manipulate matrices in this way. With some knowledge of programming, it is possible to write a program to rearrange rows and columns similar to the method shown. Thus, we can find new matrices for the graph that correspond to any relabeling of the vertices. However, if we have two matrices A and B of graphs, it is impractical for large n to see whether the graphs are isomorphic by finding all possible rearrangements of A to see whether any of them agree with B. For an n by n matrix, there are $n!$ vertex rearrangements, so we know that the order of magnitude of the number of steps is at least $n!$, and we haven't even counted assignments and comparisons. Even for n moderately small, $n!$ is extremely large. As we have seen, $2^n = o(n!)$, and we regard time 2^n as intractable. (For example, $20! \approx 2.43 \times 10^{18}$.)

However, the results of this section can be helpful by cutting down the work considerably. We saw in Definition 5.2.2 that the degree of a vertex is simply the row sum for that vertex. Therefore, we can always rearrange the rows and columns so that the smallest degrees occur first. We can then read off the degree list, which is simply the list of row sums. If two matrices A and B are given, they can be compared to see whether the degree lists are equal. If not, the matrices are not isomorphic. If the degree lists are equal, we know that an isomorphism is possible, but vertices that correspond must have the same degree. This means that if we carry out the task of considering all rearrangements of vertices, it is necessary to rearrange only vertices of the same degree.

For example, if matrices A and B were both 20 by 20 and each had 10 vertices of degree 3, 5 of degree 4, and 5 of degree 6, the number of rearrangements would be $10!5!5! \approx 5.23 \times 10^{10}$, a significant reduction from $20! \approx 2.43 \times 10^{18}$. Thus, although the process might still be time-consuming, the little preliminary work based on the degree list would reduce the cases considerably. Note that the matrix of Example 5.4.7 has degree list 1, 2, 2, 2, 3. Thus there are only $1!3!1!$ or 6 possible rearrangements that give all matrices of graphs isomorphic to this one with the rows arranged in order of row sums. This compares with 5! or 120 such matrices if row sums are ignored. Further refinements are possible, as in Exercises 5.4.12 and 5.4.13.

Some Graph-Related Algorithms

EXAMPLE 5.4.8   *Algorithm To Find the Component Σ of a Vertex in Graph G*

Given the graph G, how can we find all of the vertices in the component Σ of a vertex A? We are thinking of a graph with a large number n of elements, and would like a reasonably efficient way to find the vertices of the component. One approach is as follows. Start with $A_1 = A$. Now find a vertex A_2 that is adjacent to $\{A_1\}$. (See Definition 5.1.13.) Then we know that A_2 will be in the component of A. Now find A_3, which is adjacent to $\{A_1, A_2\}$. If it exists, A_3 must be in the component of A. If it doesn't exist, then the component must consist of A_1 and A_2. Thus we have the following method. (It will be done in more detail in the next example.)

1. $\Sigma := \{A_1\}$ [initialize]
2. WHILE there is a vertex v adjacent to some vertex of Σ
    1. Find a vertex v adjacent to Σ
    2. $\Sigma := \Sigma \cup \{v\}$

Then the resulting Σ is the required component.
   We can prove that this works as follows.

First, all vertices v of Σ are connected to A_1 [i.e., are in the component $C(A_1)$]. This can be seen by induction. In step 1, the basis, it was true. Assuming it is true for some set Σ arrived at in the course of this algorithm, the adjunction in step 2.2 adjoins v to Σ when v is adjacent to Σ. This implies that $v \approx v_1$, where $v_1 \in \Sigma$. But $v_1 \approx A_1$ by induction. Therefore $v \approx A_1$ by induction.

To show that Σ consists of *all* vertices $v \approx A_1$, we assume the contrary and will arrive at a contradiction. Suppose then that $w \approx A_1$, but $w \notin \Sigma$. Then by Theorem 5.1.14 and Definition 5.1.13, there exists a vertex v adjacent to Σ but not in Σ. Such a v would, however, have to be in Σ because of the WHILE instruction 2. This is the required contradiction.

We now consider this algorithm in detail.

EXAMPLE 5.4.9 *The Component Algorithm*

We now take the algorithm sketched in Example 5.4.8, and see how it is implemented in more detail. For convenience, we take $\mathbb{I}_n = \{1, \ldots, n\}$ as the set of vertices of G. It is convenient to replace a set Σ by its characteristic function σ, written as a sequence. Thus, σ is a sequence of length n, and

$$\sigma_i = 1 \quad \text{for} \quad i \in \Sigma, \quad \text{and} \quad \sigma_i = 0 \quad \text{for} \quad i \notin \Sigma$$

We break the algorithm up into smaller pieces. We first find a procedure to determine whether a vertex j is adjacent to the set Σ. The input is σ and j, and the output is the variable Adj $=$ Adj(σ, j), which tells whether j is adjacent to σ. Thus, Adj $= 1$ if j is adjacent to σ, and 0 otherwise.

Procedure Adjacent$(\sigma, j;$ Adj$)$
1. IF $\sigma_j = 1$ THEN
    1. Adj $:= 0$
    2. End
2. FOR $i := 1$ to n [test all vertices]
    1. IF $(\sigma_i = 1$ and $c_{ij} = 1)$ THEN [*translation: $i \in \Sigma$ and ij is an edge*]
        1. Adj $:= 1$
        2. End
3. Adj $:= 0$ [failed to find an adjacent vertex]

We next find a procedure to enlarge Σ, if possible. The procedure will yield the enlarged Σ and the information that it was possible to enlarge it. This will be coded into a variable "Possible," which is 1 if a vertex was adjoined and 0 if none could be found. The method is to search through all the vertices, find the ones adjacent to σ, and adjoin them if any are found.

Procedure Enlarge(σ; Possible)
1. Possible := 0 [will change if we find an adjacent vertex]
2. FOR i := 1 to n [check all vertices]
   1. IF Adj(σ, i) = 1 THEN [adjoin i]
      1. σ_i := 1 [adjoined]
      2. Possible := 1

We didn't use an End instruction after step 2.1.2 because while we were in the loop we decided to look for other adjacent vertices. Finally, we have the Component Algorithm, which will find the component σ of k. We may need to make $n - 1$ enlarging passes for the worst possible case.

Component(k; σ) Algorithm
1. FOR i := 1 to n
   1. σ_i := 0
2. σ_k := 1 [these steps initialize $\Sigma = \{k\}$]
3. Possible := 1
4. WHILE Possible = 1
   1. Enlarge (σ; Possible)

A time analysis shows that the routine Adjacent takes $O(n)$ time. The Enlarge procedure has n loops. Loop r must use the Adjacent procedure, and so takes $O(n)$ time. The total time for the full Enlarge procedure is therefore $O(n^2)$. Finally, the Component Algorithm uses Enlarge $O(n)$ times, so the WHILE part of this algorithm has worst time $O(n^3)$. Naturally, the first $O(n)$ assignments don't affect this estimate.

EXAMPLE 5.4.10 *Algorithm To Determine Whether G Is Connected*

G is connected if the component of a vertex has all n vertices of G. One algorithm is as follows.

Algorithm Connected(G)
1. Component(1, σ) [now test if $\Sigma = G$]
2. FOR i := 2 to n
   1. IF σ_i = 0 THEN
      1. Output is "not connected"
      2. End
3. Output is "connected"

This algorithm is also $O(n^3)$. Look at the graph in Figure 5.4.7 to get an idea that not all graphs can simply be glanced at to decide whether they are connected.

Figure 5.4.7

A Minimization Problem

We conclude this section with an optimization problem on a graph. Suppose we are given a graph whose *edges are weighted*. This means that each edge is assigned a positive number, also called the **weight of that edge.** The **weight of a path** is defined as the sum of the weights of all edges in the path. The vertices might represent cities, and the weights could be the mileages between cities. Then the weight of a path would be the total length, in miles, of that path. Similarly, the weight might represent the time to travel between locations or the cost of going from one place to another. The problem is to find a path of **least weight** that joins two given vertices.

EXAMPLE 5.4.11

In Figure 5.4.8, it is required to find the path joining A to Z having the least total weight. Each edge is assigned a weight as indicated in the figure.

Method   Instead of concentrating on getting to Z with the least weight, we look at the problem of arriving at vertices near A with a path of least weight. As we continually expand this "near" set, we will be able to include Z and thereby arrive at the solution. In Figure 5.4.9, of the two possible edges, the one with smallest weight goes to B. This immediately

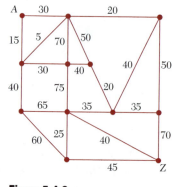

Figure 5.4.8

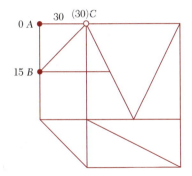

Figure 5.4.9

settles the problem of how to get from *A* to *B* with least total weight. Go directly there for a path of weight 15. To see this more formally, observe that any other path to *B* would start from *A* to *C*, and we would already be starting with a weight of more than 15. We now mark *B* permanently with the weight 15. This means that we can get to *B* with weight 15, and any other path to *B* must have weight at least 15. Similarly, mark *A* with 0, since this is our starting vertex. Mark *C* with the temporary weight 30, since we know we can get to *C* with weight at most 30. (See Figure 5.4.9.)

We now venture away from the new permanently marked vertex to the vertices adjacent to it (its neighbors). Temporarily label each of these neighbors that are not yet marked with the total weight of the path getting there. This is the permanent weight plus the weight of the additional edge to get to the neighbor. If a neighbor already has a temporary weight and the new path yields a smaller weight, take the smaller weight as the new temporary weight. This happens in Figure 5.4.10, where the weight 30 was replaced by the smaller weight 20. We thus arrive at Figure 5.4.10, in which vertex *C* has the least temporary weight. Now, *the least of the temporary weights is made permanent.* The reason is that any other way

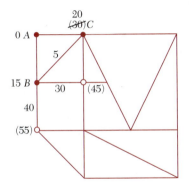

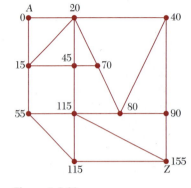

Figure 5.4.11

of arriving at this vertex would necessarily use other paths that would have a total weight at least equal to this minimum weight. Label this new vertex permanently and continue this process.

The process always adds a new permanent vertex as long as any vertices are left, and then updates the temporary weights to include the new permanent vertex in the computation. When we finish, we find that Z has weight 155. (See Figure 5.4.11.) This means that we can arrive at Z with weight 155 and cannot do better. If we take care in our labeling, we can also list the neighboring vertex and edge that give any vertex its permanent status. (We do this in our analysis later.) In this way, we can find the optimal path as in Figure 5.4.12.

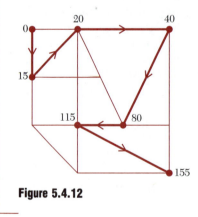

Figure 5.4.12

Remark   This is a good example of an inductive procedure. We wanted to find something about the vertex Z. But instead of starting our analysis with that vertex, we started with A (the known case) and built up, in each case arriving at a new vertex by using the previously decided cases.

The algorithm illustrated in Example 5.4.11 is due to E. Dijkstra. We now list this least-weight algorithm in pseudocode. First, we give a definition.

DEFINITION 5.4.12

Suppose C is a path $u \cdots v$, from u to v, and that D is a path $v \cdots w$, from v to w. Then the path $C + D$ is the path $u \cdots v \cdots w$ obtained by adjoining the vertices of D to the vertices of C. It is required that the end vertex of C be the initial vertex of D.

EXAMPLE 5.4.13   Dijkstra's Algorithm

This algorithm will find the minimal weight of paths joining a fixed vertex u_0 to any other vertex. It keeps track of the temporary weights found along the way. It will also keep track of the actual path used, by noting the vertex $P(v)$ *preceding* a vertex v in a minimal path. We first state some conventions.

1. The vertices V are labeled $1, 2, \ldots, n$.
2. The weight of edge ij is called $w(i, j)$.
3. If ij is not an edge, we write $w(i, j) = \infty$.
4. The initial vertex is chosen to be 1.
5. The path of least weight from 1 to i has weight $W(i)$.
6. The permanently marked vertices constitute the set Σ.
7. In the algorithm, vertices i have "temporary" weights $T(i)$. These vertices are adjacent to Σ. We write $T(i) = \infty$ if i has not yet been assigned a temporary weight.
8. The algorithm computes the path of least weight to i, as well as its weight. In this path, the vertex preceding i is called its predecessor $P(i)$.
9. We compute naturally with ∞, using $\infty + x = x + \infty = \infty$, and $n < \infty$, for all x (including ∞) and any integer n. These convenient conventions allow us to use undefined functions such as $w(i, j)$, when i and j are not adjacent, and $T(i)$, when i is not adjacent to Σ. They are always used in a minimization process, so the value ∞ will not be used. We could just as well replace ∞ by a number larger than the sum of the weights $w(i, j)$.

Dijkstra's algorithm
1. $\Sigma := \{1\}$ [initialize Σ, the set of permanently marked vertices]
2. $W(1) := 0$
3. FOR $i := 2$ to n
   1. $T(i) = \infty$ [initializing the temporary weight]
   2. $P(i) := 1$ [initializing P]

4. FOR $i := 2$ to n
    1. IF $w(1, i) < T(i)$ THEN $T(i) := w(i, 1)$ [initially recomputing temporary weights based on adjacency to 1]
5. WHILE $\Sigma \neq V$ [now the cycle—find the smallest temporary weight, at vertex s, and adjoin s to Σ]
    1. Min := least value of $T(j)$ for $j \notin \Sigma$
    2. $s :=$ a vertex $\notin \Sigma$ such that Min $= T(s)$ [s is the next vertex to be adjoined to Σ]
    3. $\Sigma = \Sigma \cup \{s\}$ [adjoin it to Σ]
    4. $W(s) :=$ Min [and give it the permanent weight]
    5. FOR $i := 2$ to n [now update temporary weights with the new permanent set Σ]
        1. IF $i \notin \Sigma$ THEN [update temporary weight and predecessor P, if necessary]
            1. IF $W(s) + w(s, i) < T(i)$ THEN
                1. $T(i) := W(s) + w(s, i)$
                2. $P(i) := s$

We can do a time analysis, using the number of vertices n as the size of the input. The initializations take $O(n)$ time. The WHILE instruction loops $n - 1 = O(n)$ times, and within each loop time $O(n)$ was taken on assignments and comparisons. Thus, the time of this algorithm is $O(n^2)$.

We now show that this algorithm terminates and that it yields the path of minimal weight.

THEOREM 5.4.14

Let G be a connected, weighted graph with vertices $\{1, \ldots, n\}$, where each edge ij has positive weight $w(i, j)$. Let the foregoing algorithm determine weights $W(i)$ and predecessor function P. Then the algorithm terminates when $W(i)$ is defined for all vertices in V, and P is defined for all vertices except 1, with $P: V - \{1\} \to V$. Further:

1. The path from 1 to u of least weight has weight $W(u)$.
2. Let $u \neq 1$ and $v = P(u)$. Suppose $C(v)$ is a path of least weight joining 1 to v. Then a path of least weight joining 1 to u is $C(v) + vu$.

Proof   We first note that, by Corollary 5.1.15, if Σ does not contain all the vertices of G, then there is a vertex v that is adjacent to Σ. By step 5.5.1.1 such a vertex will have a finite temporary weight $T(v)$. Therefore, Σ will eventually consist of all vertices of G, so the algorithm must end.

We now show by induction that $W(v)$ is the least possible weight of all paths joining 1 to v, and that a minimal weight path to v can be found so that the vertex just before v is $P(v)$. The proof of this is very similar to the heuristic reasoning in Example 5.4.11. It is clearly true at the start, since the path of least weight joining 1 to itself has weight 0, and there is nothing more to prove about $P(1)$.

Now suppose Σ_k is the value of Σ after the kth run through the WHILE loop 5. By definition, $\Sigma_0 = \{1\}$, and we have shown the result is true for the vertex 1. We assume the result for the vertices of Σ_k and prove it for the vertices of Σ_{k+1}.

Suppose the vertex s is found in step 5.2. By 5.3, s is the only new vertex in Σ_{k+1}, so we must prove the result for the vertex s. The predecessor $s_0 = P(s)$ was assigned in step 5.5.1.1, where all predecessors and temporary weights are updated. In steps 5.5.1.1.1–5.5.1.1.2, i plays the role of s and s plays the role of s_0. Thus, we have $T(s) = W(s_0) + w(s_0, s)$. Because $W(S) = T(s)$ (see 5.1, 5.2, and 5.4), we have

$$W(s) = W(s_0) + w(s_0, s), \qquad s_0 = P(s) \qquad (5.4.2)$$

If C is the path of least weight to s_0, the weight of C is $W(s_0)$ by induction. By definition, the weight of the path $C + s_0 s$ is $W(s_0) + w(s_0, s) = W(s)$. Thus (5.4.6) will prove the result once we have shown that $W(s)$ is the least weight of all paths joining 1 to s.

Now let D' be *any* path joining 1 to s. By Corollary 5.1.15, there is a first vertex t on this path that is adjacent to Σ_k. The vertex t_0 preceding it is in Σ_k. Thus, we can write $D' = C' + t_0 t + \alpha$. (See Figure 5.4.13.) But in the competition to decide which vertex would be adjoined to Σ_k, s was chosen over t. This means that the weight of $C + s_0 s$ was $\leq$ weight of $C' + t_0 t$. Thus, the weight of $C + s_0 s \leq$ weight of $C' + t_0 t + \alpha =$ weight of D'. Since D' was an arbitrary path joining 1 to s_0, we have the result. ■

Variations on this problem are possible. For example, suppose you are a salesperson and you want to find a route *going through all of these cities*, beginning and ending at A, and you want one with the least weight possible. This problem, the "traveling salesperson problem," has no known tractable solution. Similarly, if you want to visit each city once and minimize weights, there is then the question of whether a Hamiltonian path exists at all. If one does, how do you find the one of minimum weight? Or suppose you want to start at A and end at Z, passing through each city? Or suppose you don't even care which city you land in? All of these have no known tractable solution.

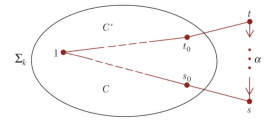

Figure 5.4.13

Exercises

In these exercises, all graphs are understood to be simple.

1. Draw a graph with degree list 1, 2, 2, 3, 4.
2. Explain why any two graphs with degree list 1, 2, 2, 3, 4 are isomorphic.
3. Draw a graph with degree list 3, 3, 3, 3, 4. Explain why any two such graphs must be isomorphic.
*4. Draw *all* graphs with degree list 2, 2, 2, 2, 2, 2.
*5. Show that there is only one connected graph that has n vertices each of degree 2 ($n \geq 3$).
6. Prove that two graphs G and G' are isomorphic if and only if their vertices can be labeled so that they have the same matrix.
7. Determine whether the graphs of Figure 5.4.14 are isomorphic. Explain your answer.

Figure 5.4.14

8. Determine whether the graphs of Figure 5.4.15 are isomorphic. Explain your answer.

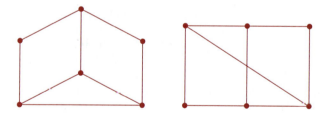

Figure 5.4.15

*9. Determine whether the graphs of Figure 5.4.16 are isomorphic. Explain your answer.

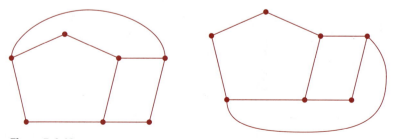

Figure 5.4.16

10. Let G and G' be isomorphic graphs. Show that the number of components of G is equal to the number of components of G'.

*11. Give an example of two graphs that have the same degree list but are not isomorphic by virtue of the previous exercise.

12. If v is any vertex of a graph G, consider the degrees of all vertices connected to v by an edge of the graph, arranged in ascending order. Call the resulting list the degree list of the neighbors of v. Show that the graphs G and G' are isomorphic under an isomorphism f, and if $f(v) = v'$ then the degree list of the neighbors of v is the same as the degree list of the neighbors of v'.

*13. Give an example where Corollary 5.4.3 does not distinguish between two graphs, but where the result of the previous exercise does.

*14. Say that a graph **contains a triangle** if it has three distinct vertices, all of which are connected. Otherwise, say that the graph is **triangle-free**. Find all triangle-free graphs having six vertices each of degree 3.

15. For the matrix of Example 5.4.7, relabel the rows and columns so that the row sums will be in ascending order, and find the matrix for the graph whose vertices are so relabeled.

16. In the Component Algorithm of Example 5.4.9, we stated that the WHILE instruction (step 4) might have to loop as much as $n - 1$ times. Explain why this is so by giving a graph where this is necessary.

17. Figure 5.4.17 gives the airline routes connecting various cities, and the cost of the tickets. Find the most economical route connecting Canyon Town and Metro City.

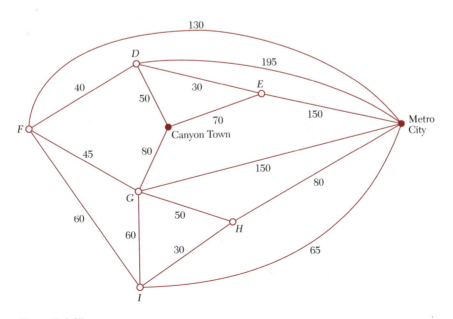

Figure 5.4.17

18. Figure 5.4.18 gives the times (including waiting and travel times) of the airline routes of the previous exercise. Find the fastest air route connecting Metro City and Canyon Town.

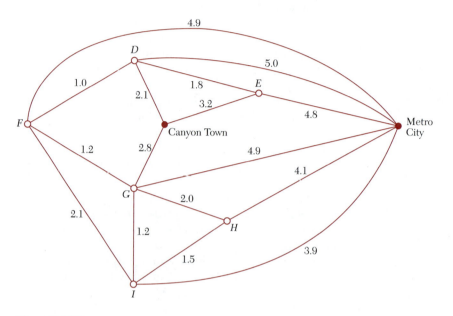

Figure 5.4.18

19. In the weighted graph of Figure 5.4.19, find a path of least weight joining A and B.

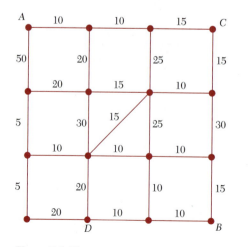

Figure 5.4.19

20. Repeat Exercise 19 for the path joining C to D.
21. Write an algorithm that interchanges rows r and s of a matrix and then interchanges columns r and s. The input is a matrix a_{ij}, $1 \leq i, j \leq n$, and the values r and s. Estimate the time for your algorithm.
22. (*Computer-related*) Convert your algorithm in the previous exercise into a program in a suitable programming language.
23. Suppose a graph has matrix a_{ij}. Write an algorithm that finds the degree list for the graph. The matrix is understood to be n by n. Estimate the time of your algorithm.
*24. In the Dijkstra Algorithm of Example 5.4.13, steps 5.1 and 5.2 found the vertex s. Find a procedure to do this. Inputs are Σ, coded by its characteristic function σ, the permanent weights W defined for vertices of Σ, and temporary weights $T(i)$, $1 \leq i \leq n$. The output is the value of Min and the required s.
25. Is there a Dijkstra problem and algorithm for weighted *directed* graphs? Explain.

5.5 Games of Strategy

We consider certain kinds of perfect information, two-person, zero-sum games.

> **DEFINITION 5.5.1**
>
> **A two-person, zero-sum game** is a game played by two people, called A and B. The game must end, and when it does each player receives an award, called the **payoff.** A's payoff is the negative of B's.

Because A's payoff is the negative of B's, the sum of the awards is 0. In a betting situation, if A's payoff is \$5, then B's is $-\$5$, and B simply gives A \$5. We usually keep track only of A's payoff. The idea of a game is that A tries to **maximize** her payoff, and B tries to maximize his. (A is Alice, and B is Bill.) This means that B tries to **minimize** A's payoff. In a two-person, zero-sum game, anything you do to improve your payoff necessarily reduces your opponent's payoff. And anything you do to reduce your opponent's payoff increases your payoff. A zero-sum, two-person game is a "dog eat dog" game.

A game like tic-tac-toe (three in a row) is an example of a two-person game. We ordinarily think of such a game in terms of winning and losing, not in terms of payoffs. However, we may assign payoff 1 for a win, payoff -1 for a loss, and payoff 0 for a tie. Each player wants to maximize his or her payoff, that is, to win. Next in order of preference is a tie with payoff 0.

EXAMPLE 5.5.3   Tic-Tac-Toe

In this game, players alternately place X's or O's in a 3 by 3 array. Player A places the X's, and she goes first. Player B places the O's. A player wins when he or she gets three in a row horizontally, vertically, or diagonally. Figure 5.5.1 gives a winning position for A (value 1), a draw (value 0), and a loss (value −1). Remember that the value is always A's payoff.

Win   Draw   Lose

Figure 5.5.1

Most people "know how to play the game well." This means that any person can force a draw, and if the other person plays carelessly it is possible to win.

EXAMPLE 5.5.4

Figure 5.5.2 gives a position in tic-tac-toe. We can see, by counting, that it must be A's move. (She started first, and, by convention, the first player uses X.) A has a choice of five moves, one of which is a "winning move." Do you see it?

A to play and win

Figure 5.5.2

EXAMPLE 5.5.5   Go-moku

Another game in the family of tic-tac-toe is Go-moku. This game is of Japanese origin and can be described simply as five in a row. However, it is not played on a 5 by 5 array. It might be played on a Go board, which is equivalent to a 19 by 19 array. Instead of X's and O's, black and white stones are put on the intersections, as indicated in Figure 5.5.3. By convention, black goes first. Of course, it can be played on a checkerboard or even with X's and O's. It can be quite complicated, even if it is a perfect information game. More familiar are the games of chess and checkers. All of these games are win, lose, or draw games.

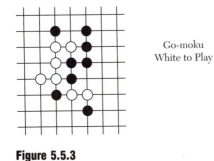

Go-moku
White to Play

Figure 5.5.3

EXAMPLE 5.5.6   The Game of Ten

The **game of ten** is sometimes taught to children when they are learning to add. Alice goes first and chooses a number 1 or 2. Bill then adds 1 or 2 to it, obtaining an answer. Then A adds 1 or 2 to it, and so on. The first person to reach 10 wins. Here, a position is simply given by an integer between 0 and 10 inclusive. It is not too hard to figure out how to win this game. We discuss it further in Example 5.5.13, and variations are given in the exercises.

EXAMPLE 5.5.7

Backgammon is a two-person, zero-sum game. Because it uses dice, a random element enters the game. It is still, however, a game of skill, and it qualifies as a perfect information game. Card games are usually not perfect information, because, most of the time, certain cards are hidden. This is true in a typical poker game, which is not normally a two-person game, and where you don't know your opponent's hand.

One of the challenges of a zero-sum, two-person, nonrandom perfect information game is that it is clearly a battle of wits—you have all the information on the table, and there are no chance elements to surprise you. We now analyze games of this type that are played sequentially, like tic-tac-toe.

DEFINITION 5.5.8

A **sequential game** is a two-person, zero-sum game played by each player moving alternately, as in tic-tac-toe. Each player moves from one **position** to another, according to the rules of the game, until no move is possible. A position from which no move is possible is called an **end position**. Player A starts the process. The game is required to end in a finite time, and the rules determine what A's payoff is.

For simplicity, we simply call this a **game** in what follows. We can put such games into a reasonable graph-theoretic setting. We represent the positions as **vertices.** If it is possible to move from one position P_1 into a position P_2, we draw a directed edge from P_1 to P_2. Thus, we can construct a **game graph** G. We have insisted that the game must end after a finite number of moves. This means that there are **no cycles** on G, for it there were the game could continue forever. We will usually assume that it is **finite**; that is, there are finitely many positions. Thus, we have the following definition.

DEFINITION 5.5.9

An **acyclic** digraph is a digraph with no cycles on it other than the trivial one-vertex ones. A **game graph** G is a finite, directed, acyclic graph. A vertex with out-degree 0 is called an **end vertex**.

The game graph of ten is illustrated in Figure 5.5.4. It is relatively easy to picture, since it has only 11 positions.

Even a simple game such as tic-tac-toe has a rather large graph. If, for example, we take as a possible position any configuration of X's, O's, and blanks in a 3 by 3 array, there are $3^9 = 19,683$ positions! If we insist on an equal number of X's as O's, or once more X than O, we can cut this down, but the number is still large. But for complicated games, the figures get astronomical. For example, playing Go-moku on a checkerboard will lead to 3^{64} or about $3.4 \cdot 10^{30}$ positions. Even with high-speed computers with enormous memories, such problems cannot be brutally attacked. It would be hard enough to store the game graph into a computer, much less analyze a proper strategy for it.

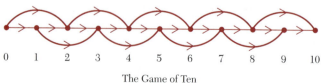

The Game of Ten

Figure 5.5.4

EXAMPLE 5.5.10

Chess is an example of a game for which, to this day, nobody knows a proper strategy. This game cannot go on forever because there are special rules that prevent this from happening. Still, a study of chess through computers, by way of its graph, has been reasonably successful. The current computer champ (called Deep Thought, at this writing) can beat most humans, including those who regard themselves as superior players. We shall say more on this later.

Win/Lose Games

EXAMPLE 5.5.11

If G is a game graph, how do you play it? We consider a game graph G as in Figure 5.5.5. A starts at vertex v and moves to one of three choices. Then B moves, and they move alternately. The first person who has no moves to make *loses;* the other person wins. Thus, players A and B each want to move into either of the end vertices marked X. It is player A's turn. What is her move?

Method   Our method is to identify each position as a winning or a losing position. You can be in a winning position and yet end up losing the game because you haven't played well. If you are in a losing position, then no matter what you do you must move into a winning position (for your opponent) and you will then lose if your opponent plays well. If you are in a winning position, your goal is to move into a *losing* position, because then it will be your opponent's turn. Thus, the positions marked

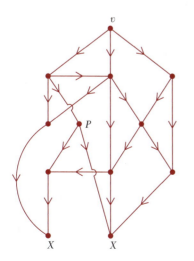

Figure 5.5.5

with an X are losing positions. You want your opponent to be there, so you will want to move into it.

We label each vertex 0 or 1; 0 is for the losing position, and 1 is for the winning position. We want the following to be true for a labeling:

Rule 1:   *All moves* from a 0-vertex go to a 1-vertex.

Rule 2:   *There is a move* from a 1-vertex to a 0-vertex.

Our method is to work backwards from the known losing positions to find winning positions and then, with this knowledge, to discover new winning or losing positions, and so on, until we have labeled all positions. The method is illustrated in Figure 5.5.6. Start labeling the end vertices 0; these are definite losers. Call this set of vertices V_0. Now find the set V_1 of all vertices from which the next move must end in V_0. That is, each edge from a vertex of V_1 leads to a vertex of V_0. Label these as 1-vertices. Now find the set V_2 of all vertices from which the next move must end in either V_0 or V_1. Label these according to Rules 1 and 2. Continuing in this way, we label the original game graph.

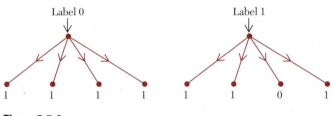

Figure 5.5.6

In short, we systematically find winning and losing positions by working backwards from the known losing positions. Note that the vertex P in Figure 5.5.5 is not in V_1, although there was a direct path from it to a vertex of V_0. There were two moves from it: one to a 0-vertex and one to a 1-vertex. By Rule 2, it must be labeled 1. Figure 5.5.7 gives the complete labeling of the game graph.

Using Figure 5.5.7, you can play the game and concentrate on finishing your reading of a major novel at the same time. You are at v, a 1-vertex. Move into a 0-vertex. This must be possible by Rule 2. Your opponent must then move into a 1-vertex by Rule 1. Keep moving into a 0-vertex. When it is your turn, you'll be in a 1-vertex, and you can keep moving into a 0-vertex indefinitely. You will win, because the game is finite and you will finally move into one of the 0-vertices marked X. The labeling solves the game and robs it of any mystery it might have had. This method will be extended to games with varying payoffs. For the present, we shall work with win/lose games.

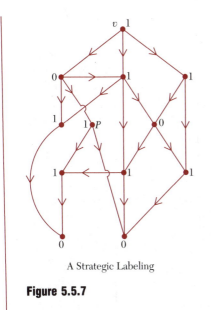

A Strategic Labeling

Figure 5.5.7

DEFINITION 5.5.12

A **labeling** of a game graph G is function L from the vertices of G into
$\{0, 1\}$, where $f(v)$ is called the **label** of v. If $f(v) = 0$, we say that v is a
0-vertex, and similarly for 1-vertices. A **strategic labeling** of G is a labeling such that

1. All edges with an initial 0-vertex have a terminal 1-vertex.
2. If v is a 1-vertex, there is an edge with initial vertex v whose terminal
   vertex is a 0-vertex.

Remark It follows from 2 that the end vertices are 0-vertices.

As in Example 5.5.11, a strategic labeling of a game graph gives a
strategy for playing the win/lose game in which you lose if you can't
make a move. If you are at a 1-vertex, simply move to a 0-vertex, and
continue doing this till you win. If you are at a 0-vertex, you are in a
losing position. Play on, hoping that your opponent doesn't know how
to play well.

It is not always necessary to use the technique given in Example 5.5.11
to find a strategic labeling. Sometimes, inspiration or insight gives a
strategic labeling immediately.

EXAMPLE 5.5.13   *The Game of Ten*

You want to get to 10 by adding 1 or 2, starting at 0. A little insight tells you that if you get to 7, you win. The same insight says that you might as well aim for 4, and similarly for 1. In short, you have a strategic labeling: Label 1, 4, 7, and 10 as 0-vertices and the rest as 1-vertices. We can then verify conditions 1 and 2 of Definition 5.5.12. Figure 5.5.8 gives the strategic labeling of the game graph of ten.

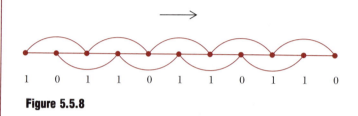

    1   0   1   1   0   1   1   0   1   1   0

Figure 5.5.8

The method used in Example 5.5.11 for a strategic labeling of a game graph is clearly recursive. We now prove that this method can always be done. The following definition is useful.

DEFINITION 5.5.14

A **lower set** L of vertices in a game graph is a set of vertices with the property that if an edge of G has its initial vertex in L, then its terminal vertex must be in L, too.

Remark   A lower set L, as a subgraph of G, is itself a game graph. An end vertex in L is necessarily an end vertex in G.

THEOREM 5.5.15

Any game graph G has a strategic labeling.

Proof   We show by induction that any lower set of G has a strategic labeling. The basis is the lower set T consisting of all the end vertices of G. We label these as 0-vertices, and we clearly have a strategic labeling of T. We now assume that a lower set L can be strategically labeled. We want to show, if $L \neq G$, how to extend this to a larger one. As a guide for the induction step, Figure 5.5.6 illustrates Rules 1 and 2 of Definition 5.5.12, and will show how to prolong the strategic labeling to one more vertex.

Suppose $L \neq G$. We claim there is a vertex $v \in G - L$ such that $\{v\} \cup L$ is a lower set of G. Let us call such a vertex v **suitable.** To find a suitable vertex, we proceed as follows. Choose any vertex $v_0 \in G - L$. If v_0 is not suitable, there is an edge starting at v_0 whose terminal vertex v_1 is also not in L. If v_1 is not suitable, there is an edge starting

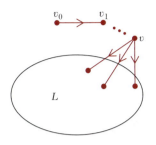

v_0 v_1

v

L

Figure 5.5.9

at v_1 whose terminal vertex v_2 is not in L. This process continues indefinitely if there are no suitable vertices. But there are no cycles in a game graph, and the graph is finite, so the process cannot continue indefinitely. Therefore at some point, we will find a suitable vertex v. Figure 5.5.9 illustrates this process.

We now use conditions 1 and 2 of Definition 5.5.12 to guide us in the labeling of a suitable vertex v. When we label v accordingly, these conditions will continue to be true on the lower set $L \cup \{v\}$, and so we have extended the lower set and its labeling to a larger one. ■

We illustrate this technique with a game that is difficult to analyze without such a strategic analysis.

EXAMPLE 5.5.16   Wytoff's Game

In the two piles of checkers in Figure 5.5.10, pile 1 has 10 and pile 2 has 17. Players A and B alternate moves as follows: Any player can remove any positive amount from pile 1, including the whole pile, *or* any positive amount from pile 2, *or* an equal amount from each pile. The last person to make a move (and remove everything) wins. It is A's turn. How does she win?

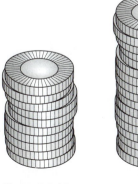

Figure 5.5.10

Method

We need a strategic labeling of the positions. Player A hopes that she is at a 1-vertex, so she will be able to move into a 0-vertex. But a labeling is required.

It is best to represent the piles (m, n) as a point in the plane or, better, as a single checker on the checkerboard of Figure 5.5.11. Some possible moves are illustrated. The rules are geometrically stated as moving left, down, or diagonally down to the left. There is one end vertex $(0, 0)$ at the lower left.

We start our labeling at the end vertex. All vertices (squares on the checkerboard) that can get to this in one move are necessarily 1-vertices. These are labeled $\times$ in Figure 5.5.12. As you can see, you want to be there. The vertices labeled ■ in this figure only lead to the $\times$ squares. They are therefore 0-vertices. For example, the pile $(1, 2)$ is a losing position if it is your move.

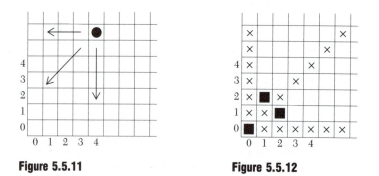

Figure 5.5.11   Figure 5.5.12

Once we label the vertices ■ as 0-vertices, we can find a host of other 1-vertices—the ones leading to them in one step. This leads to Figure 5.5.13 with new 0-vertices ■. We can obviously continue this process indefinitely. We arrive at Figure 5.5.14.

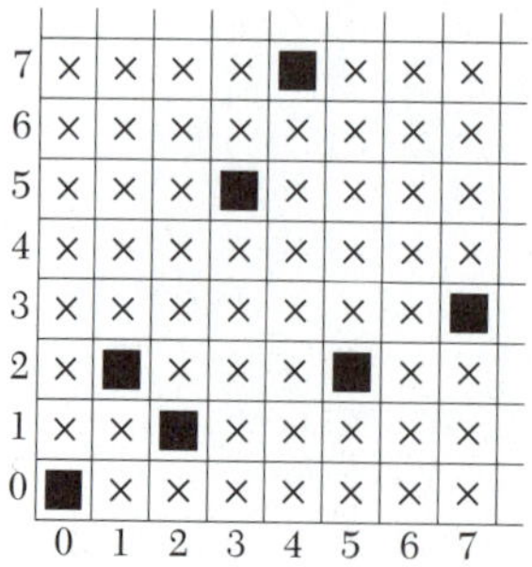

Figure 5.5.13

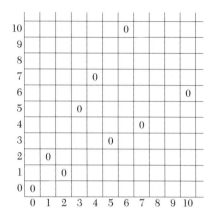

Figure 5.5.14

We can now read off the 0-vertices. By symmetry, the piles (m, n) and (n, m) are labeled the same, so we only give the 0-vertices (m, n), where $m \leq n$. These are

$$(0, 0), (1, 2), (3, 5), (4, 7), (6, 10), (8, 13), \ldots \qquad (5.5.1)$$

According to the original problem, player A is in the position $(10, 17)$ at the beginning. Since 10 occurs in the list of 0-vertices, player A need only remove 11 from the 17 pile to bring it to a $(6, 10)$ position. She then waits for B's move, perfectly confident that she is going to win if she continues to move into these 0-positions. The list of 0-vertices is the key to winning this game.

EXAMPLE 5.5.17

Is there a simple way of generating the sequence of 0-vertices (5.5.1)? By observation, the differences $n - m$ for the 0-vertex (m, n) are 0, 1, 2, 3, ... The nth term in this sequence of 0-vertices appears to be $(a_n, a_n + n)$. This is convenient to know, since a pile $(54, 50)$ (which has difference 4) can be reduced to $(6, 10)$ in a flash by pulling 44 off each pile. But how do you find a_n? Again, by observation, a_n is *the first non-negative integer not used in either the x or y component* in the sequence through the $(n - 1)$st term. In the sequence (5.5.1), for example, the next term is $(a_6, 6 + a_6)$. The first natural number not appearing so far is 9. The next 0-vertex is therefore $(9, 15)$. The next one is $(11, 18)$, since 10 has already been used.

THEOREM 5.5.18

Let the sequence (a_n, b_n) be defined recursively as follows:

1. $a_0 = b_0 = 0$.
2. a_n is the first natural number $F > a_{n-1}$ such that $F \neq b_i$ for $i = 1,$ $\ldots, n - 1$.
3. $b_n = a_n + n$.

Now define the vertex (m, n) to be a 0-vertex if (m, n) or (n, m) is one of the terms of the sequence (a_n, b_n). Otherwise define (m, n) to be a 1-vertex. Then this labeling is a strategic labeling of the game graph of Wytoff's game. ■

We leave the proof for the exercises.

EXAMPLE 5.5.19   Nim

Nim is an ancient game somewhat similar to Wytoff's game. Its analysis is not subject to a simple two-dimensional diagram as Wytoff's was, but the strategy is perhaps easier. It is played with several piles of checkers as in Figure 5.5.15. Here the players must remove a positive number of checkers (possibly all) from any *one* pile of his or her choosing. Play alternates, and the one who picks up the last checker wins. How do you play?

Figure 5.5.15

Method   The 2-pile game is fairly easy to play. Level them, if possible, and then do in one pile what your opponent does in the other. (This is the copycat strategy.) If the piles are equal when you start, you're at a 0-vertex. You'll probably lose because the copycat strategy is fairly easy to see. But what happens with more than two piles?

A strategy is given by defining the 0-positions and verifying that they satisfy Rules 1 and 2 of Definition 5.5.12. We first illustrate the correct strategy before stating it. How do you play if the numbers of checkers in the piles are 3, 5, and 8? The method is as follows. Write these numbers in binary and add each column as indicated, with no carrying:

$$\begin{array}{rl} 3: & 11 \\ 5: & 101 \\ 8: & \underline{1000} \\ & 1112 \end{array}$$

Now move so that each column sum is *even*. Thus, change the 8-pile (1000) to 110 (6). Your correct move is to pull away two checkers from the 8-pile, leaving six and the piles 3, 5, and 6. Thus, we define the strategic labeling in Nim as follows: If the numbers of each pile are written in binary notation and the sum of *each* column is even, the position is defined to be a 0-position. Otherwise, it is a 1-position. By definition, no piles constitute a 0-position.

We now show that this is a strategic labeling.

Any 0-position leads to a 1-position. For if all columns add up to an even number, then, when we lessen one column, one of the 1's must change to a zero, and hence an even sum becomes an odd one. The result is then a 1-position.

From any 1-position, it is possible to move to a 0-position. We illustrated this before. Choose the column farthest to the left that has an odd sum. Take any of the piles with a 1 in that position. Change this to a 0 and change the rest of the bits so that all column sums become even. In this way, you have changed a 1-position into a 0-position.

Games with Variable Payoffs

We now return to our original idea of a two-person, zero-sum game in which each end position determines a **value** to the player in that position. In general, an end position will have a **dual value** (P_1, P_2).

DEFINITION 5.5.20

The value (P_1, P_2) at an end position represents **the amount player A wins,** depending on whose turn it is. The amount P_1 is the payoff to A if A is in this position, and P_2 is the payoff, again to A, if B is in this position.

Since the game is zero-sum, the payoff to B is $-P_1$ or $-P_2$, according to whether player A or player B is in this end position.

For example, if the payoff is $(1, -1)$ at an end position, then each player receives a payoff 1 if he or she is in that position. (When A receives -1, B receives 1.) Note, however, that if A has just moved into this position, B receives 1, because it is B's turn. Payoffs always refer to value when it is a person's move. The payoff $(1, 1)$ at an end position means that player A receives 1 if either player is at this position. A tie game corresponds to the payoff $(0, 0)$.

EXAMPLE 5.5.21

In the win/lose game, every end position was a losing position for either player. Thus, all end vertices were 0-vertices. The payoffs were all $(-1, 1)$ at the end vertices. For win/lose games, a strategic labeling assigned only two possibilities at each vertex:

1. The 0-vertices had the payoff $(-1, 1)$.
2. The 1-vertices had the payoff $(1, -1)$.

DEFINITION 5.5.22

A **valued game graph** G is one in which each end vertex v is assigned a dual payoff $(P_1(v), P_2(v))$.

EXAMPLE 5.5.23

We now want to learn how to play the game represented by a game graph with a dual payoff function. In Figure 5.5.16, regardless of who lands at an end vertex, player A collects the money located there. (Since this is a two-person game, it is paid by B.) It is A's move, at position v. She would love to win \$5. But she is smart enough to take \$2, since by moving to r, B will surely go to the $-\$1$ payoff, and A will actually lose \$1. If it is B's move at position v, he would like A to win $-\$1$, but knows better than to move to r, because then A would take the \$5. B settles for losing \$3. (A's payoff is \$3.) Thus, we can say that the value at position v is \$2 if it is A's move, and \$3 if it is B's. The payoff at v is (\$2, \$3). Both these amounts represent A's potential winnings. The difference revolves around whose turn it is. If it is A's turn, A can guarantee at least \$2 by making the correct move. If it is B's turn, he can guarantee that A's winning will not be more than \$3. In this way, we can have a dual payoff function at *each* of the vertices v of G. Both P_1 and P_2 are A's payoff.

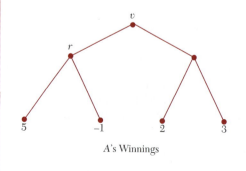

A's Winnings

Figure 5.5.16

DEFINITION 5.5.24

A **payoff evaluation function** of a valued game graph is an assignment $(P_1(v), P_2(v))$ to each vertex v of G that has the following properties:

1. $P_1(v) = \text{maximum}(P_2(v_1))$ for all v_1 such that vv_1 is a directed edge of the game graph.
2. $P_2(v) = \text{minimum}(P_1(v_1))$ for all v_1 such that vv_1 is a directed edge of the game graph.

Remark The idea is that player A looks over all positions she can go to (these are the v_1 in condition 1), and will pick the one with the largest value P_2, knowing that it will be B's move. Thus, she maximizes $P_2(v_1)$. Similarly, B minimizes $P_1(v_1)$, and this is condition 2. Once a dual payoff function is found, the strategy is straightforward. If A is moving with value P_1, she will move to the position for which P_2 is the same. (It can never be larger.) If she chooses a smaller P_2, then she has made a "bad move." Similarly, if B has value P_2, he will move to a position v with the same $P_1(v) = P_2$.

EXAMPLE 5.5.25

As soon as either of the players reaches any of the bottom positions of Figure 5.5.17, player A wins the indicated amount of money. A is at position v. What is her best move? What is the best reply?

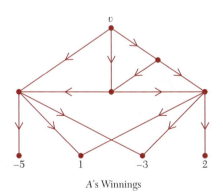

A's Winnings

Figure 5.5.17

Method The dual payoffs are indicated in the valued game graph of Figure 5.5.18. We now work ourselves up to all positions much as we did in Theorem 5.5.15, always adjoining a vertex whose only move is to a position that already has a payoff function. Using Rules 1 and 2 of

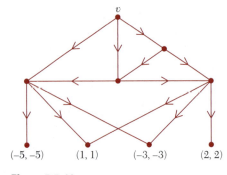

Figure 5.5.18

Definition 5.5.24, we have, at any new position, P_1 = maximum of all $P_2(v)$, where v can be reached in one move, and P_2 = minimum of all $P_1(v)$, where v can be reached in one move. In this way, we build up Figure 5.5.19.

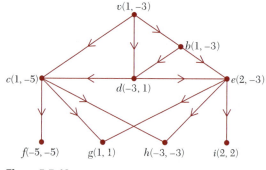

Figure 5.5.19

For example, to find the value at c, P_1 is the largest of the P_2 values of f, g, and h. This is 1, by observation. Similarly, P_2 is the least of the P_1 values, which is -5. We build up in order: c, e, d, b, and v. By observation, A's best move is to d (the one with the largest P_2 value), and B's best reply is to c (the one with the smallest P_1 value).

EXAMPLE 5.5.26

The labeling of a valued game graph can be described by the following algorithm, similar in construction to the one for the strategic labeling of G. The graph G is an input, as well as the payoff functions $P_i(v)$ ($i = 1$, 2) defined on the end vertices.

Payoff Evaluation Labeling
1. $L := $ set of end vertices [initializing the lower set on which P_i will be defined]

2. WHILE $L \neq G$ [enlarge L]
    1. $v :=$ a vertex in $G - L$ such that $L \cup \{v\}$ is a lower set [see the proof of Theorem 5.5.15, which shows the existence of such a v]
    2. $P_1(v) =$ maximum $P_2(w)$ over all w such that vw is an edge of G
    3. $P_2(v) =$ minimum $P_1(w)$ over all w such that vw is an edge of G
    4. $L := L \cup \{v\}$

EXAMPLE 5.5.27   *Bipartite Games*

In many games, positions can be occupied only by A or only by B. This is true of chess, where a position is not only a configuration of pieces on a board but a knowledge of whose move it is. In such situations, there are A-positions (positions from which A can move) and B-positions, and a move from an A-position automatically goes into a B-position, and vice versa. However, this does not alter the reasoning we have given. In this case we define $P_1(v)$ only for the A-vertices and $P_2(v)$ only for the B-vertices. The game graph may be viewed as **bipartite.**

EXAMPLE 5.5.28

An important special case of a bipartite game is a **game tree**. In this case, there is a **unique path** from A's initial position to any other position. Therefore, any position that can be arrived at after an odd number of moves is a B-position, and similarly the A-positions are the ones arrived at in an even number of moves. The resulting graph, as in Figure 5.5.20,

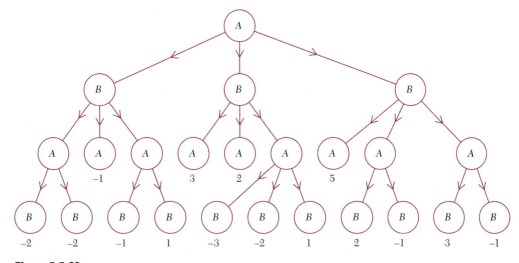

Figure 5.5.20

is called a tree, and the theory of trees will be discussed in the next chapter. We have layered the vertices so that the top is the initial position, the row below is the positions after one move, and so on, with the A-positions and B-positions alternating from the top down. In this case, the *single* payoff function P represents P_1 at an A-position and P_2 at a B-position. The value at an A-vertex is the maximum of the values it leads to. The value at a B-vertex is the minimum of the values it leads to. The full payoff function is shown in Figure 5.5.21. In this game, A's correct move is to the right. The best play is indicated by the darkened path.

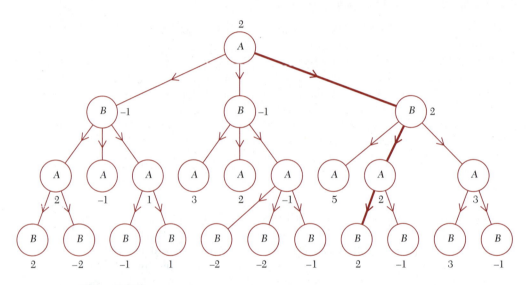

Figure 5.5.21

EXAMPLE 5.5.29   *Chess*

How are computer chess programs devised? The number of positions is so enormous and the graph so complicated that the procedure indicated in Example 5.5.26 would require far too much time and memory. For example, it involves starting with *every* mating position and working backward through *all* variations to the opening position.

However, a variation is used. Each position v can be given a working value $W(v)$ according to rules determined by the programmer in consultation with chess experts. This value is based on several considerations, mostly traditional values determined by chess masters over the years, but sometimes determined by developers of chess programs. Some of the factors going into this working value are the pieces each player has, with different pieces assigned different values; their mobility; and the possibility of an immediate capture. To decide on a move in position v, a graph of the positions that can evolve from v in D or fewer steps is found. (D is called the depth of search.) Each end position in this restricted depth game is then assigned the working value discussed earlier, and the payoffs are then computed as in the algorithm of Example 5.5.26. With W as the payoff function, the correct move is found, as in Example 5.5.25.

If the depth D is too large, the time and memory considerations overwhelm the computer. So variations are needed to cut down on the time consumed. Perhaps certain moves are eliminated immediately because they are too unpromising on the basis of the working value of the new position. (Sometimes brilliancies are not found because of this.) Certain moves may lead to situations that need more depth, and so more depth might be used for a few positions.

Thus, with some variation, the payoff evaluation labeling of Example 5.5.26 is applied, and some very fine chess programs have found themselves on the market.

Exercises

1. The game of ten (Example 5.5.6) is being played with a small change in rules. It is agreed that no one can land on the number 7. Draw a game graph for this revised game, find a strategic labeling, and describe a strategy to win.

2. Alice and Bill feel that the game of ten is too boring because they know the strategy. Therefore they change the game to twenty. The new rules allow a jump of either 2 or 3, and anyone who gets to 20 or more wins. Who wins? How does the person win?

3. Repeat Exercise 2 except that no one is allowed to land on 10.

4. Alice and Bill change the rules of ten as follows: (a) The first one to reach 20 or more wins. (b) Alice may jump 1 or 3, but Bill may jump 2 or 3. (c) Alice starts at 0. Describe Alice's strategy. (*Hint:* The game graph is bipartite.)

5. A and B are playing Nim. The piles have 4, 5, 7, and 9 chips in them. What is the winning move?

6. In a Nim game, the piles have 7, 9, and 11 chips in them. If it is your move, can you win? Explain.

7. You are playing Nim. The piles have 8, 10, and 15 chips in them. How many winning moves do you have?

8. A is playing Nim with B. The piles have 7, 14, and 21 chips in them. A agrees to let B go first, but only after she (A) adds a new pile of chips, making four piles. B agrees to this arrangement. How does A win?

9. A is playing Nim with B. They both know the winning strategy, so

they are getting bored with the game. They decide to allow a *split* of a pile as an alternative move, calling their game SplitNim. Thus, from 17, 5, and 8, by splitting the 17-pile, it would be possible to move to 11, 6, 5, and 8. How do you play this game if your position is $(1, 2)$? if your position is $(1, 3)$ or $(2, 3)$?

*10. As in the previous exercise, A and B are playing SplitNim. B is confronted with the piles $(1, 2, 3)$. Can he win? What about positions $(1, 2, 4)$ and $(2, 3, 4)$? Use the results of the previous exercise if required.

*11. Bored with Nim, players A and B change the rules so that the person who picks up the last chip *loses*. Find a strategy for this game. (*Hint:* Play the winning strategy but be careful about emptying a pile completely.)

12. You are playing Wytoff's game. There are 30 and 35 chips in the two piles. What is your winning move?

13. In Wytoff's game, what is the winning move with 11 and 43 chips in the two piles?

*14. Prove Theorem 5.5.18.

*15. You are playing Wytoff's game, with the additional rule that no more than three chips can be removed from any pile in any move. (You can remove as many as three from each pile, as long as you take an equal number from each pile.) Using the graphical procedure indicated in Figures 5.5.11–5.5.14, find enough 0-positions to play any such game with ten or fewer chips in either pile. Is $(4, 7)$ a winning position for you? If so, how many winning moves are there, and what are they?

*16. Redo Exercise 15 with the additional rule that no more than two chips may be removed from any pile at any time.

17. Explain why the game of ten is a special case of the game described in the previous exercise.

*18. Find a strategy for playing Wytoff's game where the rules are the same, except that the one who takes off the last chip *loses*. [*Hint:* Use a checkerboard and label the $(0, 0)$ position a 1-vertex, rather than a 0-vertex.]

*19. Explain why a strategic labeling of a game graph is a special case of the dual-payoff labeling of Example 26.

20. In the graph of Figure 5.5.17, assume that the first player to reach an end vertex receives the amount indicated from the other player. Find the payoffs at each vertex of the graph. What is the best move if it is your turn?

21. In the game graph of Figure 5.5.22, player A receives the indicated amount at the end vertex from player B.
   (a) Convert this into a valued game graph and compute the evaluation function at each vertex.
   (b) What is the best move for A if she is at vertex v? What is the best move for B if he is there?

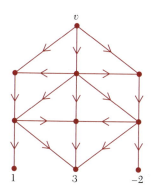

Figure 5.5.22

22. Using the algorithm of Example 5.5.26, find the payoff evaluation function at vertex v of Figure 5.5.23. In particular, if A had a choice, would she go first in this game or second?

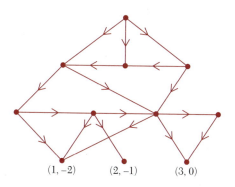

Figure 5.5.23

23. In the game tree of Figure 5.5.24, find the payoff functions at each vertex, identify A's winning move, and show how the best play proceeds.
24. In the game tree of Figure 5.5.25, find the payoff functions at each vertex, identify A's winning move, and show how the best play proceeds.
25. In the bipartite game graph of Figure 5.5.26, find the payoff functions at each vertex, and show how the best play proceeds from vertex v.

The following series of exercises introduce **Grundy numbers**, which can be used to play Nim-like games. A Grundy labeling is a generalization of a strategic labeling. Each vertex of a game graph is assigned a natural number n, called its Grundy number. In this case the vertex is called an n-vertex. The definition is recursive. As a basis, the end vertices

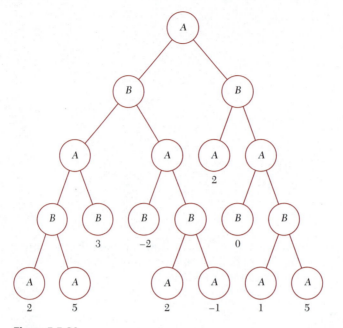

Figure 5.5.24

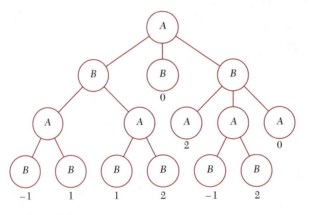

Figure 5.5.25

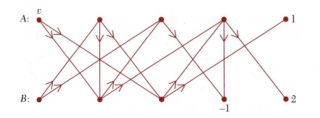

Figure 5.5.26

are all 0-vertices. If v is not an end vertex, and if the possible moves from v lead to vertices $v_1, \ldots, v_k$, with Grundy numbers $g_1, \ldots, g_k$, then the Grundy number of v is the first natural number not included among the numbers g_i, $1 \le i \le k$.

*26. Prove that any game graph G has a Grundy labeling. (*Hint:* Show that this is true for the lower sets of G by induction.)

*27. Write an algorithm that labels the vertices of a graph with its Grundy numbers. Use Example 5.5.26 as a model.

28. Grundy-label the game graph of Figure 5.5.27.

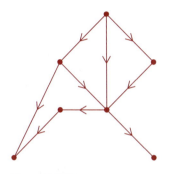

Figure 5.5.27

29. Grundy-label the game graph of the ten game of Figure 5.5.4.

30. Show that if you are at a vertex v with Grundy number n and if $0 \le m < n$, then there is a move into a position with Grundy number m.

*31. Show that if each positive Grundy number is changed to 1, the result is a strategic labeling of the graph.

*32. Let P_n be the digraph whose vertices are the integers from 0 through n, with edges joining x to y if $x > y$. Show that P_n is a game graph and that the vertex i has Grundy number i. (P_n may be thought of as a single Nim pile of size n. Thus, a Nim pile of size n has Grundy number n.)

*33. Suppose a game is played on k game graphs $G_1, \ldots, G_k$ as follows. Each digraph has a checker at one of its vertices. A player can play on any digraph and move the checker there using a legal move. Play alternates, and the player who has no move loses. Using Grundy numbers, show how to play this game. (*Hint:* Find the Grundy numbers of the positions and play Nim with them, converting your answer to a move on one of the graphs.)

*34. Slow Nim is played like Nim, except that *no more than four chips can be removed at any time*. Find a strategy for this game. In particular, what is your winning move if you have piles of 4, 6, and 12 chips? (*Hint:* Analyze an individual pile as a game graph in itself. Find its Grundy number and use Exercise 33.)

*35. Fast Nim is played like Nim except that no less than two chips can be removed at any time. If only one chip is in the pile, then it can be removed, too. Find a strategy for this game. In particular, what is your winning move if you have piles of 4, 6, and 12 chips? (*Hint:* See Exercise 34.)

5.6 Networks

A *network* is a directed, weighted graph with two vertices, called the *source* and the *sink*, singled out. In Figure 5.6.1, imagine pumping oil from depot A (the source) to city B (the sink). The figure shows the *maximum capacity* of each pipe, say in gallons per second, in the direction indicated. This might be determined by the size of pipe and the power of the pumps at the intermediate stations indicated as vertices in the figure. We assume that an intermediate station will send out as much oil as it receives so that no bottlenecks occur. A flow through any pipe is always bounded by the capacity of the pipe. In this section, we will learn how to maximize the total flow leaving the source and arriving at the sink.

Networks are a model for traffic flow problems. Although the foregoing illustration considers the flow of oil, other applications include the flow of information, traffic patterns, neural transmissions in the brain, and telephone messages. Figure 5.6.2 gives a possible flow pattern for the network of Figure 5.6.1. The label $x \mid y$ on any edge denotes a flow of x through the edge with capacity y. The notation is suggestive because the ratio x/y is the fraction of the capacity being used. As we can see, the flow in Figure 5.6.3 is larger than the one in Figure 5.6.2.

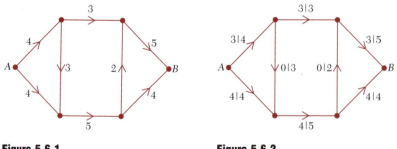

Figure 5.6.1   **Figure 5.6.2**

DEFINITION 5.6.1

A **network** N is a directed graph with two vertices, called the **source** C and the **sink** K, singled out. Each directed edge e of N is assigned a value $c(e) > 0$, called the **capacity** of the edge e. The out-degree of the sink K is 0, and the in-degree of the source C is 0. Finally, it is assumed that if v is any vertex of N, there is a path from C to v to K.

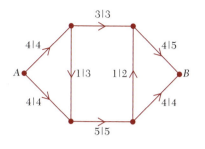

Figure 5.6.3

DEFINITION 5.6.2

Let N be a network, let E be the set of directed edges on N, and let V be the set of vertices. A **flow** f on N is a function $f: E \to \mathbb{R}$ satisfying the following conditions:

1. For each vertex v of N other than the source C and the sink K, the total flow into v is equal to the total flow out. Explicitly, for fixed vertices v_0 and w_0, we define

$$\text{Inflow}(w_0) = \sum_{vw_0 \in E} f(vw_0); \quad \text{Outflow}(v_0) = \sum_{v_0w \in E} f(v_0w)$$

By definition, $\text{Inflow}(C) = 0$ and $\text{Outflow}(K) = 0$. Thus, requirement 1 can be stated

$$\text{Inflow}(v) = \text{Outflow}(v) \tag{5.6.1}$$

for all vertices $v \in V - \{C, K\}$.

2. $0 \le f(e) \le c(e)$ for all edges $e \in E$.

DEFINITION 5.6.3

If $f: E \to \mathbb{R}$ satisfies condition 1 of Definition 5.6.2, then f is called an **algebraic** flow.

Thus, an algebraic flow might be negative or exceed the capacity of an edge, but we still have Inflow = Outflow at any vertex other than the sink or the source. It is intuitively clear that for any flow the amount leaving the source is precisely the amount entering the sink. We now prove this.

THEOREM 5.6.4

For any flow f on a network N, algebraic or not, with source C and sink K, we have

$$\text{Inflow}(K) = \text{Outflow}(C)$$

Proof   We consider the sum

$$S = \sum_{vw \in E} f(vw) \qquad (5.6.2)$$

It is convenient to define $f(vw) = 0$ when vw is not an edge. This gives the definition of S, as well as the definition of Inflow and Outflow, as an unrestricted sum over all vertices. We can thus write (5.6.2) as a double sum:

$$S = \sum_{v,w \in V} f(vw) = \sum_{v \in V} \left[\sum_{w \in V} f(vw) \right]$$

In the inner sum, v is held fixed, and the sum for each $v \in V$ is Outflow(v). Thus, $S = \Sigma \text{Outflow}(v)$, the sum extending over all v. Similarly, by first summing on w, $S = \Sigma \text{Inflow}(v)$. Thus,

$$\Sigma \text{Outflow}(v) = \Sigma \text{Inflow}(v)$$

We now let Σ' denote a sum over all vertices except for the sink K and the source C. Then we have

$$\Sigma \text{Inflow}(v) = \Sigma' \text{Inflow}(v) + \text{Inflow}(K)$$

and

$$\Sigma \text{Outflow}(v) = \Sigma' \text{Outflow}(v) + \text{Outflow}(C)$$

Therefore,

$$\Sigma' \text{Inflow}(v) + \text{Inflow}(K) = \Sigma' \text{Outflow}(v) + \text{Outflow}(C)$$

But by Equation (5.6.1), we have $\Sigma' \text{Inflow}(v) = \Sigma' \text{Outflow}(v)$. Therefore, $\text{Inflow}(K) = \text{Outflow}(C)$, which is the result. ■

DEFINITION 5.6.5

If f is any flow on a network, algebraic or not, then the common value of Inflow(K) and Outflow(C) is called the **value** of the flow f. **A maximal flow** is a flow of maximal value.

In general, we do not have a unique maximal flow. The maximal *value* of a flow is unique, but we can expect many flows with this maximal value.

The following theorem will be a useful tool for finding a maximal flow.

THEOREM 5.6.6

Let f and g be algebraic flows with values F and G. Then $f + g$ is an algebraic flow with value $F + G$.   ■

The proof is left to the reader.

The reader may wonder why algebraic flows, with possible negative values, are even considered. The answer may be seen in Figures 5.6.4 and 5.6.5. Figure 5.6.4 shows a flow of value 5 on a network N, and Figure 5.6.4 shows an algebraic flow of value 1. When we add these flows as in Theorem 5.6.6, we arrive at Figure 5.6.6, a flow whose value 6 (= 5 + 1) is larger than the flow of Figure 5.6.5. Thus, with the help of an algebraic flow, we arrived at a flow with a larger output.

A simple construction creates an algebraic flow for a path α on the network. However, we must use paths on N without considering the orientation of the edges on N. To do this, we use the following convention.

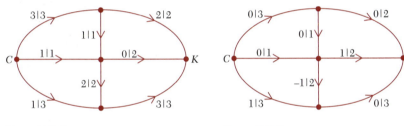

Figure 5.6.4   **Figure 5.6.5**

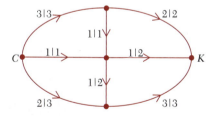

Figure 5.6.6

In short, a path on the extension of N allows edges opposite the direction given in N.

The following construction shows how a path on the extension of N, together with a real number b, determines an algebraic flow on N.

DEFINITION 5.6.8

Let α be a path on the extension of N with initial and terminal vertices at C and K, respectively, and with no repeated vertices. Further, let b be a real number. The function $g = b \cdot \alpha$ is defined on all edges e of N as follows:

1. If e is an edge of α, then $g(e) = b$.
2. If e' is an edge of α, then $g(e) = -b$.
3. If neither e nor e' is an edge of α, then $g(e) = 0$.

The definition is illustrated in Figure 5.6.7 for $b = 3$.

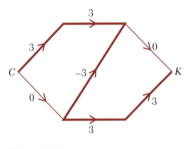

Figure 5.6.7

THEOREM 5.6.9

If α is any path as in Definition 5.6.8 and b is a real number, then $b \cdot \alpha$ is an algebraic flow with value b.

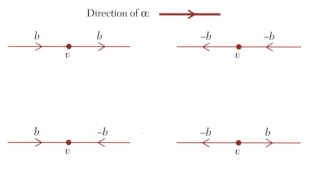

Figure 5.6.8

Proof Outflow(C) = b and Inflow(K) = b, by definition. At any vertex v not on the path α, Inflow(v) = Outflow(v) = 0, again by definition. For any vertex v other than C or K on α, there are four cases, as illustrated in Figure 5.6.8. In each case, Inflow(v) = Outflow(v). This proves the result. ∎

We now show how to find a maximal flow on a network N. The method, called the **Ford-Fulkerson Algorithm,** is done in stages. For a given flow f that does not have the maximum possible value, a certain path α, called an *augmentation path*, is found and a positive number b is found so that

$$g = f + b \cdot \alpha$$

is itself a flow. The value of g is b more than the value of f. This process is repeated until the maximum flow is achieved. The process can start with the flow that is identically 0.

In the remainder of this section, we define an augmentation path and show how to use it to increase the value of a given flow. We then show how to find an augmentation path for a given flow whose value is not maximal, and we also show that such a path must exist for a nonmaximal flow. Finally, we show that the process must terminate, at least when the capacities of the edges are integers or rationals.

DEFINITION 5.6.10

An **augmentation path** (for short, an *a*-path) for a flow f on a network N is a path on the extension of N joining C to K with no repeated vertices, with the following properties:

1. If $e \in N$ is an edge of the path α, then $f(e) < c(e)$. [Here, $c(e)$ is the capacity of the edge e.]
2. If e is an edge of the path α and e' is an edge in N, then $f(e') > 0$.

In case 1, we call $c(e) - f(e)$ the **leeway** of the edge e. In case 2, we call $f(e')$ the leeway of the edge e.

In short, α is a path that uses edges that are not used up to full capacity, but if α goes opposite the flow direction on an edge, the flow must be positive on that edge.

EXAMPLE 5.6.11

The network N of Figure 5.6.9 has a flow f indicated on the left and an *a*-path indicated with the darker lines. The leeway is observed to be ≥ 2 for all edges in the *a*-path. The adjusted flow $f + 2 \cdot \alpha$ is the flow on the right. The result is a flow whose value is 2 more than the original value.

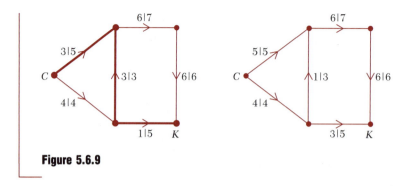

Figure 5.6.9

THEOREM 5.6.12

Let f be a flow on a network N, and let α be an a-path for f. Define b = minimum of the leeways of the edges of α. Then $g = f + b \cdot \alpha$ is a flow whose value is b more than the value of f.

Proof We know by Theorem 5.6.6 that $g = f + b \cdot \alpha$ is an **algebraic** flow whose value is b more than the value of f. Therefore, we need to show

$$0 \leq g(e) \leq c(e) \tag{5.6.3}$$

for all edges e of N. (This is Definition 5.6.2, part 2.) We do this by cases.

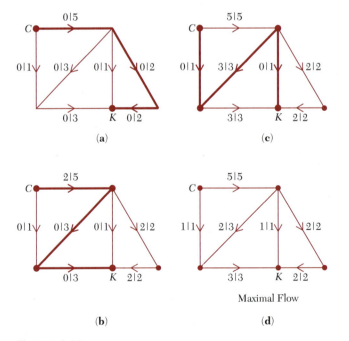

Figure 5.6.10

1. If neither e nor e' is on the path α, then $b \cdot \alpha = 0$ on e, so $g = f$ on e, and (5.6.3) is true.
2. If e is on the path α, then $b \leq c(e) - f(e)$ by the definition of b. Therefore $0 < f(e) + b \leq c(e)$. Since $g = f + b$, we have (5.6.3) in this case.
3. If e' is on the path α, then $b \leq f(e)$. Since $g(e) = f(e) - b \geq 0$ in this case, we have $0 \leq g(e) < f(e) \leq c(e)$. ∎

EXAMPLE 5.6.13

Starting with the 0 flow, a flow of maximal value is found in four stages in Figure 5.6.10 by repeated use of Theorem 5.6.11. An appropriate a-path was found, in each case, by inspection.

EXAMPLE 5.6.14   Algorithm To Find an a-Path

How do we find an augmentation path for a flow f on the extension of N or show that one does not exist? The idea is as follows:

Call an edge $e = PQ$ of the extension of N **allowable** if either (a) e is an edge of N and $f(e) < c(e)$ or (b) e' is an edge of N and $f(e') > 0$. Thus, an allowable edge is one with positive leeway. An a-path is thus a path starting at C, ending at K, with no repeated vertices and that uses only allowable edges. The idea is to search systematically for all allowable edges from the source and to continue the process until we reach the sink.

Starting at the source, we move along all allowable edges to other vertices. We label such vertices and continue moving from them to other vertices, which we again label. The process continues until we reach the sink K or run out of vertices to examine. In the latter case, there is no a-path. In the former, we can retrace our steps to find the required a-path.

If v is a labeled vertex and we find that edge vw is allowable and w is not yet labeled, we label w and define $P(w) = v$. [Think of $P(w)$ as the vertex previous to w.] We initially define $P(v) = 0$, to indicate that v is unlabeled. Thus, if $P(v) = u$, then uv is an allowable edge.

We let the vertices be labeled $v_0, v_1, \ldots, v_n$, where $C = v_0$. For convenience, we define the leeway$(vw) = 0$ when the leeway is not positive, even if vw is not an edge of the extended network. We begin with an *examining* procedure. This checks all vertices w adjacent to a fixed vertex v to see if vw is allowable. If it is, w is labeled. If w is the sink, which is our goal, we set FOUND $:= 1$. Once the process is done, we label v as examined, so we don't need to examine v again. We define $X(v) = 0$ if v is not yet examined, and $X(v) = 1$ if it has. The procedure is as follows:

Examine(v) Procedure:
1. FOR $i := 1$ to n
    1. IF ($P(v_i) = 0$ and vv_i has leeway $L > 0$) THEN
        1. $P(v_i) := v$ [labeling v_i and assigning it the previous vertex v]
        2. IF $v_i = K$ THEN
            1. FOUND := 1
            2. End [of procedure; don't bother checking any other vertices]
2. $X(v) := 1$ [we won't examine v again]
3. End [of procedure]

It is convenient to define a selection procedure to decide which vertex to examine. This procedure looks over labeled vertices that have not been examined and chooses the first one on the list to be examined. The input is the set of vertices, some of which are labeled, some of which have been examined. The output is the integer $i = $ Loc, indicating that v_i should be examined. If Loc $= 0$ is selected, then no unexamined vertices could be found. The procedure follows.

Select(Loc) Procedure:
1. FOR $i := 1$ to n [check the vertices one at a time]
    1. IF ($P(v_i) \neq 0$ and $X(v_i) = 0$) THEN [v_i is labeled and v_i is not examined]
        1. Loc := i
        2. End [don't check any more vertices]
2. Loc := 0 [could not find an unexamined, labeled vertex]

We can now give the full a-path algorithm. Its input is a flow f on a network. Its output is an a-path α or a statement that no a-path exists.

APath(f; α) Algorithm
1. FOUND := 0
2. FOR $i := 1$ to n
    1. $P(v_i) := 0$ [initialize each vertex as unselected]
    2. $X(v_i) := 0$ [initialize each vertex as unexamined]
3. Examine(C)
4. IF FOUND $= 1$ THEN
    1. $\alpha := CK$
    2. End
5. FOR $i := 1$ to $n - 1$ [we only need up to $n - 1$ passes]
    1. Select(Loc)
    2. IF Loc $= 0$ THEN
        1. Output := "There is no a-path"
        2. End
    OTHERWISE
        3. $j := $ Loc
        4. Examine(v_j)

5. IF FOUND = 1 THEN [now find path by working back-
    wards from K]
    1. $v = P(K)$
    2. $\alpha := vK$ [initializing]
    3. WHILE $v \neq C$
        1. $v := P(v)$
        2. $\alpha := v*\alpha$
    4. End [an a-path α has been found]

A flow f is given as in Figure 5.6.11a. Find an a-path for f, using the
algorithm of Example 5.6.14.

Method   Identify the vertices (other than C and K) as a through f rather
than $v_1, \ldots, v_6$. We start by examining C (step 3). This yields the labeling
of a, b, and c in Figure 5.6.11b. The labeled vertices are darkened, and
the value of P is indicated. We now select a (step 5.1), the first labeled
unexamined vertex, and examine a (step 5.2.4), which yields no labeled
vertex, since edge ae has no leeway. We next examine b, which yields
the new labeled vertex d, and then c, which yields the labeled vertex e.
At this point, we are at Figure 5.6.11c, in which a, b, and c have been
examined. Vertex d yields no new vertex (since e and b are already labeled), and vertex e yields K, making FOUND = 1. (This is the Examine
procedure step 1.1.2.1.) At this juncture (step 5.2.5), the construction of
α begins. It is found by the labeling to be $CceK$, as in Figure 5.2.11d.

Remark   The algorithm of Example 5.6.14 can be used repeatedly
in conjunction with Theorem 5.6.12 to find a maximal flow in a
network. When this is done on a large network, it is efficient to keep
track of the minimal leeway along the a-path while we do the labeling in the a-path algorithm. This is covered in the exercises.

The process illustrated in Example 5.6.13 is a greedy one. Once an a-
path is found, we use it to adjust a flow to one of larger value, but we
are then committed to the new flow, which conceivably has no augmentation path and yet might not be a maximal flow. In what follows,
we show that if any flow is not a flow of maximal value, then an a-path
can be found to increase it. It is useful to first introduce another algebraic
idea.

Let f be a flow on a network N. We extend the definition of f to the
extension N by defining $f(e) = -f(e')$ whenever e' is in N.

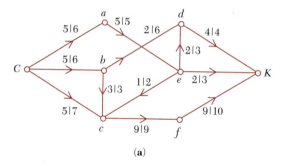

(a)

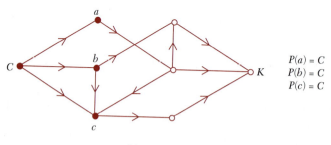

$P(a) = C$
$P(b) = C$
$P(c) = C$

(b)

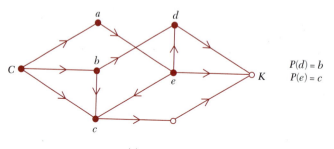

$P(d) = b$
$P(e) = c$

(c)

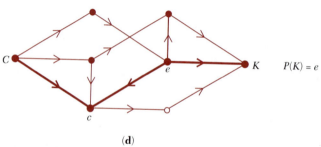

$P(K) = e$

(d)

Figure 5.6.11

Using this definition, we can say that f is a flow if Outflow(v) = Inflow(v) = 0 at any vertex v other than C or K, provided the outflow and inflow are computed on the extension of N. (See Figure 5.6.12 for an illustration.) In the same way, Definition 5.6.8 (the definition of $g = b \cdot \alpha$) may be simply stated as $f(e) = b$ for all edges e on the path, and $f(e')$ = $-f(e)$ for all edges e.

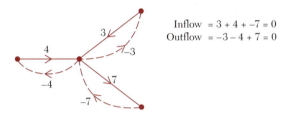

Inflow $= 3 + 4 + -7 = 0$
Outflow $= -3 - 4 + 7 = 0$

Figure 5.6.12

THEOREM 5.6.17

Let f be a flow on the network N that is not a flow of maximal value. Then there is an augmentation path α for the flow f.

Proof Since f is not maximal, there is a flow g with a larger value. The difference $h = g - f$ is an algebraic flow with positive value. Since $h + f$ is a flow, it follows that

$$0 \le h(e) + f(e) \le c(e) \qquad \text{for all edges in } N$$

Thus,

$$-f(e) \le h(e) \le c(e) - f(e) \qquad \text{for all edges } e \text{ of } N \qquad (5.6.4)$$

We now show that if $h(e) > 0$, where e is in the extension of N, then e is an allowable edge for f. (See Example 5.6.14.) There are two cases:

1. If e is an edge of N, then $0 < h(e) \le c(e) - f(e)$.
2. If e' is an edge of N, then $0 < h(e) = -h(e') \le f(e')$.

Statements 1 and 2 show that e is an allowable edge for the flow f.
    From this point on, we use only the following facts about h:

1. h is an algebraic flow on the extended network N.
2. The value of h is positive.
3. If $h(e) > 0$, then e is an allowable edge for f.

It will be necessary to change h to a new algebraic flow h', but the new h' will continue to have the same three properties.
    Since the algebraic flow h has positive value, Outflow$(C) > 0$. It follows that there is an edge e (of the extended network N) leaving the

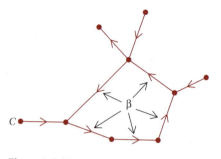

Figure 5.6.13

source C such that $h(e) > 0$. But Inflow = Outflow at all vertices other than the source or sink. Therefore, if $e = CP_1$ and $P_1 \neq K$, then there is an edge e_1 leaving P_1 with $h(e_1) > 0$. We now continue prolonging the path started at C, always choosing a departing edge with a positive h. If the sink K is reached, we have a path α whose edges have positive h and, hence, are allowable. In this case, we have arrived at an a-path α. The alternative is that α loops, as in Figure 5.6.13.

When a loop occurs, we proceed as follows. Take the cycle β as indicated in Figure 5.6.13. Let b equal the minimum value of $h(e)$ over the edges of β, and let $h' = h - b \cdot \beta$. Here, $b \cdot \beta$ is defined exactly as in Definition 5.6.8, except that β is a cycle. In this case, because β has no source or sink, the value of $b \cdot \beta$ is 0, so h' is an algebraic flow with the same value as h. Once again, we may verify that if $h'(e) > 0$, then e is allowable. We can now repeat the entire process of constructing an allowable path using h' instead of h. If necessary, we eliminate cycles as before until we arrive at an a-path that terminates at K. The process cannot continue indefinitely, since $h' = 0$ on more edges than h. ∎

THEOREM 5.6.18

If the capacities on a network N are rational numbers, a flow of maximum value will be achieved in a finite number of steps by using augmentation paths.

Proof First assume that the capacities are integers. Then the process of using a-paths, starting with the 0 flow, always increases the value of the flow by a positive integer. Because the value is always bounded by the sum of the capacities on the edges from C, this process must stop. When it does, a flow with maximal value is achieved, by Theorem 5.6.16.

If the capacities are rational, simply scale these values by multiplying them all by their least common denominator. When the maximum flow for the resulting network is found, we scale back, by dividing by the least common denominator, and the result is a flow of maximum value for the given network. ∎

Remark For real capacities, the method of augmentation paths might not yield a flow of maximal value because it might continue forever. However, this is not a critical practical problem, since a rational approximation for a real number can be used, so an arbitrary close approximation to a maximal flow can be found.

Exercises

In Exercises 1–4, find a maximal flow on the network in the figure, showing the capacities and directions of each edge. What is the value of maximum flow?

1. See Figure 5.6.14.
2. See Figure 5.6.15.
3. See Figure 5.6.16.
4. See Figure 5.6.17.

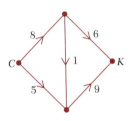

Figure 5.6.14

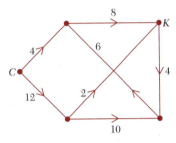

Figure 5.6.15

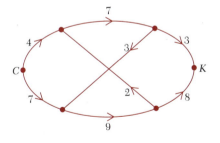

Figure 5.6.16

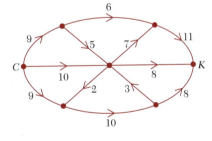

Figure 5.6.17

In Exercises 5–8, find an *a*-path for the indicated flow, using the algorithm of Example 5.6.14, or show that no *a*-path exists.

5. See Figure 5.6.18.
6. See Figure 5.6.19.
7. See Figure 5.6.20.
8. See Figure 5.6.21.

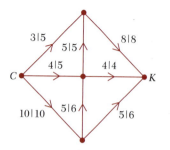

3|5 5|5 8|8 4|5 4|4 *C* *K* 10|10 5|6 5|6

Figure 5.6.18

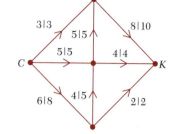

3|3 5|5 8|10 5|5 4|4 *C* *K* 6|8 4|5 2|2

Figure 5.6.19

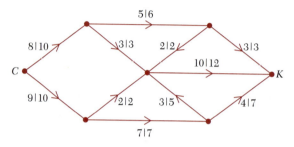

5|6 8|10 3|3 2|2 3|3 10|12 *C* *K* 9|10 2|2 3|5 4|7 7|7

Figure 5.6.20

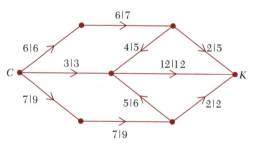

6|7 6|6 4|5 2|5 3|3 12|12 *C* *K* 7|9 5|6 2|2 7|9

Figure 5.6.21

*9. Suppose a network and a flow are defined as in Definitions 5.6.1–5.6.2, except that more than one sink and more than one source are allowed. State and prove a theorem analogous to Theorem 5.6.4.

*10. Suppose, in a practical situation, an intermediate **vertex** v has a capacity $c(v)$ so that it is only able to handle a flow of at most $c(v)$. Show how this situation can be incorporated into a standard network, where only the edges have capacities. (*Hint:* Replace v with two vertices.)

11. Explain what is meant by infinite capacity: $c(e) = \infty$. In particular, when is an edge e of the extension of N allowable?

12. Let f_1 and f_2 be maximal flows on a network N, and let c be a real

number satisfying $0 \le c \le 1$. Show that $cf_1 + (1 - c)f_2$ is also a maximal flow.

13. Prove Theorem 5.6.6.

*14. Let β be a cycle in the extension of a network N with no repeated vertex except for the beginning and ending vertex of β. Let b be a real number. If $g = b \cdot \beta$ is defined as in Definition 5.6.8, prove that g is an algebraic flow with value 0. (This was used in the proof of Theorem 5.6.17.)

Exercises 15–18 ask you to work out the full Ford-Fulkerson Algorithm by developing some further procedures and by using some of the algorithms developed so far as procedures in the final algorithm.

*15. It was noted after Example 5.6.15 that the a-path algorithm of Example 5.6.14 can keep track of the minimum leeway L for the constructed path α. Rewrite the procedure Examine(v) to do this during the labeling part of Examine(v).

*16. Write an algorithm Path(α, b) to construct the algebraic flow $g = b \cdot \alpha$. The input should be the path α given as a sequence of vertices. We assume that α starts at C and ends at K. The output should be $g(v, w)$, defined for every pair (v, w) of vertices. Define $g = 0$ when vw is not an edge of N.

*17. Using the algorithms developed in the previous two exercises, write an algorithm Increase(f) to implement Theorem 5.6.12. The input should be a flow f. The output should be $f + L \cdot \alpha$, identified as f, where α is an a-path found by the augmentation path procedure, and L is the minimum leeway of the edges of α.

*18. Using the algorithms developed, write an algorithm Ford-Fulkerson(N) that takes a given network as input and gives a maximal flow as output. Assume that the process of successively incrementing by a-paths will terminate.

19. Find two a-paths for the flow in Figure 5.6.11a other than the one found in Figure 5.6.11d.

*20. Suppose some edges of a weighted graph are undirected so that flow can go in either direction, up to a given capacity. Explain how to maximize the flow on such a weighted graph.

Summary Outline

● A **simple graph** G is a finite set V, called the vertices of G, together with a function $C: V \times V \to \mathbb{N}$ that satisfies the three conditions:

$$C(v, w) = C(w, v) \tag{5.1.1}$$
$$C(v, v) = 0 \text{ and} \tag{5.1.2}$$
$$C(v, w) = 0 \text{ or } 1 \tag{5.1.3}$$

If $C(v, w) = 1$, then v and w are **adjacent** in G. If $C(v, w) = 0$, then they are not adjacent. If v and w are adjacent, we say that vw is an **edge** of G. C is called the **counting function.** A simple graph is also called a **graph.**

- A **directed graph** (or simply a digraph) G is given by a set V of vertices and a counting function $C: V \times V \to \mathbb{N}$ that satisfies Equations (5.1.2) and (5.1.3) in the definition of a graph.

- A **general graph** allows a vertex to be adjacent to itself. Thus, the counting function in the definition of a graph must satisfy Equations (5.1.1) and (5.1.3).

- A **multigraph** omits Equation (5.1.3), so more than one edge is allowed to join two vertices.

- A **subgraph** G' of a graph G is a graph whose vertices V' are a subset of the vertices V of G and whose counting function C' satisfies $C' \le C$ for the vertices of V'. For a **spanning subgraph** G', we have $C' = C$ for the vertices of V'.

- The **matrix of a graph** G has entries $c_{ij} = C(i, j)$, the counting function of G. It is assumed that the vertices are labeled $1, 2, \ldots, n$. The matrix is a square n by n matrix.

- a **path of length k** on a graph G joining vertices P to Q is a sequence of vertices $P = P_0, P_1, P_2, \ldots, P_{k-1}, P_k = Q$, where $P_{i-1}P_i$ is an edge of the graph for $i = 1, \ldots, k$. P is called the initial vertex of the path, and Q is the end vertex. (However, see Chapter 6 for the stricter definition used in the discussion of trees.)

- A **cycle** is a path with the same starting vertex and end vertex.

- The **trivial cycle** at P is the 0 length cycle consisting of P itself.

- A graph G is **connected** if for all vertices P and Q of G, there is a path joining P to Q.

- $P \approx Q$ (P **connects** to Q) means that there is a path in the graph with P as initial vertex and Q as end vertex.

- The **component** $C(v)$ of a vertex v is the subgraph of G spanned by the equivalence class of v under the relation $\approx$.

- **Theorem** G is connected if $A \approx P$ for some fixed vertex A and all vertices P.

- A vertex v is **adjacent** to a set Σ of vertices of a graph if $v \notin \Sigma$, but vw is an edge of the graph for some vertex w in Σ.

- **Theorem**   If C is the matrix of a graph G, the number of paths of length p joining vertex i to vertex j is r_{ij}, where $R = C^p$.

- **Degrees**

  The **degree** of a vertex v, written $\deg(v)$, is the number of edges with end point at v.

The **out-degree** of a vertex v in a digraph is the number of edges with initial vertex v.

The **in-degree** of a vertex v in a digraph is the number of edges with terminal vertex v.

● **Theorem**   The degree of the vertex i in a graph or multigraph G is the sum of the terms in the ith row of the matrix of G.

● **Theorem**   In a graph or multigraph, the sum of all the degrees of the vertices is twice the number of edges.

● An **Euler path** on a multigraph G is a path in which each edge of the multigraph is used exactly once. If an Euler path begins and ends at the same vertex, then it is called an **Euler cycle.**

● **Theorem**   If there is an Euler path on a multigraph G, which begins and ends at different vertices, then the initial and end vertices have odd degree, and the rest have even degree. For an Euler cycle, all degrees must be even.

● **Theorem**   Let G be a **connected** multigraph such that **each of its vertices has even degree.** Then there is an Euler **cycle** on G.

● If all vertices of a connected graph have even degree, with the exception of P and Q that have odd degree, then there is an Euler **path** on G joining P to Q.

● A **Hamiltonian path** on G is a path in which each vertex of G appears exactly once. A **Hamiltonian cycle** is a cycle in which each vertex of G appears exactly once, except for the beginning and ending vertex, which appears just twice.

● **Ore's Theorem**   Let a graph have n vertices. If, for each pair of nonadjacent vertices having degrees d_1 and d_2, we have $d_1 + d_2 \geq n$, then there is a Hamiltonian cycle on G.

● **Theorem**   Let a graph G have n vertices. If, for each pair of nonadjacent vertices having degrees d_1 and d_2, we have $d_1 + d_2 \geq n - 1$, then there is a Hamiltonian path on the graph.

● A **coloring** of a graph is an assignment of colors to the vertices of that graph in such a way that adjacent vertices have different colors.

● The **Four-Color Theorem** states that any planar graph can be colored with four colors. Long an unsolved problem, it was settled in 1976 by Haken and Appel.

● A graph is **n-colored** if it is colored with n colors. A bipartite graph is one that is 2-colorable.

● **Theorem** G is bipartite if and only if all cycles on G have an even number of edges.

- **Theorem**   If a connected graph G is 2-colored with m vertices having color 0 and n having color 1, and if there is a Hamiltonian cycle on G, then $m = n$. If there is a Hamiltonian path on G, then $m = n$ or $m = n \pm 1$.

- A **planar graph** is one that can be drawn in the plane so that edges can meet only at vertices.

- **Euler's Theorem for a Planar Graph**   If G is a connected, planar, general multigraph, then $v - e + r = 2$. Here v, e, and r are the numbers of vertices, edges, and regions that the graph has in its plane representation.

- **Theorem**   Let G be a simple, connected, planar graph with v vertices and e edges, with $v \geq 3$. Then $e \leq 3v - 6$.

- **Theorem**   Let G be a simple, connected, planar, bipartite graph with v vertices and e edges, with $v \geq 3$. Then $e \leq 2v - 4$.

- **Theorem**   Any planar graph G is 5-colorable.

- An **isomorphism** f of the graph G into G' is a 1-1 and onto mapping f: $V \rightarrow V'$ such that $C(v, w) = C'(f(v), f(w))$ for all vertices v and w of G.

- **Theorem**   If vertices correspond under an isomorphism, their degrees are equal.

- The **degree list** of a graph is the list, in ascending order, of the degrees of vertices in the graph. Repetitions are included.

- **Theorem**   Isomorphic graphs have the same degree lists.

- **The Component Algorithm** finds the component of a vertex v in a graph G. The worst-time case is $O(n^3)$. The **Connected Algorithm** determines whether a graph is connected. Its worst-time case is $O(n^3)$.

- A **weighted graph** is a graph in which each edge is assigned a positive number, called its weight.

- **Dijkstra's Algorithm** finds the path of least weight joining two given vertices in a weighted graph.

- **Games**

  A **two-person, zero-sum game** is a game played by two people, called A and B. The game must end. When it does, each player receives an award, called the payoff. A's payoff is the negative of B's.

  A **nonrandom, perfect information game** is a game in which there is no element of chance in the play, and nothing is hidden from the players.

  A **sequential game** is a two-person, zero-sum game played by each player moving alternately, as in tic-tac-toe. Each player moves from one **position** to another, according to the rules of the game, until no move is possible. A position from which no move is possible is called an **end position**.

Player A starts the process. The game ends in a finite time, and the rules determine what A's payoff is.

● An **acyclic** digraph is a digraph with no cycles on it, other than the trivial one-vertex ones. A **game graph** G is a finite, directed, acyclic graph. A vertex with out-degree 0 is called an **end vertex.**

● A **strategic labeling** of G is a labeling such that

1. All edges with an initial 0-vertex have a terminal 1-vertex.
2. If v is a 1-vertex, there is an edge with initial vertex v, whose terminal vertex is a 0-vertex.

A strategic labeling is used to play the win/lose game of trying to reach an end vertex first.

● **Theorem**   Any game graph G has a strategic labeling.

● **Wytoff's game**   A two-pile game, in which players remove something alternately from either pile, or the same amount from both. The one who removes the last chip wins.

● **Nim**   An n-pile game, in which players remove something from one pile alternately. The one who removes the last chip wins.

● A **dual-payoff function** $(P_1(v), P_2(v))$ is given by two payoff functions P_1 and P_2. Both are A's payoff. P_1 is the payoff if it is A's turn, P_2 if it is B's.

● A **valued game graph** is one in which each end vertex v is assigned a dual payoff $(P_1(v), P_2(v))$.

● A **payoff evaluation function** of a valued game graph is an assignment $(P_1(v), P_2(v))$ to each vertex v of G that has the following properties:

1. $P_1(v) = \text{maximum}(P_2(v_1))$ for all v_1 such that vv_1 is a directed edge of the game graph.
2. $P_2(v) = \text{minimum}(P_1(v_1))$ for all v_1 such that vv_1 is a directed edge of the game graph.

● A **bipartite game** is one in which the positions are divided into A- and B-positions. Any A-position leads to a B-position, and, similarly, any B-position leads to an A-position.

● A **game tree** is the graph of a bipartite game in which there is a unique path from the initial position to any vertex.

● **Payoffs** in bipartite games or game trees are given by functions P_1, defined only on the A-positions, and P_2, defined only on the B-positions. They satisfy the same condition as in the payoff evaluation function.

● A **network** is a directed graph N with two vertices, the source C and the sink K, singled out. Each edge of N is assigned a nonnegative number $c(e)$ called the **capacity** of that edge. The in-degree of C and the out-degree of K are 0. For any vertex v of N, there is a path from C to v to K.

- If f is a function defined on the edges e of a network, then **Inflow**(v) is the sum of the terms $f(wv)$ for all edges wv in the network. **Outflow**(v) is the sum of the terms $f(vw)$ for all edges vw in the network.

- An **algebraic flow** f on a network N is a function on the edges of N such that Inflow(v) = Outflow(v) for all vertices that are not the sink or the source.

- A **flow** f is an algebraic flow with the conditions $0 \le f(e) \le c(e)$ for all edges e in the network.

- **Theorem** For any algebraic flow f on a network, Outflow(C) = Inflow(K).

- The **value** of a flow f = Outflow(C) = Inflow(K).

- A **maximal flow** on a network is a flow of maximum value.

- **Theorem** The sum of algebraic flows is an algebraic flow. The value of the sum is the sum of the values.

- The **extension** of a network N is the network obtained by introducing oppositely directed edges into N. If $e = PQ$ is an edge of N, then $e' = QP$, as well as e, is in the extension of N.

- If α is a path on the extension of N, $b\cdot\alpha$ is the algebraic flow g such that

  1. $g(e) = b$ for e in the path.
  2. $g(e) = -b$ for e' in the path.
  3. $g(e) = 0$ if neither e nor e' is in the path.

- If f is a flow, the **leeway** of an edge e of the extension of N is defined as $c(e) - f(e)$ if e is an edge of the network, and as $f(e)$ if e' is an edge of the network.

- If f is a flow on a network, an **augmentation path** or an **a-path** α is a path joining C to K on the extension of N, all of whose edges have positive leeway.

- If f is a flow on a network N, an edge e of the extension of N is **allowable** if its leeway is positive. An a-path is a path joining C to K, all of whose edges are allowable.

- **Theorem** If α is an a-path for the flow f, and if b is the least of the leeways of the edges of α, then $f + b\cdot\alpha$ is a flow whose value is b more than the flow of f.

- **Theorem** If f is a flow that is not maximal, then there is an a-path for f.

- The **Ford-Fulkerson Algorithm** finds a maximum flow from an initial flow f (possible 0) by finding an a-path α for f and adding $b \cdot \alpha$ to f, where b is the minimal leeway of the edges of α. This process is continued until no more a-paths can be found, at which time a maximal flow is obtained.

- **Theorem**   If the capacities of a network are rational, the Ford-Fulkerson algorithm terminates after finitely many passes.

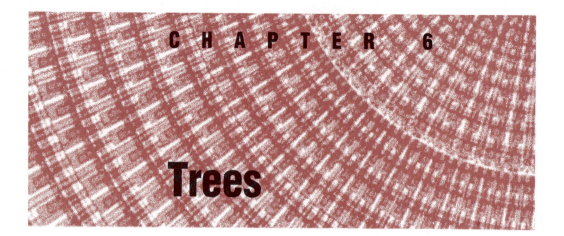

CHAPTER 6

Trees

A tree is a special kind of graph. It is usually used to show a hierarchical structure (as in a chain of command table) and to classify objects. Trees play an important role in computer science. In this chapter we give some theory and applications of this subject. Applications include the structure of algebraic and logical formulas, efficient (Huffman) coding, and binary tree searches.

6.1 Trees

A tree is a type of graph that occurs in so many different situations that a separate study of the subject is necessary. Trees are used extensively whenever it is necessary to classify or store many items. In this section, we consider several examples of trees that occur in different practical situations. We leave definitions, theory, and analysis to later sections.

EXAMPLE 6.1.1   Catalogs

Imagine a library with numerous books. It is desired to classify these books according to topic so that the user can easily find books on different topics. A simplified scheme is as follows. Let the broad classifications be (1) math/science (MS), (2) literature (Lit), and (3) biography (Biog). The MS books may be classified into (a) math (M), (b) computer science (C), (c) physics (P), and (d) other (O). The Lit books might be classified into (a) contemporary (Co), (b) classics (Cl), (c) other (O), and so forth.

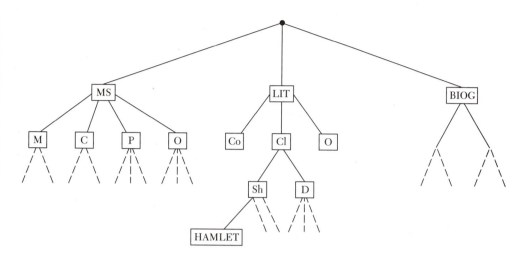

Figure 6.1.1

Further subclassifications are possible. The Lit-Cl books might be sub-divided into (a) Shakespeare (Sh), (b) Dante (D), and others. Such a scheme is realistic—one would expect a large library to be organized in a similar way. (The Dewey Decimal System is such a scheme.) Figure 6.1.1 illustrates the classification scheme just described. The graph is called a **tree** because it looks like one if turned upside down.

EXAMPLE 6.1.2   Directories

A similar classification scheme is often used to organize the directory of a floppy or hard disk. Such disks often contain hundreds or thousands of files. It is now common to arrange files in file directories that exhibit the files in a more convenient way. For example, the main classifications

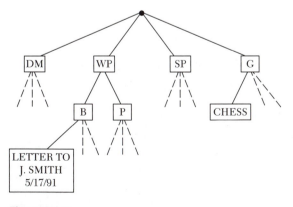

Figure 6.1.2

might be (1) word processing, (2) data management, (3) spreadsheets, and (4) games. The WP files might be subclassified into (a) business (B) and (b) personal (P), with further classifications as necessary.

One method used to designate directories or subdirectories (in personal computers that use DOS) is to use the backslash "\" character. Thus the directory \WP\Bus might be the name of the business correspondence subdirectory of the word processing directory. The file \WP\Bus\Jones refers to the file named Jones located in the directory \WP\Bus. Figure 6.1.2 illustrates a possible directory scheme to store files.

One characteristic of the graphs in Figures 6.1.1 and 6.1.2 is that each vertex (book or file) is **uniquely classifiable.** In other words, there is a unique path leading from the top of the tree to any book. Furthermore, the tree is connected because each vertex can be joined to the top by a path. We shall use these properties in the next section to define a tree.

Each of the trees in Figures 6.1.1 and 6.1.2 has a vertex with special significance, the topmost vertex, called the **root** of the tree. It is from the root that one proceeds from the general to the more specific. In Figure 6.1.2, we go from the root to the word processing directory, and then to the business directory of the word processing directory, to the letter to Mr. J. Smith.

EXAMPLE 6.1.3 Family Trees

Figure 6.1.3 illustrates a simplified **family tree.** In this example, children are given, but no marriages. The root represents the great grandparents XX of the children at the bottom. The first level below XX is their children. Below them are their children, and so on. As in the preceding examples, the resulting graph is a rooted tree. It is seen to be a **directed graph,** in a natural way: Simply draw the arrow downwards. Certain

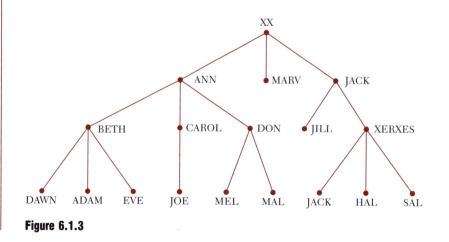

Figure 6.1.3

·people had no children. These vertices are called **leaves,** in analogy with a leaf on a tree.

Terminology for a rooted tree is often based on this example. For example, we say for any tree that A is the parent of B or that B is the child of A when the directed edge AB is in the tree. Thus, leaves are vertices that are not parents. Equivalently, a leaf is a vertex whose out-degree is 0.

Note on Relatives   We can define notions such as **cousins, nephews, second cousins, first cousins once removed,** and others, by using the terminology of graph theory. For example, A and B are **siblings** if they have the same parent(s). This is indicated in Figure 6.1.4, where A and B have C as a parent. Graph-theoretically, there is a vertex C such that CA and CB are in the same (directed graph of the) family tree. Figure 6.1.4 also indicates O as an **uncle** or **aunt** of A and of B. (There does not seem to be a sex-independent term for this.) Also in the figure, B and E are **cousins,** as are B and S. Finally, the concepts that U and V are **second cousins** and that U and W are **first cousins once removed** certainly seem to be two of the more mystifying relationships within a family, but can be easily visualized in a graph, as in Figure 6.1.4. Thus, to find first cousins once removed, simply find cousins and go one generation down from one of them.

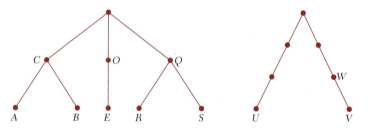

Figure 6.1.4

EXAMPLE 6.1.4   *Algebraic and Logical Expressions*

Suppose we have a fairly complicated algebraic expression such as

$$(XY - Z/W) + (ZW + XY)$$

A little experience tells us the meaning of this expression and how to compute it. Thus, this expression is the **sum** of two simpler expressions $U = XY - Z/W$ and $V = ZW + XY$, each of which can be similarly interpreted in terms of simpler expressions. For example, $XY - Z/W$ is the **difference** of two simpler expressions XY and Z/W, which are respectively the **product** and the **quotient** of two simpler expressions that can no longer be simplified. The process of going from the complicated to the simple suggests a tree operation, and we now illustrate this process.

The symbols $+$, $-$, $*$, and $/$ are **binary** operators: They each take two variables or expressions traditionally placed on either side of the operator. (Traditionally, multiplication is indicated by juxtaposition. Thus, $X \cdot Y$ and $X * Y$ are sometimes used to indicate multiplication, but it is customary in algebra to simply use XY. In what follows, however, we use $X * Y$, because we need a symbol for the operation of multiplication.) If we take the algebraic operators as **vertices** with two children as variables or expressions, we can show the structure of any algebraic expression involving these arithmetic operators as a tree. Thus, $U + V$ can be illustrated simply as in Figure 6.1.5.

Figure 6.1.5

In this illustration, $U = XY - Z/W$ and $V = ZW + XY$. We can now replace U and V with their algebraic values, expressed as a tree, to obtain a further breakdown of the formula. If we continue this process until we reach the variables X, Y, Z, and W, we arrive at Figure 6.1.6 as the full tree diagram of the algebraic expression.

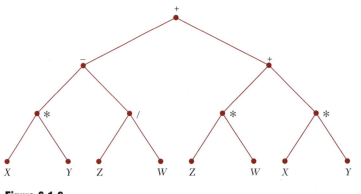

Figure 6.1.6

In the same way, a tree diagram for a logical statement can be found. For example, the statement $(p \wedge q) \vee \neg(r \vee s)$ is an OR statement connecting two simpler statements. The one is a simple AND statement, and the other is a NOT statement. If we carry out a similar procedure outlined for the algebraic statement, we arrive at Figure 6.1.7. In this figure there is only one branch coming down from the NOT connective ¬, since ¬ is a unary connective and operates on only one statement.

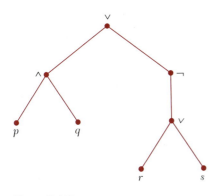

Figure 6.1.7

We can see at a glance, from Figure 6.1.6, that the algebraic expression that it represents is a sum, because + is the root. The summands can be analyzed further, and so on down the tree. Similarly, Figure 6.1.7 is the tree diagram of an OR statement, since ∨ is the root. The statements that this OR connects can be analyzed further by going down the tree.

EXAMPLE 6.1.5   Counting

When we discussed the multiplication principle for counting in Section 4.1, we warned that to use it, "there can only be one possible path to the final outcome." We can now picture this clearly with graphs and trees. For example, suppose we wanted to find out how many two-element subsets of $\{1, 2, 3, 4\}$ there are. The actual answer is $C(4, 2) = 6$. A common error is to say that there are four choices for the first element, and three for the second, and thus $4 \cdot 3 = 12$ choices in all. Figure 6.1.8

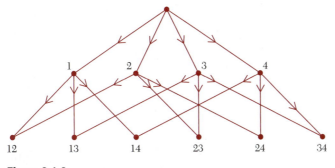

Figure 6.1.8

shows the fallacy in this reasoning. There are 12 different **paths** leading from the top (root) to the bottom. Note the two **different** paths leading to the set {1, 2}. The graph is not a tree, and the multiplication principle cannot be used.

Just as we wanted each book in the library example to be uniquely classifiable, it is necessary, in order to use the multiplication principle, that each object we are counting be chosen in a unique way. In each case, this means that there can be only one path joining the root to any leaf in the graph. Figure 6.1.8 is *not* a tree because this is clearly not true. The following example of counting can be easily done with the help of a tree.

EXAMPLE 6.1.6

How many five-letter strings contain letters from the alphabet {A, B} in which B must be followed by A or end the string?

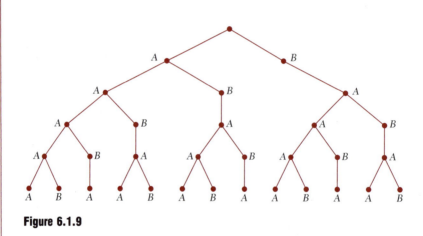

Figure 6.1.9

Method   We classify the strings according to the first letter of the string. In Figure 6.1.9, the branch to the left represents the strings that start with "A", and similarly the branch to the right represents the strings that start with "B". At the next level, we consider the second letter of the string. Because of the restriction in the problem, there is only one branch leaving the "B", whereas an "A" branch and a "B" branch leave the "A". Continuing down five levels, we stop, because the string has five letters. We count 13 leaves at the bottom, which is the answer. Any path down gives a valid string, and any valid string gives a unique path down.

Once again as in the directory example, we proceeded from the general to the specific. The first branch to the left contains all words starting with A; the one on the right contains the ones starting with B. And as we go down the graph, the word gets more and more specific because more of its beginning is known. The reader should look at each level to see how the number of such words vary with the length. At that point a guess should be made. Finally, a general statement and a proof should be given.

Exercises

1. Nathan is not sure what he is going to do today. He may go to visit Jill, in which case he might end up playing tennis or playing cards with her friends. On the other hand, if he stays home, he will either watch television (the basketball game or the science fiction movie) or study math or English. And if he studies English, he must decide whether to work on the essay, which is due next week, or to read the new assignment. Draw a tree diagram showing all of his choices. How many leaves are there?

*2. In a family, recursively define nth cousins using the terminology of graph theory.

3. If two second cousins marry, is the resulting family tree a tree in the sense of this section? Explain.

4. Mark Twain has a story in which the narrator ends with the statement "I am my own grandfather." Without going into the details of the yarn, is this possible for the trees we have been considering? Explain.

5. A three-letter string is to be made up using the alphabet $\{A, B, C\}$. The letters B and C cannot follow B, and the letter C cannot follow C. How many such strings are there? Draw a tree indicating all of the possibilities. How many ten-letter strings are there of this type?

6. A string of length 5 on the alphabet $\{A, B\}$ is to be constructed in which no letter can appear three times in a row. Draw a tree similar to Figure 6.1.9 for the set of such strings. How many such strings are there?

7. Find the algebraic expression represented by the tree of Figure 6.1.10.

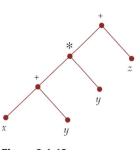

Figure 6.1.10

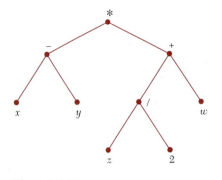

Figure 6.1.11

8. Find the algebraic expression represented by the tree of Figure 6.1.11.
9. Draw a tree diagram for the algebraic expression

$$\alpha = (X * Y) * Z - Y/Z$$

10. Draw a tree diagram for the algebraic expression

$$\beta = (a + b)/(c - a/b)$$

11. Draw the tree diagram for the algebraic expression

$$\gamma = x * (y * (x * (y * x)))$$

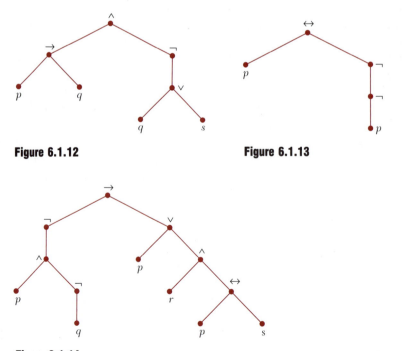

Figure 6.1.12　　　　　　　　　**Figure 6.1.13**

Figure 6.1.14

12. Figure 6.1.12 is the tree diagram of a statement. Find the statement.
13. Figure 6.1.13 is the tree diagram of a statement. Find the statement.
14. Figure 6.1.14 is the tree diagram of a statement. Find the statement.
15. Find the tree diagram for the statement $p \rightarrow (\neg q \vee r)$.
16. Find the tree diagram for the statement $(p \vee \neg q) \rightarrow (r \rightarrow (s \vee p))$.
17. Find the tree diagram for the statement $\neg p \rightarrow (p \rightarrow \neg p)$.
18. Solve Example 6.1.6, using the recursion methods of Section 4.3. In particular, compare with Example 4.3.12.
19. In Example 6.1.6, strings on the alphabet $\{A, B\}$ were considered in which B must either be followed by A or end the string. Show that the number of strings of this type of length n is the Fibonacci number f_{n+2}.

6.2 Theory of Trees

In this chapter, we consider simple graphs only. It will prove to be convenient to revise the definitions of a path and a cycle.

DEFINITION 6.2.1

A **path** in a graph G is a sequence $\alpha = v_0, v_1, v_2, \ldots, v_n$, where the v_i are all **distinct** vertices and $v_{i-1}v_i$ is an edge of G for $i = 1, \ldots, n$. The path α is said to have length n. It has $n + 1$ distinct vertices and n distinct edges.

Similarly, a **cycle** is such a sequence where the v_i are distinct except that $v_0 = v_n$. We also require $n \geq 3$.

The reason for this change of definition is as follows. We want to say that there is a unique path joining a pair of vertices. But this can never happen if we use our previous definition of path. For we can always introduce extraneous vertices and backtrack on them. Thus, in Figure 6.2.1a, path PQR joins P and R but so does $PQSQR$. When we use the stricter Definition 6.2.1, backtracking is not permitted, because the vertex Q cannot be repeated. In addition, self-intersecting paths, such as

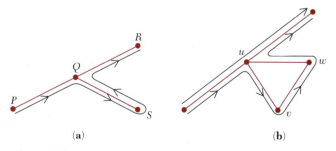

(a)　　　　　　　　(b)

Figure 6.2.1

the one in Figure 6.2.1b, are not permitted. This change of definition does not lead to any inconsistencies in our results about connectedness. For example, if two vertices are connected by a path with a loop as in Figure 6.2.1b, we can always eliminate that loop as indicated. Using our stricter definition of path and cycle, we can accurately state, for example, in the library catalog graph of Figure 6.1.1, that for any vertex in this graph there is a **unique** path joining it to the topmost vertex. We couldn't strictly say this if we allowed any backtracking. Since $n \geq 3$ in the definition of a cycle, the backtracking sequence PQP in Figure 6.2.1 is not a cycle, but $uvwu$ in Figure 6.2.1 is.

Because we use connectedness in what follows, let us show that this present definition of path leads to the same notion of connectedness as before, where our paths were allowed to have repeated vertices.

THEOREM 6.2.2

If $P \approx Q$, then there is a path connecting P and Q all of whose vertices are different.

Proof We prove that if α is a path (in the sense of Definition 5.1.7) of length n joining P to Q, then there is a path joining P to Q with no repeated vertices. We prove this by induction on n. If $n = 0$, the result is true, since the path then consists of just one vertex P (which is also Q in this case). Now assume the result is true for all paths of length $<n$. Suppose α has length n, joins P and Q, and has a vertex R repeated. Then we have $\alpha = P\beta R\sigma R\gamma Q$. (Here we think of a path as a string of vertices and β, σ, and γ are also paths or the empty string.) We now eliminate $R\sigma$, to obtain $\alpha' = P\beta R\gamma Q$ as another path joining P to Q. It has shorter length since one of the R's was eliminated. By induction, then, P and Q can be joined by a path with no repeated vertices. ■

The proof works for directed as well as nondirected graphs. As a consequence of this theorem, if a graph G has n vertices then, if a connecting path exists at all, it can be found with n or fewer vertices (or $n - 1$ or fewer edges).

DEFINITION 6.2.3

An **acyclic graph** is a graph that contains no cycles. A **tree** is a connected, acyclic graph.

The notion of acyclic was given in Definition 5.5.9. However, this was for a digraph, not a graph.

Figure 6.2.2 illustrates various trees. Note that if any extra edge is introduced into one of these trees, joining two of the vertices there, the

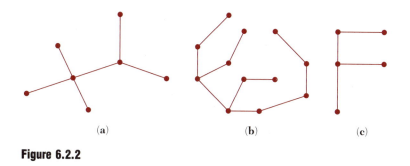

(a)   (b)   (c)

Figure 6.2.2

graph contains a cycle and, hence, is not a tree. Similarly, if any edge is removed, with the vertices kept intact, the graph is not connected and, hence, is not a tree. Somewhat loosely, we may say that a tree is on the borderline between being connected and being acyclic. It is both connected and acyclic, but barely so. Note also that any two points are connected in these graphs by only one path. We now show this in general.

THEOREM 6.2.4

Let G be a tree. Then for any vertices P and Q in G, there is a unique path joining P to Q.

Proof   If $P = Q$, the unique path is necessarily the trivial path P.

Suppose $P \neq Q$ and there are two paths α and β leading from P to Q, where

$$\alpha = PP_1P_2 \cdots P_{n-1}Q \qquad (P = P_0, Q = P_n)$$

$$\beta = PQ_1Q_2 \cdots Q_{m-1}Q \qquad (P = Q_0, Q = Q_m)$$

Now let α and β first differ in location r, so $P_i = Q_i$ for all $i < r$ but $P_r \neq Q_r$. Let $A = P_{r-1} = Q_{r-1}$. Let s be the least integer $s > r$ for which P_s is equal to one of the Q_j. Let $B = P_s$ (see Figure 6.2.3). Then we can construct a nontrivial cycle from A to B along α and then, continuing from B to A, along β. This contradicts that G is a tree.   ∎

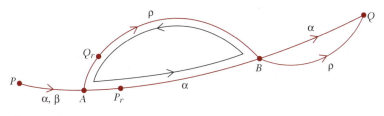

Figure 6.2.3

THEOREM 6.2.5

Let G be a tree. Then the following are true.

1. If any edge is removed from G, the resulting graph will be disconnected and thus not a tree.
2. If any additional edge is introduced joining two vertices of G, the resulting graph will contain a cycle and thus not be a tree.

Proof   1. Suppose edge PQ is removed from G. Then P and Q cannot be connected in the resulting graph because only one connecting path was possible in G, and this was clearly the edge PQ.

2. Suppose an edge PQ is introduced into the graph G, where P and Q are not adjacent. Because G is a tree, and hence connected, P and Q are joined by some path $PP_1P_2 \cdots Q$ in G. Then $PP_1P_2 \cdots QP$ would be a cycle in the resulting graph.   ■

How does one construct a tree on n vertices? A simple method is as follows. After an edge is drawn, connect a *new* vertex to one of the vertices already used, and continue this process until all vertices are used. The result is a tree. The following analysis shows that this method always works and gives all trees. We start by considering the reverse process.

THEOREM 6.2.6

Any tree with more than one vertex has a vertex of degree 1. If this vertex is removed, along with the edge joining it to its adjacent vertex, the resulting graph is also a tree.

Proof   Find a maximal path $v_1v_2 \cdots v_k$ in the tree. This is defined to be a path that cannot be prolonged at either end, since no more vertices are available to be adjoined to v_1 or v_k. Then, clearly, v_1 has degree 1, because (1) it has v_2 adjacent to it; (2) no other v_i in the path is adjacent to it (else v_1 and v_i would be connected by two paths); and (3) no other vertex in the tree is adjacent to it because the path cannot be prolonged. (The same argument also shows that v_k has degree 1.) This proves the first part of the theorem.

Now suppose v is a vertex of degree 1 in a tree, adjacent only to w with edge e. We now claim that if v is removed, along with e, the resulting graph is a tree. There can certainly be no nontrivial cycles in the resulting graph because such a cycle would be one in the original tree. In addition, the resulting graph is connected. For if P and Q are in the resulting graph, we can connect them in the original tree. The vertex v could not be used in the connecting path because it has degree 1. This proves the theorem.   ■

LEMMA 6.2.7

1. Let G be a tree, and let v be a vertex not in G. Let v be adjoined to G by one edge joining v to some vertex of G. The resulting graph G' is also a tree.

2. Further, any tree can be constructed in this way, by starting with one vertex and adjoining new vertices one at a time.

Proof (1) The graph G' is clearly connected because all vertices of G are connected to each other in G, and v is connected to some vertex of G, as in Figure 6.2.4. We further claim that G' has no cycles in it. To see this, note that the new vertex has degree 1, since, by definition, it is adjacent to only one vertex. Thus it can't appear in a cycle. Therefore, since G contains no cycles, the same is true of G'. Thus G' is a tree.

We prove (2) by induction on the number of vertices. It is clearly true for a tree with one vertex. Now suppose G is a tree with n vertices, and assume the result for trees with $n - 1$ vertices. Find a vertex in G of degree 1, and eliminate it and its edge to form a new tree. By induction, this reduced tree can be formed in the manner indicated by the theorem. Now put back the vertex and the edge to form the given tree, using this adjunction process. This completes the proof. ■

Figure 6.2.4

THEOREM 6.2.8

A tree with n vertices has $n - 1$ edges.

Proof By induction. The preceding proof shows that when we add a new vertex, a new edge is introduced. The result is true for $n = 1$, so the general result follows. ■

COROLLARY 6.2.9

If an edge of a tree G is removed and the end vertices of this edge remain intact, the resulting graph H is not connected.

Proof Let G have n vertices and $n - 1$ edges. For the new graph, the number of edges is $n - 2$, but the number of vertices is n. Thus, H is not a tree by Theorem 6.2.8. But it is clearly acyclic, because any cycle in H is a cycle in G, which is impossible. Thus H cannot be connected. ■

The following converse of Theorem 6.2.8 is true.

THEOREM 6.2.10

A **connected** graph with n vertices and $n - 1$ edges is a tree.

Proof   Suppose the graph is not a tree. Then it has a cycle. We can remove an edge from that cycle, and the resulting graph will clearly remain connected but with fewer than $n - 1$ edges. This process can be repeated until there are no more cycles and we arrive at a tree. But the resulting tree has n vertices and fewer than $n - 1$ edges. But by Theorem 6.2.8, we must have $n - 1$ edges. This is a contradiction, so the graph was a tree to start with.   ■

Remark   This theorem gives a straightforward way of verifying that a graph is a tree. As we saw in Example 5.4.10, there is a relatively simple algorithm to determine whether a graph is connected.

Minimal Spanning Trees

DEFINITION 6.2.11

If G is any graph, a **spanning subgraph** H is a subgraph of G with the same vertices V as G and some of the edges of G. (See Definition 5.1.5 for the general definition of a subgraph.) If G is connected, a **minimal connected spanning subgraph** is a connected spanning subgraph with the further property that if any edge is removed the resulting subgraph is no longer connected.

For example, suppose the vertices of a graph represent cities and the edges represent pipelines connecting the cities. Then the edges of a minimal connected spanning subgraph consist of some of these pipelines, so each city can be connected to another by using a chain of these pipelines. What makes it minimal is that the removal of any pipeline would cut off one of these cities from another. Figure 6.2.5 illustrates a minimal connected spanning subgraph of a graph. Its edges are drawn darker than the other edges of the graph.

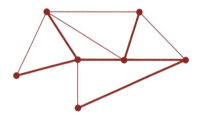

Figure 6.2.5

THEOREM 6.2.12

Let G be a connected graph, and let H be a subgraph of G with the same set of vertices. Then H is a minimal connected spanning subgraph if and only if H is a tree.

Proof If H is a minimal connected spanning subgraph, then it is connected, by definition. We now show that it is a tree. For if there were a cycle in it, any edge of that cycle can be removed, and the resulting graph would still be connected. This contradicts the definition of a minimal connected spanning subgraph.

Conversely, if H were a tree, by Corollary 6.2.9 it would not be connected if any edge were removed. Thus H is a minimal connected spanning subgraph for G. ■

Because of Theorem 6.2.12, we refer to **spanning trees** rather than minimal connected spanning subgraphs. The following two theorems give essentially different ways of finding a spanning tree for a graph. One works from the top down, the other from the bottom up.

THEOREM 6.2.13

Let G be a connected graph. Suppose edges are removed, one at a time, so that the resulting graph always remains connected. (The vertices are not removed.) If this is done until no further edge can be removed in this way, then the resulting graph is a spanning tree for G.

Proof The resulting graph is connected and has the same vertices as the original graph. Thus it is a spanning subgraph. By hypothesis, no edge can be removed to retain this property. Thus, by definition, the resulting graph is a minimal connected spanning subgraph of G. Hence it is a spanning tree. ■

THEOREM 6.2.14

Let G be a connected graph. Suppose we start with the vertices of G and no edges, and suppose edges of G are adjoined, one at a time, so that the resulting graph always has no cycles in it. If this is done until no further edge can be adjoined in this way, then the resulting graph is a spanning tree for G.

Proof The resulting subgraph H in this case has no cycles. Let us show that it is connected. To show this, suppose v and w are any two vertices. Because G is connected, they can be connected by a path in G. We now show how to find a path *in H* connecting them. To do this, we show how to replace any edge e in this path by a path in H joining the end points of e. There are two cases. If the edge is in H, we don't do any replacing. If not, then by assumption, the adjoining of that edge into H introduces a cycle in H. But this means that the end points of e can be joined by a path in H; simply take the other segment of the cycle as in Figure 6.2.6. This process therefore shows that any two vertices can be joined by a path in H with possible repeated vertices. Thus H is a tree, and this completes the proof, by Theorem 6.2.12. ■

If a connected graph G is **weighted,** then the procedures outlined in the proof of Theorem 6.2.14 can be modified to give a spanning tree with

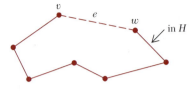

Figure 6.2.6

the least possible weight. (The weight of any subgraph is defined as the sum of the weights of its edges.) For example, if the edges represent pipelines, and the weight of any edge is the cost of building that pipeline, we can find a spanning system of pipelines costing as little as possible. Let us first illustrate the procedure.

EXAMPLE 6.2.15

For the weighted graph of Figure 6.2.7, construct a spanning tree of minimal weight.

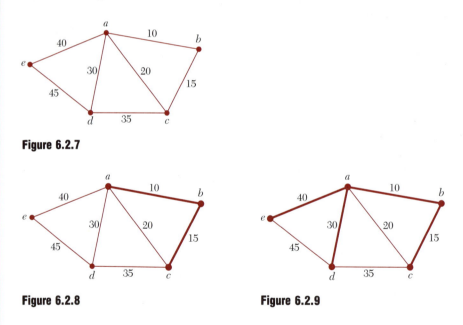

Figure 6.2.7

Figure 6.2.8   **Figure 6.2.9**

Method First choose edge *ab* of least weight 10. Then choose edge *bc* with the next lowest weight 15. This yields Figure 6.2.8, where the chosen edges are drawn in heavy lines.

Now we would like to choose *ac*, the edge of next lowest weight, but it introduces a cycle *abca*, so we ignore it and choose *ad*, the next lowest weight. When this is done, we find that we must similarly not choose *cd*; thus, the completed construction is as in Figure 6.2.9. Its weight is 95.

If the reader has qualms about this method, he or she is justified. We did not use edge ac (weight 20) but did use edge ae (weight 40). This was forced because of our method. Could we have done better if we were more clever? It turns out that this method does give a spanning tree of minimal weight. Let's first state it more explicitly, and then analyze why it works. The following theorem gives the procedure.

THEOREM 6.2.16   Kruskal's Theorem

Let G be a connected, weighted graph. The following algorithm constructs a spanning tree of G with minimal weight:

1. Order the edges of G by increasing weights. In case of a tie, any order of edges with the same weight is permissible.
2. Starting with the vertices of G and no edges, adjoin edges of G in the order fixed by step 1 to form a subgraph H. However, if the adjoining of any edge would cause a cycle to be formed in H, do not use that edge, but go on to the next edge.
3. Proceed with the adjunctions of step 2 until all edges have been tested.

Then the resulting graph is a spanning tree of G and one with the least total weight.

Proof   According to Theorem 6.2.14, this procedure yields a spanning tree H. We now consider why it is one of least weight.

For simplicity, we suppose that the weights are all different. We assume that a spanning tree H' of *smaller* weight can be found that differs from the tree obtained by the foregoing procedure. If we compare the edges of H with those of H', we find an edge e of H that is not used in H'. Choose the first such edge e in our original list. Thus, all edges of H with weight smaller than that of e are also in H'.

Now adjoin e to H'. Because H' is a spanning tree, the resulting graph contains a cycle, and this cycle must contain e. The edges of this cycle can't all be in H because H is a tree and has no cycles. Thus, there is some edge e' on this cycle that is in H' but not in H.

We claim that e' has *larger* weight than e. If not, then it has a smaller weight. The only reason e' was not chosen in H is that its introduction would have introduced a cycle of edges already chosen in H. The edges of this cycle would have weights smaller than e' and, thus, smaller than e. By our choice of e, these edges would all belong to H'. Thus, we would have a cycle in H', which is impossible.

Because the weight of e' is larger than the weight of e, replace e' by e in H', and we still get a spanning tree, but one of smaller weight. This contradicts the definition of H', as a spanning tree of least weight. ■

Remark   The algorithm is remarkable because it is relatively crude, yet works. In Dijkstra's Algorithm (Example 5.4.13), when we found a path of least weight to a vertex, we did not start by first going along the edge of least weight, then continuing on the edge of least weight, and so on. The procedure used in Kruskal's Algorithm is called a "greedy algorithm," because it doesn't look ahead and greedily tries to find the best edge one step at a time. In general, for a sophisticated construction such as this, one does not expect a greedy algorithm to work, because such an algorithm doesn't look ahead to future obstacles. Theorem 6.2.16 shows that such pessimism is not always valid.

How amenable is the Kruskal procedure to the construction of an efficient algorithm? Let us outline the procedure broadly and then see the problems, if any.

EXAMPLE 6.2.17   *Kruskal's Algorithm*

In the following algorithm, a connected, weighted graph G is given. For simplicity we arrange the edges of G in a sequence e_i, $1 \leq i \leq E$, where E is the number of edges of G. Weights $w(e_i)$ are also input with $w(e_i) > 0$ for all $i \leq E$. The weight $w(\Sigma)$, where Σ is any set of edges, is the sum of the weights of the edges in Σ. The output of the algorithm is a set Σ of edges that is a spanning tree of G and that has minimal weight compared with any other spanning tree. We proceed from the edge of least weight and adjoin edges in the manner of Theorem 6.2.16.

Kruskal Algorithm
1. Rearrange the sequence e_i so that the corresponding sequence $w(e_i)$ is in ascending order.
2. $\Sigma := \{e_1\}$ [initialize Σ]
3. FOR $i := 2$ to E [check each edge not yet considered]
    1. IF $\Sigma \cup \{e_i\}$ is acyclic THEN
        1. $\Sigma := \Sigma \cup \{e_i\}$

Step 1 consists of a sorting of the sequence $w(e_i)$.

Step 3.1 requires a check of the condition that a graph is acyclic. We haven't given an algorithm for this, but with a little extra care (hence time) we can reduce it to a connectedness problem. To see this, we note that, in step 3.1, Σ is acyclic. Therefore, its **components** are trees. (We are identifying a set of edges with a graph, namely the graph whose vertices are all the end vertices of the edges in the set and whose edges are the edges of the set.) Suppose the edge e_i under consideration has end vertices v_1 and v_2. There are now three cases.

Case 1   Neither v_1 nor v_2 is a vertex in the graph Σ. In this case, the adjunction of e_i is possible, and e_i starts a new component. See Figure 6.2.10 for this case.

Case 2   One vertex, say v_1, is not in the graph of Σ, but the other, v_2, is. In this case, no cycles are introduced when the edge e_1 is adjoined to Σ. The vertex v_1 must be adjoined to the component in which v_2 is found. See Figure 6.2.11 for this case.

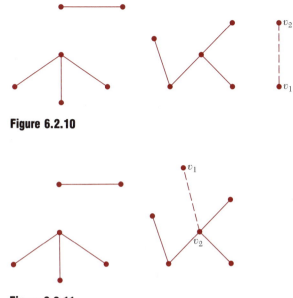

Figure 6.2.10

Figure 6.2.11

Case 3 Both v_1 and v_2 are in the graph of Σ. Here there are two subcases.

(a) The vertices are in different components of Σ. In this case, the adjunction of e_i can be done without introducing a cycle. When adjoined, the components are amalgamated into one. This can be seen in Figure 6.2.12.
(b) The vertices v_1 and v_2 are in the same component. In this case, the adjunction of e_i *will* introduce a cycle into Σ, so e_i must be *rejected*. See Figure 6.2.13 for this case.

Thus two strategies suggest themselves regarding the condition in step 3.1. One is to maintain and update a list of the components of the graph of Σ each time an edge is adjoined to it. We can then check whether case 3(b) arises. Cases 1, 2, and 3(a) show how to update the components.

The other strategy is to maintain a list of the vertices in the graph of Σ and to check whether each of the end vertices of e_i is in this list. If so,

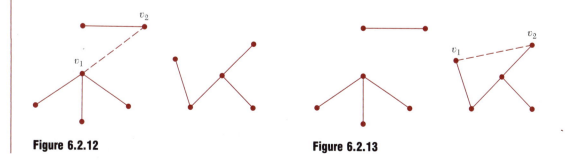

Figure 6.2.12   **Figure 6.2.13**

we find the component of v_1 in this list. The edge is rejected if v_2 is in the component of v_1.

We shall not go into a time analysis of this algorithm. There are other, better, strategies in the literature. We note, however, that the implementation of Kruskal's Algorithm is not routine. In the exercises, another algorithm (Prim's), also greedy, is introduced where the condition for adjoining an edge is not as complicated as here.

Exercises

1. Must a graph with n edges and n vertices be connected? Explain. If a graph has $n - 2$ edges and n vertices, must it be acyclic? Explain.

2. A forest is defined as an acyclic graph. Show that a component of a forest is a tree.

3. Suppose a forest has k trees and n vertices. How many edges are in the forest? (Use the definition given in Exercise 2.)

*4. Suppose G is a tree and α is a cycle of positive length in the original sense of Definition 5.1.7. Show that backtracking must occur. That is, show that there is a sequence vwv somewhere in the cycle, where v and w are adjacent vertices of G. (Hint: Do by induction on the length of α.)

5. Prove that if a connected graph G has a cycle in it, then any edge of that cycle may be removed from G and the resulting graph is still connected.

*6. Prove that if an acyclic graph G is not connected, then a new edge may be introduced into G joining two vertices of G so that the resulting graph remains acyclic.

*7. According to Theorems 6.2.8 and 6.2.10, a graph is a tree if and only if it is connected and if the number of edges is one less than the number of vertices. Write an algorithm to determine whether a graph is a tree. The input is the matrix c_{ij} of the graph and the number n of vertices. The output is either the string "tree" or the string "no tree". You may use the Connected Algorithm constructed in Example 5.4.10 as a procedure in your algorithm.

8. Suppose a graph G has v vertices and e edges. Prove that if G is connected, then $e \geq v - 1$.

9. Suppose a graph G has v vertices and e edges. Prove that if G is acyclic, then $e \leq v - 1$.

*10. Prove the analog of Theorem 6.2.10: An **acyclic** graph with n vertices and $n - 1$ edges is a tree.

11. Let H be a spanning tree of a connected graph G, and let e be an edge of G that is not in H. Show that one edge of H may be removed and replaced by e, with a resulting subgraph H' that is also a spanning tree. (Hint: Join the end points of the edge e by a path in H. Show that any edge along that path can be replaced by e.)

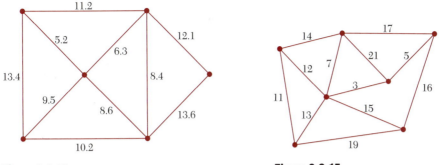

Figure 6.2.14   **Figure 6.2.15**

12. The cost of constructing a pipeline between various cities in a certain county is estimated in Figure 6.2.14. It is required to build pipelines so that each of the cities can be connected to each other by means of these pipelines, if necessary through other cities. Which pipelines should be constructed to do this task in the most economical way? Use Kruskal's Algorithm.

13. Find a spanning tree of minimal weight for the weighted graph of Figure 6.2.15.

*14. The proof of Kruskal's Theorem for the spanning tree (Theorem 6.2.16) of minimal weight made the assumption that the weights were all different. What is the situation if some of the weights are the same? In particular, prove, in this case, that the algorithm does give a spanning tree of minimal weight and show that all spanning trees can be found by this algorithm.

*15. Kruskal's Algorithm (Theorem 6.2.16) builds up a spanning tree from no edges by adding edges of least weight, making sure that no cycles are introduced. Devise an algorithm to do the same thing by starting with all edges of a connected, weighted graph and *removing* edges of the greatest weight, making sure that the graph does not get disconnected. Show that this algorithm works.

16. Write an algorithm in pseudocode for the method in Exercise 15.

17. Use the method in Exercise 15 to find a spanning tree of minimal weight for the graph in Figure 6.2.14.

18. Use the method in Exercise 15 to find a spanning tree of minimal weight for the graph in Figure 6.2.15.

*19. (*Prim's Algorithm*) Kruskal's Algorithm adjoined edges sequentially, starting with the edge of minimal weight and skipping any edge that introduced cycles. The following algorithm, due to Prim, also finds a spanning tree of minimal weight. Once again, G is taken as a connected, weighted graph. In this algorithm, the emphasis is on adjoining vertices rather than edges. The algorithm is as follows:

   1. Choose an arbitrary vertex v_0.
   2. Initialize the set Σ of chosen vertices by setting $\Sigma = \{v_0\}$. Initialize the set of chosen edges by choosing $E = \varnothing$.

3. Once Σ and E are known, and Σ does not consist of all of the vertices of G, minimize the weight of the edge vw, where v is in Σ and w is adjacent to Σ. (See Definition 5.1.13.) Assume the edge with this minimum weight is v_1w_1. Replace Σ by $\Sigma \cup \{w_1\}$ and E by $E \cup \{v_1w_1\}$.

4. Continue doing step 3 until Σ is the set of all vertices of G. At that time, the edges of the required spanning tree of least weight consist of E.

Show that Prim's Algorithm works.

20. Rewrite Prim's Algorithm in pseudocode.

21. Using Prim's Algorithm (Exercise 19), find the spanning tree of minimal weight for the graph of Figure 6.2.14.

22. Using Prim's Algorithm (Exercise 19), find the spanning tree of minimal weight for the graph of Figure 6.2.15.

23. A string of length 6 is to be formed from the letters A, B, and C. These are valued at 2, 7, and 10, respectively. The value of a string is the sum of the values of its letters. The rule of construction is that, except when it is the last letter of the string, letter "C" must be followed by "A". It is desired to form a string of maximal value. What string value will the greedy strategy give? What is the best strategy? Explain.

*24. Let G and G' be two graphs with disjoint sets of vertices. Let v and v' be vertices of G and G', respectively. Then we can **glue** these graphs at v and v' as follows. Relabel v and v' as a new vertex R, and take the union of the two graphs as the resulting glued graph. (See Figure 6.2.16.) Prove that if two trees are glued, the result is a tree.

*25. Let G and G' be two graphs with disjoint sets of vertices V and V', respectively. Let v and v' be vertices of G and G', respectively. Then

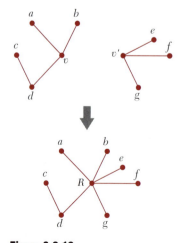

Figure 6.2.16

we can **graft** these graphs at v and v' as follows. Introduce a new vertex w and take the vertices of the grafted graph H as $V \cup V' \cup \{w\}$. Use the same edges in H together with the two additional edges wv and wv'. (See Figure 6.2.17.) Prove that if two trees are grafted the result is a tree.

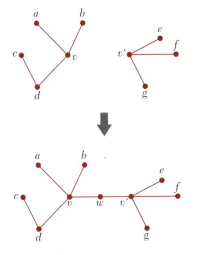

Figure 6.2.17

6.3 Rooted Trees

Each of the trees discussed in Section 6.1 had a vertex, called a root, with special significance. This was the topmost vertex in the trees illustrated in that section. In the family tree example, we were able to distinguish a parent from a child, according to the levels of the vertices. Without identifying the root, this would not be possible. We could say that two vertices were adjacent, but it would not be possible to say, given a tree, which of two adjacent vertices represents the parent and which the child. We shall see that this is possible for rooted trees.

In this section and the next, all of our trees will be understood to be rooted.

When we draw a diagram of a rooted tree, it is customary to draw the root at the top and to call that level 0. The vertices adjacent to the root are placed below it on level 1. These may be called the **children** of the root. Their children (namely the vertices adjacent to them that have not been considered so far) are called the second generation and are put on level 2. In this way, a rooted tree may be drawn so that the root is at the top, called level 0, and adjacent vertices are on levels that differ by 1. We show that this may be done uniquely for rooted trees, so it makes sense from a graph-theoretic point of view to say which of two adjacent vertices is the parent and which the child. We do this in the following definition.

DEFINITION 6.3.2

If G is a tree, with v_0 as the root, the **level** n of any vertex v is the length n of the unique path from v_0 to v. We also say that v is on level n. In particular, v_0 is on level 0.

THEOREM 6.3.3

If v and w are adjacent vertices on a tree, their levels differ by 1.

Proof   Suppose the path that joins the root v_0 to v has length n and w is not on this path. Then by extending it with the edge that joins v to w, we find that the path that joins v_0 to w has length $n + 1$. If the path joining v_0 to v *does* contain w, then w must be the vertex immediately preceding v, or else there would be two paths joining v and w. In this case, then, the level of w is $n - 1$. ■

Thus, a rooted tree may be made into a **directed** graph, called the digraph of the tree, by directing each edge from the vertex on level n to the one on level $n + 1$. Thus, in our way of drawing the diagram of a rooted tree, the direction is downward. (See Figure 6.3.1.) We can now say that a vertex v is the **parent** of another vertex w if they are adjacent and the level of v is one less than the level of w. Equivalently, v is the parent of w if the edge vw is in the digraph of the tree. We also say in this case that w is the **child** of v. As noted in the previous section, we say that a vertex v is a **leaf** if it is not a parent.

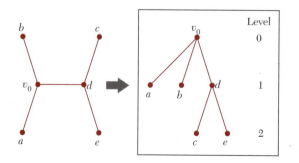

Figure 6.3.1

THEOREM 6.3.4

If v has level $n > 0$, there is one and only one vertex w adjacent to v having level $n - 1$. In brief, every vertex that is not a root has exactly one parent.

Proof If we consider the path of length n from v_0 to v, the vertex just before we reach v clearly has level $n - 1$. If there were another such vertex adjacent to v, there would be two paths from v_0 to v, contradicting the hypothesis that the graph is a tree. ■

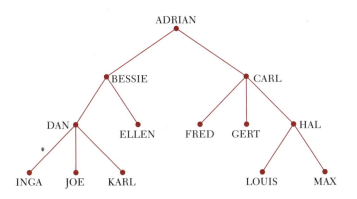

Figure 6.3.2

Figure 6.3.2 gives a simple family tree. We now wish to list the members of the family in a way that shows the various generations. The following list does the trick. Each time an indentation occurs, we go into the next generation.

ADRIAN [who married and had children:]
  BESSIE [who had children:]
    DAN [whose children are:]
      INGA
      JOE
      KARL
    ELLEN [the other child of Bessie]
  CARL [the other child of Adrian, who had children:]
    FRED
    GERT
    HAL [whose children are:]
      LOUIS
      MAX

Listing the family linearly, we obtain

ADRIAN, BESSIE, DAN, INGA, JOE, KARL, ELLEN, CARL, FRED, GERT, HAL, LOUIS, MAX

Using this method, we arrive at a full list of all the vertices in the tree. How can this be done for an arbitrary tree? We see that it is necessary to **order** the children of any parent to have a unique listing. In a family tree, the order is usually taken by choosing the oldest first. For a directory on a floppy disk, alphabetic order is usual.

DEFINITION 6.3.5

If, in a rooted tree, the children of every vertex are given a definite linear order, the tree is called an **ordered** rooted tree. If v precedes w in the ordering, we write $v < w$. If v and w are the only children of a given vertex and $v < w$, we call v and w the **left child** and **right child** of that vertex, respectively.

In what follows, we assume that all trees are ordered, rooted trees.

We can now see how linear order can be obtained in a general situation. We first list the root, then the first child, and then the full listing of each descendent of that child. We then list the second child of the root, and do the same, proceeding with all the children of the root. This is a recursive procedure, because we need a listing for each of the children of the root to obtain the full listing! To proceed, it is useful to first define a descendent.

DEFINITION 6.3.6

Let v be a vertex in a rooted tree. We recursively define the condition "w is an nth-generation descendent of v" as follows:

1. w is a 0th-generation descendent of v if and only if $w = v$.
2. If $n > 0$, w is an nth-generation descendent of v if w is the child of some vertex u that is an $(n - 1)$st-generation descendent of v.

The vertex w is **a descendent of v** if w is an nth-generation descendent of v for some n.

We can give an inductive definition of descendent without the intervention of the natural number n. Thus,

1. v is a descendent of v [the basis case].
2. w is a descendent of v if there is a vertex u such that w is the child of u and u is a descendent of v.

The definition can be rephrased in the language of relations (no pun intended) as in Section 2.1. Define aRb to mean that a is a child of b or $a = b$. Then the relation wDv (w is a descendent of v) is the transitive closure of R, as in Definition 2.1.13.)

In any (rooted) tree, the descendents of any vertex v form a rooted tree with v as the root. (See Exercise 6.3.9.) In that case, the notions of parent and children are the same in this subtree as in the original tree.

The following inductive definition shows how to list the vertices of an ordered rooted tree. It is called a *preorder traversal* of the tree.

DEFINITION 6.3.7

Let G be an ordered rooted tree with root v. The **preorder traversal** of G is a listing of the vertices of G according to the following rules:

1. If G consists of the vertex v only, then the preorder traversal of G is simply v.
2. If v has children $v_1, v_2, \ldots, v_k$, we let G_1 be the tree of descendents of v_1, and so on, through G_k. Now let L_1 be the preorder traversal of $G_1, \ldots$ and L_k the preorder traversal of G_k. Then the preorder traversal of G is the listing $v, L_1, \ldots, L_k$.

Figure 6.3.3 sketches the path of the linear listing for the family tree of Figure 6.3.2. During the course of the path outlined, we omit any vertex that has already been listed.

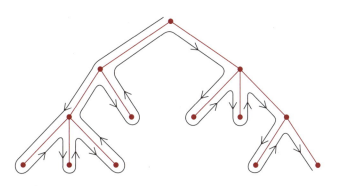

Figure 6.3.3

Polish Notation

We now illustrate a use of these listings by considering **Polish notation**.

In forming an algebraic expression involving, say, the operations $+$, $-$, $*$, $/$, we usually use parentheses to group partially calculated results. We saw in Example 6.1.4 how to draw a tree diagram for any algebraic or logic expression. The expressions $(a + b)/(c + 2d)$ and $(p \lor q) \to \neg r$ are illustrated in Figure 6.3.4. The idea is that each of the operators $+$, $-$, $*$, and $/$ are vertices with exactly two children in a given order, as are the binary connectives $\lor, \land, \to$, and $\leftrightarrow$. These are called **binary** operators, because they each accept two variables. The variables $a, b, \ldots$ or

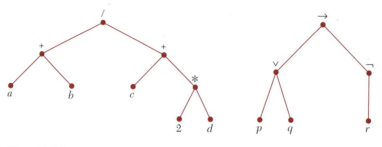

Figure 6.3.4

$p, q, \ldots$ are also vertices, but they have no children. The symbol " $-$ " is sometimes used to denote the negative of a number. As such, it is a **unary** operator, as is ¬—they accept only one variable. To distinguish the negative operator from the subtraction operator, we denote it by N. Thus, the expression $-(a + b) - c$ will be written $N(a + b) - c$. Its diagram is in Figure 6.3.5. Since N is a unary operator, it is represented by a vertex with one child.

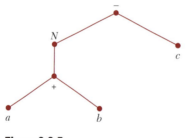

Figure 6.3.5

We can now give the definition of Polish notation.

DEFINITION 6.3.8

If α is an algebraic formula or a statement in propositional calculus whose tree diagram is T, the Polish notation for α is the preorder traversal listing for T.

EXAMPLE 6.3.9

Find the Polish notation for $(a + b)*(c - d/a)$.

Method The tree is given in Figure 6.3.6. Therefore the Polish notation for this formula is simply $* + ab - c/da$. This is the same method as listing the family tree.

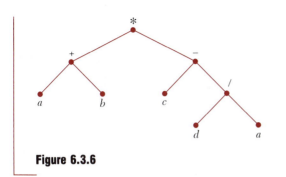

Figure 6.3.6

The List-to-Tree Algorithm

We now ask how to retrieve a tree when its vertices are given in a preorder traversal listing. In Polish notation, we have additional information about the vertices: We know the number of children for each of the vertices. For example, because + is a binary operator, it has two children. In fact, the number of children of every entry in a formula given in Polish notation is known. We shall see that this information is sufficient to retrieve the tree from the listing.

Now consider the general case of a tree in which the number of children of each vertex v is known. We first introduce a useful definition.

DEFINITION 6.3.10

The **weight** of a vertex v in a rooted tree is defined to be $1 - n$, where n is the number of children of v.

In particular, only leaves have positive weight. We show in the next theorem that **the sum of the weights of all vertices in a tree is 1.** For example, the vertices of the tree of Figure 6.3.6 have weights as follows:

$$* \quad + \quad a \quad b \quad - \quad c \quad / \quad d \quad a$$
$$\text{Weights:} \quad -1 \ -1 \ \ 1 \ \ 1 \ -1 \ \ 1 \ -1 \ \ 1 \ \ 1 \quad (\text{sum} = 1)$$

The weights are placed under the listing. Note that the binary operators $*, +, -, /$ have weight $1 - 2 = -1$, whereas the variables a, b, c, d have weight 1, since they have no children. The listing is the Polish notation (or preorder traversal listing) for this tree.

THEOREM 6.3.11

The sum of the weights of all of the vertices of a rooted tree is 1.

Proof The proof is by induction on the number of vertices in the tree. If the tree consists of one vertex, it is the root. The root has no children in this case, so its weight is 1. Now suppose the tree T has root v, which has k children $v_1, v_2, \ldots, v_k$. Thus, v has weight $1 - k$. Let the descendents of v_i form the subtree T_i. (See Exercise 6.3.9.) We may assume (our induction hypothesis) that the sum of the weights in each of the T_i is 1. Therefore, since v has weight $1 - k$, and the vertices of the tree consist of A together with the vertices of T_i, the sum of the weights of the vertices is $(1 - k) + 1 + \cdots + 1 = 1$. Here, the sum has k 1's, one for each of the T_i. ∎

THEOREM 6.3.12

Let L be the preorder traversal listing of the vertices of a rooted tree. If we keep a running count of the weights from the first entry on, this count will not be positive until the list is complete, when it will be 1.

Before proving this, let us illustrate with the Polish notation example given earlier:

*	+	a	b	−	c	/	d	a	
−1	−1	1	1	−1	1	−1	1	1	(weights)
−1	−2	−1	0	−1	0	−1	0	1	(running sums)

The general proof is by induction.

Proof   The result is true (vacuously) for a tree whose only vertex is the root. Now suppose that a tree has root v, whose children are $v_1, \ldots, v_k$. Let the descendents of v_i form the subtree T_i. As our inductive hypothesis, we may assume the result for each tree T_i. Let the listing for T_i be L_i. By the definition of preorder traversal, the listing for the given tree is $v, L_1, L_2, \ldots, L_k$. If we put weights under each vertex in this listing, we start with $1 - k$, the weight of the root v, which is ≤ 0, because we are assuming that v has at least one child. As we scan through L_1, the sum of the weights is always ≤ 0 (by induction) until we come to the end of the list, when the sum is 1, by Theorem 6.3.11. The same situation prevails for each tree L_i. Thus, as we scan through the weights, we find that each listing L_i contributes at most one 1 to the sum, and this only when we finish the listing L_i. Therefore, we add total weights of less than k to the initial $1 - k$ until we have gone through the complete listing. Thus, the running sum is always less than 1 until the last entry of L_k. This completes the proof. ∎

Theorem 6.3.12 permits an easy way of retrieving the tree structure from its preorder traversal listing. This can be done if we know the weights of each vertex or, equivalently, if we know the number of children of each vertex. Let us illustrate by retrieving the family tree of Figure 6.3.2 from its listing. We have listed the weights under the vertices:

ADR	BES	DAN	ING	JOE	KAR	ELL
−1	−1	−2	1	1	1	1

CAR	FRE	GER	HAL	LOU	MAX
−2	1	1	−1	1	1

From this listing, we see that ADR is the root and ADR has two children, the first of whom is BES. The foregoing proof shows that the listing after ADR consists of two pieces and that each piece is completed as soon as the weight 1 occurs. Thus, *starting with BES*, we count running sums:

	BES	DAN	ING	JOE	KAR	ELL
Wts	−1	−2	1	1	1	1
Sum	−1	−3	−2	−1	0	1

	CAR	FRE	GER	HAL	LOU	MAX
	−2	1	1	−1	1	1

This shows that BES's descendents end with ELL and that CAR's descendents go through MAX. The family tree thus has ADR as root, with the two children BES and CAR. The same process can then be performed on the sublists BES ... ELL and CAR ... MAX, and we ultimately retrieve the family tree of Figure 6.3.2. Note that we see immediately from the weights that BES has two children, whereas CAR has three.

This procedure is clearly recursive, because we use it again, once the lists BES ... ELL and CAR ... MAX are found. We state the procedure more formally later, but before we do so it will be useful to show how to **graft** various trees onto a given root. (Compare with Exercise 6.2.25.) See Figure 6.3.7 for an illustration of this grafting technique.

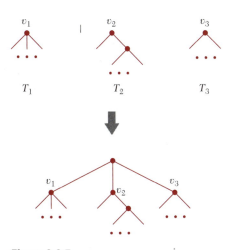

Figure 6.3.7

DEFINITION 6.3.13

Let $T_1, \ldots, T_k$ be ordered rooted trees with disjoint vertices, let v_i be the root of T_i, and let v be a vertex distinct from all of the vertices in the trees T_i. Then the **graft** of the trees T_i is an ordered, rooted tree whose root is v, with the following properties.

1. The children of v are $v_1, \ldots, v_k$ in that order.
2. Vertices other than v are adjacent if and only if they are in the same tree T_i and are adjacent in that tree.

Remark   Briefly, we graft the trees T_i onto v to form a new tree.

THEOREM 6.3.14

The graft of trees T_i $(i = 1, \ldots, k)$ is a tree.

Proof   The graph is connected, because any vertex $\neq v$ is in one of the T_i's and so can be connected to one of the v_i's, which in turn is connected to v. Further, T is acyclic, because any cycle that doesn't include v must necessarily be a cycle within one of the T_i's, which is impossible. (Otherwise it would have to jump from one of the T_i's to another, which is also not possible.) Any cycle that includes v would have to enter v from one v_i and exit through another v_j, and there could be no way of joining v_i and v_j to complete the cycle. Thus T is a tree.   ■

EXAMPLE 6.3.15   Algorithm List/Tree

We can now state the algorithm to convert a list of vertices into a tree. This algorithm is called List/Tree because it converts lists to trees. The list is supposed to be a preorder traversal list for an ordered rooted graph, and the weight of each vertex (an integer ≤ 1) is given. The algorithm also discovers if a list is spurious—that is, if no tree can give rise to this list. We give it informally.

The Algorithm List/Tree

If the list consists of a single vertex v, its weight must be 1, and the tree is the tree whose only vertex is v. If the weight of v is not 1, the list is spurious.

If the list has more than one entry, the sum of the weights must be 1, else it is a spurious list. The first entry v is the root of the required tree. If v has weight r, then v must have $k = 1 - r$ children $v_1, \ldots, v_k$, whose descendents form trees $T_1, \ldots, T_k$ with lists $L_1, \ldots, L_k$ and that can be found as follows: v_1 is the term following v in the given list, and starts the list L_1. Then L_1 terminates as soon as the running sum of its weights, starting with the weight of v_1, is equal to 1. The next term is v_2, which starts the list L_2. Similarly, L_2 terminates the first time the running sum 1 occurs starting with v_2. The procedure continues until L_k is found. The list is spurious unless it is exhausted precisely when L_k is found.

If now List/Tree is applied to the lists $L_1, \ldots, L_k$ to form trees $T_1, \ldots, T_k$, the required tree consists of the vertex v as root, with the trees $T_1, \ldots, T_k$ grafted onto v as in Definition 6.3.13.

We illustrate with a simple example.

EXAMPLE 6.3.16

Given the following list, with associated weights, reconstruct the tree.

Vertices	a	b	a	b	c	c	d	e
Weights	-1	-2	1	1	1	-1	1	1

Method   The sum of the weights is 1, so there is a possibility that this may be a list for a tree. The first vertex a has weight -1, so it must have two children, the first of which is the b following it. We now attempt to form the tree of descendents of this b. We find the running weight starting with this b as follows:

Vertices	a	b	a	b	c	c	d	e
Weights	-1	$[-2$	1	1	1]	-1	1	1
Running Weights		-2	-1	0	1			

Thus, b has three children, named a, b, c, and these children have no further children, because their weights are each 1. The list L_1 is marked with brackets, and its tree is the tree T_1 of Figure 6.3.8.

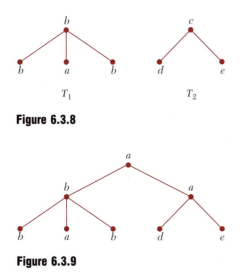

Figure 6.3.8

Figure 6.3.9

Listing L_2 follows L_1. Here, c has two children d and e, neither of which has children. Its tree is T_2 of Figure 6.3.8.

Finally, trees T_1 and T_2 are grafted onto the initial vertex a to obtain the desired tree of Figure 6.3.9.

This example illustrates that the same label may be used for more than one vertex in a tree. We take the position that the vertices are all different but that each vertex is assigned a label, where the assignment need not be 1-1. For example, the algebraic formula $(a+(b+(a+c)))+(a+b)$ has four $+$'s, three a's, and two b's. These are all names of different vertices of the tree for this formula. The tree is shown in Figure 6.3.10. The Polish notation for this formula is seen to be $++a+b+ac+ab$.

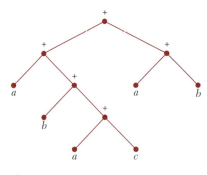

Figure 6.3.10

To reconstruct Figure 6.3.10 from the Polish notation, we need only know that $+$ has weight -1 and that a, b, and c have weight 1. The List/Tree Algorithm will do the rest.

EXAMPLE 6.3.17   Reverse Polish Notation

On certain hand calculators, a variant of Polish notation is used to good advantage. To compute $a+b$, one must enter a, then b, and then the operator $+$ to obtain an answer. If we write $a+b$ *in the order in which entries are entered into the calculator*, we obtain the list $ab+$ (reverse Polish notation) rather than the usual Polish notation $+ab$. Calculators using "algebraic logic" need parentheses or the occasional use of an equal sign to finish a calculation. This is not necessary if a calculator uses reverse Polish notation. For example, the complicated expression $(a+b)/(c*d+3)$, written $/+ab+*cd3$ in Polish notation, would be written $ab+cd*3+/$ in reverse Polish. Reading from left to right, we see that the calculator will take the entries a and b, add them, then take the entries c and d and multiply them. Then the entry 3 will be added to the most recent calculation (cd), and finally the two calculated results will be divided. Note: Once a calculation, such as $ab+$, is made, the a and the b are lost and replaced by their sum.

Exercises

1. (a) Write the following in Polish notation:

$$(x + y)^{(x - y/z)}$$

[Use the notation $E(u, v)$ as the functional way of writing u^v, so that E may be a label of a vertex with children u and v.]
   (b) Write the expression in reverse Polish notation.
2. Repeat Exercise 1 for the algebraic expression

$$(a + b \sin c)^{2/5} + \sqrt{a - b \sin c}$$

Use sqrt(x) for $\sqrt{x}$.
3. Write the statement $((p \to q) \lor q) \leftrightarrow (p \to q)$ in Polish notation.
4. Write the statement $\lnot\lnot\lnot(p \lor q) \leftrightarrow (\lnot p \land \lnot q)$ in Polish notation.
5. An algebraic expression is written $+ / + / + abcdef$ in Polish notation. Convert this to the usual algebraic notation and to reverse Polish.
6. Repeat Exercise 5 for the expression $+ + + ab/c/def$.
7. A statement is written $\lor \to p \lor q \lnot s \lnot p$ in Polish notation. Convert this to the standard way of writing this formula.
8. A statement is written $\lor\lor p \lor pq \lor pq$ in Polish notation. Convert this to the standard way of writing this formula.
*9. Show that the set of all descendents of a vertex v in a tree forms a tree.
10. The following list is a preorder traversal of a family tree:

> JOHN (3), FRED (2), ALEX, BESS (1), TED, GERRY (2),
> ZENO (1), DAWN, ALICE (3), JOHN, MARY, EPSILON,
> PRINCE

The numbers in parentheses are the number of children that person has. No number means that the person has no children. Draw the tree. Determine all of ALEX's first cousins.
11. The list in Exercise 10 contains the number of children of each vertex in the order in which these vertices appear in the preorder traversal of that tree. Listing these numbers in order, we obtain the sequence 3, 2, 0, 1, 0, 2, 1, 0, 3, 0, 0, 0, 0. Under what circumstance is a listing of integers of this sort the listing for a tree?

A sequence such as the one given in Exercise 11 is called a **tree sequence**. Exercises 12–16 concern such sequences.

*12. Prove that in any tree sequence a term 0 may be replaced by a tree sequence and the result is a tree sequence. Interpret graph-theoretically.
13. Prove that if $S_1, S_2, \ldots, S_k$ are tree sequences, then so is the sequence $k, S_1, \ldots, S_k$. This means that the integer k is followed by the terms of S_1, then by the terms of S_2, and so on. Interpret graph-theoretically.
14. Prove that if k is a term of a tree sequence S, and T is a tree sequence,

then the term k may be replaced by the terms $(k+1)$, T in S and the result is a tree sequence. Interpret graph-theoretically.

*15. Use the result of Exercise 13 to show how a tree sequence may be defined recursively. In particular, show that the tree sequence given in Exercise 11 is a tree sequence, using your definition.

*16. Use the result of Exercise 14 to show how a tree sequence may be defined recursively. In particular, show that the tree sequence given in Exercise 11 is a tree sequence, using your definition.

In Exercises 17–21, each list may or may not be a preorder traversal for a tree. The numbers in parentheses are the weights of the vertices. In each case, draw the tree if it is such a list. If it is not such a list, explain why.

17. $A(-2)$, $B(-1)$, $C(1)$, $D(1)$, $E(0)$, $F(1)$, $G(1)$
18. $A(-1)$, $B(-2)$, $A(1)$, $B(1)$, $C(1)$, $C(-1)$, $D(1)$, $E(0)$, $F(1)$
19. $A(-3)$, $B(-2)$, $C(1)$, $D(1)$, $E(-1)$, $F(1)$, $G(1)$, $G(1)$, $H(1)$
20. $A(-2)$, $B(0)$, $C(0)$, $D(0)$, $A(1)$, $B(1)$, $C(1)$
21. $A(-2)$, $B(1)$, $C(-2)$, $D(1)$, $E(1)$, $A(1)$, $B(1)$

Write the preorder traversal for each of the following trees.

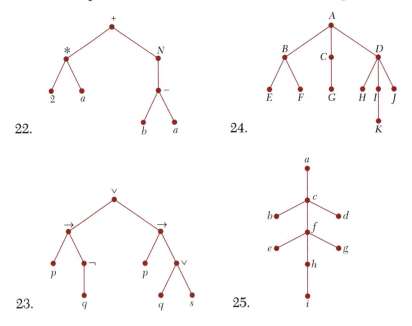

22.   24.

23.   25.

26. Is the formula $+*b/ce+*aab$ the Polish notation for some algebraic expression? Use the List/Tree Algorithm.

*27. Devise an algorithm to convert the tree of an algebraic expression, or more generally any tree, into reverse Polish notation.

*28. Devise an algorithm, PolishList/Tree, in analogy with the List/Tree Algorithm that will convert a list written in reverse Polish notation into its corresponding tree.

6.4 Some Applications

This section devotes itself to applications of some of our results on trees.

Counting Principles

We mentioned in Section 4.1 that the Multiplication Principle depends on the "uniqueness of the path to the final outcome." This is best seen in terms of rooted trees. A tree is used, because (by definition) there is a unique path from the root to any vertex. The idea is that each "stage" of a procedure is a vertex of a tree and that the "choices" are the edges leaving that vertex. The final outcome of a procedure occurs when there are no more choices (i.e., a leaf). In all cases, we are concerned with counting the number of leaves in a tree; this is simply the number of outcomes. The following result is a simple consequence of the Addition Principle.

EXAMPLE 6.4.1   *Graph-Theoretic Version of the Addition Principle*

Suppose the rooted trees $G_1, G_2, \ldots, G_k$ are grafted at the vertex v, forming a new tree G. Let n_i be the number of leaves in G_i. Then the number of leaves in G is $n_1 + n_2 + \cdots + n_k$.

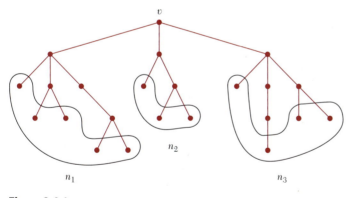

Figure 6.4.1

This is illustrated in Figure 6.4.1. In brief, the number of leaves in a tree is the sum of the number of leaves on each of its main branches. To prove it, note that any leaf of G is joined by a unique path from v to one of the v_i and then to the leaf. The roots of G_i are the disjoint cases, and there are n_i leaves for the case G_i.

As a consequence, we have the following graph-theoretic version of the Multiplication Principle.

EXAMPLE 6.4.2 *Graph-Theoretic Version of the Multiplication Principle*

Let G be a rooted tree. Suppose its root has n_0 children and each level 1 vertex has n_1 children, and so on, to the level k vertices that have n_k children. Further assume that the level $k + 1$ vertices are all leaves. Then the number of leaves is the product $n_0 \cdot n_1 \cdots n_k$.

The proof is by induction on the maximum level $k + 1$. The basis, $k = 0$, is clear. By induction, each level 1 vertex has $n_1 \cdots n_k$ leaves as descendents, and since there are n_0 such vertices there are $n_0(n_1 \cdots n_k)$ leaves in all, by the Addition Principle. (See Figure 6.4.2 for an example.)

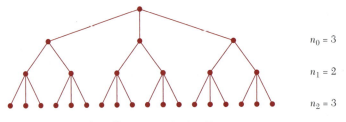

$n_0 = 3$

$n_1 = 2$

$n_2 = 3$

Number of leaves $= 3 \times 2 \times 3 = 18$

Figure 6.4.2

EXAMPLE 6.4.3

How many strings of length at most 20 on the alphabet $\{1, 2, 3\}$ are there in which each "1" must be followed by a "2" or a "3"?

Method In principle, this may be solved as in Section 4.3. Now, however, we picture the strings as paths in a tree, each vertex after the root designating the next character in the string. We use "T" to indicate that a string terminates. Let G_n be the tree for strings whose length is at most n. Figure 6.4.3 gives the main description of G_n, as well as the basis G_0 and G_1.

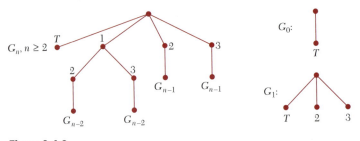

Figure 6.4.3

The required recursive definition can be read directly from Figure 6.4.3. If g_n is the number of such strings of length $\leq n$, then we can see from the diagram that

$$g_0 = 1, \quad g_1 = 3; \quad g_n = 1 + 2g_{n-2} + 2g_{n-1} \quad (n \geq 2)$$

A simple hand calculation gives $g_{20} = 667,731,285$. A formula for g_n can also be found from the methods of Section 4.3.

Binary Trees

We consider ordered, rooted trees where each vertex has at most two descendents. This notion leads to several interesting applications.

DEFINITION 6.4.4

A **binary tree** is an ordered, rooted tree whose vertices have out-degree at most 2. In a binary tree, each child must be labeled left or right, even if there is only one child. A **full binary tree** is a binary tree whose vertices have out-degree 0 or 2. Thus, if a vertex in a full binary tree is not a leaf, there are two edges leaving it. A **complete binary tree** is a full binary tree, all of whose leaves are on the same level.

A simple example of a binary tree is given in Figure 6.4.4. This represents the bit strings of length 3. The tree is complete. We can label the edges leaving any vertex as 0 or 1, using 0 for the left-hand edge and 1 for the right-hand edge. The leaves are all at level 3, and each leaf is identified by the path chosen to arrive there from the root. For example, we arrive at 010 by starting at the root and going left, then right, then left. This scheme is useful for any binary tree, complete or not. All vertices are given by the (unique) path from the root that takes you there— that is, a sequence of 0's and 1's.

This discussion gives a simple way of defining a binary tree by listing its leaves. Clearly, any vertex in a rooted tree is somewhere along the path to one of its leaves. Therefore all of the vertices of a binary tree show up as **initial substrings** of leaves.

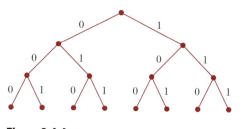

Figure 6.4.4

EXAMPLE 6.4.5

A binary tree has leaves given by 01, 111, 1010, and 1011. Sketch the tree.

Method   The vertices are the initial substrings: 0, 01; 1, 11, 111; 1, 10, 101, 1010; 1, 10, 101, 1011, and the root. (There are repetitions here.) We have agreed to use 0 for "left" and 1 for "right." The graph is given in Figure 6.4.5.

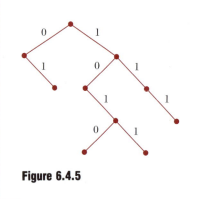

Figure 6.4.5

Huffman Codes

We observed in Section 1.1 that all printable characters are given by certain ASCII codes from 0 to 255. Equivalently, these are given by 8-bit strings, also called bytes. Standard ASCII consists of 128 characters and can be coded by 7-bit strings representing integers from 0 through 127. This means that any file containing n characters needs $8n$ bits for storage, $7n$ bits if only standard ASCII codes are used. It turns out there is a more economical procedure.

Huffman coding codes each character as a k-bit integer, where k is **variable** and depends on the character being coded. The idea is that a letter that is used frequently may be coded with fewer bits than one seldom used, so the total number of bits needed for storage is smaller than the number from the ASCII, fixed-width approach.

EXAMPLE 6.4.6   A Huffman Code

We take a simple example, using only the letters A, B, C, and D. We can code these as 00, 01, 10, and 11. In this way, the string

$$A\,A\,A\,A\,B\,C\,A\,A\,B\,A\,A\,A\,B\,A\,D\,A\,A\,A\,C\,C\,B$$

would be coded as

00 00 00 00 01 10 00 00 01 00 00 00 01 00 11 00 00 00 10 10 01

or, condensing and using groups of eight for readability,

00000000 01100000 01000000 01001100 00001010 01

Thus, a total of 42 bits are used for a string of length 21. Now it is easy to see that A occurs quite frequently (13 times), whereas B, C, and D occur 4, 3, and 1 time, respectively. Now use the following code, in which A is coded by only 1 bit:

A: 0
B: 11
C: 100
D: 101

(We'll learn later how to construct this code.) Then the original string

$A\ A\ A\ A\ B\ C\ A\ A\ B\ A\ A\ A\ B\ A\ D\ A\ A\ A\ C\ C\ B$

is coded as

0 0 0 0 11 100 0 0 11 0 0 0 11 0 101 0 0 0 100 100 11

which, when grouped in strings of size 8, becomes

00001110 00011000 11010100 01001001 1

a reduction from 42 to 33, a decrease of about 21%.

This string can be uniquely decoded. Because only A has a code starting with 0, the decoded string *must* start with $AAAA$. Then only B starts with 11, so the next letter must be B. This procedure continues until we obtain a complete, unique decoding.

This example leads to two problems, both of which can be solved by the Huffman coding procedure.

1. What method of coding characters by strings of bits allows a unique decoding?
2. Of the allowable coding methods, which one uses the least number of bits to code a string of characters?

EXAMPLE 6.4.7 *The Binary Tree of a Code*

Let us examine the code of Example 6.4.6 by using a binary tree. Figure 6.4.6 places each character at the point in a tree that is given by its code. That is, we can read the code for any letter by going down to it from the root and reading off the path as a sequence of 0's and 1's. The level

Code
A: 0
B: 1 1
C: 1 0 0
D: 1 0 1

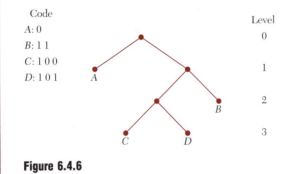

Level
0
1
2
3

Figure 6.4.6

column gives information on the length of the path used to get to any vertex; it is the number of bits used in the code for a character.

We can now answer question 1. In Figure 6.4.6, the characters to be coded must be placed at vertices in such a way that **the path from the root to any character does not meet any other character.** Equivalently, **no code for any character is the beginning string of the code for any other character.** This allows for unique decoding. In the previous illustration, once we arrive at the string 11 we knew it was the code name for B. On the other hand, if C, for example, were coded as 110, we couldn't decode 11 as B because it might be the beginning of the code for C. Because of the preceding condition, such ambiguities never occur.

DEFINITION 6.4.8

Suppose some of the vertices of a binary tree are labeled with characters $A_1, A_2, \ldots, A_k$ in such a way that a path from the root to any of these characters does not meet any other character. Then this labeling is called a **decodable labeling** of the characters.

Thus, we can summarize the foregoing examples by saying that a proper (variable length) encoding of characters is given by a decodable labeling.

To answer question 2, we must have some idea of the **frequency** with which each character is used in a file. (In practice, an estimate might be used.) Intuitively, to decrease the number of bits we use, we want the most frequent symbols to be coded with the fewest symbols. The preceding example worked so well because the character A was quite frequent and was coded with only one bit.

DEFINITION 6.4.9

Suppose that in a string of characters of length N the character x occurs n times. Then n is called the **frequency** of the character x in the string. Let the characters in the string be A_i, with frequency n_i, $i = 1, 2, \ldots$.

Now suppose we have a coding system in which A_i is coded as a string of bits of length L_i. Then the total number of bits needed to encode the entire string of characters is

$$W = \sum n_i L_i$$

Here, W is called the **weight** of the code with respect to the frequency data n_i. In addition, since N is the number of characters in the string, we have

$$N = \sum n_i$$

The **average number of bits per character** is W/N.

In Example 6.4.6, we had the symbols, frequencies, and weight as indicated in Table 6.4.1.

Table 6.4.1

A_i	n_i	L_i	$n_i L_i$
A	13	1	13
B	4	2	8
C	3	3	9
D	1	3	3
			33: Weight

In this illustration, there were $N = \sum n_i = 21$ characters. Thus, the average number of bits used per character is $33/21 \approx 1.571$, significantly better than 2, the number of bits used for a code with a fixed number of bits per character.

Question 2 thus becomes a minimization problem. Of all the decodable labeling of the vertices, which one has the minimum weight? (The resulting labeling is called a **Huffman code**.) We can answer this question by a series of lemmas, which verify the idea that the characters with the lowest frequency have the longest code.

In what follows, we assume that we are given a sequence of characters $A_1, A_2, \ldots, A_n$ with corresponding frequencies n_i.

LEMMA 6.4.10 Switching Lemma

Suppose variables are encoded in such a way that a character A has a smaller frequency than the character B, and the length of the code of A is smaller than the length of the code of B. (Graphically, a character is higher on the graph than another character and has smaller frequency.) Then by switching the codes for A and B, we obtain a coding with a smaller weight.

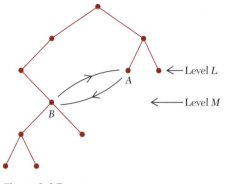

Figure 6.4.7

Proof Suppose A has length L and frequency a, and B has length M and frequency b. (See Figure 6.4.7.) The contributions to the weights before and after the switch are $aL + bM$ and $aM + bL$, respectively. The difference is

$$(aL+bM) - (aM+bL) = a(L-M) + b(M-L) = (a-b)(L-M) > 0$$

Thus the switch gives a smaller weight. The calculation also shows that if $<$ is replaced by $\le$, a switch yields a code whose weight is not larger than the original weight. ■

LEMMA 6.4.11

Suppose the characters A_i have frequencies n_i, where the n_i are in increasing order. Then a code with minimal weight can be found in which the two characters A_1 and A_2 of *least* frequency are both leaves with the same parent. In short, the configuration of Figure 6.4.8 occurs in a Huffman code.

Proof Find any character at the lowest level of the tree. If it is not A_1, find A_1, and use the Switching Lemma to switch it to the lowest level. By the Switching Lemma, the resulting graph has a smaller weight, contradicting the assumption that the graph is one of minimal weight. (If there are several characters at the lowest level, or if there are several characters with the same minimal weight, the same argument shows that a switch leads to another minimal graph.) Thus, we may assume that A_1 is at the lowest level.

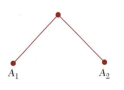

Figure 6.4.8

Now try to move A_1 up to where its parent vertex is. If we could, the resulting tree would have smaller weight, which is impossible. But what prevents this move? The only possibility is that the parent (necessarily unlabeled) has yet another child, also labeled. (Moving A_1 up would then result in a path to the other child, which would pass through it.) Now, take this other child and switch it with A_2, obtaining a graph whose weight is at least as small as the given one. We thus arrive at Figure 6.4.8. ■

The importance of Lemma 6.4.11 is that it replaces two vertices by one tree, which we now know is part of the ultimate Huffman code. This is the beginning of an inductive scheme, whereby we consider how such clusters occur in minimal coding trees.

DEFINITION 6.4.12

A **cluster** is a decodable labeling of some of the characters A_i. (In particular, we regard any vertex, by itself, as a cluster, with itself as the only vertex.) The **frequency** of a cluster is defined as the sum of the frequencies of the characters in the cluster. Clusters are called **disjoint** if they have no characters in common.

Thus, Lemma 6.4.11 shows that in any Huffman coding of a sequence A_i of characters, the two characters A_1 and A_2 of least frequency may be replaced by one cluster as in Figure 6.4.8. We now show that this procedure may be continued for clusters.

DEFINITION 6.4.13

The **graft** of disjoint clusters C_1 and C_2, written $C_1 \oplus C_2$, is defined as the cluster shown in Figure 6.4.9. If n_i is the frequency of C_i, the frequency of $C_1 \oplus C_2$ is $n_1 + n_2$.

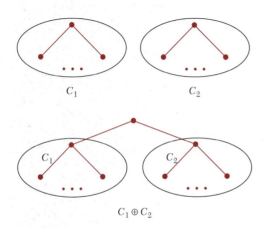

C_1 C_2

$C_1 \oplus C_2$

Figure 6.4.9

Lemma 6.4.11 shows that to find the code of minimal weight for the sequence $A_1, A_2, \ldots, A_n$ listed in order of increasing frequency, we need to find the code of minimal weight for the sequence of clusters: $A_1 \oplus A_2, A_3, \ldots, A_n$.

LEMMA 6.4.14

Let the clusters $C_1, C_2, \ldots, C_k$ be used in a Huffman code, and let them be listed in order of frequency. Then a Huffman code can be formed from the clusters $C_1 \oplus C_2, C_3, \ldots, C_k$. ■

The proof is deferred to the exercises. It is quite similar to the proof of Lemma 6.4.11 (which is a special case).

We illustrate with a simple example.

EXAMPLE 6.4.15

Characters A, B, C, D, E have frequencies as in Table 6.4.2. Find a Huffman code for this sequence of characters.

Table 6.4.2

Character:	A	B	C	D	E
Frequency:	13	31	17	5	21

Method   We follow the previous lemmas. First arrange the characters in order of increasing frequency:

Character:	D	A	C	E	B
Frequency:	5	13	17	21	31

Now graft the clusters of least frequency to form

Cluster:	$D \oplus A$	C	E	B
Frequency:	18	17	21	31

And rearrange the clusters again in order of frequency:

Cluster:	C	$D \oplus A$	E	B
Frequency:	17	18	21	31

And graft the clusters of least frequency:

Cluster: $C \oplus (D \oplus A)$ E B

Frequency: 35   21   31

Continuing, we obtain

Cluster: $E \oplus B$. $C \oplus (D \oplus A)$

Frequency: 52   35

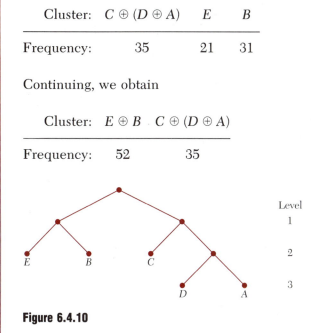

Level
1

2

3

Figure 6.4.10

And so we finally arrive at the Huffman code $(E \oplus B) \oplus (C \oplus (D \oplus A))$, whose graph is given in Figure 6.4.10.

The Huffman code for this system is given in Table 6.4.3.

Table 6.4.3

Symbol	Code
A	111
B	01
C	10
D	110
E	00

In this case, the frequencies did not warrant that the most frequent character, B, be coded by a 1-bit string.

Binary Search Trees

Binary trees can be used effectively to index a group of records. Suppose the data for a company's employees consists of several thousand employees. These records might be stored in a computer file, and it is necessary to occasionally check whether a certain person is an employee and, if so, to find his or her record. Also, it is necessary to add records as new employees show up. How does one proceed?

We encountered this problem in the time analysis of the InsertSort Algorithm of Example 3.3.7. Locating a record in a sorted file of size n could be done efficiently by using a binary search, taking $O(\log n)$ time. But inserting it into a sequence of n records had a worst-time case of order of magnitude n. Thus, if "Jones, John E." joins the company, his record would have to be put in between "Jones, Jane R." and "Jong, Kai L." This means that all records from "Jong" on would have to be pushed down to make room for Jones's record. Each push down, as we have seen, is an assignment and takes time. On a computer this could be a time-consuming job. Of course, if we put Jones's record at the end, we don't have this pushing-down problem, but we lose the alphabetizing, so we would have difficulty locating a record or finding whether someone is an employee.

There is a way of proceeding by using binary trees. The idea is to "index" the records of the employees by labeling each one as the vertex of a binary tree. Let us illustrate by doing this on the following list:

Jones, Johnson, Smith, Arden, Hausner, Cogan, Perry, Tucker

Start with Jones and put her at the root. The idea is that all names preceding it alphabetically go to the left and the ones after it go to the right. In this case, Johnson goes left and Smith goes right, leading to Figure 6.4.11. Continuing, Arden is to the left of Jones and Johnson, leading to Figure 6.4.12. Hausner is next on the list. He is left of Johnson but right of Arden, leading to Figure 6.4.13.

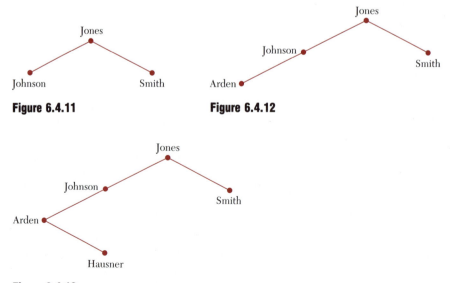

Figure 6.4.11

Figure 6.4.12

Figure 6.4.13

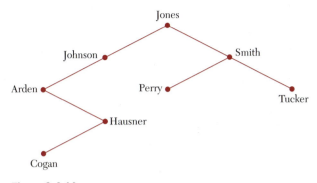

Figure 6.4.14

Now continue, following down the graph. Cogan is to the left of Jones and Johnson, to the right of Arden, and to the left of Hausner. We can similarly fill in Perry and Tucker to obtain Figure 6.4.14.

This method continues when new people join the company. A new employee gets adjoined to the binary tree as a new leaf. All that we do is check whether the record is alphabetically before or after the root. Go left for before, and right for after. Do the same at each named vertex until there is no labeled vertex. This will locate the place where the new employee belongs on the binary tree. For example, Innis and Roberts get adjoined as in Figure 6.4.15.

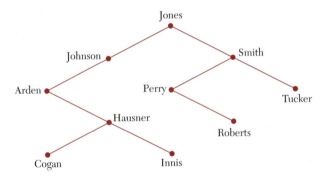

Figure 6.4.15

It is not necessary to actually draw this graph. We can label the vertices as in Example 6.4.5, and thus we have the following labeling of the employees:

Jones: Root	Johnson: 0	Smith: 1	Arden: 00
Hausner: 001	Cogan: 0010	Perry: 10	Tucker: 11
Innis: 0011	Roberts: 101		

These vertex labels remain with the employees as new ones are introduced. Thus, a physical realphabetizing of the names is never done.

EXAMPLE 6.4.16

Let us see how to assign a vertex to Ms. Lanham in Example 6.4.15. She is alphabetically after the root, Johnson, so her code would have to start with 1. Smith is labeled 1, and Lanham is before Smith, so Lanham must begin with 10. Perry is labeled 10, and Lanham is before her, so Lanham must begin with 100. But there is no 100 (yet) in the list. Give it to Lanham. The method finds Lanham if she is there; if she is not, it assigns a vertex to her.

Exercises

1. Find a formula in closed form for the number of strings of length at most *n* of the type described in Example 6.4.3.
2. Do Example 6.4.3 with the proviso that the string must have length exactly 20.
3. How many strings of length at most 20 on the alphabet {1, 2, 3} are there in which each "1", except the last, must be followed by a "2" or a "3"?
4. How many strings of decimal digits are there of length at most 10 in which each even digit is followed by an odd digit?

Exercises 5–9 ask for a count of trees. The graphical method of Example 6.4.3 can be applied, but we mut be careful because we are asking not for a count of leaves but for a count of trees.

*5. How many binary trees are there in which each leaf is on level *n*? Recall that a single child must be identified as left or right, so the answer is 3 when *n* = 1. (*Hint:* Find a recursion that is not linear. Then experiment and prove your guess by induction.)
*6. How many ordered, rooted trees are there in which all leaves are on level 4 and each vertex other than a leaf has either one or two children?
*7. Define a ternary tree as a rooted tree in which each vertex, other than a leaf, has at most three children. In analogy with the binary trees, each child must be labeled left, middle, or right, even if there are fewer than three children. How many ternary trees are there in which each leaf is at level *n*? (*Hint:* Look at Exercise 5 and generalize.)
*8. How many ordered, rooted trees are there in which all leaves are on level 4 and each vertex other than a leaf has either two or three children? (*Hint:* Find a recursion.)
*9. How many ordered, rooted trees are there in which all leaves are on level 4 and each vertex other than a leaf has either one or two children? It is further required that each "only child" that is not a leaf must have two children. By definition, an only child is the child of a vertex whose out-degree is 1. (*Hint:* Find a recursion.)

10. The only leaves on a binary tree are 001, 010, 1110, and 10. Sketch the tree and state how many vertices it has.

11. The vertices of a binary tree consist of all strings of 0's and 1's of length ≤5 in which no three consecutive bits are the same. How many leaves are there? How many vertices?

12. Find the average number of bits per letter for the code developed in Example 6.4.15.

13. In Example 6.4.15, you have a hunch that by coding A as the 1-bit string 0, you will do better than the text coding. Find a decodable labeling in which A is coded in this manner and in which the total weight is minimized subject to this condition. Compare weights with the text solution to see if your hunch was right. (*Hint: A* must be a cluster that is not combined with other clusters until the final cluster is obtained. Therefore treat the symbols B, C, D, and E separately to find their best cluster K. Your answer is $A \oplus K$.)

14. If characters a, b, c, d, and e are the only characters in an alphabet, and they have frequencies 5, 20, 12, 8, and 1 respectively, find a Huffman code for this system. Use the algorithm outlined in the text.

15. For the code developed in Exercise 14, find the average length used to code a character.

16. For the decodable labeling of Figure 6.4.16, decode the message

$$11101011 \quad 11010110 \quad 10000111 \quad 10001110$$

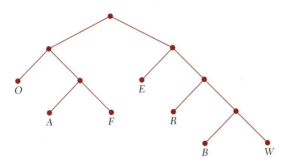

Figure 6.4.16

17. The characters in the following table occur with the indicated frequency (in percent). Devise a Huffman code and determine the average number of bits used per character. Compare this for the best fixed-width code.

Character:	A	R	C	T	E	I	S	X
Frequency (percent):	6	20	4	15	25	10	17	3

18. Using the code created in Exercise 17, code the phrase "ARCTIC SEAS." (Ignore the space.)

19. Suppose variables A_i have frequency n_i and are encoded in some decodable tree. Suppose two characters have the same frequency or

are on the same level. Show that they can be interchanged and the resulting encoding has the same weight.

*20. Suppose a cluster C has frequency N and weight W. Let this cluster be attached to a tree at level L. Prove that the contribution of this cluster to the weight of the tree is $W + LN$.

*21. State and prove the Switching Lemma for clusters. Use the result in Exercise 20. Hence, prove Lemma 6.4.14. Explain why Lemma 6.4.11 is a special case of this result.

22. Let C be a two-leaved cluster as in Figure 6.4.8. Describe the clusters $C \oplus C$, $C \oplus (C \oplus C)$, $C \oplus (C \oplus (C \oplus C))$, and $(C \oplus C) \oplus (C \oplus C)$. (Here we obviously do not assume that the vertices of the clusters are distinct.)

23. Describe the graph of all 3-bit strings in terms of the cluster C of the previous exercise and the operation $\oplus$.

24. Using the binary search tree method indicated in the text, place the words in the following quotation on a binary tree, using alphabetic order to decide precedence: "For every atom belonging to me as good belongs to you." (Ignore capitalization.)

25. The following names are placed on a binary tree in the order given:

> Leonard, George, Carol, Harry, Bill, Tom,
> Alice, John, Frieda, Ellen, Marie, Anthony

Each name is assigned a vertex on the tree. List the names and list each vertex (given as a string of bits) next to the name.

26. Place the following words on a binary tree.

> EVERY THING IS COMING UP ROSES
> EXCEPT FOR THIS ROSE

Find the vertex for each of these words, expressed as a string of 0's and 1's.

27. Suppose two words are put on a binary search tree and each is located at a vertex given by a string of bits. Give an algorithm to determine, in terms of their bit strings, when a word alphabetically precedes another.

28. The following are vertices for names on a binary search tree. Place them in "alphabetic order." Namely, list them in the same order that the names would be listed if they were in alphabetic order:

> 0101, 11010, 0010, 1001, 1, 01, 110, 1110, 111

Summary Outline

- A **path** in a graph is a sequence $v_0, v_1, \ldots, v_n$ of vertices such that the v_i are distinct and $v_{i-1}v_i$ is an edge of G for $i = 1, 2, \ldots, n$.

- A **cycle** in a graph G is a sequence $v_0, v_1, \ldots, v_n$ of vertices, with $n \geq 3$, such that the v_i are distinct ($i = 0, 1, \ldots, n-1$), $v_0 = v_n$, and $v_{i-1}v_i$

is an edge of G for $i = 1, 2, \ldots, n$. (See Chapter 5 for the weaker definitions of path and cycle used there.)

● An **acyclic graph** is a graph that contains no cycles.

● A **tree** is a connected, acyclic graph.

● **Theorem** For any tree G,

1. There is a unique path joining any vertices P and Q in G.
2. If any edge is removed from G, the resulting graph is disconnected and hence is not a tree.
3. If any additional edge is introduced joining two vertices of G, the resulting graph contains a cycle and hence is not a tree.

● **Theorem** A tree with n vertices has $n - 1$ edges.

● **Corollary** If an edge of a tree is removed, the resulting graph is not connected.

● **Theorem** A connected graph with n vertices and $n - 1$ edges is a tree.

● A **spanning subgraph** H in a graph G is a subgraph G with the same vertices of G.

● A **minimal** connected spanning subgraph of a connected graph G, or a **spanning tree,** is a connected spanning subgraph with the further property that if any edge is removed from it the resulting graph is not connected.

● **Theorem** Let G be a connected graph and let H be a subgraph of G with the same set of vertices. Then H is a minimal spanning subgraph if and only if H is a tree.

● **Two methods for constructing spanning trees** Let G be a connected graph.

1. Suppose edges are removed, one at a time, so that the resulting graph always remains connected. (The vertices are not removed.) If this is done until no further edges can be removed in this way, then the resulting graph is a spanning tree for G.
2. Suppose we start with the vertices of G and no edges and adjoin edges of G, one at a time, so that the resulting graph always has no cycles in it. If this is done until no further edges can be adjoined in this way, then the resulting graph is a spanning tree for G.

● **Kruskal's Algorithm** for constructing a spanning tree of minimal weight on a connected, weighted graph G:

1. Order the edges of G in order of increasing weights. In case of a tie, any order of edges with the same weight is permissible.
2. Starting with the vertices of G and no edges, adjoin edges of G in the order fixed by step 1 to form a subgraph H. However, if adjoining

any edge causes a cycle to be formed in H, do not use that edge, but go on to the next edge.

3. Proceed with the adjunctions of step 2 until all edges have been tested.

● **Greedy algorithm**   A many-step algorithm that attempts to optimize each step without regard for the consequences on subsequent steps.

● A **rooted tree** is a tree with one of its vertices singled out. This vertex is called the **root** of the tree.

● The **level** of any vertex v of a rooted tree G is the length n of the unique path from the root v_0 to v. We also say that v is on level n. In particular, v_0 is on level 0.

● **Theorem**   Adjacent vertices on a rooted tree have levels that differ by 1.

● In a rooted tree, a vertex v is the **parent** of another vertex w if they are adjacent and the level of v is 1 less than the level of w. In this case w is the **child** of v.

● A vertex v is a **leaf** if it is not a parent; that is, it has no children.

● **Theorem**   If v has level $n > 0$, there is one and only one parent of v.

● An **ordered rooted tree** is one in which the children of every vertex are given a definite ordering. If v precedes w in the ordering, we write $v <$ w. In this case, if v and w are the only children, we call v and w the **left child** and **right child**, respectively.

● Recursive definition of an **nth-generation descendent** of a vertex A in a rooted tree:
  (a) v is a 0th-generation descendent of A if and only if $v = A$.
  (b) If $n > 0$, v is an nth-generation descendent of A if v is the child of some vertex w that is an $(n - 1)$st-generation descendent of A.

● The vertex v is **a descendent of** A if v is an nth-generation descendent of A for some n.

● The **preorder traversal** of an ordered rooted tree G is a listing of the vertices of G, defined recursively as follows:
  1. If G consists of the vertex A only, then the preorder traversal of G is simply A.
  2. If A has children $B_1, B_2, \ldots, B_k$, we let G_1 be the tree of descendents of B_1, and so on, through G_k. If L_i is the preorder traversal of G_i ($i = 1, \ldots, k$), then the preorder traversal of G is the listing $A, L_1, \ldots, L_k$.

● **Polish notation** for an algebraic formula α is the preorder traversal listing for the tree diagram T of α.

● The **weight** of a vertex v in a rooted tree is $1 - n$, where n is the number of children of v.

- **Theorem**   The sum of the weights of all of the vertices of a rooted tree is 1.

- The **graft** of rooted trees T_i onto a root v: Let $T_1, \ldots, T_k$ be ordered rooted trees with disjoint vertices, let v_i be the root of T_i, and let v be a vertex distinct from all of the vertices in the trees T_i. Then the graft of the trees T_i is the ordered, rooted tree whose root is v such that

  1. v has the children $v_1, \ldots, v_k$ in that order.
  2. The descendents of v_i consist of the tree T_i.

- **Algorithm List/Tree**   An algorithm that converts a preorder traversal list to its tree. The weight of each vertex (an integer ≤ 1) is given.

- **Reverse Polish notation**   A listing of the vertices of an algebraic formula similar to Polish notation, except that operators appear on the right instead of the left.

- **Addition Principle for Counting** (graph-theoretic)   Suppose the rooted trees $G_1, G_2, \ldots, G_k$ are grafted at the vertex A, forming a new tree G. Let n_i be the number of leaves in G_i. Then the number of leaves in G is $n_1 + n_2 + \cdots + n_k$.

- **Multiplication Principle for Counting** (graph-theoretic)   Let G be a rooted tree. Suppose its root has degree n_0 and level 1 vertices each have degree n_1, level 2 vertices have degree n_2, $\ldots$, level k vertices have degree n_k, and level $k + 1$ vertices are all leaves. Then the number of leaves is the product $n_0 \cdot n_1 \cdots n_k$.

- A **binary tree** is an ordered, rooted tree whose vertices have out-degree at most 2. In a binary tree, each child must be labeled left or right, even if there is only one child.

- A **full binary tree** is a binary tree whose vertices have out-degree 0 or 2. Thus, in a full binary tree either a vertex is a leaf or there are two edges leaving it.

- A **complete binary tree** is a full binary tree, all of whose leaves are on the same level.

- **Huffman coding** codes each character of an alphabet as a k-bit integer, where k is variable and depends on the character being coded. It is used to minimize the number of bits needed to code a message.

- A **decodable labeling** is the labeling of the vertices of a binary tree with characters $A_1, A_2, \ldots, A_k$ in such a way that a path from the root to any of these characters does not meet any other character.

- The **weight** of a code is

$$W = \sum n_i L_i$$

where n_i is the frequency of the ith character and L_i is the length of the bit string needed to code it.

- **Switching Lemma** If variables are encoded in such a way that a character A has a smaller frequency than the character B, and the length of the code of A is smaller than the length of the code of B, then by switching the codes for A and B we obtain a coding with a smaller weight.

- A **cluster** is a decodable labeling of some of the characters A_i.

- The **frequency** of a cluster is defined as the sum of the frequencies of the characters in the cluster.

- Clusters are called **disjoint** if they have no characters in common.

- The **graft** of disjoint clusters C_1 and C_2, written $C_1 \oplus C_2$, is obtained by grafting the trees of the clusters at a root and keeping the character assignment in C_1 and C_2. The **frequency** of the graft of C_1 and C_2 is the sum of their frequencies.

- **Lemma** (used to create a Huffman coding) Let clusters $C_1, C_2, \ldots, C_k$ be listed in order of frequency. Then a Huffman code can be formed from the clusters $C_1 \oplus C_2, C_3, \ldots, C_k$.

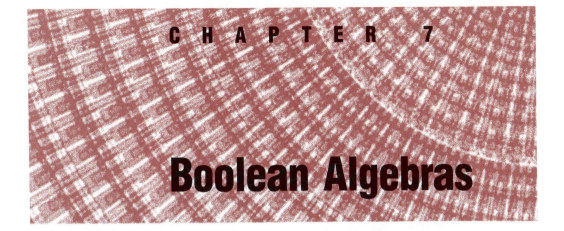

CHAPTER 7

Boolean Algebras

Boolean algebra is a tool for studying and applying logic. As its name implies, it is an algebraic method. Switching circuits, which are used in the design of computer chips, are introduced with a discussion of the techniques needed to simplify their design. We also show how a purely algebraic approach to some parts of logic is possible by using Boolean algebras, and how this technique can also be used to study set theory.

7.1 Switching Circuits

A **switch** or a **gate** is a device with several inputs and one output. Each **input** can be on or off, which are identified with T and F, or 1 and 0, respectively. In what follows, we usually use 1 and 0. The **output** of a switch is either 0 or 1, depending on the values of the inputs and on the switch in question. Thus, a switch with n inputs may be thought of as a black box, with n independent inputs that may be on or off, and one

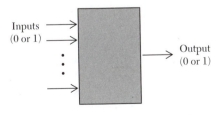

Figure 7.1.1

output, which is on or off depending on the inputs. (See Figure 7.1.1 for a schematic switch.) In more abstract terms, a switch is a function $f(x_1, \ldots, x_n)$ whose value is 0 or 1, where each x_i is either 0 or 1:

$$f: \mathbb{Z}_2 \times \mathbb{Z}_2 \times \cdots \times \mathbb{Z}_2 \to \mathbb{Z}_2$$

EXAMPLE 7.1.1   AND, OR, NAND, NOR, and NOT Gates

Any truth table on the atomic statements $p, q, \ldots$ may be regarded as a switch by simply regarding $p, q, \ldots$ as inputs. For example, the compound statement $p \wedge q$ is a switch that is 1 only when both p and q are 1. This is shown schematically in Figure 7.1.2. This switch is naturally called an **AND gate**. Similarly, we have an **OR gate** and a **NOT gate**, as illustrated. The little circle is used as a NOT gate, and so **NAND** (not and) and **NOR** (not or) are drawn as illustrated. Since the connective $\neg$ is unary, the NOT gate has only one input, as in Figure 7.1.2. The symbols used to represent these gates in the figure are standard.

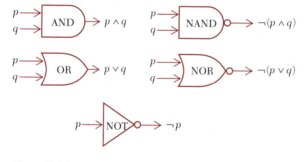

Figure 7.1.2

When several of these AND, OR, and NOT gates are connected in a specific way, the result corresponds to a compound statement. For example, the statement

$$\neg p \wedge (q \vee r) \tag{7.1.1}$$

corresponds to the switching device of Figure 7.1.3. In a similar manner, *any compound statement $P(p, q, \ldots)$ corresponds to a switch with inputs $p, q, \ldots$*. We shall soon see that, conversely, any switch comes from some compound statement.

Figure 7.1.3 is essentially the tree diagram of Figure 7.1.4 of the statement (7.1.1). The method of drawing the tree diagram was described in Example 6.1.4. However, the point of view we are now considering is somewhat different. Besides the cosmetic change (the tree is now lying on its side), we are regarding the leaves p, q, r as inputs that may be

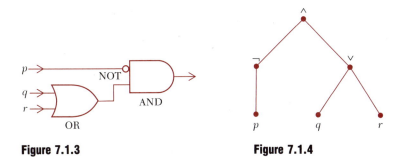

Figure 7.1.3   **Figure 7.1.4**

either 0 or 1. Any parent except for the root is a gate, here either a NOT, an AND, or an OR gate, that determines, depending on the input (its children), whether the output is 0 or 1. Finally, the root corresponds to the output of the switch.

In a switch such as in Figure 7.1.3, we generally omit the arrows for the input and outputs. They are understood to go from left to right.

EXAMPLE 7.1.2   *Logical Equivalence and Tautologies*

From the point of view of switches, the logical equivalence of statements has the following meaning: $P \equiv Q$ if the corresponding switches have the same output for the same inputs.

Similarly, P is a tautology if the corresponding switch is always on, regardless of the inputs.

EXAMPLE 7.1.3

Design a switch for the statement $p \leftrightarrow q$. Thus, the switch is on if and only if the inputs are either both on or both off.

Method   We use the equivalence

$$p \leftrightarrow q \equiv p \wedge q \vee \neg p \wedge \neg q$$

Thus, the required switch is in Figure 7.1.5. The junctures for p and q allow the inputs to flow into the various gates as needed.

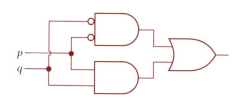

Figure 7.1.5

The following example gives a procedure to construct a statement with a given truth table. The statement will use only the connectives $\vee$, $\wedge$, and $\neg$.

EXAMPLE 7.1.4

Find a statement with the truth table of Table 7.1.1 that uses only the connectives $\vee$, $\wedge$, and $\neg$.

Table 7.1.1
A Switch

p	q	r	S
0	0	0	0
0	0	1	0
0	1	0	1 ←
0	1	1	0
1	0	0	1 ←
1	0	1	0
1	1	0	0
1	1	1	1 ←

Method We look at each of the rows where the output is 1. (These are indicated by the arrow.) The first such row is where $(p, q, r) = (0, 1, 0)$. We now consider the statement $\neg p \wedge q \wedge \neg r$. In this string of ANDs, the NOT ($\neg$) is used on those atomic statements that were 0 in that row. Note that $\neg p \wedge q \wedge \neg r$ is 1 (true) *only when* $(p, q, r) = (0, 1, 0)$ (i.e., for the indicated row only) since these are the only values that make each part of the compound AND statement true. We now do this for each of the other rows to obtain Table 7.1.2.

Table 7.1.2

(p, q, r)	Corresponding Compound AND Statement
$(0, 1, 0)$	$\neg p \wedge q \wedge \neg r$
$(1, 0, 0)$	$p \wedge \neg q \wedge \neg r$
$(1, 1, 1)$	$p \wedge q \wedge r$

The construction is such that each of the compound AND statements *is 1 only for the indicated* (p, q, r); *otherwise it is 0.* But this shows that

$$S = (\neg p \wedge q \wedge \neg r) \vee (p \wedge \neg q \wedge \neg r) \vee (p \wedge q \wedge r)$$

is the required statement. It is 1 for the inputs of each of the indicated rows, because one of the terms in this OR expression is 1 for each row. Further, it is 0 for the other inputs, because all of the terms in this OR statement are 0 in this case.

This method clearly generalizes to any switch. We therefore have the following theorem.

THEOREM 7.1.5

Any switch with n inputs p, q, ... may be expressed as a statement S involving p, q, ..., and the connectives $\vee$, $\wedge$, and $\neg$ only. ■

The method of Example 7.1.4 actually does more because it gives an algorithm to find S, and it finds a specific (and unique) representation of S. We review these in the following definition and theorem.

DEFINITION 7.1.6

The **conjunction** of statements P_1, P_2, ... is the statement $P_1 \wedge P_2 \wedge \cdots$.
The **disjunction** of statements P_1, P_2, ... is the statement $P_1 \vee P_2 \vee \cdots$.
An **atomic statement** on p_1, p_2, ..., p_n, or simply an **atom**, is the conjunction of statements P_1, P_2, ..., P_n, where each P_i is either p_i or $\neg p_i$.

THEOREM 7.1.7

Any switch on the inputs p_1, ..., p_n may be uniquely represented as the disjunction of atoms on p_1, ..., p_n. This representation is called the **disjunctive normal form** (dnf). ■

Thus, the switch of Example 7.1.4 has the dnf

$$S = (\neg p \wedge q \wedge \neg r) \vee (p \wedge \neg q \wedge \neg r) \vee (p \wedge q \wedge r)$$

Each of the three terms $\neg p \wedge q \wedge \neg r$, $p \wedge \neg q \wedge \neg r$, and $p \wedge q \wedge r$ is an atom on p, q, r.

Theorem 7.1.5 may be restated in terms of gates. Suppose we have an unlimited number of AND, OR, and NOT gates. Then any switch can be constructed by using only these gates. This is a practical result. In the construction of electronic equipment, complicated switches must often be designed. It is useful to be able to depend on these basic gates to design any switch.

This result is formalized by the following notion of completeness.

DEFINITION 7.1.8

A set of connectives is called **complete** provided any switch can be expressed as a statement using only these connectives.

In the terminology of this definition, we may restate Theorem 7.1.5 as follows: *The set* $\{\vee, \wedge, \neg\}$ *of connectives is complete.* We can strengthen this result:

THEOREM 7.1.9

The set $\{\neg, \vee\}$ of connectives is complete as is the set $\{\neg, \wedge\}$.

Proof We already know that any switch may be expressed as a statement using $\vee$, $\wedge$, and $\neg$. We now show how $\wedge$ can be eliminated. By De Morgan's Law we have

$$P \wedge Q \equiv \neg(\neg P \vee \neg Q)$$

This shows that any use of $\wedge$ can be replaced by a use of an equivalent form involving $\neg$ and $\vee$. Hence, any statement involving $\vee$, $\wedge$, and $\neg$ is logically equivalent to one involving just $\vee$ and $\neg$. This shows that $\vee$ and $\neg$ form a complete set of connectives. Similarly, the logical equivalence (again, De Morgan's Law)

$$P \vee Q \equiv \neg(\neg P \wedge \neg Q)$$

shows that $\{\wedge, \neg\}$ is a complete set of connectives. ■

Remark It is also true that the connective $\vee$ by itself, or $\wedge$ by itself, is *not* complete. Can you prove this?

EXAMPLE 7.1.10   NAND and NOR Gates

There *are* single connectives that are complete. The NAND gate (for NOT AND), which is indicated by "|," has the truth table given in Table 7.1.3. The "|" is called the **Scheffer stroke.** We may easily verify that

$$p \mid q \equiv \neg(p \wedge q)$$

Sometimes | is written $\uparrow$, which is suggestive of an AND symbol with a line through it.

Table 7.1.3   The NAND Connective |

p	q	$p \mid q$
F	F	T
F	T	T
T	F	T
T	T	F

To show that NAND is complete, we need only show how to form $\wedge$ and $\neg$ by using | alone. But

$$\neg p \equiv p \mid p$$

and

$$p \wedge q \equiv \neg(p \mid q) \equiv (p \mid q) \mid (p \mid q)$$

A similar argument applies to the NOR gate (for NOT OR). This is indicated by $\downarrow$ and is called the **Peirce arrow.** Here,

$$p \downarrow q \equiv \neg(p \vee q)$$

The truth table for the NOR operator $\downarrow$ is given in Table 7.1.4.

Table 7.1.4   The NOR Connective $\downarrow$

p	q	$p \downarrow q$
F	F	T
F	T	F
T	F	F
T	T	F

It will be convenient to make some typographical abbreviations. First, we shall not write the symbol $\wedge$, so a switch such as $p \wedge q \wedge r$ is simply written pqr. (This is similar to algebra, where multiplication signs are dropped by convention.) Next, we write $\neg p$ as p'. This makes the expression for a dnf more legible. Instead of $p \wedge q \wedge \neg r \wedge s$, we simply write $pqr's$. Thus, the switch S of Example 7.1.4, previously listed as

$$S = (\neg p \wedge q \wedge \neg r) \vee (p \wedge \neg q \wedge \neg r) \vee (p \wedge q \wedge r)$$

will now be written as

$$S = p'qr' \vee pq'r' \vee pqr$$

EXAMPLE 7.1.11   A Diagram for a dnf

The truth table for $p \vee q$ has three rows with output 1, and this leads to the dnf for $p \vee q$:

$$p \vee q \equiv (p \wedge \neg q) \vee (\neg p \wedge q) \vee (p \wedge q) \tag{7.1.2}$$

or

$$p \vee q \equiv pq' \vee p'q \vee pq \tag{7.1.3}$$

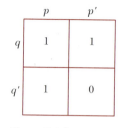

Figure 7.1.6

The truth table for $p \vee q$ may be diagrammed as in Figure 7.1.6. The boxes under p correspond to the inputs where p is on, or 1. Similarly, the boxes under p' correspond to the inputs where p is 0, with a similar convention for the boxes to the right of q and q'. Therefore, these boxes represent all four possible inputs for p and q. The four rows of the truth table for $p \vee q$ now appear as four boxes. In each box, we put the output (0 or 1) corresponding to the inputs represented by the box. Thus, since $p \vee q$ is true, for example, when p and q are, we put the value 1 in the upper left-hand box.

The dnf for $p \vee q$ can be read off directly from Figure 7.1.6. Using this figure, our procedure for finding the dnf, as illustrated in Example 7.1.4, can be neatly summarized as follows: For each box holding a 1, multiply the row and the column labels it is in to find an atom (as in Definition 7.1.6). The dnf is then the disjunction of the atoms found in this way. Of course, the multiplication is really a conjunction, as we have agreed to omit the $\wedge$ sign and use the ordinary multiplication notation. We shall soon show how this procedure works for more complicated switches.

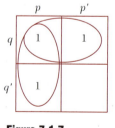

Figure 7.1.7

Figure 7.1.7 shows how to retrieve $p \vee q$ from this figure. The entries below p correspond to the atoms for which p is true. These are one of the ovals. Similarly, the atoms for which q is true are the other oval. Thus, the two ovals correspond to p and to q, and we simply take the disjunction of the statements corresponding to the ovals. If we were to construct a switch for $p \vee q$, it would be quite silly to use its dnf— $p \vee q$ itself is much simpler.

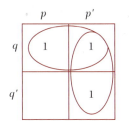

Figure 7.1.8

In much the same way, the switch $p \rightarrow q$ has the dnf

$$pq \vee p'q \vee pq'$$

as in Figure 7.1.8. We usually omit the 0 entries to avoid the clutter. The logical equivalence

$$(p \rightarrow q) \equiv \neg p \vee q$$

can be seen in the figure by noting the two ovals, one under p' and the other to the right of q. The union of these ovals is the set of boxes where 1 appears.

This diagram method works for more than two variables.

EXAMPLE 7.1.12

Construct a simple switch S for Table 7.1.5, using ANDs, NOTs, and ORs.

Table 7.1.5
A Switch

p	q	r	S
0	0	0	1
0	0	1	0
0	1	0	1
0	1	1	1
1	0	0	1
1	0	1	0
1	1	0	1
1	1	1	0

Method   Construct Figure 7.1.9. Each rectangle corresponds to exactly one atom on p, q, and r, as indicated in the figure. The q and the q' columns have each been split in half. Each half has an r and an r' component. We try to place these components next to each other, but we succeed only with r'. It will sometimes be useful to think of the extreme

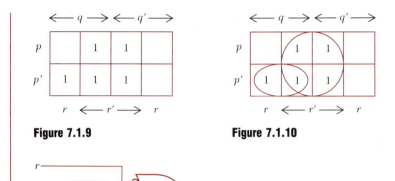

Figure 7.1.9   **Figure 7.1.10**

Figure 7.1.11

right and left columns as being adjacent. (Think of pasting the left and right edges together to form a cylinder.) In this case, the two r columns (actually the rq column and the rq' column) can also be regarded as adjacent. We have put the number 1 in each rectangle corresponding to the row in which $S = 1$. If we now group the rectangles as indicated in Figure 7.1.10 (this is where inspiration comes in), then $S = 1$ on the union of the two indicated ovals. These ovals correspond to the statements r' and $p'q$. That is, the large oval with four boxes consists precisely of the boxes where r is false; it corresponds to r'. The smaller 1 by 2 oval corresponds to the statement $p'q$. Thus, $S \equiv r' \vee p'q$, and we can use the switch as indicated in Figure 7.1.11.

It is possible to extend this method to any number of inputs. We can always introduce another input by doubling the number of rows or col-

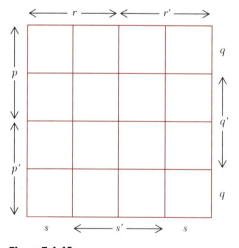

Figure 7.1.12

umns. However, if we have more than four inputs, it becomes impractical to simplify a switch by simply drawing a figure, and it is convenient to use appropriate software to do this work. Figure 7.1.12 gives an appropriate figure for four inputs. If we imagine the first and last columns as adjacent, and similarly the first and last rows, then we again have a situation where each input x and its complement x' occupy adjacent rows or columns. The resulting figures are called **Karnaugh maps**, in honor of their inventor.

EXAMPLE 7.1.13

Figure 7.1.13 shows the boxes corresponding to the inputs p' and s'. The boxes for p' are all the boxes to the right of the label p'. These correspond to all inputs where $p' = 1$ (or $p = 0$). All of these boxes constitute a 2 by 4 rectangle. Similarly, the 4 by 2 rectangle for s' lies above the label s'. The **intersection** of these rectangles is a square (i.e., the 2 by 2 rectangle). These are the boxes corresponding to inputs where p' and s' are true or, equivalently, where $p's'$ is true.

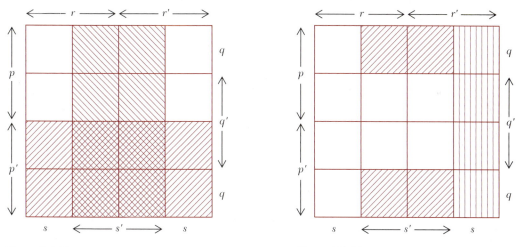

Figure 7.1.13   **Figure 7.1.14**

Figure 7.1.14 similarly gives a square corresponding to qs'. The "square" is actually split into two parts, the upper and lower 1 by 2 rectangles; it is a wrapped-around square. The figure corresponding to $r's$ is not a square; it is a 4 by 1 rectangle.

An intersection of *three* inputs, p, r', and s', is shown in Figure 7.1.15. It consists of the two boxes in the indicated 2 by 1 rectangle. These correspond to all inputs where $p = 1$, $r = 0$, and $s = 0$. (The two possible values of q give the two boxes in the rectangle.) Finally, an individual box corresponds to an atom. The atom $p'q'rs$ is shown in Figure 7.1.15.

In optimizing a switch by cutting down the number of gates used to construct it, we try to write a switch as a union of as few of these rec-

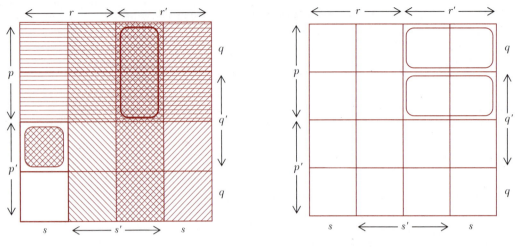

Figure 7.1.15

Figure 7.1.16

tangles as possible, with each rectangle as large as possible. In Figure 7.1.16, the two rectangles corresponding to $pq'r'$ and pqr' are adjacent. If they both showed up in a dnf, we would combine them into the square representing pr'. In general, adjacent rectangles of the same size can be combined into a larger one.

EXAMPLE 7.1.14

Let the dnf for a four-input (p, q, r, and s) switch be

$$pqrs \lor pqr's \lor pq'rs \lor pq'r's' \lor pq'r's$$

Find a simple switch for this expression, using $\land$, $\lor$, and $\lnot$.

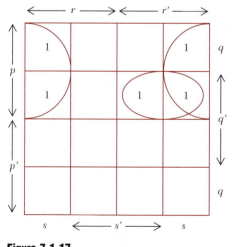

Figure 7.1.17

Method   Diagram this switch as in Figure 7.1.17. By observation, we see that this switch can be constructed with two groupings, one of which is split into two 2 by 1 rectangles representing ps. The other is a 1 by 2 rectangle representing $pq'r'$. Therefore, a simple representation for the given switch is $ps \vee pq'r'$ or, using the unabbreviated notation,

$$(p \wedge s) \vee (p \wedge \neg q \wedge \neg r)$$

A switching arrangement is given in Figure 7.1.18.

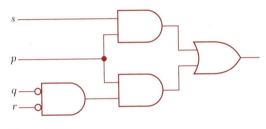

Figure 7.1.18

Exercises

In Exercises 1–5, put the given statement into disjunctive normal form.

1. $p \rightarrow \neg q$
2. $p \wedge (\neg q \vee r)$
3. $p \leftrightarrow \neg r$
4. $(p \wedge q) \vee (r \wedge s)$
5. $p \vee q \vee (\neg p \wedge r)$

6. Find a statement logically equivalent to $p \leftrightarrow \neg r$ that uses only the connectives $\neg$ and $\wedge$.
7. Repeat Exercise 6 for $(p \vee q) \rightarrow r$.
8. Find a statement logically equivalent to $p \leftrightarrow \neg r$ that uses only the connectives $\neg$ and $\vee$.
9. Repeat Exercise 8 for $(p \wedge q) \leftrightarrow r$.
*10. Show that the set of connectives $\{\vee, \wedge\}$ is *not* complete.
*11. Show that the set consisting of $\vee$ alone is not complete.
*12. Find a statement using NANDs ($|$) only that is logically equivalent to $\neg p \wedge q$.
*13. Repeat Exercise 12 for NOR gates.
14. Design a switch, using AND, OR, and NOT gates, that has three inputs and whose output is on if and only if the majority of the inputs are on. Use as few gates as you can.

A switch is known to have the Karnaugh map in each of the following diagrams. An asterisk (*) indicates that any output, 0 or 1, is allowed. A blank box represents the 0 output. For each diagram, design a switch, using as few AND, OR, and NOT gates as possible. It is not necessary to diagram the switch if there are more than three inputs; just give the statement that represents it.

15.

16.

17.

18.

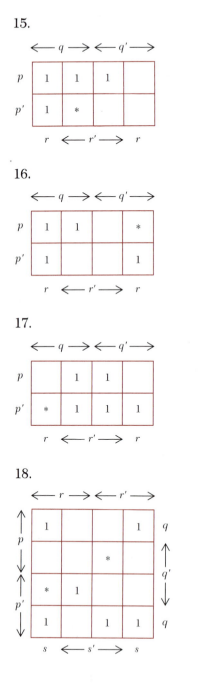

19.

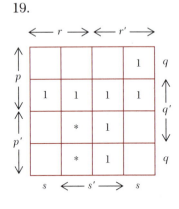

20.

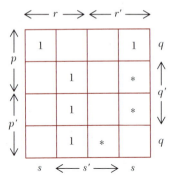

21.

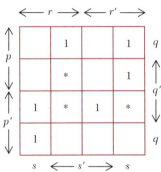

In Exercises 22–28, use Karnaugh maps to find a statement representing the desired switch using as the connectives $\wedge$, $\vee$, and $\neg$. Use as few connectives as possible. It is not necessary to diagram it.

22. Design a switch with inputs p, q, r whose output is 1 if and only if the value of (p, q, r) is one of $(0, 0, 0)$, $(1, 0, 0)$, $(1, 1, 0)$, or $(1, 1, 1)$.
23. Repeat Exercise 22 with four inputs (p, q, r, s). The output is 1 only when the inputs are one of $(0, 0, 1, 0)$, $(0, 0, 1, 1)$, $(1, 0, 1, 0)$, or $(1, 1, 1, 0)$.
24. A switch is on if p and q are on or if q and r are on, and it is off when p, q, and r are off. Design such a switch.
25. Design a switch whose values are given by the dnf $p'qr \vee p'q'r \vee pq'r$.
26. Design a switch whose values are given by the dnf $p'q'rs \vee p'q'rs' \vee p'q'r's' \vee pq'r's'$.
27. Design a switch whose values are given by $p'qr \vee pq'r' \vee p'qr' \vee pq'r$.
28. Design a switch whose values are given by the dnf $pqrs \vee pqrs' \vee p'qrs \vee p'qrs'$.
29. What is the dnf for the statement p? (Assume inputs p, q, and r.)
30. What is the dnf for the NAND operator $p \mid q$? for the NOR operator $p \downarrow q$?
31. What is the dnf for the exclusive or (XOR)? The exclusive or of two statements is true if and only if one and only one of the statements is true.
32. A switch is to be designed as simply as possible with AND, OR, and NOT gates. Its output is on if two of the inputs are on and the other two are off. Its output is off if all the outputs are on or all are off. Otherwise, it is not important what the output is. Find a simple design.
*33. Determine whether NAND is associative. Namely, is $(p \mid q) \mid r \equiv p \mid (q \mid r)$?
*34. Repeat Exercise 33 for NOR ($\downarrow$).

7.2 Boolean Algebra

Mathematical logic was originated by the English mathematician George Boole (1815–1864). It was thought that it might be possible to develop an **algebra** of statements that would help in logic much as ordinary algebra helps in arithmetic. This more algebraic treatment of the subject is now called **Boolean algebra** in honor of its inventor. This algebra is also useful for the treatment of sets as well as for the design of switches, as discussed in the previous section. The version we give was developed in this century and differs somewhat from the original. Its advantage is that the resulting algebra is similar to the school algebra we have all learned. The methods developed in this section give a distinctly more algebraic approach to the propositional calculus.

The logical equivalence $p \wedge q \equiv q \wedge p$ is so reminiscent of the commutative law of algebra ($xy = yx$), that an ordinary algebraic approach to the propositional calculus seems quite reasonable. What we shall do is replace *logical equivalence* with *equality* and decide which of the connectives will act as addition and multiplication. A "natural" selection might be $\vee$ for $+$, $\wedge$ for $\cdot$, and $\neg$ for $-$ (the negative). However, this doesn't lead to ordinary algebraic usage. For example, De Morgan's Law becomes $-(p + q) = (-p)(-q)$. As we shall see, there is another solution. It turns out that the most useful connective to use for addition is XOR.

DEFINITION 7.2.1

The binary connective $p + q$ is defined to be true if and only if p and q have opposite truth values. That is, one must be true, and the other false. This connective is called the **exclusive or** or **XOR** connective. We denote this connective by $+$, in anticipation of its algebraic use. Its truth table is given in Table 7.2.1.

Table 7.2.1   XOR

p	q	$p + q$
F	F	F
F	T	T
T	F	T
T	T	F

The only difference between the connectives OR and XOR is in the last line. An XOR is true only if one *and only one* of the statements it connects is true.

To make the connection with algebra clearer, we continue to use the conventions introduced in the last section. In particular, we use 0 for F and 1 for T, and we use the notation pq or $p \cdot q$ for $p \wedge q$. In addition, if two statements P and Q are logically equivalent, we call them equal and write $P = Q$.

With these conventions, many of our tautologies become similar to some standard algebraic identities. For example, the "commutative law" tautology $p \wedge q \leftrightarrow q \wedge p$, which is equivalent to the equivalence $p \wedge q \equiv q \wedge p$, becomes, simply, $pq = qp$. The equivalence $T \wedge p \equiv p$ now becomes the familiar $1 \cdot p = p$.

Table 7.2.1 itself may be regarded as an "addition table" for 0 and 1. The last line is $1 + 1 = 0$; the other lines are more familiar: $1 + 0 = 0 + 1 = 1$, and $0 + 0 = 0$. Thus, addition in this table is taken mod 2.

We now state the axioms for a general Boolean algebra. We will then be able to use properties of this algebra to obtain results about the propositional calculus, switches, and other topics.

DEFINITION 7.2.2

A **Boolean algebra** B is a set B with at least two distinct elements, 0 and 1, and two operations, $+$ and $\cdot$, from $B \times B \to B$ that satisfies the following axioms. We usually abbreviate $x \cdot y$ as xy.

Axioms for a Boolean Algebra

For all x and y in B:

1. $x + y = y + x$
2. $xy = yx$ (Commutative Laws)

For all x, y, and z in B:

3. $(x + y) + z = x + (y + z)$
4. $(xy)z = x(yz)$ (Associative Laws)
5. $x(y + z) = xy + xz$ (Distributive Law)

For all x in B:

6. $0 + x = x$ and $0x = 0$ (Properties of 0)
7. $1x = x$ (Property of 1)
8. $x + x = 0$; $xx = x$ (Boolean Conditions)

Axioms 1 through 7 constitute some of the most familiar rules of algebra. For example, they are satisfied by the natural numbers $\mathbb{N}$, the integers $\mathbb{Z}$, and the real numbers $\mathbb{R}$. However, nothing in Axioms 1 through 7 guarantees subtraction (the possibility of solving the equation $a + x = b$ for x, given values a and b in B). Nor do they guarantee the possibility of division (the possibility of solving the equation $ax = b$ for x, given $a \neq 0$ and b in B). We can see this since $\mathbb{N}$ does not have unlimited subtraction, and $\mathbb{Z}$ (which does) does not have unlimited division by nonzero elements. Thus, although Axioms 1 through 7 are familiar laws of algebra, we must be careful when using them. Any rule of algebra that uses subtraction or division either makes no sense or might be false.

EXAMPLE 7.2.3

An example where a usual law of algebra fails for a Boolean algebra is as follows. In "ordinary algebra," when $ab = 0$, we can conclude that $a = 0$ or $b = 0$. This fact makes sense in B, but it does not follow from the axioms and is not necessarily true for B. An ordinary algebraic proof would take the equation $ab = 0$ and, assuming $a \neq 0$, divide by a to conclude that $b = 0$. However, we cannot necessarily divide by a nonzero element of B. (See the remark following Theorem 7.2.8.)

EXAMPLE 7.2.4

Subtraction *is* possible in a Boolean algebra. In algebra, when $x + y = 0$, we say that y is the negative of x or $y = -x$. Since $x - y = x + (-y)$, once we have a negative, we will have subtraction. The Boolean condition $x + x = 0$ implies that any x has a negative and, in fact, is its own negative: $-x = x$. The equation

$$x + a = b$$

can be solved for x by adding $-a$ to both sides of the equation. Here, $-a = a$, so we *add a* to both sides:

$$x + a + a = b + a$$

and, simplifying,

$$x = b + a$$

If $+$ were taken as OR $(\vee)$ in the propositional calculus, all of the axioms would be true, with the exception of the first Boolean condition $x + x = 0$. It would have to be replaced by the condition $x + x = x$, and there would be no subtraction in our algebra.

In algebra, any algebraic system satisfying Axioms 1 through 7, and in which any element has a negative, is called a **commutative ring with unity 1.** Thus, a Boolean algebra is, from an algebraic viewpoint, a commutative ring with a unity 1, and it also satisfies the Boolean conditions 8. It is often called a **Boolean ring.**

If the elements $x, y, \ldots$ of the Boolean algebra are interpreted as statements, how are $\neg x$ and $x \vee y$ to be retrieved algebraically? We can restate this question in terms of the propositional calculus. How can we define $\neg p$ and $p \vee q$ in terms of "$+$" and "$\cdot$"? There is a simple answer: $\neg p$ has the opposite truth value of p; but so does $1 + p$ ($1 + 1 = 0$; $1 + 0 = 1$). Therefore,

$$\neg p \equiv 1 + p$$

Similarly, using truth tables, we can verify that

$$p \vee q \equiv p + q + pq$$

We continue to use the notation p' for $\neg p$. We therefore give the following definitions, *valid in any Boolean algebra.*

DEFINITION 7.2.5

If B is a Boolean algebra, we define

$$p' = 1 + p \tag{7.2.1}$$

$$p \vee q = p + q + pq \tag{7.2.2}$$

We have thus algebraically retrieved our other two connectives.

The power of the axioms for a Boolean algebra and the definitions of p' and $p \vee q$ is that they allow us to derive facts about the propositional calculus in a completely familiar algebraic way. The Boolean conditions allow further simplifications.

We now show some of this power by deriving De Morgan's Laws algebraically.

THEOREM 7.2.6   De Morgan

In any Boolean algebra,

$$(xy)' = x' \vee y' \tag{7.2.3}$$

$$(x \vee y)' = x'y' \tag{7.2.4}$$

Proof   To prove (7.2.3), we start with the right-hand side and work algebraically:

$$
\begin{aligned}
x' \vee y' &= x' + y' + x'y' &&\text{[from (7.2.2)]}\\
&= x + 1 + y + 1 + (x + 1)(y + 1) &&\text{[from (7.2.1)]}\\
&= x + y + 1 + 1 + xy + x + y + 1 &&\text{(algebra)}\\
&= xy + 1 &&\text{(Boolean conditions)}\\
&= (xy)' &&\text{[from (7.2.1)]}
\end{aligned}
$$

Similarly, to prove (7.2.4), we start with the right side again and proceed:

$$
\begin{aligned}
x'y' &= (x + 1)(y + 1) &&\text{[from (7.2.1)]}\\
&= xy + x + y + 1 &&\text{(algebra)}\\
&= (x \vee y) + 1 &&\text{[from (7.2.2)]}\\
&= (x \vee y)' &&\text{[from (7.2.1)]}
\end{aligned}
$$

This proves the result.   ■

In a similar way, we can easily prove the theorem on the negation of a negation.

THEOREM 7.2.7

If x is in a Boolean algebra B, then $x'' = x$ and $xx' = 0$.

Proof

$$x'' = (x + 1) + 1$$
$$= x + 1 + 1$$
$$= x + 0 = x$$

We leave the proof of $xx' = 0$ to the reader. ■

Boolean algebras are sometimes defined in terms of the operations xy, $x \vee y$, and x'. In this case, the previous theorems are taken as axioms.

It is also possible to express some of the concepts of **set theory** by using Boolean algebra. Suppose the set I is taken as the universal set, and we consider all subsets A of I. We have defined the intersection $A \cap B$ and the union $A \cup B$ of two subsets. It is convenient to define the **symmetric difference** $A + B$ of two sets A and B by the formula

$$A + B = A \cup B - A \cap B$$

Thus, x is in $A + B$ if and only if x is in A or in B but not in both. This set is illustrated in Figure 7.2.1. (This set operation was also considered in Exercises 4.1.30–4.1.34, where it was called $A \triangle B$.)

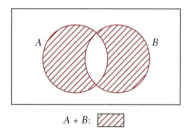

$A + B$: ▨

Figure 7.2.1

Now, if we simply write $A \cap B$ as AB, we see that $A \cap B = A + B + AB$, and we have a strong analogy of $\cup$ with $\vee$, and $\cap$ with $\wedge$. More precisely, we have the following theorem.

THEOREM 7.2.8

Let I be a set and let B be the subsets of I. Then B is a Boolean algebra (called the Boolean algebra of subsets of I) where addition and multiplication are taken as the symmetric difference and the intersection, respectively, and where 0 and 1 are taken as the null set $\varnothing$ and the universal set I, respectively.

Proof   The proof is a verification of Axioms 1 through 8 of Definition 7.2.2. Note, in particular, that the Boolean conditions are, in the language of set theory,

$$A + A = \varnothing; \qquad A \cap A = A$$

The properties of 1 and 0 are simply

$$A + \varnothing = A; \qquad A \cap \varnothing = \varnothing; \qquad A \cap I = A$$

We leave some further verifications to the exercises. ■

Theorem 7.2.8 shows that Boolean algebra is powerful enough to discuss set theory as well as the propositional calculus. For example, the definition $x' = x + 1$ interprets x' as $\neg x$, when x is a statement. If A is a *set*, how is A' interpreted? By definition, $A' = A + I$ is the set of points of I that are not in A—the **complement** of A. Similarly, the set $A \vee B$ is defined as $A + B + AB$, and this is the **union** of the sets A and B. Thus, our one algebraic system is able to interpret and analyze set theory as well as the propositional calculus.

Example 7.2.3 can be illustrated for sets. If $AB = \varnothing$, then A and B are disjoint sets. We clearly can't conclude that either A or B is the null set $\varnothing$.

We have discussed the notion of **inclusion** of sets: $A \subseteq B$. For sets we can show that $A \subseteq B$ if and only if $A \cap B = A$. This can be generalized to any Boolean algebra.

DEFINITION 7.2.9

In any Boolean algebra B, we define $x \subseteq y$ to mean that $xy = x$.

We can relate this definition to the propositional calculus.

THEOREM 7.2.10

Let P and Q be statements. Then $P \rightarrow Q$ is a tautology if and only if $P \subseteq Q$.

Proof   First suppose that $P \rightarrow Q$ is a tautology. This means that we can never have P true (1) and Q false (0). But then $P \wedge Q$ must always have the same truth value as P. This is certainly true if P is 0, because then $P \wedge Q$ is 0. It is equally true if P is 1, for then Q must be 1 and so $P \wedge Q$ is 1. Thus, $P \wedge Q \equiv P$ and, by our convention, $PQ = Q$. But in Boolean algebra terms, this says that $P \subseteq Q$.

Conversely, if $P \subseteq Q$, we have $P \wedge Q = P$, by Definition 7.2.9. Thus, we can never have $P = 1$ and $Q = 0$. Thus $P \rightarrow Q$ must always be 1 and is a tautology, by definition. ■

Remark   By Theorem 1.4.4, we have $P \subseteq Q$ if and only if P logically implies Q.

We now illustrate how Definition 7.2.9 can be used to prove, algebraically, some familiar facts about sets.

EXAMPLE 7.2.11

In a Boolean algebra B, prove

$$x \subseteq x \vee y \tag{7.2.5}$$

$$\text{If } x \subseteq y, \text{ then } y' \subseteq x' \tag{7.2.6}$$

$$\text{If } x \subseteq y \text{ and } y \subseteq z, \text{ then } x \subseteq z \tag{7.2.7}$$

Method   To prove (7.2.5), we compute

$$x(x \vee y) = x(x + y + xy)$$
$$= xx + xy + xxy$$
$$= x + xy + xy = x$$

by algebra. But this proves (7.2.5).
    To prove (7.2.6), suppose $x \subseteq y$. Then

$$xy = x$$

by definition. But then we have, by algebra,

$$x'y' = (x + 1)(y + 1)$$
$$= xy + x + y + 1$$
$$= x + x + y + 1$$
$$= y + 1 = y'$$

This proves (7.2.6). We now prove (7.2.7). Since $x \subseteq y$, we have $xy = x$. Similarly, $yz = y$. Therefore $xz = xy{\cdot}z = x{\cdot}yz = xy = x$. Therefore $x \subseteq z$.

Remark   Paraphrasing (7.2.6), complementation reverses inclusions. This is a set-theoretical version of the contrapositive law, which states that $\neg(p \rightarrow q) \equiv \neg q \rightarrow \neg p$. Condition (7.2.7) shows that $\subseteq$ is a transitive relation in a Boolean algebra.

The following example is taken from "A Selection from Symbolic Logic," by Lewis Carroll, the famous author of *Alice in Wonderland*. The example is given exactly as stated.

EXAMPLE 7.2.12

(1) Babies are illogical;
(2) Nobody is despised who can manage a crocodile;
(3) Illogical persons are despised.
   Univ. "persons"; a = able to manage a crocodile; b = babies; c = despised; d = logical.

Explanation   Carroll gives three premises and desires a conclusion (which is unstated). He suggests taking, as the universal set, the set of persons. Similarly, he suggests taking a as the set of persons who are able to manage a crocodile, b as the set of babies, and so on.

Method   We write (1)–(3) as statements of set theory. They become

$$b \subseteq d' \tag{7.2.8}$$

$$a \subseteq c' \tag{7.2.9}$$

and

$$d' \subseteq c \tag{7.2.10}$$

Although there are several consequences of these premises, presumably the one Carroll wants is the one involving a and b. Thus, eliminate d from (7.2.8) and (7.2.10) and then eliminate c. This is straightforward. From (7.2.8) and (7.2.10) (using transitivity), we obtain

$$b \subseteq c \tag{7.2.11}$$

Taking the complement of (7.2.9), we get $c \subseteq a'$. Combining with (7.2.11) by using transitivity, we get $b \subseteq a'$. Freely translating into English, the required answer is "Babies are not able to manage a crocodile." (Carroll's answer is "Babies cannot manage crocodiles.")

Carroll gives 60 such exercises, of which this is the first. Many of them involve more than transitivity and complementation. He also gives answers. More of his problems are given in the exercises.

EXAMPLE 7.2.13

It is not surprising that Boolean algebra covers both set theory and propositional calculus, since (as can be seen by the preceding example) there is a relation between the subjects. Thus,

x is in $A \cup B$ if and only if x is in A OR x is in B.

x is in $A \cap B$ if and only if x is in A AND x is in B.

x is in $I - A$ if and only if x is NOT in A.

x is in $A + B$ (in one but not the other) if and only if x is in A XOR x is in B.

We conclude this section by looking at the disjunctive normal form in an algebraic way and interpreting this for sets.

DEFINITION 7.2.14

If $x_1, x_2, \ldots, x_n$ are elements in a Boolean algebra, an **atom** on these elements is a **nonzero product** $x = z_1 z_2 \cdots z_n$, where z_i is equal to either x_i or x_i'.

This is consistent with our previous use of the term in Definition 7.1.6. The atoms (on $x_1, \ldots, x_n$) satisfy some elementary properties, given in the following theorem.

THEOREM 7.2.15

Let $x_1, \ldots, x_n$ be elements of a Boolean algebra. Then there are at most 2^n atoms on these elements. Further, suppose $x = z_1 z_2 \cdots z_n$ and $y = w_1 w_2 \cdots w_n$, where each z_i and w_i is either x_i or x_i'. Then, unless $z_i = w_i$ for each $i = 1$ to n, we have $xy = 0$.

Proof By definition, an atom x is a product of factors z_i, where each z_i can be one of two choices, x_i or x_i'. Thus there can be at most 2^n atoms. (There might be less if one of these products is 0.)

Different atoms x and y have product 0. For if x and y are as before, and not identical, they must differ in one factor, z_i. One of the products must have a factor x_i, the other x_i'. Thus, xy will have the factor $x_i x_i' = 0$, so $xy = 0$.

The *nonzero* products $x = z_1 z_2 \cdots z_n$, for different z_i, are all distinct. For if x and y were two such nonzero products with $x = y$, then if they differed in one factor, the preceding result shows that $xy = 0$. But if $x = y$, then, multiplying by x, we obtain $xx = 0$, so x must equal 0, which is a contradiction. ■

Figure 7.2.2 gives the familiar Venn diagram for three sets A, B, and C. In this "general" situation, the universe is divided into $8 (= 2^3)$ pieces. As illustrated in the figure, each of these pieces is an atom on the sets A, B, and C. In this diagram, $A \cap B$ is the union of the two atoms ABC and ABC'. This is the "disjunctive normal form" for AB. (We omit the intersection sign $\cap$ according to our previous convention.)

It is easy to consider three sets where there are fewer than eight atoms on them. Figure 7.2.3 gives some of the possibilities.

The following theorem gives an algebraic setting for the disjunctive normal form.

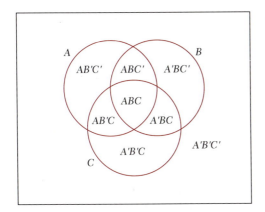

Figure 7.2.2

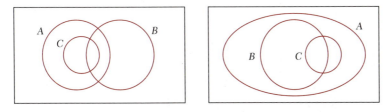

Figure 7.2.3

THEOREM 7.2.16

Let $x_1, x_2, \ldots, x_n$ be elements of a Boolean algebra B. Let P be any algebraic expression in the x_i. That is, P can be formed from the x_i using multiplications and additions, as well as the constants 0 and 1. Then P can be written uniquely as the sum of different atoms in the x_i. This is called the disjunctive normal form for P.

Proof   Using algebra and the Boolean conditions, we can simplify P to a sum of terms, each of which is a product of some of the x_i. In such a product, if any variable x_j is missing, simply multiply it by $1 = x_j + x_j'$. In this manner, each term is a sum of products $y_1 y_2 \cdots y_n$, where each y_i is either x_i or x_i'. If any such product appears twice in the sum, it can be eliminated because $a + a = 0$. Finally, we can eliminate the terms that are 0. We thus obtain P as the sum of distinct atoms.

Uniqueness is obtained as follows. Suppose

$$P = a_1 + a_2 + \cdots = b_1 + b_2 + \cdots \qquad (7.2.12)$$

where each of the a_i and b_i are atoms, and the a_i and the b_i are all distinct. We now show that a_1 must equal one of the b_j. To do this, suppose none of the b_i are equal to a_1. Multiply this equation by a_1 and use the fact that different atoms have product 0. This yields the equation $a_1 = 0$, which is a contradiction, because a_1 is an atom. Thus, a_1 — one of the b_i. By rearranging terms, we may suppose $a_1 = b_1$. Now add a_1 to Equation (7.2.12) to obtain

$$a_2 + \cdots = b_2 + \cdots$$

Continuing in this way, we find that the a_i's are in 1-1 correspondence with the b_j's. (A more formal proof uses an induction on the number of a_i's.) ■

In the propositional calculus, the disjunctive normal form expresses a proposition as a **disjunction** of atoms, that is, as the union of atoms. But a union is the same as an addition for distinct atoms a and b. Because $ab = 0$, we have

$$a \vee b = a + b \;\mid\; ab = a + b + 0 = a + b$$

Similarly, a sum of distinct atoms is the same as their union.
We illustrate Theorem 7.2.16 by an example.

EXAMPLE 7.2.17

Write $(x \vee yz')'$ in dnf.

Method   We work algebraically:

$$
\begin{aligned}
(x \vee yz')' &= x \vee yz' + 1 \\
&= x + yz' + xyz' + 1 \\
&= x' + yz' + xyz' \\
&= x'(y + y')(z + z') + (x + x')yz' + xyz' \\
&= x'yz + x'yz' + x'y'z + x'y'z' \\
&\quad + xyz' + x'yz' + xyz' \\
&= x'yz + x'y'z + x'y'z'
\end{aligned}
$$

Other methods are possible. For example, after the third equal sign, we could simplify: $yz' + xyz' = (1 + x)yz' = x'yz'$. We have the full power of ordinary algebra at our disposal, so shortcuts are often available.

Exercises

1. Verify, using truth tables, that in the propositional calculus,

$$p \wedge (q + r) \equiv (p \wedge q + p \wedge r)$$

[This translates into the Boolean identity $p(q + r) = pq + pr$.]

2. The law of the excluded middle $(p \vee \neg p)$ is tautology 1 on our list of tautologies (Table 1.3.1). What is the corresponding algebraic statement in a Boolean algebra B?

3. Repeat Exercise 2 for the tautology $\neg(p \wedge \neg p)$.

4. In any Boolean algebra, show that if $a + b = a + c$, then $b = c$.

5. In any Boolean algebra, show that $a(b \vee c) = ab \vee ac$.

6. In any Boolean algebra, show that $a \vee bc = (a \vee b)(a \vee c)$.

*7. In any Boolean algebra, show that $a \vee (b \vee c) = (a \vee b) \vee c$.

*8. In any Boolean algebra, show that $a \subseteq b$ if and only if $b = a \vee b$.

9. What algebraic expression in a Boolean algebra corresponds to the statement $p \rightarrow q$ in the propositional calculus?

10. Consider the tautologies $p \rightarrow$ T and F $\rightarrow p$. What are the corresponding algebraic identities for these in a Boolean algebra?

*11. What algebraic expression in a Boolean algebra corresponds to the statement $p \leftrightarrow q$?

*12. Consider the tautology $p \wedge q \rightarrow p$. What is the corresponding algebraic identity in a Boolean algebra?

The following exercises are numbered 13, 30, and 40 in Lewis Carroll's "A Selection from Symbolic Logic."

13. (1) All humming-birds are richly coloured;
    (2) No large birds live on honey;
    (3) Birds that do not live on honey are dull in colour.
    Univ. "birds"; a = humming-birds; b = large; c = living on honey; d = richly coloured.

14. (1) Things sold in the street are of no great value;
    (2) Nothing but rubbish can be had for a song;
    (3) Eggs of the Great Auk are very valuable;
    (4) It is only what is sold in the streets that is really *rubbish*.
    Univ. "things"; a = able to be had for a song; b = eggs of the Great Auk; c = rubbish; d = sold in the street; e = very valuable.

15. (1) No kitten, that loves fish, is unteachable;
    (2) No kitten without a tail will play with a gorilla;
    (3) Kittens with whiskers always love fish;
    (4) No teachable kitten has green eyes;
    (5) No kittens have tails unless they have whiskers.
    Univ. "kittens"; a = green-eyed; b = loving fish; c = tailed; d = teachable; e = whiskered; h = willing to play with a gorilla.

16. Prove that the Boolean algebra Axiom 1 is valid for the subsets of a universal set I, using our interpretation of "$+$" and "$\cdot$."

17. Verify Axiom 5 for the subsets of a universal set I.

18. Let A, B, C be sets. Give a Venn diagram for $(A + B) + C$. Complete this sentence as simply as possible: "x is an element of $(A + B) + C$ if and only if x is . . . ."

19. Verify Axiom 3 for sets. (*Hint:* Use the result of the previous exercise.)

20. Prove in any Boolean algebra that if $ab = 1$, then $a = b = 1$.
21. Prove in any Boolean algebra that $ab \subseteq a$.
22. Prove in any Boolean algebra that $xx' = 0$.
23. Write $xy' + z$ in disjunctive normal form.
24. Write $xy' + yz' + zx'$ in disjunctive normal form.
25. Write $x \vee y'z \vee x'yz$ in disjunctive normal form.
*26. (*Algebra of Inequalities*) In a Boolean algebra, prove the following:
    (a) If $a \subseteq b$, then $ax \subseteq bx$.
    (b) If $a \subseteq b$ and $bx = 0$, then $a + x \subseteq b + x$.
    (c) If $a \subseteq b$, then $a \vee x \subseteq b \vee x$.
27. Prove: In a Boolean algebra, if $a \subseteq b$, then $a \vee (a + b) = b$.
*28. Prove: In a Boolean algebra, if $a \subseteq b$, then there is a unique x such that $a \vee x = b$ and $ax = 0$.
*29. Show that the equation $0 \cdot x = 0$ of Axiom 6 follows from the rest of the axioms. (*Hint:* Use the Boolean conditions.)
*30. Show that the Boolean algebra Axiom 1 (the commutative law for addition) follows from the others. (*Hint:* Use the Boolean conditions.)

Exercises 31–40 discuss a Boolean algebra B as a partially ordered set $(B, \subseteq)$. You may use the results in the previous set of exercises to do any one of these exercises.

31. Prove that a Boolean algebra B is a partially ordered set, using the relation $\subseteq$.
*32. Show that in a Boolean algebra, any two elements a and b have a least upper bound c. This is an element c satisfying

$$a \subseteq c \quad \text{and} \quad b \subseteq c$$

$$\text{If } a \subseteq x \text{ and } b \subseteq x \text{ then } x \subseteq c$$

(*Hint:* Try the union $c = a \vee b$.)
*33. Show that in a Boolean algebra any two elements a and b have a greatest lower bound d. This is an element d satisfying

$$d \subseteq a \quad \text{and} \quad d \subseteq b$$

$$\text{If } x \subseteq a \text{ and } x \subseteq b \text{ then } x \subseteq d$$

*34. If a and b are in a Boolean algebra, write $a < b$ for the covering relation "a is covered by b." (See Definition 2.1.28.) What is the set-theoretic description of the statement that $A < B$ for sets A and B? What is the set-theoretic description of the statement that $\varnothing < A$?
*35. Prove that if $a < b$ in a Boolean algebra, then $0 < a + b$.
*36. Prove that if $0 < a$ and $ab = 0$, then $b < a + b$.

In the following exercises, B is a finite Boolean algebra. Say that an element a is **small** if $0 < a$. (Check the meaning of "small" for sets, as a guide in the following exercises.)

*37. Prove that if a and b are small, then either $a = b$ or $ab = 0$.

*38. Let $a_1, \ldots, a_n$ be the set of small elements of B. Show that the elements

$$a_{n_1} + a_{n_2} + \cdots + a_{n_k}, \quad 1 \le n_1 < n_2 < \cdots < n_k \le n$$

are all distinct.

*39. Show that the number of elements in a finite Boolean algebra is a power of 2. (Use the results of the previous exercises.)

*40. Suppose a Boolean algebra is generated by elements $x_1, \ldots, x_n$. This means that every element of B is an algebraic expression in the x_i's. (See Theorem 7.2.16 for a definition.) Show that the *small* elements are the same as the atoms on the elements x_i.

Summary Outline

● A **switch** is a device with several inputs and one output. Each **input** can be on or off or, equivalently, T and F, or 1 and 0, respectively. The **output** of a switch is 0 or 1, depending on the values of the inputs and on the switch in question. Any compound statement $P(p, q, \ldots)$ of the propositional calculus corresponds to a switch with inputs $p, q, \ldots$.

● **Theorem**   Any switch on with n inputs $p, q, \ldots$ may be expressed as a statement S involving $p, q, \ldots$ and the connectives $\vee, \wedge$, and $\neg$ only.

● The **conjunction** of statements $P_1, P_2, \ldots$ is the statement $P_1 \wedge P_2 \wedge \cdots$. The **disjunction** of statements $P_1, P_2, \ldots$ is the statement $P_1 \vee P_2 \vee \cdots$.

● An **atomic statement** on $p_1, p_2, \ldots, p_n$, or simply an **atom,** is the conjunction of statements $P_1, P_2, \ldots, P_n$, where each P_i is p_i or $\neg p_i$.

● **Theorem**   Any switch on the inputs $p_1, \ldots, p_n$ may be uniquely represented as the disjunction of atoms on $p_1, \ldots, p_n$. This representation is called the **disjunctive normal form** (dnf).

● A set of connectives is called **complete** if any switch can be expressed as a statement by using only these connectives.

● **Theorem**   The connectives $\neg$ and $\vee$ are complete; the connectives $\neg$ and $\wedge$ are complete.

● A **Karnaugh map** is a diagram of a switch that shows the atomic parts of its disjunctive normal form. It can sometimes be used to rewrite the switch into a simpler form.

● The **exclusive OR,** denoted by $+$ or XOR, is a binary logical connective. The statement $p + q$ is true if and only if p and q have opposite truth values.

● A **Boolean algebra** B is a set B with at least two distinct elements, 0 and 1, and two operations, $+$ and $\cdot$, from $B \times B \to B$ that satisfies the following axioms. (The expression xy is an abbreviation of $x \cdot y$.)

1. $x + y = y + x$
2. $xy = yx$ (Commutative Laws)
3. $(x + y) + z = x + (y + z)$
4. $(xy)z = x(yz)$ (Associative Laws)
5. $x(y + z) = xy + xz$ (Distributive Law)
6. $0 + x = x$ and $0 \cdot x = 0$ (Properties of 0)
7. $1x = x$ (Property of 1)
8. $x + x = 0; xx = x$ (Boolean Conditions)

- **Negation** and **disjunction** are defined in any Boolean algebra:

$$p' = 1 + p$$

$$p \lor q = p + q + pq$$

- **Inclusion** is defined in any Boolean algebra:

$$x \subset y \text{ means that } xy = x.$$

- **De Morgan's Laws**   In any Boolean algebra,

$$(xy)' = x' \lor y'$$

$$(x \lor y)' = x'y'$$

- Some **properties of inclusion** in a Boolean algebra:

$$x \subseteq x \lor y$$

$$\text{If } x \subseteq y \text{ then } y' \subseteq x'$$

$$\text{If } x \subseteq y \text{ and } y \subseteq z \text{ then } x \subseteq z \quad (transitivity)$$

- **Double negative**   If x is in a Boolean algebra B, then $x'' = x$.

- The **symmetric difference** of sets A and B is $A \cup B - A \cap B$.

- **Theorem** (algebra of sets)   The collection B of all subsets of a given set I is a Boolean algebra (called the Boolean algebra of subsets of I. Addition and multiplication are taken as the symmetric difference and the intersection, respectively. The elements 0 and 1 are the null set $\emptyset$ and the universal set I, respectively.

- An **atom** on the elements $x_1, x_2, \ldots, x_n$ in a Boolean algebra is a nonzero product $x = z_1 z_2 \cdots z_n$, where z_i is equal to x_i or x_i'.

- **Theorem**   Let $x_1, \ldots, x_n$ be elements of a Boolean algebra. Then there are at most 2^n atoms on these elements. Further, suppose $x = z_1 z_2 \cdots z_n$ and $y = w_1 w_2 \cdots w_n$, where each z_i and w_i is x_i or x_i'. Then, unless $z_i = w_i$ for each $i = 1$ to n, we have $xy = 0$.

- **Theorem**   Let $x_1, x_2, \ldots, x_n$ be elements of a Boolean algebra B. Let P be any algebraic expression in the x_i; that is, P can be formed from the x_i using multiplications and additions, as well as the constants 0 and 1. Then P can be written uniquely as the sum of different atoms in the x_i. This representation is called the **disjunctive normal form** for P.

CHAPTER 8

Number Theory

What used to be the extreme example of "pure" mathematics has turned out to be useful in many fields, including cryptography, applied mathematics, and computer science. This chapter introduces the theory of divisibility, congruence, prime numbers, orders of numbers modulo a prime, and the Chinese Remainder Theorem. Some applications include random number generation, cryptography, error-correcting techniques, and parallel computation.

8.1 Divisibility and the Euclidean Algorithm

In this section, we develop some divisibility properties of the integers. This theory is quite old, and much of it appears in Euclid's *Elements*. The Greeks regarded the integers as a way of measuring lengths. Once a unit u of length was found, any length nu, where n is a positive integer, was conceived in geometric terms much as a foot is 12 "units of inch." Thus, divisibility (of lengths) was important, because it showed which lengths could be used as a unit relative to another length.

As usual, we let $\mathbb{Z} = \{0, 1, 2, \ldots, -1, -2, \ldots\}$ denote the set of integers. Unless otherwise stated, all variables such as x and y are integers. If b/a is an integer, then we say that a *divides* b.

DEFINITION 8.1.1

For a and b in $\mathbb{Z}$, a **divides** b, or b is a **multiple of** a, if $b = aq$ for some integer q. We write $a \mid b$. In the language of predicate logic ($\mathbb{Z}$ = the universe),

$$a \mid b \equiv \exists q (b = aq)$$

The following lemma summarizes some elementary properties of divisibility.

LEMMA 8.1.2

1. $1 \mid a$ for all a
2. $a \mid 0$ for all a
3. $a \mid a$ for all a
4. If $a \mid b$ and $b \mid c$, then $a \mid c$
5. If $a \mid b$, then $ax \mid bx$
6. If $xa \mid xb$, $x \neq 0$, then $a \mid b$
7. If $a \mid b$ and $a \mid c$, then $a \mid (bx + cy)$
8. If $a \mid b$, then either $b = 0$ or $\mid a \mid \, \leq \, \mid b \mid$

Proof The proofs are all straightforward. For example, to prove statement 7, if $a \mid b$ and $a \mid c$, then $b = aq_1$ and $c = aq_2$. Thus, $bx + cy = a(xq_1 + yq_2)$, which shows that $a \mid bx + cy$. ■

To illustrate,

1. Since $3 \mid 6$ and $3 \mid 27$, we have by statement 7 of Lemma 8.1.2 that $3 \mid (6x + 27y)$.
2. If x is even ($2 \mid x$), then we have by statement 5 of Lemma 8.1.2 that $10 \mid 5x$ and $8 \mid 4x$.

According to the definition, $0 \mid x$ provided $x = 0 \cdot q$ for some q. But this is possible when and only when $x = 0$. Thus,

$$0 \mid x \quad \text{iff} \quad x = 0$$

This property of 0 is put in for completeness. In fact, most of our results will concern $a \mid b$, where $a \neq 0$.

The next theorem is the very well known result that any division yields a quotient and a remainder. Geometrically, if we take away enough integer b units from a, we arrive at a remainder that is less than b.

THEOREM 8.1.3   The Division Algorithm

Let a and b be integers, with $b > 0$. Then there are two numbers q and r (called the quotient and the remainder) such that

$$a = bq + r, \qquad 0 \leq r < b \tag{8.1.1}$$

The numbers q and r are unique. ■

Remark   Since $a/b = q + r/b$ and $0 \leq r/b < 1$, we have the formulas $q = [a/b]$ and $r = a - bq = a - b[a/b]$.

The Division Algorithm will be the basis for much of our study of the theory of numbers. It was given as Theorem 2.5.2, where it was used to find the base b representation of an integer. See also Examples 3.2.24 and 3.2.25. Some proofs of this theorem are outlined in Exercises 8.1.5, 8.1.6, and 8.1.7.

The condition $0 \leq r < b$ can be written $r = 0, 1, \ldots,$ or $b - 1$, or simply $r \in \mathbb{Z}_b$. These numbers are also called the remainders modulo b. A good part of grade school arithmetic was, alas, devoted to finding q and r for given a and b. Most computer languages have built-in functions to find the quotient and the remainder.

EXAMPLE 8.1.4

Some examples of division are

(a) Divide 13 by 3: $13 = 3{\cdot}4 + 1$, so $q = 4$ and $r = 1$.
(b) Divide 91 by 7: $91 = 7{\cdot}13 + 0$, so $q = 13$ and $r = 0$. Here, $7 \mid 91$, because $r = 0$.
(c) Divide -7 by 3: $-7 = 3(-3) + 2$: $q = -3$ and $r = 2$ (*not* $-7 = 3(-2) + -1$, because -1 is not in $\mathbb{Z}_3$).

DEFINITION 8.1.5

Let a and b be integers, not both 0. The greatest common divisor d of a and b, written $d = \text{GCD}(a, b)$, is the largest integer x that divides both a and b.

The notation (a, b) is also used for $\text{GCD}(a, b)$.

It follows, by definition, that if $d = \text{GCD}(a, b)$ and if $c \mid a$ and $c \mid b$, then $c \leq d$. (By definition, d is the greatest common divisor, and since c is a common divisor we must have $c \leq d$.) Later we show more, namely that if $c \mid a$ and $c \mid b$, then $c \mid d$. Some simple illustrations follow:

$$\text{GCD}(15, 24) = 3; \quad \text{GCD}(31, 43) = 1;$$

$$\text{GCD}(16, 48) = 16; \quad \text{GCD}(0, 35) = 35$$

How does one go about finding $\text{GCD}(a, b)$? If $a = 0$, this is simply b, and similarly if $b = 0$. So we may assume that a and b are both nonzero. One way of proceeding (the most direct and time-consuming way) is to test the numbers from 1 through the smaller of $\mid a \mid$ and $\mid b \mid$, keeping track of the largest of the common divisors of a and b. When the search is over, $\text{GCD}(a, b)$ will be found. This method simply uses the definition. This algorithm is as follows.

EXAMPLE 8.1.6   Long Search Algorithm to Find GCD

With inputs a and b, $0 < b \le a$, find output GCD $=$ GCD(a, b).

LongSearch$(a, b;$ GCD$)$
1. IF $a = b$ THEN [GCD is obviously a]
    1. GCD $:= a$
    2. End
2. GCD $:= 1$ [initializing]
3. FOR $i := 2$ to b [check all possible common divisors]
    1. IF $i \mid a$ and $i \mid b$ THEN
    2. GCD $:= i$ [a common divisor]

This algorithm is never used, though it is a direct implementation of the definition. Its time has order of magnitude b, assuming that division takes no time. We shall see at the end of this section that the worst-time case for the Euclidean Algorithm has time $O(\log b)$.

A more effective and speedier method for finding GCD(a, b) is based on the Division Algorithm. (This method is due to Euclid and was published about 350 B.C.) At each step, we divide the larger number by the smaller. We then replace the larger of the numbers by the smaller, and the smaller of the numbers by the remainder. The process terminates when the remainder is 0. We first illustrate with a numerical example.

EXAMPLE 8.1.7

Find GCD$(851, 1147)$.

Method

$$\underline{1147} = \underline{851} \cdot 1 + \underline{296}$$
$$\underline{851} = \underline{296} \cdot 2 + \underline{259}$$
$$\underline{296} = \underline{259} \cdot 1 + \underline{37}$$
$$\underline{259} = \underline{37} \cdot 7 + \underline{0}$$

The last divisor is the GCD. Thus,

$$\text{GCD}(851, 1147) = 37$$

The algorithm ignores the quotients that are found along the way.

The proof that this method works is based on the following lemma.

LEMMA 8.1.8

Suppose $a = bq + r$. Then

1. Any common divisor of a and b is also a common divisor of b and r.
2. Any common divisor of b and r is also a common divisor of a and b.
3. $GCD(a, b) = GCD(b, r)$.

Proof

1. If $d \mid a$ and $d \mid b$, then $d \mid (a - bq)$, or $d \mid r$.
2. If $d \mid b$ and $d \mid r$, then $d \mid (bq + r)$, or $d \mid a$.
3. By parts 1 and 2, the set of common divisors of a and b is equal to the set of common divisors of b and r. Thus, the largest integer in each set is the same: $GCD(a, b) = GCD(b, r)$. ■

Thus, in Example 8.1.7 we have

$$
\begin{aligned}
GCD(1147, 851) &= GCD(851, 296) \\
&= GCD(296, 259) \\
&= GCD(259, 37) \\
&= GCD(37, 0) \\
&= 37
\end{aligned}
$$

The process thus reduces the problem to finding $GCD(37, 0)$, and this is clearly 37. In fact, the *only* positive divisors of 37 and 0 are 1 and 37, and by Lemma 8.1.8 the only positive common divisors of 1147 and 851 are 1 and 37.

THEOREM 8.1.9   The Euclidean Algorithm

Let $b \neq 0$ and

$$
\begin{aligned}
a &= bq_1 + r_1 \\
b &= r_1 q_2 + r_2 \\
r_1 &= r_2 q_3 + r_3 \\
&\vdots \\
r_{n-1} &= r_n q_{n+1} + r_{n+1} \\
r_n &= r_{n+1} q_{n+2} + 0
\end{aligned}
$$

Then $GCD(a, b) = r_{n+1}$.

Proof By Lemma 8.1.8,

$$GCD(a, b) = GCD(b, r_1)$$
$$= GCD(r_1, r_2)$$
$$= \cdots$$
$$= GCD(r_n, r_{n+1})$$
$$= GCD(r_{n+1}, 0) = r_{n+1} \quad\blacksquare$$

COROLLARY 8.1.10

Let $d = GCD(a, b)$. Let $c \mid a$ and $c \mid b$. Then $c \mid d$. In words, any common divisor of a and b must divide their greatest common divisor.

Proof By Lemma 8.1.8 and the proof of Theorem 8.1.9, the set of common divisors of a and b equals the set of common divisors of d and 0. But a common divisor of d and 0 is the same as a divisor of d, because all nonzero integers divide 0. $\quad\blacksquare$

EXAMPLE 8.1.11

The Euclidean Algorithm of Theorem 8.1.9 can be put into pseudocode. The input is a and b, with $0 \leq a, b$. The output is $GCD = GCD(a, b)$.

Euclid(a, b; GCD)
1. IF $b = 0$ THEN GCD $:= a$ [this also sets $GCD(0, 0) = 0$, which is a standard convention]
2. WHILE $b > 0$ [do a long division by b]
    1. $q := [a/b]$ [the quotient]
    2. $r := a - bq$ [the remainder]
    3. GCD $:= b$ [temporarily, if the division is exact]
    4. $a := b$
    5. $b := r$ [now repeat, if $b > 0$]

Another corollary of the Euclidean Algorithm is that $d = GCD(a, b)$ can be written $d = ax + by$. This expression is called a **linear combination** of a and b.

COROLLARY 8.1.12

Let $d = GCD(a, b)$. Then for some integers x and y, we have

$$d = ax + by$$

Proof We prove, by induction, that each remainder occurring in the Euclidean Algorithm can be expressed as a linear combination of a and *b.* Since d is the last nonzero remainder, this will prove the result. For a basis, the first remainder is so expressible: Because $a = bq + r$, we have $r = a - bq = ax + by$ with $x = 1$ and $y = -q$. For the induction step, we have at the kth step in the Euclidean Algorithm the division $r_k = r_{k+1}q_{k+2} + r_{k+2}$. Clearly, if r_k and r_{k+1} are expressed as a linear combination of a and b, then r_{k+2} can be so expressed from this equation. ∎

This proof can be used to find x and y algorithmically while the GCD is being computed. Explicitly, if

$$r_k = x_k a + y_k b$$

for each k, the proof shows that

$$r_{k+2} = r_k - r_{k+1}q_{k+2}$$
$$x_{k+2}a + y_{k+2}b = x_k a + y_k b - (x_{k+1}a + y_{k+1}b)q_{k+2}$$
$$= (x_k - q_{k+2}x_{k+1})a + (y_k - q_{k+2}y_{k+1})b$$

Thus, we can use the recursions

$$x_{k+2} = (x_k - q_{k+2}x_{k+1}); \qquad y_{k+2} = (y_k - q_{k+2}y_{k+1}) \qquad \text{(8.1.2)}$$

during the Euclidean Algorithm. Since $\text{GCD}(a, b)$ is the last nonzero r_n, the corresponding x_n and y_n give the required x and y. We can imagine starting the Euclidean Algorithm with the superfluous zeroth equation $b = 0 \cdot a + b$ to get $r_0 = b$ and $x_0 = 0$, $y_0 = 1$. Then the first real division yields $r_1 = a - bq_1$, so we can use $x_1 = 1$, $y_1 = -q_1$. Thus, the initial conditions are

$$x_0 = 0; \quad y_0 = 1; \quad x_1 = 1; \quad y_1 = -q_1$$

These initial conditions for x_k and y_k ($k = 0$ and 1) together with the recursion equations (8.1.2) generate the final values of x and y.

In a short computation, we can work backwards from the Euclidean Algorithm to express d as a linear combination of a and b. We illustrate this technique with a numerical example.

EXAMPLE 8.1.13

(a) Find $d = \text{GCD}(1128, 342)$.
(b) Express d as a linear combination of 1128 and 342.
(c) Find all positive common divisors of 1128 and 342.

Method We first use the Euclidean Algorithm to find the GCD. (We underline the remainders to separate them from the quotients, which do not enter this algorithm.)

$$\underline{1128} = \underline{342}\cdot 3 + \underline{102}$$

$$\underline{342} = \underline{102}\cdot 3 + \underline{36}$$

$$\underline{102} = \underline{36}\cdot 2 + \underline{30}$$

$$\underline{36} = \underline{30}\cdot 1 + \underline{6}$$

$$\underline{30} = \underline{6}\cdot 5$$

Thus, GCD(1128, 342) = 6. To express 6 as a linear combination of the given integers, work backwards. We have, using the equation with $\underline{6}$ as the remainder:

$$\underline{6} = \underline{36}\cdot 1 - \underline{30}$$

Now eliminate $\underline{30}$, using the equation with $\underline{30}$ as remainder:

$$\underline{6} = \underline{36}\cdot 1 - (\underline{102} - \underline{36}\cdot 2)$$

$$= \underline{36}\cdot 3 - \underline{102}\cdot 1$$

Now eliminate $\underline{36}$:

$$\underline{6} = (\underline{342} - \underline{102}\cdot 3)3 - \underline{102}\cdot 1$$

$$= \underline{342}\cdot 3 - \underline{102}\cdot 10$$

And finally eliminate $\underline{102}$:

$$\underline{6} = \underline{342}\cdot 3 - (\underline{1128} - \underline{342}\cdot 3)10$$

$$\underline{6} = \underline{342}\cdot 33 - \underline{1128}\cdot 10$$

This is the required linear combination.

To find all common divisors of 342 and 1128, we simply apply Corollary 8.1.10 and find all positive divisors of 6. These are 1, 2, 3, and 6.

DEFINITION 8.1.14

Two numbers a and b are **relatively prime** if GCD$(a, b) = 1$.

The following theorem gives a useful algebraic equivalent.

THEOREM 8.1.15

Two numbers a and b are relatively prime if and only if there are integers x and y such that

$$1 = ax + by \qquad\qquad (8.1.3)$$

Proof If $\text{GCD}(a, b) = 1$, Equation (8.1.3) follows from Corollary 8.1.12. On the other hand, if (8.1.3) is true, and if d is any common divisor of a and b, then $d \mid 1$ by statement 7 of Lemma 8.1.2. Thus $d = \pm 1$ and $\text{GCD}(a, b) = 1$. This proves the theorem. ■

The following results are learned early in grade school, and for this reason they are sincerely believed by many to be obvious. But they are relatively deep.

THEOREM 8.1.16

1. If $a \mid bc$ and $\text{GCD}(a, b) = 1$, then $a \mid c$.
2. If $a \mid c$ and $b \mid c$ and $\text{GCD}(a, b) = 1$, then $ab \mid c$.

For example, if $7 \mid 19c$ then $7 \mid c$. In grade school, we might say that $19c/7$ is an integer, and since no cancellations occur between 19 and 7 (they are relatively prime), all cancellations must occur between c and 7, so $7 \mid c$. Statement 1 of Theorem 8.1.16 justifies this line of reasoning. Careful: If $4 \mid c$ and $6 \mid c$, you don't necessarily have $24 \mid c$. Why? Can you prove that $12 \mid c$, using Theorem 8.1.16?

Proof of Theorem 8.1.16

1. If $a \mid bc$ and $\text{GCD}(a, b) = 1$, by Theorem 8.1.15, we have

$$1 = ax + by$$

Multiply by c to obtain

$$c = acx + bcy$$

But since $a \mid bc$, a divides the right side. Therefore $a \mid c$.
2. We have $a \mid c$ and $b \mid c$ and $\text{GCD}(a, b) = 1$. Write $c = aq$. Since $b \mid c$, we have $b \mid aq$. By part 1, $b \mid q$. Thus, $q = br$, so $c = aq = abr$, proving that $ab \mid c$. ■

We now have some of the machinery to prove some results about prime numbers.

DEFINITION 8.1.17

A **prime number** p is a number $p > 1$ whose only positive divisors are 1 and p. A number that is >1 and not a prime is called **composite**.

Thus, the positive integers are partitioned into three sets: the primes, the composites, and 1. The first few primes are

$$2, 3, 5, 7, 11, 13, 17, 19, 23, 29, 31, 37, \ldots$$

Many divisibility results are true for primes but not for other numbers. Thus, we have the next result.

THEOREM 8.1.18

Let p be a prime and suppose $p \mid ab$. Then $p \mid a$ or $p \mid b$.

Proof   If p does not divide a, p and a must be relatively prime. (Their only possible positive common divisors are 1 and p, and we have eliminated p.) Therefore, since $p \mid ab$ and $\text{GCD}(a, p) = 1$, we must have $p \mid b$ by statement 1 of Theorem 8.1.16.   ■

Remark   The theorem may be stated in its contrapositive form: If $p \nmid a$ and $p \nmid b$, then $p \nmid ab$. ($p \nmid a$ is the notation for "p does not divide a.") Theorem 8.1.18 generalizes to more than two factors by a simple induction: If p is a prime and if $p \mid a_1 a_2 \cdots a_n$, then $p \mid a_1$ or $p \mid a_2, \ldots$, or $p \mid a_n$. It is necessary to have p as a prime. For example, $6 \mid 10{\cdot}3$, yet $6 \nmid 3$ and $6 \nmid 10$.

The sequence of primes has been intensively studied. It exhibits certain irregular patterns. For example, it is possible to have consecutive primes arbitrarily far from each other (as far as we want provided we go out into the sequence far enough), yet there appear to be infinitely many "twin prime" pairs—primes that differ by 2, such as 5, 7 or 29, 31. (Nobody has proved this latter conjecture.) On the other hand, it also exhibits certain statistical patterns. For example, if n is large, the number of primes from 1 to n has order of magnitude $n/\log n$. (This is the natural logarithm.) This latter result, called the Prime Number Theorem, is quite deep. The following result shows that we can never run out of primes.

THEOREM 8.1.19

There are infinitely many primes.

Proof (Euclid)   We show that if $p_1, p_2, \ldots, p_N$ are N primes, then it is possible to find a prime that is not one of the p_i. We let

$$n = p_1 p_2 \cdots p_N + 1$$

Let q be *any prime divisor* of n. Then clearly, $q \neq p_i$, for if it were we would have $q \mid p_1 \cdots p_n$. Therefore, since $q \mid n$, we would have $q \mid 1$, a contradiction. Thus we have shown that any prime divisor of n is different from each of the p_i, which gives the result.   ■

THEOREM 8.1.20   *Unique Factorization Theorem*

Any $n > 1$ can be written as the product of primes in only one way, except for the order of factors. (For example, $60 = 6 \cdot 10 = 2 \cdot 3 \cdot 2 \cdot 5$ and $60 = 4 \cdot 15 = 2 \cdot 2 \cdot 3 \cdot 5$. The factorization into primes is the same in both cases.)

Proof   We have already shown (Theorem 2.3.15) that any $n > 1$ is either a prime or a finite product of primes. The proof of unique factorization is by induction. It is clearly true for $n = 2$, the basis for this induction. Now suppose the result is true for all $m < n$. Suppose

$$n = p_1 p_2 \cdots p_r = q_1 q_2 \cdots q_s \qquad (8.1.4)$$

are both factorizations of n into primes. If $r = 1$, n is a prime p_1 that has the unique factorization $n = p_1$. If $r > 1$, then, since $p_1 \mid n$, we have $p_1 \mid q_1 \cdots q_s$. By Theorem 8.1.18, $p_1 \mid q_i$ for some i. Reordering the q's if necessary, we may suppose $p_1 \mid q_1$. Since q_1 is prime, we must have $p_1 = q_1$. Dividing (8.1.4) by p_1, we obtain

$$n/p_1 = p_2 \cdots p_r = q_2 \cdots q_s$$

But since $n/p_1 < n$, it has a unique factorization. Thus the rest of the p's and the q's are the same, except for their order. This is the result. ∎

Remark   This result offers a convenient theoretical way of coding ordered pairs of natural numbers as one positive integer. The method is simply to code the pair (a, b) as the positive number $2^a 3^b$. (We choose 2 and 3 because they are the first two primes.) Unique factorization tells us that if $2^a 3^b = 2^c 3^d$, we must have $a = c$ and $b = d$. This allows us to uniquely decode a number. For example, the pair $(2, 1)$ is coded as the number $2^2 3^1 = 12$. Similarly, the pair $(0, 2)$ is coded as $2^0 3^2 = 9$. We can decode by factoring. The number 18 can be decoded by factoring $18 = 9 \cdot 2 = 2^1 3^2$, so 18 is the code name for $(1, 2)$. By extending the number of primes, we can code triples of natural numbers and, in fact, all finite sequences of natural numbers. We use this technique in Chapter 10.

   The disadvantage of the preceding method of coding pairs of integers is that the code numbers get very large even for small pairs. The pair $(1, 5)$ is coded as $2 \cdot 3^5 = 486$, and the pair $(9, 9)$ is coded as a number larger than 10 million. Thus, this technique is not a practical way to code ordered pairs of positive integers.

   How can we determine whether a number n is prime? For example, is 9929 a prime? If we used the definition directly, we would have to check the possible divisors from 2 through 9928. The following theorem shortens this considerably.

THEOREM 8.1.21

Let $n > 1$. If there are no divisors d of n between 2 and $\sqrt{n}$ inclusive, then n is a prime.

Proof If n were not a prime, then we would have $n = rs$, where both r and s are >1. We could not have *both* r and s larger than $\sqrt{n}$ because, in that case, their product would be greater than n and, hence, could not be n. Therefore, one of r and s would be $\leq \sqrt{n}$, so n would have a divisor $\leq \sqrt{n}$, and thus a prime divisor $\leq \sqrt{n}$. ■

For example, to check whether 97 is a prime, we need to check for divisors between 2 and 10. The only primes in this range are 2, 3, 5, and 7. Since none of these are divisors of 97, it follows that 97 is a prime. Similarly, to check 9929, only the primes between 2 and 97 need be tested as divisors. In a prime testing program, one method is to test the divisor $d = 2$ and then all odd divisors from $d = 3$ through $\sqrt{n}$.

We now return to our discussion of the time taken by the Euclidean Algorithm. In Example 8.1.7, we found GCD(851, 1147), using only four long divisions. A little experimentation shows that the remainders decrease most rapidly when the quotients are large. This suggests that the slowest time (the worst-time case) occurs when the quotients are 1.

EXAMPLE 8.1.22

How many divisions are needed to find GCD(13, 8) with the Euclidean Algorithm?

Method

$$13 = 8 \cdot 1 + 5$$
$$8 = 5 \cdot 1 + 3$$
$$5 = 3 \cdot 1 + 2$$
$$3 = 2 \cdot 1 + 1$$
$$2 = 1 \cdot 2$$

Thus, five divisions were necessary, including the last division by 1. There are actually *more* divisions than in Example 8.1.7.

The remainders are, of course, the Fibonacci numbers. The equation $r_n = r_{n+1}q + r_{n+2}$ becomes $r_n = r_{n+1} + r_{n+2}$ when q is 1, and this is the Fibonacci recursion if we read these remainders backwards. (The remainders decrease. Read up in the divisions in Example 8.1.22 to find the sequence of remainders in increasing order.) We know that the Fibonacci numbers increase exponentially (Example 4.3.7). If we used the same technique as in the example to find GCD(f_{n+1}, f_n), we would perform $n - 1$ divisions. (In the example, $8 = f_6$.) Thus, we suspect logarithmic time.

THEOREM 8.1.23

Suppose $0 < b < a$ and the Euclidean Algorithm takes n long divisions. Then $b \geq f_{n+1}$.

Proof   We may assume $n \geq 2$. The sequence of remainders r_k in the Euclidean Algorithm satisfies the equations

$$r_k = r_{k+1}q_{k+2} + r_{k+2}, \qquad k = 0, 1, \ldots, n - 2 \qquad \textbf{(8.1.5)}$$

Here, since n divisions were used, we have

$$r_n = 0, \quad r_{n-1} \geq 1, \quad \text{and} \quad r_{n-2} \geq 2 \qquad \textbf{(8.1.6)}$$

In this setup $r_0 = b$. Because q_k is a positive integer, we have $q_k \geq 1$ for all k, so (8.1.5) yields the inequalities

$$r_k \geq r_{k+1} + r_{k+2}, \qquad k = 0, 1, \ldots, n - 2 \qquad \textbf{(8.1.7)}$$

Thus, we have $b = r_0 > r_1 > \cdots > r_{n-1} > 0 = r_n$. We prefer to reverse the sequence (reading up the sequence of remainders). This amounts to taking the sequence s_k, where $s_k = r_{n-k}$. The result is the sequence $0 = s_0 < s_1 < \cdots < s_n = b$. The inequality (8.1.7) becomes

$$s_{k+2} \geq s_{k+1} + s_k, \qquad k = 0, \ldots, n - 2 \qquad \textbf{(8.1.8)}$$

The inequalities (8.1.6) become

$$s_1 \geq 1 \qquad \text{and} \qquad s_2 \geq 2 \qquad \textbf{(8.1.9)}$$

We can now apply Theorem 4.3.18 on solving linear inequalities. The inequalities (8.1.8) and (8.1.9) are compared with the equalities

$$F_{k+2} = F_{k+1} + F_k, \qquad k = 0, \ldots, n - 2, \quad \text{with } F_1 = 1 \text{ and } F_2 = 2$$

This is the Fibonacci equation, starting with terms 1 and 2, which are f_2 and f_3. Thus, its solution is $F_n = f_{n+1}$. By Theorem 4.3.18, we have $b = s_n \geq f_{n+1}$.   ■

Remark   The result can also be proved by induction by using Equations (8.1.8) and (8.1.9).

COROLLARY 8.1.24

In the Euclidean Algorithm, where $0 < b < a$ are the inputs, take b as the size. Then the worst time for this algorithm has order of magnitude $\log b$.

Proof We assume that the time has order of magnitude equal to the number of long divisions. If n divisions are needed, then $b \geq f_{n+1}$. We know that f_{n+1} has the same order of magnitude as Ψ^n, where $\Psi = (1 + \sqrt{5})/2$. (See Example 4.3.7.) Therefore, $b \geq K\Psi^n$. Taking logarithms gives $\log b \geq \log K + n \log \Psi$. But this shows that $n = O(\log b)$. The proof shows that the Fibonacci numbers give the worst case. ∎

Remark An alternative approach for obtaining information on the inequalities (8.1.8) and (8.1.9), without the intervention of Theorem 4.3.18 and the knowledge of the properties of the Fibonacci sequence, is given in Exercise 8.1.34.

Exercises

1. Find GCD(72, 138), using the Euclidean Algorithm. Express your answer as a linear combination of 72 and 138. Find all positive divisors of 72 and 138.

2. Repeat Exercise 1 for the integers 1481 and 4317.

*3. Prove that any two consecutive Fibonacci numbers are relatively prime.

4. Let $d = \text{GCD}(a, b)$. Suppose $c = ax + by$ for some x and y. How are c and d related?

5. (*Proof of the Division Algorithm*) For fixed $b > 0$, any $n \geq 0$ can be written $n = bq + r$, where $r \in \mathbb{Z}_b$. Prove this by induction on n. Using this result for $n \geq 0$, prove that it is also true for negative n.

*6. (*Alternative Proof of the Division Algorithm*) For fixed a and b, let r be the *least* natural number of the form $a - bq$. Prove that $0 \leq r < b$. Assume that both a and b are positive.

*7. (*Uniqueness in the Division Algorithm*) Prove that if $b > 0$ and

$$a = bq + r = bq_1 + r_1, \qquad 0 \leq r, r_1 < b$$

then $r = r_1$ and $q = q_1$.

8. Let $d = \text{GCD}(a, b)$. Show that a/d and b/d are relatively prime.

9. Prove $\text{GCD}(xa, xb) = x\text{GCD}(a, b)$, where $x > 0$.

10. Prove statements 4 and 8 of Lemma 8.1.2.

11. Using Theorem 8.1.15, prove that if $\text{GCD}(a, b) = \text{GCD}(a, c) = 1$, then $\text{GCD}(a, bc) = 1$.

*12. Suppose that $a^2 = bc$, where b and c are positive and relatively prime. Prove that b and c are squares. (*Hint:* Use unique factorization.)

13. A proper divisor of an integer $n > 1$ is a positive divisor that is less than n. Prove: If $n > 1$, the largest proper divisor of n is $\leq n/2$. Using this fact, show how the time of the LongSearch Algorithm of Example 8.1.6 can be cut approximately in half. (The algorithm is still highly inefficient.) Write the modified LongSearch Algorithm.

*14. A clock is observed at 12 noon. The minute hand is observed every 47 minutes. Show that at some point it will be observed at one minute after 12. If observed long enough, will it be observed at one minute after every hour?

15. Find the prime factorization of (a) 288, (b) 448.

16. Find the prime factorization of (a) 109, (b) 833. (*Hint:* For (b) first multiply by 3.)

17. Find the prime factorization of (a) 1001, (b) 221.

18. Find the prime factorization of (a) 1610, (b) 22,100.

19. Using the coding device given after the proof of Theorem 8.1.20, code (2, 3) as an integer. Similarly code (5, 0).

20. Repeat Exercise 19 for (a) (2, 2), (b) (2, 4).

21. Using the coding device given after the proof of Theorem 8.1.20, code (0, 1, 1) as an integer. Similarly, for (1, 2, 2).

22. *Decode* the integer 54 into an ordered pair of natural numbers. The coding procedure is the one given after the proof of Theorem 8.1.20. If no decoding is possible, state why. Similarly, decode 50.

23. Repeat Exercise 22 for (a) 72, (b) 81.

24. *Decode* the integer 500 into an ordered *triple* of natural numbers. The coding procedure is the one given after the proof of Theorem 8.1.20. If no decoding is possible, state why. Similarly, decode 100.

25. Repeat Exercise 24 for (a) 175, (b) 98.

*26. Prove that the number of positive divisors of a positive integer is *odd* if and only if the integer n is a perfect square.

27. Prove that $\sqrt{7}$ is an irrational number. (*Hint:* Assume that it is a fraction, square both sides, clear of fractions, and use unique factorization.)

28. Write the numbers from 1 through 100 in a 10 by 10 table as indicated in Figure 8.1.1. Now circle 2 and cross out the multiples of 2 (4, 6, 8, and so on). Next circle 3, the next uncrossed number, and cross out multiples of 3. (Some numbers such as 6 will be crossed out twice.) Continue with the next uncrossed number (5 in this case)

	1	2	3	4	5	6	7	8	9	10
00	1	②	③	4	5	6	7	8	9	10
10	11	12	13	14	15	16	17	18	19	20
20	21									
30	31									
40	41									
50	51									
60	61									
70	71									
80	81									
90	91									

Figure 8.1.1

and cross out its multiples. Do this until you reach 10 (actually, 7 will be the last number used since 8, 9, and 10 will have been crossed out). Show that all of the uncrossed numbers are primes. This method of generating the primes is called the **Sieve of Eratosthenes.**

29. Using the sieve method of Exercise 28, find all of the primes between 1 and 200.

*30. It was assumed during the proof of Theorem 8.1.19 that any integer >1 has a prime divisor. Prove this by induction.

31. Using Theorem 8.1.21 and the remarks following the proof, write an algorithm that decides whether a given positive integer n is a prime.

*32. Using the technique indicated after the proof of Corollary 8.1.12, adjust the Euclidean Algorithm of Example 8.1.11 to find GCD $=$ GCD(a, b), and integers x and y such that GCD $= ax + by$.

33. According to Theorem 8.1.23, if $0 < b < a$ and the Euclidean Algorithm needs n divisions, then $b \geq f_{n+1}$. Show that this is so if the Euclidean Algorithm needs n or more divisions.

34. Prove by induction that if s_k satisfies the inequalities (8.1.8) and (8.1.9) then $b = s_n \geq (1.6)^{n-1}$. Hence, show that the number of divisions in the Euclidean Algorithm is $O(\log b)$.

35. Using Lemma 8.1.2, show that the positive integers $\mathbb{N}^+$ form a partially ordered set $(\mathbb{N}^+, |)$ under the division relation.

*36. If C is the covering relation for $(\mathbb{N}^+, |)$ as defined in Exercise 35, describe what is meant by aCb in arithmetic terms. Prove your result.

Some Programming Exercises

The following exercises are for readers who are familiar with a programming language.

37. Write a program to compute GCD(a, b) using the Euclidean Algorithm. Assume that a and b are positive. Take a and b as inputs, and let $d = $ GCD(a, b).

38. Using the technique indicated after the proof of Corollary 8.1.12, write a program to find integers x and y such that GCD$(a, b) = d = xa + yb$. The program should accept positive numbers a and b as input and give their GCD d and numbers x and y such that $d = xa + yb$. (*Hint:* It is a good idea to do Exercise 32 first.)

39. Write a program to determine whether a positive integer is a prime. Use Theorem 8.1.21. The input should be n, and the output should be true (or 1) if n is a prime, false (or 0) if n is not. (*Hint:* It is a good idea to do Exercise 31 first.)

40. Using the program constructed in Exercise 39 as a basis, write a program that counts all of the primes through a given input n and compares the count to $n/\log(n)$. (The natural log is understood.) The comparison should be the ratio of the count to $n/\log(n)$, as well as the difference.

8.2 Congruence

If a light switch is on, and if the switch is toggled 143 times, the light switch will then be off. Few of us need to actually toggle the switch 143 times—it is enough to know that an odd number of toggles is equivalent to one, and an even number is equivalent to none. In the world of toggles, there are only two possibilities: 0 (even) and 1 (odd). If the minute hand of a clock starts at 38 minutes after the hour, then after 155 minutes the minute hand will be at 13 minutes after the hour. In the world of the minute hand, there are only 60 possibilities: 0, 1, 2, . . . , 59.

Each example works because, after adding 1 for some time, we proceed cyclically and come back to 0. For the toggle, 2 "=" 0 and for the minute hand, 60 "=" 0. The following definition, considered briefly in Section 2.1, generalizes these examples and permits us to handle many questions of divisibility and cyclic structure in a straightforward way. We fix an integer n in this definition. In our examples, we had $n = 2$ and $n = 60$.

DEFINITION 8.2.1

For any integers a and b,

$$a \equiv b \bmod n \tag{8.2.1}$$

(read "a is congruent to b modulo n") is defined to mean

$$n \mid (a - b) \tag{8.2.2}$$

By definition, (8.2.2) is equivalent to the equation $a - b = qn$ for some q, so (8.2.1) and (8.2.2) are also equivalent to

$$a = b + qn \qquad \text{for some } q \tag{8.2.3}$$

For example (see the first paragraph),

$$143 \equiv 1 \bmod 2 \qquad \text{and} \qquad 38 + 155 \equiv 13 \bmod 60$$

We sometimes simplify (8.2.1) by writing $a \equiv b \ (n)$ or even $a \equiv b$, if n is understood. The use of the symbol $\equiv$ anticipates that congruence is an equivalence relation. We shall see that congruence behaves very much like equality in many respects, and this fact enables us to convert various properties of divisibility into an *algebraic* setting.

THEOREM 8.2.2

Congruence mod n is an equivalence relation.

1. For any a, $a \equiv a$.

2. For any a and b, if $a \equiv b$ then $b \equiv a$.
3. For any a, b, and c, if $a \equiv b$ and $b \equiv c$ then $a \equiv c$.

(All congruences are taken mod n for some fixed value of n.)

Proof We prove 3, leaving the others to the reader. If $a \equiv b$ and $b \equiv c$, then $n \mid (a - b)$ and $n \mid (b - c)$. Therefore the sum $(a - b + b - c)$ is divisible by n, or $n \mid (a - c)$. Thus $a \equiv c$. ∎

Theorem 8.2.2 allows us to string congruences much as we string equalities. Thus, we may write $a \equiv b \equiv c \equiv d$ to mean $a \equiv b$, $b \equiv c$, $c \equiv d$, and we may conclude that $a \equiv d$.

We now show that congruences can be added and multiplied.

THEOREM 8.2.3

(Again, all congruences are mod n.) Suppose $a \equiv b$ and $c \equiv d$. Then

1. $a + c \equiv b + d$.
2. $ac \equiv bd$.
3. $a^k \equiv b^k$ for $k = 1, 2, \ldots$.

Proof 1. $n \mid (a - b)$ and $n \mid (c - d)$. Therefore $n \mid (a - b + c - d)$. But this is simply $a + c \equiv b + d$ by Definition 8.2.1.
2. Again, $n \mid (a - b)$ and $n \mid (c - d)$. Therefore,

$$n \mid [c(a - b) + b(c - d)]$$

or, simplifying, $n \mid (ac - bd)$. This is the result.
Statement 3 comes from repeated application of statement 2. We multiply the congruence $a \equiv b$ by itself k times to get statement 3. Equivalently, a simple induction does the job. ∎

We can relate congruence to the Division Algorithm.

THEOREM 8.2.4

Let $n > 1$, $r \in \mathbb{Z}_n$. Then a leaves the remainder r when divided by n if and only if $a \equiv r$ mod n.

Proof If $a = nq + r$, then $a \equiv r$ mod n by (8.2.3). Conversely, if $a \equiv r$ mod n, then, by (8.2.3), $a = qn + r$. Since $r \in \mathbb{Z}_n$, r is the remainder when a is divided by n. ∎

COROLLARY 8.2.5

$a \equiv b$ mod n if and only if a and b leave the same remainder when divided by n.

Proof If a and b each leave the same remainder r, then $a \equiv r$ and $b \equiv r$. Thus, $a \equiv b$. Conversely, if $a \equiv b$ and a leaves the remainder r, we have $a \equiv r$. Therefore, $b \equiv r$, so b also leaves the remainder r. ∎

This corollary gives another way of defining congruence through remainders. It shows that any integer a is congruent to exactly one of the numbers $0, 1, \ldots, n - 1$ in $\mathbb{Z}_n$. (Compare the introductory paragraph with $n = 2$ and $n = 60$.) Thus, the set $\mathbb{Z}_n$ may be regarded as a **set of representatives** modulo n. There are exactly n equivalence classes modulo n. Each integer in $\mathbb{Z}_n$ has a different equivalence class, and any integer is in the equivalence class determined by its remainder mod n. The following theorem introduces a notation for the remainder when a is divided by n and summarizes Theorem 8.2.4 and Corollary 8.2.5.

THEOREM 8.2.6

For any $a \in \mathbb{Z}$, write $(a)_n$ as the remainder when a is divided by n. Then

1. $(a)_n \in \mathbb{Z}_n$
2. $a \equiv (a)_n \bmod n$
3. $a \equiv b \bmod n$ if and only if $(a)_n = (b)_n$
4. $n \mid a$ if and only if $a \equiv 0 \bmod n$ if and only if $(a)_n = 0$ ∎

EXAMPLE 8.2.7

The Euclidean Algorithm can be succinctly stated by using congruences. Lemma 8.1.8, the basis for the Euclidean Algorithm, can be restated as follows:

If $a \equiv r \bmod b$, then $\mathrm{GCD}(a, b) = \mathrm{GCD}(b, r)$.

The Euclidean Algorithm can now be restated as follows. Suppose

$$a \equiv r_1 \bmod b$$

$$b \equiv r_2 \bmod r_1$$

$$r_1 \equiv r_3 \bmod r_2$$

$$\vdots$$

$$r_{n-1} \equiv 0 \bmod r_n$$

Then $\mathrm{GCD}(a, b) = |r_n|$. The absolute value sign allows the possibility of arbitrary integers as the r's. They need not be in $\mathbb{Z}_n$ for this result to be valid.

We now illustrate how congruences can help to find remainders.

EXAMPLE 8.2.8

Find $(29^7)_{13}$.

Method   Work mod 13 to find

$$29 \equiv 3 \bmod 13$$

Cubing, we find

$$29^3 \equiv 3^3 = 27 \equiv 1 \bmod 13$$

Squaring this last congruence, we find

$$29^6 \equiv 1 \bmod 13$$

Multiplying by the first congruence, we find

$$29^7 \equiv 3 \bmod 13$$

or $(29^7)_{13} = 3$.

We should not think of actually computing 29^7 and dividing by 13. In the world modulo 13, there are exactly 13 possible answers, so it is not appropriate to do a calculation of this size.

EXAMPLE 8.2.9

Find all numbers x such that $(3x)_{19} = 5$.

Method   Rewrite as a congruence:

$$3x \equiv 5 \bmod 19$$

Our strategy is to *multiply* this equation by an integer large enough to bring the coefficient of x near the modulus 19, and then reduce this coefficient mod 19. Therefore we multiply by 6, bringing the coefficient of x as near to 19 as we can:

$$18x \equiv 30 \bmod 19$$

Now, using $18 \equiv -1$ and $30 \equiv 11$, we obtain

$$-x \equiv 11 \bmod 19$$

Multiply by -1 and use $-11 \equiv 8$ to get

$$x \equiv 8 \bmod 19$$

This is the answer. Numerically, the possible positive x's are 8, 27, 46, ..., and the negative x's are -11, -30, ..., which we get by adding 19's to and subtracting 19's from 8. We can check our result. Multiply the congruence $x \equiv 8$ by 3 to obtain $3x \equiv 24 \equiv 5 \bmod 19$, so $(3x)_{19} = 5$.

In the next section we show that the operation of multiplying by 6 to obtain $18x \equiv 30$ is reversible—the factor 6 can be canceled. This is possible because 6 and 19 are relatively prime. Therefore, checking the solution is unnecessary, because all steps in the derivation are reversible. Note also the useful technique for finding a negative modulo n:

$$-a \equiv n - a \bmod n$$

Thus, for example, $-11 \equiv 19 - 11 = 8 \bmod 19$.

Cryptography

It is an ancient idea to code important messages so that an enemy cannot read them. Almost as ancient, of course, is the art of decoding. The use of coding today, for governments as well as businesses, is no less important for the same reason, especially given the enormous amount of computer information available at the touch of a finger.

EXAMPLE 8.2.10

One of the simplest codes replaces each letter with one that comes a fixed number of digits after it in the alphabet, "wrapping around" when we reach the end of the alphabet. We illustrate by replacing a letter with the letter that comes five digits after it:

```
LETTER:  A B C D E F G H I J K L M N O P Q R S T U V W X Y Z
  CODE:  F G H I J K L M N O P Q R S T U V W X Y Z A B C D E
```

Thus, the message "I WILL COME" gets compacted into IWILLCOME and gets coded into NBLQQ HTRJ, hopefully totally confusing and confounding the enemy. (For ease in reading, the code is usually broken into strings of five.) The decoder, who has the message, simply goes to the decoding table:

```
  CODE:  A B C D E F G H I J K L M N O P Q R S T U V W X Y Z
DECODE:  V W X Y Z A B C D E F G H I J K L M N O P Q R S T U
```

For this code, once the enemy has found that the code letter for I is N, the rest of the alphabet can be decoded, since all that is involved is a push of the alphabet a certain number of units to the right, with wraparound. This type of code can be simply analyzed by using congruence modulo 26. First let us code each letter with a number from $\mathbb{Z}_{26}$. The simplest is probably A → 1, B → 2, ..., Y → 25, and Z → 0, where we have converted the code number 26 to 0 (since $26 \equiv 0$ in $\mathbb{Z}_{26}$). Now if the number for a letter is x and the number for the code of that letter is y, the coding table is given simply by the equation

$$\text{CODE: } y \equiv x + 5 \bmod 26$$

Decoding merely means to solve this for x. Thus,

$$\text{DECODE: } x \equiv y - 5 \bmod 26$$

Thus, to decode J, whose number is $y = 10$, we find $x \equiv 10 - 5$ or 5, which is the number for E. The tables have been replaced by simple algebraic formulas.

EXAMPLE 8.2.11

We can illustrate a more complicated version of this type of code. Code A as F, then code B as I (3 after F), C as L (3 after I), and so forth. (In general, after we decide how to code A, the code for B would be a fixed number, here 3, after the code for A, and so on, cycling around the alphabet as before.) In this case the code is

```
LETTER:  A B C D E F G H I J K L M N O P Q R S T U V W X Y Z
  CODE:  F I L O R U X A D G J M P S V Y B E H K N Q T W Z C
```

How would we decode the message OVSVK LVPR? We can go backwards in this table, but this involves a search for each of the code letters. Better to make a decoding table, especially if we intend to code and decode a few messages:

```
  CODE:  A B C D E F G H I J K L M N O P Q R S T U V W X Y Z
DECODE:  H Q Z I R A J S B K T C L U D M V E N W F O X G P Y
```

The decoded message can now be read off as DONOT COME, presumably "Do not come."

What is involved algebraically? Every increase in x by 1 yields an increase in y of 3. This yields a linear equation. The wraparound feature simply tells us that we are working modulo 26. Such a linear equation is of the form $y \equiv mx + b$, where m is the increase in y due to an increase in 1 of x, and b is a constant depending on where we start. Here, $m = 3$ so $y \equiv 3x + b$. But we know that $y = 6$ when $x = 1$ (since A has code F). Thus, substituting in, we obtain $6 \equiv 3 + b$, and so $b \equiv 3$. Thus we have

$$\text{CODE: } y \equiv 3x + 3 \bmod 26$$

To solve for x, we do not divide. As in Example 8.2.9, we multiply by 9 (to bring the 3 closer to 26), and obtain

$$9y \equiv 27x + 27 \equiv x + 1$$

Therefore, we can solve to get

$$\text{DECODE: } x \equiv 9y - 1 \bmod 26$$

This gives the decoding table in a succinct manner. To decode E, for example, we have $y = 5$, $x \equiv 9 \cdot 5 - 1 = 44 \equiv 18$, so E is the code letter for the eighteenth letter R, agreeing with the previous decoding table.

The simple linear equation scrambles the letters pretty well and has the nice property that you need only remember m and b for the equation $y \equiv mx + b$, rather than the whole coding table. You can then compute the inverse (decoding) equation $x \equiv m'y + b'$, or else remember m' and b' also.

DEFINITION 8.2.12

A **linear substitution code** is one of the form $y \equiv mx + b$, with a unique decoding $x \equiv m'y + b'$, all congruences taken mod 26.

The number m has to be relatively prime to 26 for this to work. For example, if we jumped two letters for the code for every jump of one, we would run out of places after 13 letters. Later we shall see that we can uniquely solve $y \equiv mx + b \bmod n$ only when m and n are relatively prime. In the previous case, $n = 26$, so m must be odd and not divisible by 13.

EXAMPLE 8.2.13

The message TYOYN AQYVA S has been intercepted. It is known to be a linear substitution code, and it is also known that the codes for B and C are L and Q, respectively. Decode this message.

Method   Because the code is linear, we have the congruence $y \equiv mx + b$. We must find m and b. We are given that when $x = 2$ (the number for B), then $y = 12$, the number for L. Similarly, when $x = 3$, $y = 17$, the number for Q. Thus, both (2, 12) and (3, 17) satisfy the coding equation $y \equiv mx + b$. Putting in these known values, we obtain

$$12 \equiv 2m + b \ (26)$$

$$17 \equiv 3m + b \ (26)$$

We can solve for m and b mod 26. Subtracting, we find

$$5 \equiv m$$

Substituting into the first equation, we get $12 \equiv 10 + b$, so $b \equiv 2$. The full equation is $y \equiv 5x + 2$ mod 26.

To solve $y \equiv 5x + 2$, we write $5x \equiv y - 2 \ (26)$. Once again, our strategy is to multiply the coefficient of x to bring it closer to 26. Multiplying by 5, we obtain $25x \equiv 5y - 10$. Using $25 \equiv -1$, this yields $-x \equiv 5y - 10$ or

$$x \equiv 10 - 5y$$

We can now decode the message. The first symbol is T, numbered $y \equiv 20 \equiv -6$. Therefore, $x \equiv 10 - 5y \equiv 10 + 30 \equiv 14$, the number for N. The message starts with N. We leave the full decoding to the reader.

The codes we have given are simple substitution codes. Although the linear substitution $y \equiv mx + b$ mixes the letters pretty well, codes of these sorts can be cracked fairly easily. Because of their relative simplicity, other methods are used. One way is to code each *pair* of letters, thus giving $26^2 = 676$ entries in the coding table. Thus, the phrase "I WILL COME" would be broken into the pairs IW IL LC OM EX, and each pair is coded differently according to some substitution scheme. Another method might be to use different substitutions after, say, 20 letters or after the letter T occurs, and so on. Of course, the more complicated the method, the more complicated the decoding process will be.

Many newspapers have cryptography puzzles daily. The reader should refer to Edgar Allen Poe's short story *The Gold Bug* for an early literary example.

Random Number Generators

Random numbers are intended to be a sequence of real numbers between 0 and 1 that are chosen randomly. They are useful because they help simulate random processes. Once we know a few numbers in the sequence, we should not be able to predict the next number or even in what range the next number will be. For example, a calculator might yield 0.072, 0.799, and 0.112 when the RAN # key is pressed three times in a row. All that we are able to say about the next number is that it has a chance of being anywhere between 0 and 1 with equal probability. By definition, if you can say more than this, the numbers are not random. Yet these numbers are obviously generated by some method. Many languages have built-in random number generators. How is this done?

EXAMPLE 8.2.14   *Random Number Generators*

The following method is known to satisfy many of the tests for randomness and is in common usage. Choose a large prime p and suitable numbers a and b. Now working mod p, let $f(x) \equiv ax + b$ with $f(x) \in \mathbb{Z}_p$; that is, $f(x) = (ax + b)_p$. Now choose any integer s, called "the seed." The generation of random numbers is given by the sequence s_0, s_1, s_2, and so on, where $s_0 = s$, $s_1 = f(s_0)$, $s_2 = f(s_1)$, and in general $s_{n+1} = f(s_n)$ for $n \geq 0$. Thus, the first term after the seed s is $s_1 = f(s)$, at which time the new seed is regarded as s_1, and the process continues with this new seed. At any step, the last previous computation becomes the new seed. The numbers s_i are integers x satisfying the inequality $0 \leq x < p$. Therefore, if we divided by p, the numbers x_i all satisfy the inequality $0 \leq x_i < 1$, and it is these x_i that are the sequence of "random numbers."

If we work mod p, we expect the sequence x_i to cyclically repeat, so we may well ask what kind of randomness this is. However, if the prime p is large and a is chosen properly, the repetitions will occur every $p - 1$ times, a large period. For example, if p is order of magnitude 100,000, then we expect repetition after roughly 100,000 times. If we use, say, 100 random numbers, this repetition shouldn't be important.

We shall illustrate with a very small and impractical p in order to clarify the method. Take $p = 7$, $a = 3$, $b = 2$, so $f(x) = (3x + 2)_7$. If the seed is $s = 0$, the sequence is generated as in Table 8.2.1.

Table 8.2.1

n	$s_n = (3s_{n-1} + 2)_7 \ (n > 0)$
0	0 (initial seed)
1	$(3 \cdot 0 + 2)_7 = 2$
2	$(3 \cdot 2 + 2)_7 = 1$
3	$(3 \cdot 1 + 2)_7 = 5$
4	$(3 \cdot 5 + 2)_7 = 3$
5	$(3 \cdot 3 + 2)_7 = 4$
6	$(3 \cdot 4 + 2)_7 = 0$

Thus, the sequence is 0, 2, 1, 5, 3, 4, and 0, after which it cycles, because we have arrived at the original seed. The number 6 is left out of this sequence. What happens if we choose $s = 6$? $f(6) = 6$, so we get nothing but 6's. In this case, 6 is called a **fixed point** of the function $f(x)$, and it will always be the case for such functions that there is a fixed point that must be avoided as a seed. In general, a **fixed point** for any function $f: A \to A$ is an element x such that $f(x) = x$.

EXAMPLE 8.2.15   An Algorithm for a Random Number Generator

To take a somewhat more realistic example, let

$$f(x) = (37x + 57)_{97}$$

Starting with the seed 77, we can compute the sequence

$$77, 93, 6, 85, 1, 94, 43, \ldots$$

How are these numbers computed? Since the seed is chosen as 77, the next term is $f(77) = (37 \cdot 77 + 57)_{97}$. This can be done by hand, but it does take time because it involves two-digit multiplication and long division by 97, and it is no doubt error-prone. It is indispensable to use a computer or a programmable calculator for such a calculation. The computation of $(n)_{97}$ is a one-step routine: $(n)_{97} := n - 97[n/97]$. Calculating $y = f(x)$ is accomplished simply by the algorithm

1. $y := 37x + 57$
2. $y := (y)_{97}$

Once this is implemented as a working program, simply feed in 77 as the input and read the output 93. Run again with 93 as input, and read output 6, and so on.

The reader may check that all the numbers between 0 and 96, inclusive, are obtained when we arrive at the seed, with the exception of 55, which is the fixed point. (Later, we show how to determine these facts without actually carrying out all of these computations.) By dividing these integers by 97, we generate the random sequence

$$0.79, 0.96, 0.06, 0.88, 0.01, 0.97, 0.44, \ldots$$

When such a method is used, we say that the random numbers are generated by **iteration** of the function $f(x) = (ax + b)_p$. In an actual application, the prime would be much larger than 97, to avoid cycling for a sufficiently long time.

DEFINITION 8.2.16

Let $f: X \to X$. We define the **nth iterate** of a function recursively as follows:

$$f^0(x) = x$$
$$f^{n+1}(x) = f(f^n(x))$$

The sequence $f^n(s)$, $n = 0, 1, \ldots$, is called the **sequence of iterates of** $f(x)$ with seed s. If $f(x_0) = x_0$, then x_0 is called a **fixed point** of f.

The reader should compare with Exercises 2.3.28 through 2.3.31. Equivalently, the sequence a_n is a sequence of iterates of $f(x)$ with seed s when $a_0 = s$ and $a_{n+1} = f(a_n)$ for $n \geq 0$.

Thus, the method of Example 8.2.15 generates the sequence of iterates of the function $f(x) = (37x + 57)_{97}$ with seed $s = 77$.

EXAMPLE 8.2.17   *Finding a Fixed Point*

How can we find the fixed point for $f(x) = (ax + b)_p$ in advance? This amounts to solving the equation $f(x) = x$. In Example 8.2.15 this is

$$37x + 57 \equiv x \ (97) \quad \text{or} \quad 36x + 57 \equiv 0 \ (97)$$

The reader may attempt this according to the methods illustrated in Example 8.2.9. As noted, the solution is $x \equiv 55$. In the next section, we learn more of the theory and the practicalities behind solving linear equations. We also consider what conditions are needed for a linear function $(ax + b)_p$ and a seed s in order that the values of the sequence generated will go through all values mod p except for the fixed point.

EXAMPLE 8.2.18

Chaos theory considers iterations of a function $f(x)$ where x is a real number strictly between 0 and 1, and shows how these iterates may exhibit random-like properties. In this theory, x is real and not in $\mathbb{Z}_p$. For example, the very simple function $f(x) = 4x(1 - x)$, with seed $x = 0.361$, yields the sequence

$$0.361, 0.923, 0.285, 0.816, 0.602, \ldots$$

with random-like properties. We will not pursue this topic any further. For further information, see *Symposia in Applied Mathematics Proceedings*, Vol. 39, *Chaos and Fractals: The Mathematics behind the Computer Graphics*," edited by Robert DeVaney and Linda Keen (Providence, Rhode Island, American Mathematical Society, 1989).

Error-Catching Methods

Some arithmetical errors are quite easy to catch. For example, the computation $674 \times 893 = 601,828$ is clearly wrong, because the last digit in the product has to be 2. The last digit of any positive number, however, is simply the number mod 10. Thus, this check, using the last digit, amounts to a check mod 10, because if $ab = c$ we must have $ab \equiv c$ mod 10. One of the simplest arithmetic checking devices is to see if the arithmetic works mod 10. Similarly, a check mod 2 would be to see whether, for example, odd plus odd = even (which is $1 + 1 \equiv 0$ mod 2). A check mod 2 is called a **parity check.** The foregoing computational error was not caught by a parity check, but by a check mod 10.

Two simple arithmetic curiosities can catch errors of this sort. They are based on a simple criterion for finding a number mod 9, and similarly for finding it mod 11.

THEOREM 8.2.19   Casting Out Nines

Let

$$n = a_0 + 10a_1 + 100a_2 + \cdots + 10^k a_k$$

where the $a_i \in \mathbb{Z}_{10}$ are the decimal digits of n. Let

$$S(n) = a_0 + a_1 + \cdots + a_k$$

$$(= \text{the sum of the digits of } n)$$

Then

$$n \equiv S(n) \bmod 9$$

For example, $27{,}843 \equiv 2 + 7 + 8 + 4 + 3 \bmod 9$ (8.2.4)

$$\equiv 24 \equiv 2 + 4 \equiv 6 \bmod 9,$$

so $(27{,}843)_9 = 6$.

Proof   Start with

$$1 \equiv 1 \bmod 9$$

$$10^1 \equiv 1 \bmod 9$$

$$\vdots$$

$$10^k \equiv 1 \bmod 9$$

which are all obtained from raising $10 \equiv 1$ to the powers $0, 1, \ldots, k$. Now multiply, respectively, by $a_0, a_1, \ldots, a_k$ and sum the resulting congruences. The result is (8.2.4). ■

Some arithmetic errors can be caught with this result. For example, consider the alleged product

$$1381 \times 4713 = 6{,}507{,}653$$

We can work mod 9, using Theorem 8.2.19, to find

$$1381 \equiv 1 + 3 + 8 + 1 \equiv 4 \ (9)$$

and

$$4713 \equiv 4 + 7 + 1 + 3 \equiv 6 \ (9)$$

and

$$6{,}507{,}653 \equiv 6 + 5 + \cdots \equiv 5 \ (9)$$

But $4 \times 6 = 24 \equiv 6 \bmod 9$, and not 5. Therefore the product is incorrect, since it is not even correct mod 9.

In this case, the correct answer is 6,508,653. Thus, this method caught an error in one digit. It won't catch a transposition of digits such as 6,506,853, because it has the same digit sum as the correct answer. Anyone who has ever made an incorrect telephone call knows that transposition errors occur commonly. The following theorem helps to catch such transpositions.

THEOREM 8.2.20 Mod 11

Let

$$n = a_0 + 10a_1 + 100a_2 + \cdots + 10^k a_k$$

where the $a_i \in \mathbb{Z}_{10}$ are the decimal digits of n. Let

$$A(n) = a_0 - a_1 + a_2 - \cdots = \sum_{j=0}^{k} (-1)^j a_j$$

$$(= \text{alternating sum of the digits})$$

Then

$$n \equiv A(n) \bmod 11 \qquad\qquad (8.2.5)$$

For example,
$$293{,}467 \equiv 7 - 6 + 4 - 3 + 9 - 2 \equiv 9 \bmod 11$$

Proof As in the previous theorem, we work with $10 \equiv -1 \ (11)$ and raise it to the powers from 0 to k:

$$1 \equiv 1 \bmod 11$$

$$10 \equiv -1 \bmod 11$$

$$\vdots$$

$$10^k \equiv (-1)^k \bmod 11$$

Now, multiply these by $a_0, a_1, \ldots, a_k$, and sum. The result is the required congruence (8.2.5). ∎

The alleged product $1381 \times 4713 = 6{,}506{,}853$ can be checked as follows:

$$1381 \equiv 1 - 8 + 3 - 1 \equiv -5 \,(11)$$
$$4713 \equiv 3 - 1 + 7 - 4 \equiv 5 \,(11)$$
$$6{,}506{,}853 \equiv 3 - 5 + \cdots \equiv 1 \,(11)$$

Yet $-5 \cdot 5 = -25 \equiv 8$ mod 11, not 1, so the product is incorrect, because it is not correct mod 11.

Storage of Integers

In Section 2.5, we discussed how integers can be stored in a computer by using the binary number system. Typically, an integer is stored by using 16 or 32 bits, depending on the storage space allocated. Since $2^{16} = 65{,}536$, only 65,536 integers are distinguishable in the 16-bit case. (In an 8-bit machine, two cells are used so that a 16-bit integer is stored in two 8-bit cells.) Working with the 16-bit case, we see that we can only distinguish the integers $0, 1, \ldots, 65{,}535$, and these will be given in binary as a sequence of 16 zeros and ones. If we try to add 1 to 65,535 using base 2, we get the computation

$$1111\ 1111\ 1111\ 1111 + 1 = 1\ 0000\ 0000\ 0000\ 0000$$

In the more convenient Hex notation, this is

$$\text{FFFF} + 1 = 1\ 0000$$

Because we have only 16 bits, or 4 Hex digits, it is necessary to drop the leading digit and get $\text{FFFF} + 1 = 0000$. This is precisely addition mod 2^{16} (65,536), and the Hex equation $\text{FFFF} + 1 = 0$ is correct mod 65,536: in decimal, $65{,}535 + 1 \equiv 0$ mod 65,536.

EXAMPLE 8.2.21   Negative Number Storage

It is useful for computer applications to work mod 65,536 and to perform additions and multiplications mod 2^{16}. In this way, *negative* integers can be stored; we merely work mod 2^{16}. Thus, -681 is stored as $65{,}536 - 681 = 64{,}855$. Using this system, -1 is stored as $2^{16} - 1$, which in binary is $1111\ 1111\ 1111\ 1111$. Because of the ambiguity involved in mod 2^{16}, *we treat any number as negative if the leading digit is 1*; otherwise it is positive. In this system, the highest *positive* integer is stored as $0111\ 1111\ 1111\ 1111$, or $2^{15} - 1 = 32{,}767$, and the lowest negative integer is stored as $1000\ 0000\ 0000\ 0000$ (32,768), which gets converted mod 2^{16} to $32{,}768 - 65{,}536 = -32{,}768$.

We can relate this system with finding the "one's complement." Suppose x is any integer written in binary, using 16 digits. The one's complement changes all 0's into 1's and all 1's into 0's. If y is the one's complement of x, then we see that $x + y = 1111\ 1111\ 1111\ 1111$, which has been identified as $-1 \pmod{2^{16}}$. Thus, the equation $x + y = -1$ gives the relationship between a number x and its one's complement y. In Example 8.2.21, note that 32,767 and $-32,768$ are one's complements, and this relation holds for them.

It is a simple matter to design circuits to add mod 2^{16}, because no carrying is done in the leftmost column. In many computer languages, addition and multiplication are actually done as congruence computations modulo a power of 2. It is not uncommon for a "computer error" to give a very large answer such as, say, -562. In this case, reading the fine print, the programmer will discover that it is his or her responsibility to be sure that the range of values for an integer remains within legal limits. An alternative, taken in some computer languages, is to give an OVERFLOW message if an integer exceeds the appropriate limits in any given program.

We end by noting that this way of storing and computing with negative and large integers is not standard. Another approach takes a leading 1 in a 16-bit integer as standing for a negative sign, and the other 15 digits as the absolute value of the integer. Also, instead of dropping the leading 1, a programming language might simply refuse to do a computation if an integer is out of range. Ultimately, because there are infinitely many integers and a finite amount of space, some compromise is required, and a perfect solution is not available.

Exercises

1. Find $(3^{81})_7$. Similarly, find $(3^{81})_{19}$.
2. Solve the congruence $3x \equiv 4 \bmod 11$.
3. Solve the congruence $7x \equiv 2 \bmod 33$.
4. Solve the congruence $8x \equiv 3 \bmod 19$.
5. Solve the congruence $11x \equiv 7 \bmod 35$.
6. Show that if $x_1, x_2, \ldots, x_{n+1}$ are integers, then we have $x_i \equiv x_j \bmod n$ for some pair i, j ($i \neq j$).
*7. Let $x_1, x_2, \ldots, x_m$ be distinct integers in $\mathbb{Z}_n$ with $m > n/2$. Show that for some i and j, we have $x_i + x_j \equiv 0 \bmod n$.
*8. Give an example of a multiplication of positive integers that is *incorrect*, but that is correct mod 9, 10, and 11.
9. Suppose a number n is written in octal: $n = \sum a_i 8^i$, where a_i is in $\mathbb{Z}_8$. Give simple rules for finding numbers in $\mathbb{Z}_8$ that are congruent to n mod 7, mod 8, and mod 9. Illustrate by finding $(n)_7$, $(n)_8$, and $(n)_9$ for the octal number 543612.
*10. Is the number $(1101\ 0100\ 1110\ 0111\ 0011)_2$ divisible by 5? Answer without converting to base 10. (*Hint:* Convert to base 4 and reconsider Theorem 8.2.19 or 8.2.20 in this base.)

*11. Is the number $(1101\ 0100\ 1110\ 0111\ 0011)_2$ divisible by 7? Answer without converting to base 10. (*Hint:* Convert to octal and reconsider Theorem 8.2.19 or 8.2.20 in this base.)

*12. Is the number $(1101\ 0100\ 1110\ 0111\ 0011)_2$ divisible by 3? Answer without converting to base 10. (*Hint:* Convert to base 4 and reconsider Theorem 8.2.19 or 8.2.20 in this base.)

13. Give a simple rule for determining whether an integer written in decimal notation is divisible by 3.

14. Show that $1001 \equiv 0 \bmod 7$. Hence, find $350{,}352 \bmod 7$ using this congruence.

15. Let a and n be greater than 1. Show, using congruences, that $a^n - 1$ is divisible by $a - 1$, and $a^n + 1$ is divisible by $a + 1$ (n odd). [*Hint:* $a \equiv 1 \bmod (a - 1)$, and $a \equiv -1 \bmod (a + 1)$.]

*16. Using the results of Exercise 15, prove that if $2^n - 1$ is a prime, then n must be a prime. Remark: Primes P of this type ($P = 2^p - 1$ for some prime p) are called **Mersenne primes.** Many of the largest known primes are of this type, because there are special methods available to determine whether they are prime without going through the hopeless process of checking all divisors through it's square root. For example, a known prime number of this type is $2^{216{,}091} - 1$, an integer with over 65,000 digits. It is conjectured that there are infinitely many Mersenne primes, but no proofs, either way, have been found.

*17. Using the results of Exercise 15, prove that if $2^n + 1$ is a prime, then n must be a power of 2. Remark: Primes of this type ($P = 2^{2^k} + 1$) are called **Fermat primes**, after Fermat, who conjectured that *all* numbers of this form were primes. (Here, he did not claim to have a proof.) The numbers $2 + 1 = 3$, $2^2 + 1 = 5$, $2^4 + 1 = 17$, $2^8 + 1 = 257$, and $2^{16} + 1 = 65{,}537$ are all primes. Euler computed and showed that $2^{32} + 1$ is *not* a prime, since 641 is a factor. No further Fermat primes are known to this date, and no one knows if one exists.

18. Prove that 641 divides $2^{32} + 1$, as noted in Exercise 17.

19. Solve the simultaneous congruences:

$$4x + 7y \equiv 2 \bmod 13$$

$$2x + 5y \equiv 8 \bmod 13$$

20. Solve the simultaneous congruences:

$$3x + 5t \equiv 1 \bmod 11$$

$$4x - t \equiv 2 \bmod 11$$

21. Solve the simultaneous congruences:

$$x + 2y + 4z \equiv 1 \bmod 9$$

$$2y + 7z \equiv 2 \bmod 9$$

$$5z \equiv 3 \bmod 9$$

22. Prove: If $a \equiv b$ mod n and if $d \mid n$, then $a \equiv b$ mod d. Using this result, show that the congruence $3x \equiv 7$ mod 60 has no solution.

23. Prove: If $a \equiv b$ mod n and if d is a common divisor of a, b, and n, then $a/d \equiv b/d$ mod n/d, and conversely.

24. Solve for x: $21x \equiv 15$ mod 60. (See Exercise 23.)

25. A linear coding system on the 26 letters of the alphabet uses $m = 5$ and $b = 11$ (the code is $y \equiv 5x + 11$ mod 26). (a) Using this code, what is the code message for "DO NOT DO ANY HOMEWORK"? (b) Find the decoding numbers m' and b'.

26. Using the code in Exercise 25, decode the message ULGDV PCGGH.

27. A linear coding system on the 26 letters of the alphabet codes the letter F as H and codes the letter A as K. Write the code table and the decode table. (*Hint:* Assuming the equation is $y \equiv mx + b$ mod 26, put in the two specific values for x and the corresponding values for y. Now solve for m and b as the unknowns, as in Example 8.2.13.)

28. Using the function $f(x) = (2x + 8)_{13}$ and seed $s = 0$, list the iterates $f^n(s)$ for $n = 0$ through 12. What is the fixed point for f?

*29. You want to generate random numbers as in the text by iterating the function $(16x + 7)_{113}$.

    (a) Find the fixed point for $f(x)$.

    (b) Compute the powers of 16 modulo 113 and determine the first positive power n such that $16^n \equiv 1$.

    (c) Now start with the seed 50 and find the first 10 or so generated digits. Compare your observation with the result in part (b).

    (d) Explain why this function won't work to give all the integers in $\mathbb{Z}_{113}$ except the fixed integer

30. An integer is stored by using 16 bits as in the text. If a computation has the result $n = 2^{20}$, what is the decimal output?

31. What is the one's complement of the integer 8789. Assume that the integer is stored with 16 bits. Express your answer as a decimal integer.

32. Devise an arithmetic rule to find $(n)_{101}$. Illustrate by finding a number in $\mathbb{Z}_{101}$ that is congruent to 12,409,362.

33. Devise an arithmetic rule to find $(n)_{37}$. (*Hint:* Factor 999.) Illustrate by finding 12,409,362 mod 37.

34. Devise an arithmetic rule to find $(n)_{13}$ and $(n)_7$ and illustrate on the number 14,802,221. (Here, factor 1001.)

35. By Example 8.2.7, $d = \text{GCD}(29, 52)$ can be computed as follows:

$$52 \equiv 23 \equiv -6 \text{ mod } 29$$

$$29 \equiv 5 \equiv -1 \text{ mod } 6$$

and so $d = \mid -1 \mid$. Here, remainders, *possibly negative*, are found that are less than or equal to half the modulus in absolute value. Use this method to see how many divisions are needed to find GCD(89,

55) and compare it with the number used in the usual Euclidean Algorithm when nonnegative remainders were used.

*36. Using the method outlined in Exercise 35, suppose n divisions are needed to find GCD(a, b), where $a \geq b$. Show that $b \geq 2^{n-1}$.

8.3 Further Theory of Congruences

If we fix an integer n and do arithmetic modulo n, it is convenient to work within $\mathbb{Z}_n$ so that all numbers and results of computations are in $\mathbb{Z}_n$. Thus, we simply take $a + b$ to mean $(a + b)_n$, and similarly for ab and $-a$. For example, if $n = 7$, we write

$$5 \cdot 2 = 3, \quad -3 = 4, \quad 5^3 = 6 \quad \text{in } \mathbb{Z}_7$$

In brief, any congruence

$$ab \equiv r \bmod n \qquad (r \text{ in } \mathbb{Z}_n)$$

is replaced by the statement

$$ab = r \qquad \text{in } \mathbb{Z}_n$$

or simply

$$ab = r$$

once it is understood that we are working in $\mathbb{Z}_n$.

Since $\mathbb{Z}_n$ is *finite*, we have a finite algebraic system with rules much like the rules of algebra. We illustrate the arithmetic tables for $\mathbb{Z}_6$ in Table 8.3.1.

Table 8.3.1   Operations in $\mathbb{Z}_6$

+	0	1	2	3	4	5
0	0	1	2	3	4	5
1	1	2	3	4	5	0
2	2	3	4	5	0	1
3	3	4	5	0	1	2
4	4	5	0	1	2	3
5	5	0	1	2	3	4

$\cdot$	0	1	2	3	4	5
0	0	0	0	0	0	0
1	0	1	2	3	4	5
2	0	2	4	0	2	4
3	0	3	0	3	0	3
4	0	4	2	0	4	2
5	0	5	4	3	2	1

x	$-x$
0	0
1	5
2	4
3	3
4	2
5	1

The usual laws of algebra apply to $\mathbb{Z}_n$ insofar as they involve addition and multiplication. Briefly, these are, for all choices of a, b, and c:

Commutative Laws:	$ab = ba; a + b = b + a$
Associative Laws:	$a(bc) = (ab)c; a + (b + c) = (a + b) + c$
Identity Laws:	$a + 0 = a; a \cdot 1 = a; 0 \cdot a = 0$
Negative Law:	$a + (-a) = 0$
Distributive Law:	$a(b + c) = ab + bc$

The reader should compare these laws with the axioms for a Boolean algebra (Definition 7.2.2). As observed there, these axioms are valid in a commutative ring with unity. Of course, we do not have the Boolean condition $a + a = 0$, but we have the Negative Law. Thus, we can subtract freely. The equation $x + a = b$ has the unique solution $x = b - a = b + (-a)$. The law of algebra that does *not*, in general, apply is the possibility of division. For example, the equation $2x = 1$ in $\mathbb{Z}_6$ has no solution. As we shall see, we can divide a by b ($b \neq 0$) in $\mathbb{Z}_n$ if n is prime. This has nothing to do with whether a is divisible by b in the usual sense (in $\mathbb{Z}$).

EXAMPLE 8.3.1

Solve $3x = 2$ in $\mathbb{Z}_{11}$.

Method Multiply (in $\mathbb{Z}_{11}$) by 4 to get $12x = 8$, or $x = 8$. Conversely, checking the solution, if $x = 8$, then $3x = 24 = 2$ in $\mathbb{Z}_{11}$.

Remark As in the previous section, we multiplied by 4 in order to get the coefficient of x nearer the modulus 11. Then, by reducing it modulo 11, we found a smaller number, in this case 1 since $12 \equiv 1 \bmod 11$.

We may rephrase the question and the method of solution by using fractional notation. The question was to find $\frac{2}{3}$ in $\mathbb{Z}_{11}$. The answer: $\frac{2}{3} = \frac{8}{12} = \frac{8}{1} = 8$.

The following result gives a necessary and sufficient condition for a reciprocal to exist in $\mathbb{Z}_n$.

DEFINITION 8.3.2

A number a in $\mathbb{Z}_n$ is **invertible** in $\mathbb{Z}_n$ if there exists a number a^{-1} such that $aa^{-1} = 1$ in $\mathbb{Z}_n$. The number a^{-1} is called the **inverse** of a in $\mathbb{Z}_n$. If a is invertible, we write $ba^{-1} = b/a$. In particular, we write $a^{-1} = 1/a$.

Remark The number a^{-1} must be unique. For if a has two inverses b and c, then $ab = ac = 1$ in $\mathbb{Z}_n$. Multiplying by b, we obtain $bab = bac$. Since $ab = ba = 1$, this gives $b = c$.

THEOREM 8.3.3

The integer a is invertible in $\mathbb{Z}_n$ if and only if a and n are relatively prime.

Proof If a and n are relatively prime, then, by Theorem 8.1.15, $ax + ny = 1$ for some integers x and y. Thus, $ax \equiv 1 \bmod n$ and, hence, $ax = 1$ in $\mathbb{Z}_n$. By definition, this implies that a is invertible and that $x = a^{-1}$. Conversely, if a is invertible in $\mathbb{Z}_n$ there is an inverse x: $ax = 1$ in $\mathbb{Z}_n$. But this means that $ax \equiv 1 \bmod n$, or that $ax = 1 + qn$ for some q. Since $ax - qn = 1$, it follows that a and n are relatively prime, again by Theorem 8.1.15. ∎

This result not only states when a has an inverse in $\mathbb{Z}_n$ but also shows how to get it. Using the Euclidean Algorithm, we can find integers x and y such that $ax + ny = 1$. Then $a^{-1} = x$ in $\mathbb{Z}_n$. Other methods of finding inverses are discussed later in this section.

Once we have inverses, division is not far behind.

THEOREM 8.3.4

Let a and n be relatively prime. Then the equation $ax = b$ has the unique solution $x = ba^{-1}$ in $\mathbb{Z}_n$. We also write $x = b/a$.

Proof Suppose a^{-1} is an inverse of a in $\mathbb{Z}_n$.

(Uniqueness) If $ax = b$ in $\mathbb{Z}_n$, then multiplying by a^{-1} we find that $x = ba^{-1}$ in $\mathbb{Z}_n$. Thus, the value of x must be ba^{-1}.

(Existence) Setting $x = ba^{-1}$, we have $ax = aba^{-1} = b$ in $\mathbb{Z}_n$ because $aa^{-1} = 1$ in $\mathbb{Z}_n$. Thus, $x = ba^{-1}$ is a solution. ∎

We illustrate these results with an example.

EXAMPLE 8.3.5

Solve $17x = 23$ in $\mathbb{Z}_{58}$.

Method First find $1 = \mathrm{GCD}(17, 58)$ linearly in terms of 17 and 58. We use the Euclidean Algorithm:

$$58 = 17 \cdot 3 + 7$$
$$17 = 7 \cdot 2 + 3$$
$$7 = 3 \cdot 2 + 1$$

Now, working backwards through these equations, we obtain

$$1 = 7 - 3 \cdot 2$$
$$= 7 - (17 - 7 \cdot 2) \cdot 2$$
$$= 7 \cdot 5 - 17 \cdot 2$$
$$= (58 - 17 \cdot 3) \cdot 5 - 17 \cdot 2$$
$$= 58 \cdot 5 - 17 \cdot 17$$

This shows that $17^{-1} = -17 = 41$ in $\mathbb{Z}_{58}$. Therefore, the solution of $17x = 23$ in $\mathbb{Z}_{58}$ is $17^{-1} \cdot 23 = 41 \cdot 23 = 15 \; [=(41 \cdot 23)_{58}]$. Thus, $x = 15$. As a check, we can directly compute that $17 \cdot 15 \equiv 23 \bmod 58$. Strictly speaking, a check was not necessary, in view of Theorem 8.3.4.

The preceding results are significantly strengthened when n is prime. In what follows, we fix a prime p and work in $\mathbb{Z}_p$.

THEOREM 8.3.6

If $a \neq 0$ in $\mathbb{Z}_p$, then a^{-1} exists in $\mathbb{Z}_p$. The equation $ax = b$ has a unique solution $x = ba^{-1} = b/a$ provided $a \neq 0$ in $\mathbb{Z}_p$.

Proof   This is a restatement of Theorems 8.3.3 and 8.3.4 when n is a prime. For in this case, if an integer a is not divisible by p $(a \neq 0)$, then it is relatively prime to p, and conversely.   ∎

This theorem gives us the usual situation in algebra: we can divide, except by 0. The systems $\mathbb{Z}_p$ are therefore miniature algebraic systems in which the usual properties of algebra relating to addition, multiplication, subtraction, *and* division are valid. They are called **finite fields**. (The term **field** refers to an algebraic system where these usual laws of algebra apply.)

The following cancellation laws follow from our results on inverses. As usual, n is arbitrary and p is assumed to be prime.

THEOREM 8.3.7   *Cancellation Laws*

1. If $ab = 0$ in $\mathbb{Z}_n$ and $\text{GCD}(a, n) = 1$, then $b = 0$ in $\mathbb{Z}_n$.
2. If $ab = ac$ in $\mathbb{Z}_n$ and $\text{GCD}(a, n) = 1$, then $b = c$ in $\mathbb{Z}_n$.
3. If $ab = 0$ in $\mathbb{Z}_p$, then $a = 0$ or $b = 0$ in $\mathbb{Z}_p$.
4. If $ab = ac$ in $\mathbb{Z}_p$ and $a \neq 0$ in $\mathbb{Z}_p$, then $b = c$ in $\mathbb{Z}_p$.

Proof   These are all obtained by multiplying the equalities by a^{-1}. The reader should compare with Theorems 8.1.16 and 8.1.18.   ∎

Finite Fields

The theory of finite fields is extensively studied in algebra and has many practical and theoretical applications. We have noted that $\mathbb{Z}_p$ is a finite field for any prime p. However, there are other finite fields.

Table 8.3.2 gives the arithmetic operations of a field F_4 with four elements: 0, 1, α, and β. Some of the properties of F_4 are given in the exercises.

Do not confuse F_4 with $\mathbb{Z}_4$: $\mathbb{Z}_4$ is not a field. In it, $2 \cdot 2 = 0$, yet $2 \neq 0$. In F_4, $2 = 1 + 1 = 0$.

The table itself does not guarantee that we have a field. The various algebraic laws noted after Table 8.3.1 need to be verified. It turns out there is a finite field with p^n elements for any prime p and $n \geq 1$. Table

Table 8.3.2 Operations in the Field F_4

+	0	1	α	β
0	0	1	α	β
1	1	0	β	α
α	α	β	0	1
β	β	α	1	0

·	0	1	α	β
0	0	0	0	0
1	0	1	α	β
α	0	α	β	1
β	0	β	1	α

x	$-x$
0	0
1	1
α	α
β	β

8.3.2 illustrates this field for $n = 2$ and $p = 2$. The case $n = 1$ gives the field $\mathbb{Z}_p$.

Computing Inverses

The method given in Theorems 8.3.3 and 8.3.4 is an effective way of solving the congruence $ax \equiv b \bmod n$. However, it is not the only way. In Example 8.3.1, we dealt with such an equation by multiplying it through by some number c in order to bring the coefficient of x closer to 1. Since we want to divide by c to justify this, we have to choose c relatively prime to n. However, if n is a prime p, all we need is $c \neq 0$. We illustrate with a simple example, mod 61.

EXAMPLE 8.3.8

Solve the equation $23x = 51$ in $\mathbb{Z}_{61}$.

Method Write $x = 51/23$ and multiply numerator and denominator by a number that brings the denominator closer to 1. All numbers are taken in $\mathbb{Z}_{61}$:

$$
\begin{aligned}
x = 51/23 &= -10/23 \quad (\times 3/3) \\
&= -30/69 \\
&= -30/8 = -15/4 \quad (\times 15/15) \\
&= -225/60 \\
&= -42/(-1) = 42
\end{aligned}
$$

Remarks (1) Using negatives, we cut 51 down to -10, making the hand calculation easier. By working with negatives, it is always possible to replace a number by one congruent to it mod p that is smaller than $p/2$ in absolute value, assuming that p is an odd prime. (2) We converted $30/8$ to $15/4$ by canceling a common factor 2. This was a good idea because it cut down the denominator in size. It is legitimate because the law of algebra $ax/bx = a/b \ (x \neq 0)$ is true. All fractions are understood to be elements of $\mathbb{Z}_{61}$ by Definition 8.3.2.

We can summarize this method for computing a/b in $\mathbb{Z}_p$ as follows. If $b > 1$, divide p by b to obtain $p = bq + r$, with $0 < r < b$.

Therefore $bq = -r$ in $\mathbb{Z}_p$, and $a/b = aq/bq = -aq/r$. Thus, the denominator is reduced in size and the process can be continued until it is 1. The numerator in each case is reduced mod p. The method of cutting down size by taking negatives is done by replacing any integer a by $a - p$ if a is greater than $p/2$. Thus, in the previous example, 51 was replaced by its equivalent $51 - 61 = -10$.

EXAMPLE 8.3.9   A Quotient mod p Algorithm

The following algorithm gives the foregoing method for evaluating quotients mod p. For simplicity, we have not introduced "negatives" here. The input consists of the integers a and b in $\mathbb{Z}_p$, and p is a fixed prime number throughout. The output is $c = a/b$ in $\mathbb{Z}_p$.

QuotientMod$p(a, b; c)$ algorithm
1. IF $b = 0$ THEN
    1. Output is stern warning "You can't divide by 0".
    2. End
2. WHILE $b \neq 1$
    1. $q := [p/b]$ [greatest integer function]
    2. $a := (-aq)_p$
    3. $b := p - bq$
3. $c := a$ [the denominator b is now 1]

Fermat's Little Theorem, Powers, and Inverses

In what follows, we let p be a prime, and we work mod p, or equivalently in $\mathbb{Z}_p$. Fermat's Little Theorem is a pretty result about powers mod p.

THEOREM 8.3.10   Fermat's Little Theorem

Let $a \neq 0$ in $\mathbb{Z}_p$. Then $a^{p-1} = 1$ in $\mathbb{Z}_p$. In congruence form

$$\text{If } a \not\equiv 0 \bmod p, \text{ then } a^{p-1} \equiv 1 \bmod p$$

Proof   The following proof is based on the Box Principle. We consider the $p - 1$ distinct nonzero elements of $\mathbb{Z}_p$: $1, 2, \ldots, p - 1$. Multiply each of these by a to get the elements $a, 2a, \ldots, (p - 1)a$. Now these are all distinct and not 0. To see this, note that if $ra = sa$ in $\mathbb{Z}_p$, then $r = s$ by the Cancellation Law. Therefore, the elements $a, 2a, \ldots, (p - 1)a$ are just a rearrangement of $1, 2, \ldots, p - 1$. Therefore they have the same product: $1 \cdot 2 \cdots (p - 1) = a \cdot 2a \cdots (p - 1)a$. This simplifies to

$$(p - 1)! = (p - 1)! a^{p-1} \qquad \text{in } \mathbb{Z}_p$$

The result is obtained by canceling $(p - 1)!$, which is not 0.   ∎

Theorem 8.3.10 gives another method for finding a^{-1}.

COROLLARY 8.3.11

In $\mathbb{Z}_p$, $a^{-1} = a^{p-2}$. Thus, the solution of $ax \equiv b \bmod p$ is $x \equiv ba^{p-2}$.

Proof   Note that $a \cdot a^{p-2} = 1$.   ∎

This formula for a^{-1} can be used practically. What is needed is a method for efficiently computing powers of $a \bmod p$. This will be discussed in Examples 8.3.14 and 8.3.15.

Theorem 8.3.10 is useful for finding large powers in $\mathbb{Z}_p$. For example, to find $3^{547} \bmod 31$, we can use Fermat's Little Theorem to obtain $3^{30} = 1$ in $\mathbb{Z}_{31}$. If we divide 547 by 30, we obtain $547 = 30 \cdot 18 + 7$, so $3^{547} = (3^{30})^{18}3^7 = 3^7$. A simple computation for 3^7 gives the answer 17. In effect, we have reduced the exponent 547 modulo 30 and replaced it by 17. This method works in general.

COROLLARY 8.3.12

Suppose $a \neq 0$ in $\mathbb{Z}_p$ and $n \equiv r \bmod (p-1)$. Then $a^n \equiv a^r \bmod p$.   ∎

In brief, exponents may be reduced $\bmod (p-1)$ when exponentiating $\bmod p$. In Example 8.3.11, $3^{547} = 3^7 = 17$ in $\mathbb{Z}_{31}$. Thus, a theory of logarithms in $\mathbb{Z}_{31}$ would have $\log_3 17 = 7$, or 547. In the preceding computation, 547 was reduced to 7 modulo 30. Thus logarithms should be taken $\bmod 30$ so that we can say $\log_3 17 = 547 = 7$ in $\mathbb{Z}_{30}$. In general, any theory of logarithms in $\mathbb{Z}_p$ will locate the logarithm in $\mathbb{Z}_{p-1}$.

Corollary 8.3.12 allows us to find certain **roots** of numbers $\bmod p$ in an especially simple way—by raising to a power.

THEOREM 8.3.13

Suppose p is a prime and e is relatively prime to $p-1$. Let e' be the inverse of $e \bmod p-1$. Then

$$y \equiv x^e \bmod p \quad \text{iff} \quad y^{e'} \equiv x \bmod p$$

Proof   To see this, first note that $ee' \equiv 1 \bmod (p-1)$. Therefore, if $y \equiv x^e \bmod p$, we can raise both sides to the e' power to obtain, by Corollary 8.3.12, $y^{e'} \equiv x^{ee'} \equiv x \bmod p$. Similarly, if $y^{e'} \equiv x$, we obtain $y \equiv x^e$.   ∎

For example, to solve the congruence $x^3 \equiv 4 \bmod 11$, we first find an inverse of $3 \bmod 10$. This is 7. Therefore, $x \equiv 4^7 \bmod 11$, which gives $x \equiv 5 \bmod 11$.

Remarks   (1) The equation $x \equiv y^{e'}$ is more familiar from algebra as $x \equiv y^{1/e} \bmod (p-1)$ [fractional exponents!].
(2) If e is not relatively prime to $p-1$, the situation is much more complicated. Thus, the congruence $x^2 \equiv 5 \bmod 11$ has *two* solutions,

$x \equiv 4$ and $x \equiv 7$, but the congruence $x^2 \equiv 2 \bmod 11$ has no solutions. Corollary 8.3.12 does not give us any help in solving these simple congruences.

How can we compute $a^n \bmod p$ in an efficient way? Corollary 8.3.12 shows that we may take $n \bmod (p - 1)$, so we can always get away with fewer than $p - 2$ multiplications $\bmod p$. To keep the numbers small, once a multiplication is performed, it is reduced $\bmod p$. There is an effective, practical way to cut down on the number of such multiplications, which is based on representing n in binary. We illustrate with two examples, both done on a programmable hand calculator using a simple program to evaluate a number $\bmod p$.

EXAMPLE 8.3.14

Compute $17^{200} \bmod 677$.

Method   First we compute 17^2, then its square, and so forth, always squaring the previous answer. This leads to Table 8.3.3.

Table 8.3.3   Certain Powers of 17 mod 677

n	17^n
1	17
2	289
4	250
8	216
16	620
32	541
64	217
128	376
256	560
512	149

Now write $n = 200$ in binary as a sum of powers of 2:

$$200 = 128 + 64 + 8$$

Therefore,

$$17^{200} = 17^{128 + 64 + 8} = 17^{128}\,17^{64}\,17^8 = 376 \cdot 217 \cdot 216 = 208$$

The method is quite efficient. Instead of 199 multiplications, we formed the table using eight multiplications (squares), and the final answer used three more. The table can then be used to find any power of 17 mod 677 with only a few multiplications. In particular, this method is effective for finding the inverse of an integer mod p. For, by Corollary 8.3.11, $17^{-1} = 17^{675}$ in $\mathbb{Z}_{677}$. But $675 = 512 + 128 + 32 + 2 + 1$. [This is equivalent to $675 = (1010100011)_2$.] Therefore

$$17^{-1} = 149 \cdot 376 \cdot 541 \cdot 289 \cdot 17 = 239$$

using Table 8.3.3.

Another efficient approach to computation of powers, not using Table 8.3.3, is also based on binary notation. First observe that if $a^k = b$, then, by squaring, we obtain $a^{2k} = b^2$. Also, multiplying by a gives $a^{k+1} = ab$. If, now, k is expressed in binary, as $k = (b_1 \cdots b_r)_2$, then $2k$ is simply $(b_1 \cdots b_r 0)_2$ and $2k + 1 = (b_1 \cdots b_r 1)_2$. Thus, when $a^k = b$, we have the following:

1. Squaring b has the effect of adjoining 0 to the right end of the binary expansion of k.
2. Squaring b and multiplying by a has the effect of adjoining 1 to the right end of the binary expansion of k.

We use this to advantage in the next example.

EXAMPLE 8.3.15

Compute 34^{146} mod 633.

Method   First, find $k = 146$ in binary: $146 = (10010010)_2$. Now we start with $34^1 = 34$ in $\mathbb{Z}_{633}$. We are going to change the exponent of 34 to 10010010 by using the preceding rules: Adjoining 0 corresponds to squaring (S); adjoining 1 corresponds to squaring and then multiplying by 34 (SM). Thus, the symbols 0010010, which follow the initial 1 in the exponent of 34, correspond to the operations S S SM S S SM S, or simply

Table 8.3.4

Operation	Power	Computation	Exponent (binary)
	34^1	34	1
S	34^2	523	10
S	34^4	73	100
S	34^8	265	1000
M	34^9	148	1001
S	34^{18}	382	10010
S	34^{36}	334	100100
S	34^{72}	148	1001000
M	34^{73}	601	1001001
S	34^{146}	391	10010010

SSSMSSSMS. Thus, we square 34, square again, and again, then multiply by 34, and so on. This yields the calculator computation (all modulo 633) of Table 8.3.4 (S = square, M = multiply by 34).

The computation yields the answer 391. We see that we needed nine operations (rather than 145 multiplications) because of the judicious use of squaring.

Random-Like Sequences

In the previous section, we showed how to generate a "random" set of integers mod p by choosing a function $f(x) = ax + b$ in $\mathbb{Z}_p$, a seed s, and computing the sequence recursively defined by $s_0 = s$ and $s_{n+1} = f(s_n)$. Under certain circumstances (explained in what follows), this sequence will go through all $p - 1$ values mod p, except the fixed point.

If $f(x) = ax + b$ in $\mathbb{Z}_p$ and $a \neq 1$, there is one and only one fixed point. For the equation $f(x) = x$ is simply $ax + b = x$, or $(a - 1)x = -b$, which has exactly one solution $-b/(a - 1)$ in $\mathbb{Z}_p$.

DEFINITION 8.3.16

Let $f(x) = ax + b$ be a linear function in $\mathbb{Z}_p$, where $a \neq 0$ and $a \neq 1$. Then the sequence of iterates $f^n(s)$ is called **random-like** if this sequence goes through all $p - 1$ values mod p except the fixed point.

We now determine when we have a random-like sequence. In what follows we assume that $f(x) = ax + b$ in $\mathbb{Z}_p$ and that $a \neq 0$ and $a \neq 1$.

EXAMPLE 8.3.17

We cannot expect a random-like sequence for any given linear $f(x)$. For example, take $p = 11$ and $f(x) = 3x + 7$. Then

$$f^2(x) = f(f(x)) = 3(3x + 7) + 7 = 9x + 6$$
$$f^3(x) = f(f^2(x)) = 3(9x + 6) + 7 = 5x + 3$$
$$f^4(x) = 3(5x + 3) + 7 = 4x + 5$$
$$f^5(x) = 3(4x + 5) + 7 = x$$

Therefore, independently of the seed s, s will appear again after five iterates.

LEMMA 8.3.18

Let $f(x) = ax + b$ in $\mathbb{Z}_p$, $a \neq 0$ and $a \neq 1$, and with fixed point x_0. Then for all n, the nth iterates of f are given by

$$f^n(x) = a^n(x - x_0) + x_0 \qquad (8.3.1)$$

Proof   First, note that Equation (8.3.1) is true for $n = 0$. We need to verify it for $n = 1$ also. By definition,

$$x_0 = ax_0 + b$$

Also,

$$f(x) = ax + b$$

Subtracting, we get $f(x) - x_0 = a(x - x_0)$, so

$$f(x) = a(x - x_0) + x_0 \qquad (8.3.2)$$

This is Equation (8.3.1) for $n = 1$. The rest is an induction. Assuming (8.3.1), we have

$$
\begin{aligned}
f^{n+1}(x) &= f(f^n(x)) \\
&= a(f^n(x) - x_0) + x_0 \qquad \text{[using (8.3.2)]} \\
&= a \cdot a^n(x - x_0) + x_0 \qquad \text{(by induction)} \\
&= a^{n+1}(x - x_0) + x_0 \quad \blacksquare
\end{aligned}
$$

COROLLARY 8.3.19

The iterate $f^n(x)$ has leading coefficient a^n and fixed point x_0. If $a^n = 1$, then $f^n(x) = x$.   ■

We can now analyze when a sequence of iterates of a linear function $f(x) = ax + b$ is random-like.

DEFINITION 8.3.20

If a is not 0 in $\mathbb{Z}_p$, the **order of a** is defined as the least positive integer k such that $a^k \equiv 1$.

Remark   By Fermat's Little Theorem (Theorem 8.3.10), the order of a is $\leq p - 1$.

EXAMPLE 8.3.21

Find the order of 3 mod 11.

Method   Find the powers of 3 in $\mathbb{Z}_{11}$ and see which power is 1. These powers are 3, $3^2 = 9$, $3^3 = 27 = 5$, $3^4 = 15 = 4$, and $3^5 = 12 = 1$. Thus the order of 3 mod 11 is 5.

THEOREM 8.3.22

Let a have order $p - 1$. Then if s is any seed unequal to the fixed point x_0, the sequence of iterates

$$f(s), f^2(s), \ldots, f^{p-1}(s) = s \qquad (8.3.3)$$

is random-like. Conversely, if the sequence of iterates is random-like, then a has order $p - 1$.

Proof Assume that a has order $p - 1$. We have to show that the numbers of (8.3.3) are distinct mod p. If $f^m(s) = f^n(s)$, with $1 \le m < n \le p - 1$, we have, by (8.3.1), $a^m(s - x_0) + x_0 = a^n(s - x_0) + x_0$. But since $s - x_0$ is not equal to 0, this implies $a^m = a^n$. Using cancellation, since $m < n$, we have $1 = a^{n-m}$. But this is impossible, because we have assumed that $p - 1$ is the least exponent of a that yields 1, and $n - m$ is smaller.

Conversely, if a has order $n < p - 1$, $f^n(s) = s$ by Equation (8.3.1). Thus, repetitions occur after n iterates. ∎

The proof shows that if a has order n ($a \ne 0$ and $a \ne 1$), then there are exactly n distinct terms in the iterates of $f(x) = ax + b$. In Example 8.3.17, we chose $f(x) = 3x + 7$ in $\mathbb{Z}_{11}$. We can now see why there were repetitions after five iterates. The order of 3 mod 11 was computed to be 5 in Example 8.3.21. Thus, the sequence of iterates has only five distinct terms.

We can prove a few more facts about the order of an element in $\mathbb{Z}_p$.

THEOREM 8.3.23

Let a be a nonzero element of $\mathbb{Z}_p$, and let k be the order of a. Then k divides $p - 1$.

Proof We have $a^{p-1} = 1$, by Fermat's Little Theorem, and $a^k = 1$. Divide $p - 1$ by k to obtain $p - 1 = kq + r$, $0 \le r < k$. Thus, we have

$$1 = a^{p-1} = a^{kq+r} = (a^k)^q \cdot a^r = a^r$$

By definition, k is the least positive exponent for which $a^k = 1$. Since $r < k$, r cannot be positive, so that $r = 0$. This is the result: $p - 1 = kq$. ∎

Remark This result makes it relatively easy to find the order of an element a mod p. The only exponents we need to check are the divisors of $p - 1$.

It is true that for any prime p there are elements a of order $p - 1$. (These are called **primitive roots mod p**.) The proof of this would take us too far afield. By the foregoing results, any such a and arbitrary b yield a function $f(x) = ax + b$ whose iterates in $\mathbb{Z}_p$ are random-like provided the seed is not the fixed point.

Cryptography Again

In the last section, we considered the simple code system given by $y = ax + b$ in $\mathbb{Z}_p$. Here x was the number of a letter to be coded, and y was the number of the code. We pointed out that this is a relatively simple linear substitution code. Because we only used 26 letters (though obviously the alphabet could be extended), these codes could be cracked relatively easily. We now illustrate another substitution code, given by a more intricate formula, as well as a larger alphabet. Again, we number A by 1, B by 2, . . . , Y by 25, and Z by 0.

EXAMPLE 8.3.24   A Superalphabet

We can introduce a larger alphabet, by pairing letters, and regard, for example, BR as a single superletter. We can assign a number to this in several ways. For example, we can use 0218 as its number, taking the number for B and juxtaposing with the number for R. Instead, let us consider BR as a number written in base 26, so B is in the "26 column" and R is in the units column. (The digits in base 26 are taken as the full alphabet of 26 letters.) Thus BR is, in decimal notation, $2 \cdot 26 + 18 = 70$. We can retrieve the 2 and 18 by dividing 70 by 26 to get its quotient and remainder. The largest number appearing is the number for YY, which is 675. We have thus created a relatively large alphabet. (When decoding a number larger than 675, we simply take that number modulo 676, which is 26^2.)

The code we use is created as follows. Find a prime p and integers a, b, c, and e, and let

$$y = a + b(x - c)^e \qquad \text{in } \mathbb{Z}_p \qquad (8.3.4)$$

Here $b \neq 0$ in $\mathbb{Z}_p$ and e is relatively prime to $p - 1$. Using Theorem 8.3.13, we can solve this for x:

$$x = c + ((y - a)/b)^{e'}$$

where e' is the inverse of e mod $(p - 1)$. We call such a code an **exponential** code.

EXAMPLE 8.3.25

We illustrate an exponential code with a simple example. We need to choose a prime larger than 675 so that the superletters have distinct codes. We choose $p = 677$. The exponent e must be relatively prime to 676. We choose $e = 49$ and use the relatively simple code

$$\text{CODE: } y = x^{49} \quad \text{in } \mathbb{Z}_{677}$$

Since $\frac{1}{49} = 69$ in $\mathbb{Z}_{676}$, the inverse is

$$\text{DECODE: } x = y^{69} \quad \text{in } \mathbb{Z}_{677}$$

Using the methods of Examples 8.3.14 and 8.3.15 and a decent hand calculator or computer program, this is not difficult to compute. For example, to code DONT COME, we proceed as follows:

```
TEXT:   DO NT CO ME
   x:   119 384 093 343
   y:   641 473 623 572      (y = x^49 in Z_677)
CODE:   XQ RE WY VZ
```

Thus the coded message is XQREW YVZ. Decoding is the reverse process, where we use $x = y^{69}$ in $\mathbb{Z}_{677}$. Codes of the type (8.3.4) are relatively difficult to crack because of the numerous parameters. Even if we use a pure power $y = x^e$ in $\mathbb{Z}_p$, it is still necessary to know p and e to code and decode.

Exercises

1. Construct addition and multiplication tables for $\mathbb{Z}_7$. Explain why each row and column in the addition table is a rearrangement of the elements of $\mathbb{Z}_7$. Similarly, explain why this is so for each row and column of the multiplication table except for the row corresponding to multiplication by zero, and for the similar column.
2. Redo Exercise 1 for $\mathbb{Z}_5$.
3. Solve the congruence $41x \equiv 31 \bmod 73$.
4. Solve the congruence $48x \equiv 37 \bmod 75$.
5. Find the inverse of 14 in $\mathbb{Z}_{33}$.
*6. Find the inverse of 35 in $\mathbb{Z}_{61}$ by two methods: those of Theorem 8.3.3 and Corollary 8.3.11.
*7. Find the inverse of 41 in $\mathbb{Z}_{47}$ by two methods: the method of Theorem 8.3.3, and the algorithm QuotientMod$p$ of Example 8.3.9.
8. Prove in $\mathbb{Z}_n$: If a and b have inverses, then so does ab, and $(ab)^{-1} = a^{-1}b^{-1}$. Translate this into a statement concerning relatively prime numbers.
9. Let $\mathbb{Z}_n^*$ be the set of elements in $\mathbb{Z}_n$ that have inverses. List the elements of $\mathbb{Z}_8^*$ and of $\mathbb{Z}_{15}^*$. Construct a multiplication table for the

elements of $\mathbf{Z}_8^*$ and explain why each row and column is a rearrangement of the elements.

*10. Find a number n and an element a in $\mathbf{Z}_n$ such that $a \neq 0$ and $a^2 = 0$ in $\mathbf{Z}_n$. For which n is it possible to find such an a? For which n is it impossible?

11. If $x \equiv 5$ mod 9 and $(x - 5)/9 \equiv 2$ mod 9, find x mod 81.

12. Prove in $\mathbf{Z}_{125}$: If a does not have an inverse, then some power of a must be 0.

13. Prove: If n is a power of a prime p and if x does not have an inverse in $\mathbf{Z}_n$, then some power of x must be 0 in $\mathbf{Z}_n$. (Compare with Exercise 12.) Conversely, show that if every x that does not have an inverse in $\mathbf{Z}_n$ has the property that some power of it is 0 in $\mathbf{Z}_n$, then n is a power of some prime.

14. In the finite field F_4 defined in Table 8.3.2, compute
    (a) $\alpha + \alpha^2$ (b) $\alpha/(1 + \alpha)$ (c) $(1 + \alpha)(1 + \beta)$
    Your answer must be one of 0, 1, α, or β.

15. In the finite field F_4 defined in Table 8.3.2, compute
    (a) $1/(\alpha + \beta)$ (b) $\alpha/(1 + \beta)$ (c) $\beta^2 + \alpha^2$

16. In the finite field F_4 defined in Table 8.3.2, verify the associative law $[\beta(1 + \beta)]\alpha = \beta[(1 + \beta)\alpha]$.

17. In the finite field F_4 defined in Table 8.3.2, verify that $x^3 = 1$ for all $x \neq 0$ in this field.

*18. Prove that in *any* finite field F with s elements, $x^{s-1} = 1$ for all $x \neq 0$ in F. (*Hint:* Look at the proof of Theorem 8.3.10.)

*19. Suppose $f(x) = 3x + 14$ in $\mathbf{Z}_{19}$. Find $f^{14}(x)$, the fourteenth iterate of $f(x)$. [*Hint:* Use Equation (8.3.1).]

*20. Repeat Exercise 19, finding a linear equation for $f^9(x)$, where $f(x) = 7x + 3$ in $\mathbf{Z}_{13}$.

*21. Let $\mathbf{Z}_n^*$ be defined as in Exercise 8.3.9. Define $\varphi(n)$ as the number of elements in $\mathbf{Z}_n^*$. [$\varphi(n)$ is the **Euler φ function,** equivalently defined and evaluated in Definition 4.1.20 and Theorem 4.1.21.] Prove that if a is relatively prime to n then

$$a^{\varphi(n)} = 1 \quad \text{in } \mathbf{Z}_n$$

This is a generalization of Fermat's Little Theorem, because, when $n = p$, $\varphi(n) = p - 1$. (*Hint:* List the elements of $\mathbf{Z}_n^*$ as $a_1, a_2, \ldots, a_{\varphi(n)}$ and multiply each of them by a. Show that they are a rearrangement of the elements and now use the technique in the proof of Theorem 8.3.10.)

22. Show that $p - 1$ is the only element of order 2 in $\mathbf{Z}_p$ (p an odd prime).

23. Using Theorem 8.3.13, show that the equation $x^7 = 2$ mod 13 has a unique root. Find this root, using this theorem.

24. Prove: If $x^5 \equiv y^5$ mod 29, then $x \equiv y$ mod 29.

25. Solve the equation $x^2 = 2$ completely in $\mathbf{Z}_7$.

*26. Prove: If $f: \mathbf{Z}_p \to \mathbf{Z}_p$, with $f(x) = x^e$, and if e and $p - 1$ are relatively prime, then f is a 1-1 function.

*27. Show that if the congruence $x^4 \equiv a$ mod 29 has a solution x then either $a \equiv 0$ or $a \equiv 1$ or a has order 7 mod 29.

28. In $\mathbb{Z}_{13}$, list all the elements of orders 1, 2, and 4. Verify that the square of every element of order 4 has order 2 and that the square of every element of order 2 has order 1. Explain, theoretically, why this is so.

29. Generalize Exercise 28 by proving that in $\mathbb{Z}_p$ the square of any element of order $2r$ has order r.

*30. Prove that in $\mathbb{Z}_n$ if $a^r = a^s = 1$, then the order of a divides $\mathrm{GCD}(r, s)$.

*31. Prove in $\mathbb{Z}_p$ that if a has order $p - 1$ and n is relatively prime to $p - 1$, then a^n also has order $p - 1$. Illustrate for $p = 11$ and $a = 2$.

*32. Using the result of Exercise 31, find ten elements of order 22 in $\mathbb{Z}_{23}$.

*33. Construct a table for 2^n mod 677 similar to Table 8.3.3. Take n to consist of the powers of 2 through 512 and get each term by squaring the previous term and reducing mod 677.

*34. Using the table constructed in the previous exercise and the method used in Example 8.3.14 of the text, find 2^{133} mod 677.

*35. Repeat Exercise 34, but use the technique of Example 8.3.15.

*36. (*Logarithms*) Suppose p is a prime and a has order $p - 1$. Show that $a^x \equiv a^y$ mod p if and only if $x \equiv y$ mod $p - 1$. Now if y is in $\mathbb{Z}_p$, $y \neq 0$, define $x = \log_a y$ provided $a^x = y$ in $\mathbb{Z}_p$. Show that $\log_a y$ is 1-1 onto $\mathbb{Z}_{p-1}$. Prove that $\log_a yz = \log_a y + \log_a z$ in $\mathbb{Z}_{p-1}$.

37. Revise the algorithm QuotientMod$p$ of Example 8.3.9 to use the shortcut of taking negatives. Thus, if $n \in \mathbb{Z}_p$ and $n > p/2$, replace it with $n - p$.

Computer Exercises

(Intended only for readers who have some programming knowledge. *Warning:* some of these programs might take some time to develop.)

38. Write a program to compute a/b in $\mathbb{Z}_p$. Take p as given and a and b as variable integer inputs. Use the algorithm QuotientMod$p$ of Example 8.3.9. Use the built-in functions that compute the quotient and remainder when any integer is divided by a positive integer.

39. Write a reasonably efficient program to compute a^n mod p. Take p as a fixed prime, and a and n as inputs.

40. Write a program that encodes a pair of letters into another pair of letters as in Example 8.3.25. The program should first find the numerical equivalent x mod 676 of the pair, then compute $y = x^{49}$ mod 677, and then represent y as a pair of letters, as in the illustration preceding Example 8.3.25. For example,

$$\mathrm{DO} \to x = 4{\cdot}26 + 15 = 119 \to y = x^{49} = 641 = 26{\cdot}24 + 17 \to \mathrm{XQ}$$

41. Write a program that encodes and decodes messages, using the code $y = x^{49}$ in $\mathbb{Z}_{677}$. Use this to code the message THIS IS EASY. It will

be helpful to use the programs developed in Exercises 39 and 40 as procedures for this program. The input and the output of the program should be a finite sequence of capital letters. Ignore blanks, if any.

42. Using the program developed in Exercise 41, decode the message QGKXR SUKWR RFVHQ IPYWG.

8.4 The Chinese Remainder Theorem

In this section, we work simultaneously with congruences on different moduli. We first recall some results on divisibility from Lemma 8.1.2:

If $a \mid b$ and $b \mid c$, then $a \mid c$
If $a \mid b$, then $ax \mid bx$
If $xa \mid xb$, $x \neq 0$, then $a \mid b$

These divisibility results may be written in congruence form.

LEMMA 8.4.1

1. If $a \equiv b \bmod n$ and $d \mid n$, then $a \equiv b \bmod d$.
2. If $a \equiv b \bmod n$, then $xa \equiv xb \bmod xn$ where $x \neq 0$, and conversely.
3. If $a \equiv b \bmod n$ and if d is a common divisor of a, b, and n, then $a/d \equiv (b/d) \bmod (n/d)$. ■

The proofs are left to the reader.

THEOREM 8.4.2

Let $d = \text{GCD}(a, n)$. Then the congruence

$$ax \equiv b \bmod n$$

has a solution if and only if $d \mid b$. In this case, all solutions are the solutions of the congruence $(a/d)x \equiv (b/d) \bmod (n/d)$, which has a unique solution mod (n/d).

Proof Suppose $ax \equiv b \bmod n$ has a solution x. Since $d \mid n$, by statement 1 of Lemma 8.4.1, we have $ax \equiv b \bmod d$. Thus, $0 \equiv b \bmod d$, because $d \mid a$. But this means that $d \mid b$.

Conversely, if $d \mid b$, the given congruence is, by Lemma 8.4.1, equivalent to the congruence $(a/d)x \equiv (b/d) \bmod (n/d)$. But a/d and n/d are relatively prime because $d = \text{GCD}(a, n)$. Hence, this congruence has a unique solution mod n/d by Theorem 8.3.4. ■

EXAMPLE 8.4.3

Solve the congruences (a) $6x \equiv 9 \bmod 15$ and (b) $6x \equiv 10 \bmod 15$.

Method In (a), $\text{GCD}(6, 15) = 3$, so the first congruence is equivalent to $2x \equiv 3 \bmod 5$. This has the unique solution $x \equiv 4 \bmod 5$. The second congruence has no solution because 3 does not divide the constant term 10.

The Chinese Remainder Theorem is so named because it is a theorem about remainders that has been known to the Chinese since antiquity. Many applications have been discovered recently. The theory starts with the following lemma, again a restatement of a divisibility result.

LEMMA 8.4.4

Let m and n be relatively prime. Then if

$$a \equiv b \bmod m \qquad \text{and} \qquad a \equiv b \bmod n$$

we have

$$a \equiv b \bmod mn$$

and conversely.

Proof This is a restatement of statement 2 of Theorem 8.1.16. By hypothesis, $m \mid (b - a)$, and $n \mid (b - a)$, and $\mathrm{GCD}(m, n) = 1$. It follows by that result that $mn \mid (b - a)$. Thus $a \equiv b \bmod mn$. The converse is true by Lemma 8.4.1, because both m and n divide mn. ■

THEOREM 8.4.5 *The Chinese Remainder Theorem*

Let m and n be relatively prime, $N = mn$, and let a and b be given. Then the congruences

$$x \equiv a \bmod m \qquad x \equiv b \bmod n \tag{8.4.1}$$

have a unique solution mod N.

Proof The congruence $x \equiv a \bmod m$ has the solution

$$x = a + mt$$

where t is an arbitrary integer. This satisfies $x \equiv b \bmod n$ provided $a + mt \equiv b \bmod n$. But this is a linear congruence for t mod n, and its coefficient m is relatively prime to n. Therefore, it has a unique solution t mod n: $t \equiv t_0 \bmod n$. This can be written

$$t = t_0 + ns \qquad (s \text{ an arbitrary integer})$$

Therefore, substituting this value for t in the expression for x, we obtain

$$x = a + mt = a + m(t_0 + ns) = c + Ns$$

with s arbitrary and $c = a + mt_0$. But this is the same as $x \equiv c$ mod N. ∎

The following alternative proof is based on the Box Principle. For any x in $\mathbb{Z}_N$, let $a_1 = (x)_m$ and $a_2 = (x)_n$. Then a_1 and a_2 are in $\mathbb{Z}_m$ and $\mathbb{Z}_n$, respectively, and $x \equiv a_1$ mod m and $x \equiv a_2$ mod n. Thus, each x in $\mathbb{Z}_N$ gives rise to a pair (a_1, a_2), where a_1 is in $\mathbb{Z}_m$ and a_2 is in $\mathbb{Z}_n$. We now show that this correspondence is 1-1. If x and y are in $\mathbb{Z}_N$, and they both correspond to the same pair (a_1, a_2), then $x \equiv a_1 \equiv y$ mod m and similarly $x \equiv y$ mod n. Therefore, by Lemma 8.4.4, $x \equiv y$ mod N. Since x and y are in $\mathbb{Z}_N$, it follows that $x = y$. But there are $mn = N$ such pairs and N values of x in $\mathbb{Z}_N$. Therefore the correspondence is onto. But this is the result in question.

The first proof shows how to compute a solution of (8.4.1). Ultimately, we are reduced to finding an inverse of m mod n or of n mod m. The second proof tries all values x in $\mathbb{Z}_N$ until a solution of (8.4.1) is found.

EXAMPLE 8.4.6

Find the general solution of the system of congruences $x \equiv 4$ mod 9 and $x \equiv 26$ mod 35.

Method   As in the proof of Theorem 8.4.5, write $x = 26 + 35t$ and substitute into the first congruence. This gives

$$26 + 35t \equiv 4 \bmod 9$$

$$-t \equiv 5 \bmod 9$$

$$t \equiv 4 \bmod 9$$

Now take $t = 4$ to find $x = 26 + 35t = 166$. Here $N = 9 \cdot 35 = 315$. Hence, the solution is $x \equiv 166$ mod 315. (Equivalently, we could write $t = 4 + 9s$, to find all solutions. However, we know there is a unique solution mod 315. Therefore, once we found one, corresponding to $t = 4$, all other solutions were congruent to it mod 315.)

Why did we use the second congruence to substitute into the first? We preferred to find an inverse mod 9 rather than mod 35.

We now generalize to more than two congruences.

DEFINITION 8.4.7

Integers $n_1, n_2, \ldots, n_k$ are said to be **relatively prime in pairs** if any two of them are relatively prime.

COROLLARY 8.4.8

Suppose $n_1, n_2, \ldots, n_k$ are integers that are relatively prime in pairs. Let $a_1, \ldots, a_k$ be integers. Then the k congruences $x \equiv a_i \bmod n_i$ have a unique solution modulo the product $N = n_1 \cdots n_k$. ∎

This may be proved by induction on k.

One consequence of this corollary is that a **sequence** of nonnegative integers $a_1, \ldots, a_k$ may be replaced by *one* integer x. The number x may be regarded as the **code number** for the sequence. Equivalently, we may think of the a_i as the **coordinates** of the number x. Here the a_i are given mod n_i, and x is determined modulo their product N. More formally, we have Definition 8.4.9.

DEFINITION 8.4.9

Suppose m and n are relatively prime. For any integer x, let $x \equiv a_1 \bmod m$ and $x \equiv a_2 \bmod n$, with a_1 in $\mathbb{Z}_m$ and a_2 in $\mathbb{Z}_n$. Then x is said to have coordinates (a_1, a_2) with respect to the pair (m, n). More generally, suppose the integers $n_1, \ldots, n_k$ are relatively prime in pairs. Then x has coordinates $(a_1, \ldots, a_n)$ with respect to the integers n_i, provided $x \equiv a_i \bmod n_i$, with a_i in $\mathbb{Z}_{n_i}$.

If $N = n_1 \cdots n_k$, then by the Chinese Remainder Theorem the value of any integer x is uniquely determined mod N from its coordinates.

THEOREM 8.4.10

Let x and y have coordinates $(a_1, \ldots, a_k)$ and $(b_1, \ldots, b_k)$, respectively. Then the coordinates of $x + y$ are $a_i + b_i$, and the coordinates of xy are $a_i b_i$. These latter computations are taken mod n_i.

Proof Since $x \equiv a_i$ and $y \equiv b_i$, both mod n_i, we add and multiply to obtain the result. ∎

We have seen a hidden use of this theorem in Section 8.2, when we considered numbers mod 9, 10, and 11. We now reconsider that technique from the point of view of Theorem 8.4.10.

EXAMPLE 8.4.11

Choosing the moduli 9, 10, and 11, find the coordinates of the numbers 735 and 543 and the coordinates of their product.

Method Using the methods of Section 8.2, we can find a number mod 9 by taking digit sums. By using the last digit and the alternating sum of the digits, we can similarly find a number mod 10 and 11. Thus, we find that 735 is congruent to 6, 5, and 9 modulo 9, 10, and 11, respectively. This means that the coordinates of 735 are $(6, 5, 9)$. Similarly, the coordinates of 543 are (3, 3, 4). The coordinates of the product can be found, according to Theorem 8.4.10, by computing coordinate by coordinate. (The first coordinate is computed mod 9, the second mod 10, and the third mod 11.) Thus, the coordinates of 735·543 are (0, 5, 3). We used this computation for error checking in Section 8.2. Any alleged product of these numbers had to have the coordinates (0, 5, 3). The method was not perfect, because a number with these coordinates is determined mod N, which is 990 in this case. [Note that 990 has coordinates (0, 0, 0) as does the number 0.]

EXAMPLE 8.4.12   Parallel Computation

The techniques for 9, 10, and 11 can be used to do computations with very large numbers. The idea is that we do the computation on each component, which is reasonable in size, and at the end we put the answers together. Unlike working in a base b, no carrying is involved, so each computation can be done independently of the others. The technique is useful for parallel computation, where many smaller computations can be done at the same time. It is also useful for computing with very large numbers because we can work independently on each of its smaller components. We summarize this technique.

Suppose we wish to do many computations on very large integers, and these computations involve additions and multiplications only. Choose integers $n_1, \ldots, n_k$ that are relatively prime and set $N = n_1 \cdots n_k$. The computation proceeds as follows:

1. Replace each integer r in the computation with its components $(r_1, \ldots, r_k)$. This amounts to finding r mod n_i.
2. Independently do the computations on the ith component, working mod n_i. In this way, the answer A_i can be found mod n_i for $i = 1, \ldots, k$. Thus, the final answer A has components $(A_1, \ldots, A_k)$.
3. Finally, reconstitute the answer A by solving the congruences $A \equiv A_i$ mod n_i. This can be done by the Chinese Remainder Theorem. Further techniques are given later. The answer A, in any case, can only be found mod N.

As for the practicalities, step 2 breaks up the computation into simple, more manageable computations that can be done independently on different machines, or by different people, and at different times. Step 1 takes some time, but finding numbers mod N_i is relatively straightforward. The time-consuming part is the reconstitution 3 of a number A from its components. Even though the answer A is found mod N, we can arrange N to be much larger than the answer, so $(A)_N$ will be the same as A.

We illustrate this technique with an example that uses small numbers.

EXAMPLE 8.4.13

Using moduli 7, 9, 10, and 11, compute $A = 35 \cdot 73$.

Method We find coordinates by taking the factors mod 7, 9, 10, and 11. We have

$$A \equiv 0 \cdot 3 = 0 \bmod 7$$

$$A \equiv 8 \cdot 1 = 8 \bmod 9$$

$$A \equiv 5 \cdot 3 \equiv 5 \bmod 10$$

$$A \equiv 2(-4) \equiv 3 \bmod 11$$

Thus, A has coordinates $(0, 8, 5, 3)$ with respect to these moduli. We can now solve these congruences to "reconstitute" A. We obtain $A \equiv 2555$ mod 6930. (Here $6930 = 7 \cdot 9 \cdot 10 \cdot 11$.) Finally, a rough estimate shows that A is no larger than $40 \cdot 80 = 3200$. Therefore, A must equal 2550, because a larger A mod 6930 would take us well over this estimate.

For a large computation, a similar method reduces most of the computation to computations modulo smaller numbers, saving for the end the problem of finding A from its components. Exercise 25 shows how to get large relatively prime n_i that are near powers of 2.

We now give a useful technique for solving the simultaneous congruences $x \equiv a_i \bmod n_i$. We first illustrate it for two relatively prime moduli m and n.

DEFINITION 8.4.14

Let m and n be relatively prime integers. A **basis** for solutions modulo (m, n) is a set of two integers α_1 and α_2 such that α_1 has the components $(1, 0)$ and α_2 has the components $(0, 1)$, both with respect to the moduli (m, n).

Similarly, for the moduli (m, n, p), a basis consists of three integers α_1, α_2, and α_3 with components $(1, 0, 0)$, $(0, 1, 0)$, and $(0, 0, 1)$. The definition generalizes for any number of moduli.

The importance of this concept is that once a basis is found, the congruences

$$x \equiv a \bmod m \qquad \text{(8.4.2)}$$
$$x \equiv b \bmod n$$

can be immediately explicitly solved.

THEOREM 8.4.15

Let α_1 and α_2 be a basis for the moduli (m, n). Then the system of congruences (8.4.2) has the solution

$$x \equiv a\alpha_1 + b\alpha_2 \bmod N \qquad (8.4.3)$$

Proof The proof is an immediate consequence of Theorem 8.4.10. Working two components at a time, we see that the components of the right side of the congruence (8.4.3) are $a(1, 0) + b(0, 1)$ or (a, b). But this means that $a\alpha_1 + b\alpha_2$ satisfies the congruences (8.4.2). ■

We illustrate with a simple example.

EXAMPLE 8.4.16

Choose moduli 9 and 10. Find a basis for the moduli 9 and 10. Using this basis, show how to solve the system

$$x \equiv a \bmod 9$$

$$x \equiv b \bmod 10$$

Method We first find a basis. Now, α_1 must satisfy $\alpha_1 \equiv 1 \bmod 9$ and $\alpha_1 \equiv 0 \bmod 10$. Writing $\alpha_1 = 10t$, the first of these congruences becomes $10t \equiv 1 \bmod 9$, or $t \equiv 1 \bmod 9$. Taking $t = 1$, we find $\alpha_1 = 10$. Similarly, to find α_2, we have $\alpha_2 \equiv 0 \bmod 9$ and $\alpha_2 \equiv 1 \bmod 10$. Thus, $\alpha_2 = 9t$, and $9t \equiv 1 \bmod 10$. This gives $t = -1$, so $\alpha_2 = -9$. Thus, the basis is 10 and -9.

The solution of the given congruences is $x \equiv 10a - 9b \bmod 90$. For example, we can immediately find that the system $x \equiv 4 \bmod 9$ and $x \equiv 6 \bmod 10$ has the solution $x \equiv 10 \cdot 4 - 9 \cdot 6 = -14 \equiv 76 \bmod 90$.

This technique thus gives an effective method for reconstituting a number from its components. In the next theorem, we discover a simple method for finding a basis for a finite number of moduli that are relatively prime in pairs.

Theorem 8.4.15 applies with obvious modifications to three or more moduli that are relatively prime in pairs. For example, if α_1, α_2, and α_3 are a basis modulo (m, n, p), then the system of congruences

$$x \equiv a \bmod m, \qquad x \equiv b \bmod n, \qquad \text{and} \qquad x \equiv c \bmod p$$

has the solution $x \equiv a\alpha_1 + b\alpha_2 + c\alpha_3 \bmod N$ (where $N = mnp$).

How do we find a basis for the moduli (m, n, p)? To find α_1, for example, we need to find a number x whose components are $(1, 0, 0)$. This is a solution of

$$x \equiv 1 \bmod m, \qquad x \equiv 0 \bmod n, \qquad \text{and} \qquad x \equiv 0 \bmod p$$

The last two congruences have the solution $x = npt$ for arbitrary t. The first congruence becomes $npt \equiv 1 \bmod m$. Thus, t must be the inverse of $np \bmod m$, so α_1 is npt_1, where t_1 is an inverse of $np \bmod m$. Similarly, $\alpha_2 = mpt_2$, where t_2 is an inverse of $mp \bmod n$, and so on. This method clearly generalizes to any number of moduli as in the following theorem.

THEOREM 8.4.17

To find the ith member α_i of a basis for a system of moduli, multiply all the moduli except the ith and then multiply this product by its inverse modulo the ith modulus. ∎

EXAMPLE 8.4.18

Using the moduli 2, 3, 5, and 7, find numbers with components (1, 1, 2, 2) and (0, 2, 4, 1).

Method   We use Theorem 8.4.17. Here, $\alpha_1 = 3\cdot5\cdot7t = 105t$, where t is an inverse of 105 mod 2. Thus, we may choose $t = 1$, so $\alpha_1 = 105$.
   Similarly $\alpha_2 = 70t$, where t is the inverse of 70 mod 3. Thus, $t = 1$ is appropriate here and $\alpha_2 = 70$.
   In the same way, $\alpha_3 = 42t = 42\cdot3 = 126$, and $\alpha_4 = 30\cdot4 = 120$, because the inverse of 30 mod 7 is $t = 4$. The basis is 105, 70, 126, and 120. Thus, the solutions are as follows:
   For the components (1, 1, 2, 2), the required number is

$$105 + 70 + 252 + 240 = 667 \equiv 37 \pmod{210}$$

For components (0, 2, 4, 1), the required number is

$$0 + 140 + 504 + 120 = 764 \equiv 134 \pmod{210}$$

Exercises

1. Solve the congruence $16x \equiv 40 \bmod 28$.
2. Solve the system of congruences $x \equiv 5 \bmod 6$ and $x \equiv 3 \bmod 7$.
3. Solve the system of congruences $x \equiv 1 \bmod 10$, $x \equiv 1 \bmod 11$, and $x \equiv 2 \bmod 13$.
4. Solve the system of congruences $2x \equiv 3 \bmod 7$ and $5x \equiv 7 \bmod 13$.
5. Find a basis for the moduli (9, 10, 11).
6. Using your result in Exercise 5, solve the system
   (a) $x \equiv 3 \bmod 9$, $x \equiv 4 \bmod 10$, and $x \equiv 5 \bmod 11$.
   Similarly, solve the system
   (b) $x \equiv 1 \bmod 9$, $3x \equiv 2 \bmod 10$, and $5x \equiv 7 \bmod 11$.
7. Repeat Exercises 5 and 6 for the moduli (7, 8, 9) and the systems
   (a) $x \equiv 2 \bmod 7$, $x \equiv 3 \bmod 8$, and $x \equiv 5 \bmod 9$.
   (b) $x \equiv 1 \bmod 7$, $3x \equiv 5 \bmod 8$, and $5x \equiv 2 \bmod 9$.

8. Let m, n, and p be relatively prime in pairs. What number has components (a, a, a) with respect to the moduli (m, n, p)? Using this fact, find a number that has remainders 6, 7, and 8 modulo 7, 8, and 9, respectively.

9. Ten people on a desert island discover a pile of coconuts. They try to divide them evenly, but eight are left over. At that point, one of the people announces that she doesn't want any because she's allergic to coconuts. They attempt to divide the pile evenly among the remaining nine people but seven are left over. Two more people drop out, leaving seven, but five are left over when they try to divide the coconuts evenly among the seven. There are not more than 1000 coconuts in the pile. How many coconuts are there? (*Hint:* Exercise 8 will help solve this problem in an easy manner.)

10. Suppose r and $r + 1$ are consecutive numbers. Show that they are relatively prime. Show that $\alpha_1 = r + 1$ and $\alpha_2 = -r$ form a basis for this pair of moduli.

11. Using the basis result obtained in Exercise 10, solve the systems
    (a) $x \equiv 31 \bmod 60$, $x \equiv 45 \bmod 61$.
    (b) $x \equiv 4 \bmod 60$, $x \equiv 30 \bmod 61$.

12. Using the basis result of Exercise 10, solve the systems
    (a) $x \equiv 31 \bmod 102$, $x \equiv 50 \bmod 103$.
    (b) $x \equiv 20 \bmod 102$, $x \equiv 31 \bmod 103$.

*13. Suppose $r - 1$, r, and $r + 1$ are consecutive positive numbers. Show that these numbers are relatively prime in pairs if and only if the middle term r is even. (*Hint:* If d is a divisor of any two of them, then d must divide their difference.)

*14. Suppose r is even, and moduli $r - 1$, r, and $r + 1$ are chosen. (See Exercise 13.) Find a basis α_1, α_2, α_3 for this set of moduli. (*Hint:* Use Theorem 8.4.17 and the formula for the inverse of a product given in Exercise 8.3.8.)

15. Using your result in Exercise 14, find a basis for the moduli 77, 78, and 79. Using this basis, solve the simultaneous congruences $x \equiv 30 \bmod 77$, $x \equiv 31 \bmod 78$, and $x \equiv 32 \bmod 79$.

16. Using the result in Exercise 14, find a basis for the moduli 777, 778, 779. Hence solve the congruences $x \equiv 30 \bmod 777$, $x \equiv 31 \bmod 778$, and $x \equiv 50 \bmod 779$.

17. Show why no four consecutive integers can be relatively prime in pairs.

*18. In analogy with Exercise 14, find conditions on r in order for the integers $r - 2$, $r - 1$, r, and $r + 2$ to be relatively prime in pairs.

19. Compute $410^{89} \bmod 990$. Use moduli 9, 10, and 11, and work with the coordinates of 410.

*20. Using the technique of Exercise 19, compute $431^{245} \bmod 504$.

21. Let m, n, and p be relatively prime in pairs. Prove that if α_1, α_2, and α_3 are a basis for the moduli m, n, and p, then $\alpha_1 + \alpha_2 + \alpha_3 \equiv 1 \bmod mnp$.

The following exercises are interesting theoretically and are useful for finding very large numbers that are relatively prime in pairs.

*22. Suppose $(a)_n = r$. Define $P(x) = 2^x - 1$. Define $A = P(a)$, $N = P(n)$, and $R = P(r)$. Prove $(A)_N = R$. For example, since $(11)_3 = 2$, this result yields $(2047)_7 = 3$. [*Hint:* Write $a = nq + r$. Since $N = 2^n - 1$, we have $2^n \equiv 1$ (N). Now raise both sides to the power q and proceed from there.]

*23. Generalize Exercise 22 for the function $P_b(x) = b^x - 1$.

*24. Using the result of Exercise 22, show that if $GCD(r, s) = d$, and $R = P(r) = 2^r - 1$, $S = 2^s - 1$, and $D = 2^d - 1$, then $GCD(R, S) = D$. (*Hint:* Use the Euclidean Algorithm.)

*25. Prove: If r and s are relatively prime, then $2^r - 1$ and $2^s - 1$ are also relatively prime. (This result is useful for finding very large numbers that are relatively prime in pairs. For example, 8, 9, and 11 are relatively prime in pairs, and this result immediately shows that 255, 511, and 2047 are also relatively prime in pairs.)

*26. Generalize Exercise 24 by finding $GCD(a^r - 1, a^s - 1)$ in terms of $GCD(r, s)$. Here, a is an integer ≥ 2.

27. Using the result of Exercise 24, find $GCD(a, b)$, where

$$a = (111\ 1111\ 1111\ 1111)_2 \text{ and } b = (11\ 1111)_2$$

Summary Outline

- For a and b in $\mathbb{Z}$, a **divides** b, or b **is a multiple of** a, if $b = aq$ for some integer q. We write $a \mid b$.

- **Some properties of divisibility**

  (a) $1 \mid a$ for all a.
  (b) $a \mid 0$ for all a.
  (c) $a \mid a$ for all a.
  (d) If $a \mid b$ and $b \mid c$, then $a \mid c$.
  (e) If $a \mid b$, then $ax \mid bx$.
  (f) If $xa \mid xb$, $x \neq 0$, then $a \mid b$.
  (g) If $a \mid b$ and $a \mid c$, then $a \mid (bx + cy)$.
  (h) If $a \mid b$, then either $b = 0$ or $|a| \leq |b|$.

- **The Division Algorithm** Let a and b be integers, with $b > 0$. Then there are two numbers q and r (called the quotient and the remainder) such that

$$a = bq + r, \qquad 0 \leq r < b$$

  The numbers q and r are unique.

- The **greatest common divisor** d of a and b, written $d = GCD(a, b)$ is the largest integer that divides both a and b. Here, a and b are integers, not both 0.

- **The Euclidean Algorithm**   Let $b \neq 0$ and

$$a = bq_1 + r_1$$
$$b = r_1q_2 + r_2$$
$$r_1 = r_2q_3 + r_3$$
$$\vdots$$
$$r_n = r_{n+1}q_{n+2} + r_{n+2}$$
$$r_{n+1} = r_{n+2}q_{n+3} + 0$$

where $r_{n+2} > 0$. Then $\text{GCD}(a, b) = r_{n+2}$.

- **Corollary**   Let $d = \text{GCD}(a, b)$. Let $c \mid a$ and $c \mid b$. Then $c \mid d$.

- **Corollary**   Let $d = \text{GCD}(a, b)$. Then $d = ax + by$ for some integers x and y.

- Two numbers a and b are **relatively prime** if $\text{GCD}(a, b) = 1$.

- **Theorem** a and b are relatively prime if and only if there are integers x and y such that $1 = ax + by$.

- **Theorem**

   (a) If $a \mid bc$ and $\text{GCD}(a, b) = 1$, then $a \mid c$.
   (b) If $a \mid c$ and $b \mid c$ and $\text{GCD}(a, b) = 1$, then $ab \mid c$.

- **A prime number** p is a number $p > 1$ whose only positive divisors are 1 and p. A **composite number** is one that is >1 and not a prime.

- **Theorem**   If p is a prime and $p \mid ab$, then $p \mid a$ or $p \mid b$.

- **Theorem**   There are infinitely many primes.

- **Unique Factorization Theorem**   Any $n > 1$ can be written as the product of primes in only one way, except for the order of factors.

- **Theorem**   If $n > 1$ and if there are no divisors of n between 2 and $\sqrt{n}$ inclusive, then n is a prime.

- **Congruence** $a \equiv b \bmod n$ is defined to mean $n \mid (a - b)$.

- **Theorem**   Congruence mod n is an equivalence relation.

- **Theorem**   If $a \equiv b$ and $c \equiv d$, then

   (a) $a + c \equiv b + d$
   (b) $ac \equiv bd$
   (c) $a^k \equiv b^k$ for $k = 1, 2, \ldots$

   (All congruences are taken mod n.)

- **Theorem**   Let $n > 1$, $r \in \mathbb{Z}_n$. Then a leaves the remainder r when divided by n if and only if $a \equiv r \bmod n$.

- **Corollary** $a \equiv b \bmod n$ if and only if a and b leave the same remainder when divided by n.

- A **fixed point** for any function $f: A \rightarrow A$ is an element x such that $f(x) = x$.

- The sequence a_n is a **sequence of iterates** of $f(x)$ with seed s, when $a_0 = s$ and $a_{n+1} = f(a_n)$ for $n \geq 0$.

- **Theorem** (Casting Out Nines)   Let

$$n = a_0 + 10a_1 + 100a_2 + \cdots + 10^k a_k$$

where the $a_i \in \mathbb{Z}_{10}$ are the decimal digits of n. Let

$$S(n) = a_0 + a_1 + \cdots + a_k$$

$$(= \text{the sum of the digits of } n)$$

Then

$$n \equiv S(n) \bmod 9$$

- **Theorem**   Let

$$n = a_0 + 10a_1 + 100a_2 + \cdots + 10^k a_k$$

where the $a_i \in \mathbb{Z}_{10}$ are the decimal digits of n. Let

$$A(n) = a_0 - a_1 + a_2 - \cdots = \sum_{j=0}^{k} (-1)^j a_j$$

$$(= \text{the alternating sum of the digits})$$

Then

$$n \equiv A(n) \bmod 11$$

- The **one's complement** of a string of bits changes all 0's into 1's and all 1's into 0's.

- A number a in $\mathbb{Z}_n$ is said to be **invertible** in $\mathbb{Z}_n$ if there exists a number a^{-1} of $\mathbb{Z}_n$ such that $aa^{-1} = 1$ in $\mathbb{Z}_n$. The number a^{-1} is called the **inverse** of a in $\mathbb{Z}_n$. If a is invertible, we write $ba^{-1} = b/a$. In particular, $a^{-1} = 1/a$.

- **Theorem**   The integer a is invertible in $\mathbb{Z}_n$ if and only if a and n are relatively prime.

- **Theorem**   Let a and n be relatively prime. Then the equation $ax = b$ has the unique solution $x = ba^{-1}$ (or b/a) in $\mathbb{Z}_n$.

- **Theorem**   If $a \neq 0$ in $\mathbb{Z}_p$ (p a prime), then a^{-1} exists in $\mathbb{Z}_p$. The equation $ax = b$ has a unique solution $x = ba^{-1} = b/a$ provided $a \neq 0$ in $\mathbb{Z}_p$.

- **Theorem**   Cancellation Laws

   1. If $ab = 0$ in $\mathbb{Z}_n$ and $\mathrm{GCD}(a, n) = 1$, then $b = 0$ in $\mathbb{Z}_n$.
   2. If $ab = ac$ in $\mathbb{Z}_n$ and $\mathrm{GCD}(a, n) = 1$, then $b = c$ in $\mathbb{Z}_n$.

3. If $ab = 0$ in $\mathbb{Z}_p$, then $a = 0$ or $b = 0$ in $\mathbb{Z}_p$.
4. If $ab = ac$ in $\mathbb{Z}_p$ and $a \neq 0$ in $\mathbb{Z}_p$, then $b = c$ in $\mathbb{Z}_p$.

● **Theorem** Fermat's Little Theorem   Let $a \neq 0$ in $\mathbb{Z}_p$ (p a prime). Then $a^{p-1} = 1$ in $\mathbb{Z}_p$. In congruence form: If $a \not\equiv 0 \bmod p$, then $a^{p-1} \equiv 1 \bmod p$.

● **Corollary**   In $\mathbb{Z}_p$, $a^{-1} = a^{p-2}$. Thus, the solution of $ax \equiv b\ (p)$ is $x \equiv ba^{p-2}$.

● **Corollary**   If $a \not\equiv 0\ (p)$ and $n \equiv r \bmod (p-1)$, then $a^n \equiv a^r\ (p)$.

● **Theorem**   Suppose p is a prime and e is relatively prime to $p-1$. Let e' be the inverse of $e \bmod (p-1)$. Then

$$y \equiv x^e \bmod p \quad \text{iff} \quad y^{e'} \equiv x \bmod p$$

● **Lemma**   Let $f(x) = ax + b$ in $\mathbb{Z}_p$, $a \neq 0$ or 1, with fixed point x_0. Then for all n, the nth iterates of f are given by

$$f^n(x) = a^n(x - x_0) + x_0$$

● **Corollary**   Let $f(x) = ax + b$ have fixed point x_0. Then the iterate $f^n(x)$ has leading coefficient a^n and fixed point x_0. If $a^n = 1$, then $f^n(x) = x$.

● The **order of a** in $\mathbb{Z}_p$ ($a \neq 0$) is the least positive integer k such that $a^k \equiv 1 \bmod p$.

● **Theorem**   The order of a nonzero element of $\mathbb{Z}_p$ divides $p - 1$.

● A **primitive root** mod p is an element of order $p - 1$.

● If $f(x) = ax + b$ is a linear function in $\mathbb{Z}_p$, where $a \neq 0$ and $a \neq 1$, the sequence of iterates $f^n(s)$ is called **random-like** if this sequence goes through all $p - 1$ values mod p except the fixed point. The number s is called the **seed**.

● **Theorem**   If $f(x) = ax + b$ in $\mathbb{Z}_p$, the sequence of iterates is random-like if and only if the seed s is not the fixed point of f and a has order $p - 1$.

● **Some properties of congruences**

1. If $a \equiv b \bmod n$ and $d \mid n$, then $a \equiv b \bmod d$.
2. If $a \equiv b \bmod n$, then $xa \equiv xb \bmod xn$ where $x \neq 0$, and conversely.
3. If $a \equiv b \bmod n$, and if d is a common divisor of a, b, and n, then $a/d \equiv (b/d) \bmod (n/d)$.

● **Theorem**   Let $d = \text{GCD}(a, n)$. Then the congruence

$$ax \equiv b \bmod n$$

has a solution if and only if $d \mid b$. In this case, all solutions are the solutions of the congruence $(a/d)x \equiv (b/d) \bmod (n/d)$.

- **Lemma**   Let m and n be relatively prime. Then $a \equiv b \bmod m$ and $a \equiv b \bmod n$ if and only if $a \equiv b \bmod mn$.

- Integers are **relatively prime in pairs** if any two of them are relatively prime.

- **Chinese Remainder Theorem**   Suppose $n_1, n_2, \ldots, n_k$ are relatively prime in pairs. Let $a_1, \ldots, a_k$ be arbitrary integers. Then the k congruences $x \equiv a_i \bmod n_i$ have a unique solution modulo the product $N = n_1 \cdots n_k$.

- If m and n are relatively prime, x is said to have **coordinates** (a_1, a_2) with respect to the pair (m, n) if $x \equiv a_1 \bmod m$ and $x \equiv a_2 \bmod n$. More generally, if the integers $n_1, \ldots, n_k$ are relatively prime in pairs and $x \equiv a_i \bmod n_i$, with a_i in $\mathbb{Z}_{n_i}$, the elements a_i are the **coordinates** of x with respect to n_i.

- **Theorem**   Let x have coordinates $(a_1, \ldots, a_k)$ and let y have coordinates $(b_1, \ldots, b_k)$. Then the coordinates of $x + y$ are $a_i + b_i$, and the coordinates of xy are $a_i b_i$. These latter computations are taken mod n_i.

- Let m and n be relatively prime integers. A **basis** for solutions mod (m, n) is a set of two integers x_1 and x_2 such that x_1 has the components $(1, 0)$ and x_2 has the components $(0, 1)$ with respect to the moduli (m, n). A similar definition applies for three or more moduli that are relatively prime in pairs.

- **Theorem**   Let x_1 and x_2 be a basis for the moduli (m, n). Then the system of congruences $x \equiv a \bmod m$ and $x \equiv b \bmod n$ has the solution $x \equiv a x_1 + b x_2 \bmod N$.

- **Theorem**   To find the ith member x_i of a basis for a system of moduli, multiply all the moduli except the ith and then multiply this product by its inverse modulo the ith modulus.

C H A P T E R   9

Grammars and Automata

What is a language, and how is it constructed? This question, considered by linguists, is not an easy one. In this chapter we consider more simplified artificial languages. In particular, the language of the propositional calculus is analyzed by two methods. An automaton is a simple computing machine that performs limited, language-related tasks. These are defined in this chapter, together with some applications. Their limitations are also illustrated.

9.1 Syntax of Propositional Calculus—Type-0 Grammars

In Chapter 1, we were concerned with the truth or falsity of compound statements. We took for granted the **readability** of statements, using parentheses when necessary and dropping them for easy readability, according to certain conventions given in Section 1.2. By readability, we mean the ability to take a formula and understand its structure and composition. We now concern ourselves with the question of readability. Although this may be easy for a statement such as $(p \vee q)$, it becomes much more complicated for very long strings, or even for a moderately long string such as

$$((((p \vee (\neg\neg q \to (q \to \neg p))) \vee ((r \wedge (\neg q \to p)) \wedge \neg r)) \vee (\neg q \to r)) \leftrightarrow (p \leftrightarrow q))$$

One reason for such an analysis is that in a computer analysis of language, the computer must be given precise rules to determine whether a string such as this is meaningful and, if so, to determine its truth value.

Similarly, linguists study the grammar of a language without concerning themselves with the truth or falsity of statements in that language.

One way we analyze such strings, to determine whether they represent statements and to find their structure, is by using trees. We did some of this for algebraic expressions in Examples 6.1.4 and 6.3.16.

Until now, we concentrated mostly on **semantics**—the question of when one of these compound statements was true. We now concentrate on **syntax**—the question of what strings qualify to be called formulas. The general situation, which we discuss later, is that we have some set of characters, called the **alphabet**, and a certain subset of strings on this alphabet called a **language**. The problem is to give a method to define or identify the language. The method used to define the language is called a **grammar**. It is in the nature of the subject that many of the definitions and techniques of the grammar often involve much symbol manipulation, without regard for the meaning of the symbols.

We now give a formal definition of a compound statement, or *formula*.

DEFINITION 9.1.1

The **alphabet** for the propositional calculus consists of the following symbols:

$p, q, r, \ldots$	(infinitely many **variable symbols**)
T, F	(two **constant symbols**)
$\wedge, \vee, \rightarrow, \leftrightarrow$	(four **binary connective symbols**)
$\neg$	(a **unary connective symbol**)
$(\,,)$	(left and right **parentheses**)

These symbols are called the **symbols of propositional calculus.**

The string of symbols representing a statement is called a formula.

DEFINITION 9.1.2

A **formula** is a string on the symbols of the propositional calculus that can be obtained by the use of the following rules:

1. Any constant or variable symbol is a formula. Thus $p, q, r, \ldots$ are formulas, as are the constant symbols T and F.
2. If α and β are formulas, so are
   - (a) $(\alpha \wedge \beta)$
   - (b) $(\alpha \vee \beta)$
   - (c) $(\alpha \rightarrow \beta)$
   - (d) $(\alpha \leftrightarrow \beta)$
   - (e) $\neg\alpha$

Since a formula α is a string, its length, written length(α), is also defined.

Note that parentheses are introduced in statements 2(a)–(d) but not in statement 2(e). These are part of the definition of a formula, and their introduction, as we shall see, allows us to uniquely decompose (parse) a formula into its components. Parentheses are *not* used informally at the discretion of the user. For example, $(\neg p)$ is not a formula according to the definition, whereas $\neg p$ is.

Definition 9.1.2 can be rephrased by recursively defining a formula of length n, as follows.

EXAMPLE 9.1.3   *Alternative Version of Definition 9.1.2*

This definition defines a formula of length n by recursion.

1. *The Basis*   A formula of length 1 is either a variable symbol $p, q, r,$ ... or a constant symbol T or F.
2. *The Recursion*   A formula of length $n > 1$ is a string of length n on the symbols of the propositional calculus that is one of the following:
   (a) $(\alpha \wedge \beta)$, where α and β are formulas of length $<n$.
   (b) $(\alpha \vee \beta)$, where α and β are formulas of length $<n$.
   (c) $(\alpha \to \beta)$, where α and β are formulas of length $<n$.
   (d) $(\alpha \leftrightarrow \beta)$, where α and β are formulas of length $<n$.
   (e) $\neg \alpha$, where α is a formula of length $n - 1$.

In statements 2(a)–(d), $\text{length}(\alpha) + \text{length}(\beta) = n - 3$.

EXAMPLE 9.1.4

Some examples of formulas using these definitions are

$$p; \quad q; \quad (p \wedge q) \;\; [\text{but } not \; p \wedge q]; \quad \neg((p \to q) \vee (q \wedge r))$$

Let us verify that this latter string is a formula by Definition 9.1.2. By statement 1, p, q, and r are formulas. Hence, by statement 2, so are $(p \to q)$ and $(q \wedge r)$. Hence, again by statement 2, so is $((p \to q) \vee (q \wedge r))$. Finally, using statement 2 of Definition 9.1.2, we arrive at the result. In short, we can use this definition to build up a formula. It is more difficult to determine that a string is *not* a formula, because we have to show that a string *cannot* be built up from the rules in Definition 9.1.2. Some of the following results show how to do this.

Definition 9.1.2 leads to the following inductive procedure, which is useful to determine whether a collection $\mathbb{C}$ of formulas consists of *all* formulas.

THEOREM 9.1.5

Let $\mathbb{C}$ be a collection of formulas of the propositional calculus. Suppose

1. All variable and constant symbols are in $\mathbb{C}$.

2. If α and $\beta \in \mathbb{C}$, and if λ is any of the four binary connectives, then $(\alpha\lambda\beta) \in \mathbb{C}$.
3. If $\alpha \in \mathbb{C}$, then $\neg\alpha \in \mathbb{C}$.

Then $\mathbb{C}$ consists of all formulas of the propositional calculus.

Proof We show, by induction, that all formulas of length $\leq n$ are in $\mathbb{C}$. This is clearly true for $n = 1$, because, by hypothesis, all variable and constant symbols are in $\mathbb{C}$.

Now assume this result for all formulas with length $\leq n$. Suppose now that σ is a formula of length $n + 1$. Since the length of σ is greater than 1, σ is not a variable or constant symbol. Thus, by Definition 9.1.2, $\sigma = (\alpha\lambda\beta)$ or $\sigma = \neg\alpha$ for some formulas α and β and some binary connective λ. In each case, α and β are formulas of length $\leq n$ and, thus, are in $\mathbb{C}$ by our induction hypothesis. By hypotheses 2 and 3, this forces σ to be in $\mathbb{C}$, and the result is proved for $n + 1$. This proves the result for all formulas, by induction. ■

Theorem 9.1.5 gives an inductive procedure that is useful for proving results about formulas. It can be restated in a more familiar inductive form, which is the way we use this theorem:

> If a statement about formulas is true for the variable or constant symbols, and if, whenever it is true for formulas α and β it is also true for $(\alpha\lambda\beta)$ for any binary connective λ, and it is also true for $\neg\alpha$, then the statement is true for all formulas.

We can see this by letting $\mathbb{C}$ be the set of formulas for which the statement is true, and using Theorem 9.1.5.

We shall now see how this theorem can be applied. We start by proving that any formula has an equal number of left and right parentheses. Of course, this result seems obvious from the definition, because parentheses are always introduced in pairs. Still, we offer the formal proof to show how this induction works.

THEOREM 9.1.6

Any formula has an equal number of left and right parentheses.

Proof This statement is clearly true for the variable and constant symbols, since these contain no parentheses as a substring. This is the basis part of the theorem. If α and β have balanced parentheses, then so do $\neg\alpha$ and $(\alpha\lambda\beta)$ (for any binary connective λ) by a count of left and right parentheses. Thus the result is true by induction (Theorem 9.1.5). ■

EXAMPLE 9.1.7

The following is another simple illustration. Let us show that any formula begins with $\neg$, or "(", or a variable or constant symbol. This is clearly true for variable or constant symbols. And if it is true for α and β, it is true for ¬α and for (αλβ) for any binary connective λ. Thus, the result is true for all formulas, again by Theorem 9.1.5. We leave it to the reader to prove that any formula of length ≥2 starts with either ¬ or "(".

These simple results indicate why certain strings are not formulas. For example, ((p ∧ q) is not one because it has more left parentheses than right.

Before showing that any formula is uniquely readable (Theorem 9.1.10), we must first prove an important lemma.

DEFINITION 9.1.8

If α = στ (the concatenation of σ and τ), where σ and τ are *nonempty* strings, we say that σ is a **proper initial segment** of α and that τ is a **proper final segment**.

For example, if α = ¬((p → q) ∨ r), then "¬((p" is a proper initial segment and ") ∨ r)" is a proper final segment.

LEMMA 9.1.9

Let α be a formula, and let β be a proper initial segment of α. Then β has more left parentheses than right, or β consists only of a string of ¬'s. ∎

Remark   Try this on the formula ¬¬((p → q) ∧ T).

Proof   We let ℂ be the set of all formulas with this property, and show that the conditions of Theorem 9.1.5 are satisfied.

1. All variable or constant symbols are in ℂ. This is true vacuously, because any such symbol has length 1 and no proper substrings.
2. Now suppose α and β are in ℂ and λ is any of the four binary connectives. Let us show that (αλβ) is in ℂ. To do this, we find all of the proper initial segments of this string and verify that they have the required property. The segments are of the following form (here α' and β' are proper initial segments of α and β, respectively):
   (a) (
   (b) (α'
   (c) (α
   (d) (αλ
   (e) (αλβ'
   (f) (αλβ

In each case, we can see that there is an excess of left parentheses over right. [For cases 2(c)–(f), we need to use Theorem 9.1.6 on the number of parentheses in a formula.]

3. Finally, the case $\neg\alpha$ is similarly verified for α in **C**. This proves the result. ■

We can now state and prove the unique readability theorem alluded to earlier.

THEOREM 9.1.10   Unique Readability

Let α be a formula. Then α is precisely one and only one of the following types. In each case, α' and/or β' are uniquely determined formulas.

1. α is a variable or constant symbol; or
2. $\alpha = (\alpha' \wedge \beta')$; or
3. $\alpha = (\alpha' \vee \beta')$; or
4. $\alpha = (\alpha' \rightarrow \beta')$; or
5. $\alpha = (\alpha' \leftrightarrow \beta')$; or
6. $\alpha = \neg\alpha'$.

Proof   Case 1 is the unique case where the first (and only) symbol is a variable or constant symbol. Case 6 is the case in which α starts with $\neg$. In that case, α' is the unique string that follows the leading $\neg$ in this string. We therefore concern ourselves with a formula α that starts with "(". To prove uniqueness, we suppose that α has two representations:

$$\alpha = (\alpha'\lambda\beta') = (\alpha^*\mu\beta^*)$$

where α', α^*, β', and β^* are formulas, and λ and μ are binary connectives. Then, as strings,

$$\alpha'\lambda\beta' = \alpha^*\mu\beta^*$$

But this shows that, unless $\alpha' = \alpha^*$, one of the formulas α' or α^* is a proper initial substring of the other. But this is impossible by Theorem 9.1.6 and Lemma 9.1.9, because one string would have an equal number of left and right parentheses and the other would not. Thus, $\alpha' = \alpha^*$, and, hence, $\lambda = \mu$ and, finally, $\beta' = \beta^*$. This is the result. ■

We can use the techniques used in this proof to determine whether a string is a formula and, if it is, to find the tree representation of that formula. We now illustrate the method on a simple string before we explicitly give the algorithm in general. The method is similar to the one developed for Polish notation in Example 6.3.15.

EXAMPLE 9.1.11

Determine, using the method of Theorem 9.1.10, whether the following string α is a formula, and find its tree if it is one.

$$\alpha = (((p \rightarrow q) \wedge r) \wedge \neg(q \rightarrow p))$$

Method   We keep track of the excess of left parentheses over right parentheses, following the first parentheses. When matching occurs (excess = 0), we get the initial string

$$(\, (\, (\, p \rightarrow q \,) \wedge r \,) \wedge \neg \, (\, q \rightarrow p \,))$$
$$1 \; 2 \, 2 \quad 2 \quad 2 \, 1 \; 1 \; 1 \; 0$$

According to the algorithm, this leads to a formula whose partially completed tree is given in Figure 9.1.1.

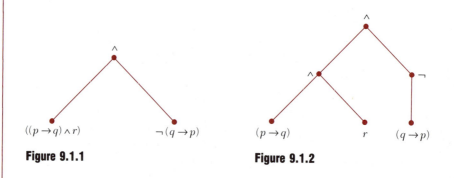

Figure 9.1.1   **Figure 9.1.2**

We now work similarly on each branch to obtain Figure 9.1.2. For example, the work on the left branch is

$$(\, (\, p \rightarrow q \,) \wedge r \,)$$
$$1 \; 1 \quad 1 \; 1 \; 0$$

We finally arrive at the tree diagram of Figure 9.1.3.

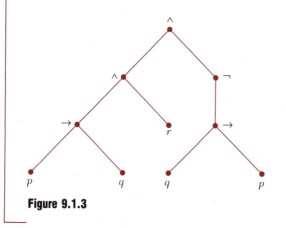

Figure 9.1.3

The following example shows how to discover when a string is not a formula.

EXAMPLE 9.1.12

Is the string $((p \lor \neg q) \land r \to (q \to s))$ a formula?

Method We use the algorithm on the string as indicated:

$$((p \lor \neg q) \land r \to (q \to s))$$
$$1\ 1\ 1\ 1\ 1\ 0$$

This leads to the tree of Figure 9.1.4.

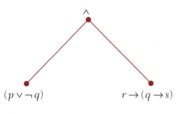

Figure 9.1.4

But clearly (again as in the proof of Theorem 9.1.10), $r \to (q \to s)$ is not a formula because it starts with a variable symbol and has length >1.

EXAMPLE 9.1.13   *String/Tree Algorithm*

The preceding examples show how to obtain the tree of a formula, and how to determine whether a given string is a formula. It is clearly recursive, because we had to use it over and over as the tree developed. The following is an informal algorithm based on this proof of Theorem 9.1.10. The pseudocode version is left to the exercises.

String/Tree Algorithm
1. If α starts with $\neg$, then $\alpha = \neg \alpha'$, and α is of type 6 in Theorem 9.1.10. Apply this algorithm to α' for further decomposition. If α' is a formula, the tree for α has a root labeled $\neg$, with the tree S for α' grafted onto it. See Figure 9.1.5.

Tree of α'

Figure 9.1.5

2. If α starts with a variable or constant symbol, then α must have length 1, and it is a variable or constant symbol, or else it isn't a formula. Its tree is a single vertex, labeled α.
3. If α starts with "(", then it must end with ")", or else it isn't a formula. Write α = (σ). Now scan σ from left to right through all initial ¬'s. Following these ¬'s there must be either a variable or constant symbol p or a left parenthesis.
   (a) In the first case, we have σ = ¬ ⋯ ¬pλσ', where λ is a binary connective. In this case, we must have α = (α'λβ') where α' = ¬ ⋯ ¬p and β' = σ'. This algorithm must be applied to β' to obtain the tree S of β' for further decomposition. The tree for α given by a root v, labeled λ, with the tree for ¬ ⋯ ¬p and the tree S of β' grafted onto it, as in Figure 9.1.6.

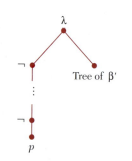

Figure 9.1.6

   (b) In the second case, scan σ from left to right from its first left parenthesis, keeping track of the excess of left parentheses over right parentheses until the number of left parentheses equals the number of right parentheses. If this occurs for the string α', we have σ = α'δ. Then we must have δ = λβ', where λ is a binary connective, or else α is not a formula. So finally, we obtain

$$α = (α'λβ')$$

   and the algorithm must then be applied to α' and β' for further decomposition. If α' and β' have trees R and S, the tree for α is given by a root v, labeled λ, with the trees for R and S grafted onto it as in Figure 9.1.7.

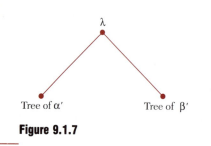

Figure 9.1.7

Type-0 Languages

We now give another method for generating languages on an alphabet Σ. This method is useful for the study of artificial and natural languages, such as English. Here, Σ is an arbitrary, nonempty set of symbols, which we also call the letters. We use the notation Σ^* to denote all strings using the letters of Σ, and Σ^+ denotes the set of all nonempty strings. The symbol ϵ denotes the empty string. The technique is first illustrated for the formulas of propositional calculus.

EXAMPLE 9.1.14   *Productions for the Language of the Propositional Calculus*

Let Σ be the alphabet of all the symbols of the propositional calculus. We also introduce two new special symbols S and B. To orient yourself, think of S as "start," as well as the symbol for a generic formula. Think of B as a "binary connective." Now consider the following transformations, called **productions**:

$$S \Rightarrow (SBS) \tag{9.1.1}$$

$$S \Rightarrow \neg S \tag{9.1.2}$$

$$S \Rightarrow x \qquad (x \text{ any variable or constant symbol}) \tag{9.1.3}$$

$$B \Rightarrow \wedge;\ B \Rightarrow \vee;\ B \Rightarrow \rightarrow;\ B \Rightarrow \leftrightarrow \tag{9.1.4}$$

Now, starting with S and using these productions, we arrive at a formula, provided (9.1.3) and (9.1.4) are used to eliminate the "variables" S and B. For example, here's how to produce the formula $((\neg p \rightarrow q) \vee r)$ using these productions:

$$
\begin{aligned}
S &\Rightarrow (S\ B\ S) && [\text{by (9.1.1)}]\\
&\Rightarrow (S \vee S) && [\text{by (9.1.4)}]\\
&\Rightarrow ((S\ B\ S) \vee S) && [\text{by (9.1.1)}\ \ (\text{first occurrence of } S \text{ only})]\\
&\Rightarrow ((S \rightarrow S) \vee S) && [\text{by (9.1.4)}]\\
&\Rightarrow ((\neg S \rightarrow S) \vee S) && [\text{by (9.1.2)}\ \ (\text{first occurrence of } S)]\\
&\Rightarrow ((\neg p \rightarrow q) \vee r) && [\text{using three applications of (9.1.3)}]
\end{aligned}
$$

Other routes are possible. For example, we could have waited till the end to substitute in the connectives.

We now generalize this technique for producing a language (i.e., a set of strings). Let Σ be an alphabet, and let V be another set, called the variable set. The set V is assumed to contain an element S, called the starter, and V and Σ are disjoint. (In Example 9.1.14, $V = \{B, S\}$.) The extended alphabet is defined as the set $\Gamma = \Sigma \cup V$.

DEFINITION 9.1.15

A **production** is an ordered couple (α, β), written

$$\alpha \Rightarrow \beta \qquad (9.1.5)$$

where $\alpha \in \Gamma^+$ and $\beta \in \Gamma^*$. If σ is any string of Γ^+, we say that the production (9.1.5) **transforms σ into τ** provided

$$\sigma = \sigma_1\alpha\sigma_2 \quad \text{and} \quad \tau = \sigma_1\beta\sigma_2$$

In this case, we write

$$\sigma \Rightarrow \tau \quad \text{or} \quad \sigma_1\alpha\sigma_2 \Rightarrow \sigma_1\beta\sigma_2 \qquad (9.1.6)$$

via the production (9.1.5). Finally, if there is a sequence of strings S, σ_1, $\sigma_2, \ldots, \sigma_n = \alpha$ **starting with** S and ending with α, such that $S \Rightarrow \sigma_1$; $\sigma_1 \Rightarrow \sigma_2$; $\sigma_2 \Rightarrow \ldots$; $\sigma_{n-1} \Rightarrow \sigma_n = \alpha$ via a set of productions, we say that the string α is **produced** by this set of productions. (This is written $S \Rightarrow \sigma_1 \Rightarrow \sigma_2 \Rightarrow \cdots \Rightarrow \alpha$.)

Remark   Note that (9.1.6) is not a new production. Rather, we say that τ is produced from σ via the production (9.1.5). Thus, in Example 9.1.14, we had $(S\ B\ S) \Rightarrow (S \vee S)$ via the production (9.1.4). It was not a new production.

DEFINITION 9.1.16

A **type-0 grammar** is given by the alphabet Σ, the variable set V, the starter S, and a finite set of productions. The set of all strings α on the alphabet Σ produced using the productions of a type-0 grammar according to Definition 9.1.15 is called a **type-0 language.**

As might be anticipated by the notation, there are also type-1, type-2, and type-3 grammars. These are all special cases of type-0 grammars and are defined by putting additional simplifying conditions on the productions allowed in the definition of the grammar.

We illustrate Definition 9.1.16 with a few examples.

EXAMPLE 9.1.17

Let $\Sigma = \{0, 1\}$. Let G be the set of all strings for which m 0's are followed by n 1's (m and n arbitrary and ≥ 0). Show that G is a type-0 language.

Method   We think of such a string as one that starts with S and is built up by adjoining 0's at the beginning and 1's at the end. Finally, we eliminate S altogether. This is accomplished by the three productions

$$S \Rightarrow 0S$$

$$S \Rightarrow S1 \qquad\qquad \textbf{(9.1.7)}$$

$$S \Rightarrow \epsilon$$

Thus, the first two productions allow the introduction of 0's at the beginning and 1's at the end, and the third eliminates S to produce a string of the required type.

In this example, the grammar consists of the alphabet $\{0, 1\}$, the variable set $\{S\}$ with the starter S as its sole member, and the three productions (9.1.7).

EXAMPLE 9.1.18

Let $\Sigma = \{a, b\}$ and $G = \{a^n b^n \mid n \geq 0\}$. ($a^n$ is a sequence of n letter a's.) Show that G is a type-0 language. Similarly, let $H =$ the set of strings with an equal number of a's and b's (any order). Show that H is also a type-0 language.

Method Since we want an equal number of a's and b's, we must adjoin a and b at the same time. This leads to the two productions $S \Rightarrow aSb$ and $S \Rightarrow \epsilon$, and these produce the language G. For H, our approach is to use the productions for G and then introduce a scrambling production, which can take the letter b before an a. This is accomplished by the production $ab \Rightarrow ba$. Finally, then, H is produced by the productions $S \Rightarrow aSb$, $S \Rightarrow \epsilon$, and $ab \Rightarrow ba$. Details are left to Exercise 9.1.15.

A similar example is as follows.

EXAMPLE 9.1.19

Let $\Sigma = \{a, b\}$ and let G be the set of words that have at least one a and at least one b. Find a set of productions for these words.

Method Think of the variable A as representing a word with at least one a in it, and similarly for the variable B. Also think of X as any word. Then the following scheme works:

$$S \Rightarrow aB \qquad S \Rightarrow bA$$

$$B \Rightarrow aB \qquad B \Rightarrow bX$$

$$A \Rightarrow bA \qquad A \Rightarrow aX$$

$$X \Rightarrow aX \qquad X \Rightarrow bX \qquad X \Rightarrow \epsilon$$

In this example, the variable set is $V = \{S, A, B, X\}$, and the nine productions of the grammar are as listed.

EXAMPLE 9.1.20   *English Grammar*

We conclude with a more familiar "language," the English language. For the purposes of this example, think of the letters of the alphabet Σ as the words in the dictionary. A string is then simply a stream (i.e., string) of words, maybe grammatically correct. The following is a very simplified version of English grammar. Think of a sentence as a string of the form

$$\text{NOUNPHRASE VERBPHRASE}$$

The sentence "the boy is walking" is of this form, with "the boy" as the noun phrase and "is walking" as the verb phrase. Thus, the following productions give some simple English sentences:

$$S \Rightarrow NV$$

$$N \Rightarrow \text{the boy}; \ N \Rightarrow \text{she}; \ N \Rightarrow \text{johnny}; \ N \Rightarrow \text{shirley}$$

$$V \Rightarrow \text{likes to bike}; \ V \Rightarrow \text{is walking}; \ V \Rightarrow \text{runs fast}$$

Obviously, more complicated grammars are possible, but we shall leave this to the linguists.

Exercises

1. Find the syntactically correct formulas for each tree of Figure 9.1.8.

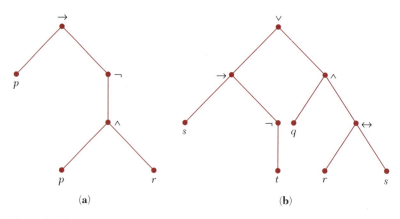

(a)   (b)

Figure 9.1.8

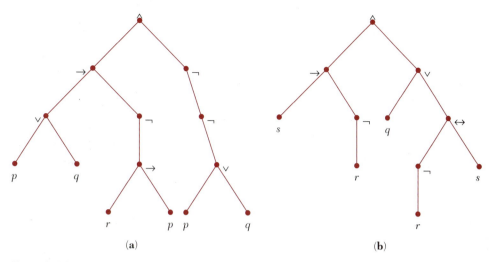

Figure 9.1.9

2. Find the syntactically correct formulas for each tree of Figure 9.1.9.
3. For each of the following, determine whether the string is a formula. If so, give the tree structure. Use the algorithm given in this section.
   (a) $(p \rightarrow \neg p)$
   (b) $(((p \lor q) \rightarrow (p \lor p)))$
   (c) $((\neg p \leftrightarrow q) \land (r \lor (p \land \neg q)))$
4. Repeat Exercise 3 for the following strings.
   (a) $\neg \neg \neg \neg (p)$
   (b) $(p \land \neg \neg \neg p)$
   (c) $\neg (((p \lor q) \rightarrow r) \land r)$
   (d) $((((p \rightarrow \neg q) \lor (r \lor s)) \lor (t \lor s)))$
5. Draw a tree diagram for the formula $(((p \rightarrow p) \rightarrow p) \rightarrow p)$.
6. Determine whether the large string given in the introductory paragraph of this section is a formula. If it is, give its tree structure.
7. Determine whether the string

   $$((r \rightarrow s) \land (p \lor ((((p \rightarrow q) \rightarrow r) \lor (r \land s) \rightarrow (t \lor \neg s))))$$

   is a formula. If it is, give its tree structure.
8. Prove by induction: In any formula, the number of binary connectives is equal to the number of left parentheses.
*9. Let a formula have P variable or constant symbols and C binary connectives. Find, by experiment or otherwise, the relationship between P and C. Prove it by induction.
10. Prove by induction that no formula ends with "(".
*11. Prove by induction that no formula has two consecutive binary connectives in it.
*12. Prove by induction that, in any formula, any variable symbol x may be replaced by $\neg x$, and the result is a formula.

*13. Prove by induction that no formula contains the consecutive symbols ")(" as a substring.

14. Following Definition 9.1.2, give a definition for a formula of the propositional calculus in Polish notation. No parentheses are in the alphabet.

*15. (a) Example 9.1.18 of the text stated that the language H consisting of all strings on a and b with an equal number of a's and b's is given by the productions $S \Rightarrow aSb$, $S \Rightarrow \epsilon$, and $ab \Rightarrow ba$. Explain why any string in H can be obtained from these productions.

   (b) Show that any string in a and b obtained from these productions is in the language H.

16. What language is generated from the following productions?

$$S \Rightarrow aSbb, \qquad S \Rightarrow \epsilon$$

Describe it as simply as you can.

17. What language is generated from the following productions?

$$S \Rightarrow a, \qquad S \Rightarrow aSa, \qquad S \Rightarrow bSb$$

Describe it as simply as you can.

18. What language is generated from the following productions?

$$S \Rightarrow \epsilon; \quad S \Rightarrow acS; \quad S \Rightarrow bcS; \quad S \Rightarrow Sc;$$

$$ab \Rightarrow ba; \quad ac \Rightarrow ca; \quad bc \Rightarrow cb$$

Describe it as simply as you can.

*19. A language is generated from the productions

$$S \Rightarrow aaabbb, \quad aba \Rightarrow ba, \quad ab \Rightarrow bab, \quad ba \Rightarrow bab$$

   (a) Show that any string in the language ends with three or more b's.

   (b) Show that any string in the language has at least one a.

*20. For the language of Exercise 19,

   (c) show that $bbbbabbbbb$ is in the language.

   (d) show that any string in the language, with the exception of $aaabbb$, has more b's than a's.

*21. A language is generated from the productions

$$S \Rightarrow ab, \qquad ab \Rightarrow aba, \qquad ba \Rightarrow bab$$

   (a) Show that ba is not in the language.

   (b) Show that $abababb$ is in the language.

   (c) Show that all strings in the language start with ab. (Use induction on the number of productions needed to arrive at the string.)

*22. For the language of Exercise 21 and the results proved there,

   (a) show that if α is any string on a and b, then $aba\alpha$ is in the language.

   (b) describe the language generated by the foregoing productions. That is, give an algorithm to determine whether a string is in this language.

In each of the following, take Σ as the two-letter alphabet $\{a, b\}$. Show that each of the languages G is a type-0 language, by exhibiting suitable productions.

23. G = strings with exactly three a's.
24. G = strings with twice as many a's as b's.
25. G = strings that read the same when reversed. (These are called palindromes.)
26. G = strings of the form $a^n b^m$ or $b^n a^m$, with $m, n \geq 0$.
27. G = strings with an odd number of a's.
28. G = $\{ab, ba\}$.
*29. G = set of words of the form $a^n b^n a^n$, for some $n \geq 1$.
*30. G = set of words that are squares—the set of α such that $\alpha = \beta\beta$.
*31. It was stated without proof that the four production types (9.1.1) through (9.1.4) of Example 9.1.14 have the formulas of the propositional calculus as their language. Prove this.

The proof of Theorem 9.1.10 was the basis for the informal String/Tree Algorithm of Example 9.1.13. The following exercises ask you to put some of this algorithm into pseudocode.

*32. Step 3 of String/List scans a string α and breaks it up into a string $\alpha = \alpha'\lambda\beta'$, where α' is a string of $\neg$'s, followed either by a constant or variable symbol p or by a formula with an equal number of left and right parentheses, starting with "(" and ending with ")". λ is a binary connective. Write a procedure Scan in pseudocode that does this. The input is the string α, assumed of length $n \geq 3$, which may be assumed given as a sequence in the alphabet of the propositional calculus. The output is α', λ, and β' or a statement that the procedure fails. You might make this recursive to take care of the initial $\neg$'s.
*33. Using the procedure outlined in Exercise 32, write the String/List Algorithm in pseudocode. It should be recursive, with inputs α and output the tree T of α, if α is a formula, or a statement that α is not a formula.

9.2 Automata

If Σ is an alphabet, a language L on Σ has been defined as a subset L of Σ^*. In the previous section we were concerned with methods of producing and identifying the strings of certain languages L in special cases. We now study an especially simple kind of algorithm that determines whether a given string $\alpha \in \Sigma^*$ is in a given language L.

An *automaton* (plural, *automata*) is a method that does this for especially simple languages. We first define an automaton heuristically as a machine subject to simple rules that performs this task. In what follows, we take $\Sigma = \{a, b\}$, a two-letter alphabet. This simplifies the discussion considerably. If an alphabet contains more letters, the various definitions, results, and discussion change accordingly, but no new complications are introduced.

DEFINITION 9.2.1

An **automaton** A is a machine with finitely many states $q_1, q_2, \ldots, q_k$. One of these states, q_1, is called the **initial state**. Some of the states are called **terminal states**. At each moment of time, the automaton is in one of the states q_i and scanning a letter of a given string α. The automaton works by scanning the string α from left to right, starting at the first letter of the string. When it is in any state, and viewing a given letter, the machine changes its state, depending only on the letter it is viewing and its current state, and moves one letter to the right to scan the next letter. When there are no more letters to view, the machine stops. If it is then in a terminal state, the string α is **accepted** by A, and α **is in the language of A.** If it is not in a terminal state, α is **rejected** by the automaton, and it is not in the language of A.

DEFINITION 9.2.2

More formally, an automaton can be defined as follows. A set $\mathbf{Q} = \{q_1, q_2, \ldots, q_k\}$ of states and a set $\Sigma = \{a_1, \ldots, a_n\}$ of letters must be given. In addition, one state, say q_1, is called the initial state, and some subset $T \subseteq \mathbf{Q}$, called the terminal states, must be given. In addition, a **transition function** $f \colon \mathbf{Q} \times \Sigma \to \mathbf{Q}$ must be given. An automaton is the list of all of these items: $A = [Q, \Sigma, q_1, T, f]$.

Remark   When the machine is in state q viewing the symbol x, it moves into state $f(q, x)$.

EXAMPLE 9.2.3

Suppose an automaton has states q_1 and q_2 and the letters $\Sigma = \{a, b\}$. Let the only terminal state be q_2, and let the transition function f be given by

$$f(q_1, a) = q_1; \quad f(q_1, b) = q_2; \quad f(q_2, a) = q_1; \quad f(q_2, b) = q_1$$

Decide whether the string "*abba*" is in the language of this automaton. Similarly decide whether "*babbb*" is accepted.

Method   We first consider the string "*abba*". Starting with the first letter "*a*", we are in state q_1. Since $f(q_1, a) = q_1$, we move to state q_1 and view the second letter b. This is indicated diagrammatically as follows:

$$
\begin{array}{ll}
\text{String:} & a \quad b \quad b \quad a \\
\text{State:} & q_1 \, q_1
\end{array}
$$

Now we are in state q_1, viewing b. Since $f(q_1, b) = q_2$, we move to state q_2, and view the next letter, leading to the diagram

$$\begin{array}{llll} \text{String:} & a & b & b & a \\ \text{State:} & q_1 & q_1 & q_2 \end{array}$$

Finally, we arrive at the diagram

$$\begin{array}{lllll} \text{String:} & a & b & b & a \\ \text{State:} & q_1 & q_1 & q_2 & q_1 & q_1 \end{array}$$

The procedure stops here. We are in state q_1 with no more letters to scan, and this state is not designated as a terminal state. Therefore, the string "*abba*" is rejected by the automaton, and the string *abba* is not in its language.

The situation for "*babbb*" is indicated in the diagram

$$\begin{array}{llllll} \text{String:} & b & a & b & b & b \\ \text{State:} & q_1 & q_2 & q_1 & q_2 & q_1 & q_2 \end{array}$$

We end in the terminal state q_2. This string is accepted by the automaton and is in its language.

This procedure gives a recursive method to decide whether a string is accepted. If α is the string $a_1 \cdots a_n$, then the sequence of states s_k (viewing the kth letter a_k) is defined by the condition

$$s_1 = q_1 \qquad \text{(initial state)}$$

$$s_{k+1} = f(s_k, a_k) \qquad \text{for } 1 \leq k \leq n$$

The string is accepted if s_{n+1} is a terminal state. Otherwise, it is rejected.

DEFINITION 9.2.4

If **A** is an automaton, the language $L(A)$ is the set of strings that are accepted by **A**.

The method illustrated in Example 9.2.3 effectively determines whether a string α is in the language of an automaton, but it gets more cumbersome as the number of states increases. In addition, it gives very little insight into what is happening. We now introduce a useful graphical description of these computations.

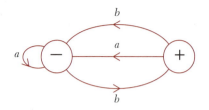

Figure 9.2.1

DEFINITION 9.2.5   *The Digraph of an Automaton*

It is most convenient to illustrate what an automaton does by means of a (general) **directed graph.** The method is as follows. Each **state** of A is taken as a **vertex** of the graph. The out-degree of each vertex is the number of letters of the alphabet. Since the alphabet Σ is assumed to have the two letters a and b, we have two directed edges, labeled "a" and "b", exiting from each vertex. If $f(q, a) = q'$ in the definition of A, we join q to q' via a directed edge labeled "a", and similarly for the directed edge labeled "b". The initial state is labeled $\ominus$, and the terminal states are labeled $\oplus$. The symbol $\overline{\oplus}$ is used for an initial vertex that is also terminal.

Figure 9.2.1 illustrates the digraph for the automaton of Example 9.2.1.

The way an automaton scans a string α can now be graphically interpreted. The string α is scanned starting at the initial vertex. The first letter determines the edge that leads to another vertex. From this vertex, the next letter determines an edge that leads to a next vertex. Continuing to scan, a string α leads to a **path**† in the directed graph. The string α is in $L(A)$ if and only if the path ends on a terminal vertex. Furthermore, any path starting at $\ominus$ leads to a string α, so $L(A)$ is in 1-1 correspondence with all paths starting at $\ominus$ and ending at a vertex of the type $\oplus$.

EXAMPLE 9.2.6

Consider the simple automaton whose digraph is given in the left half of Figure 9.2.2. Starting at the initial vertex, any string leads to a path that cycles around the triangle. The paths that end at the terminal vertex are precisely those whose lengths are multiples of 3. Thus, the language $L(A)$ consists of all strings whose lengths are multiples of 3. Note that the empty string ϵ is also accepted.

For convenience in what follows, we will use a single arrow labeled "a, b" instead of two arrows, labeled "a" and "b", when both letters cause the same transition. This is illustrated on the right in Figure 9.2.2.

---

† In this chapter, paths and cycles may have repeated vertices.

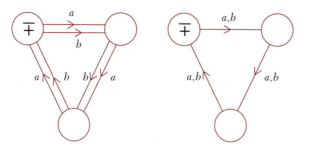

Figure 9.2.2

Digraphs of automata are easily identified.

THEOREM 9.2.7

Let **G** be a general digraph with the following additional properties.

1. Each directed edge is labeled either "*a*" or "*b*".
2. The out-degree of each vertex is 2. Of the two edges leaving any vertex *v*, one is labeled *a* and one is labeled *b*.
3. A certain vertex is labeled ⊖.
4. Certain vertices are labeled ⊕.

Then **G** is the digraph of an automaton. Automata and digraphs of this type are in 1-1 correspondence. ■

Because of Theorem 9.2.7, we usually identify an automaton with its directed graph.

We now illustrate with some examples.

EXAMPLE 9.2.8

Find an automaton whose language consists exactly of the two strings *ab* and *ba*.

Method   The automaton of Figure 9.2.3 clearly works. How is it constructed?

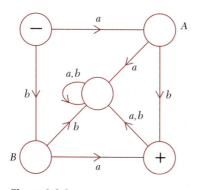

Figure 9.2.3

Note that the center state is a "sink." Once there, the string must necessarily be rejected. A good technique for constructing diagrams of automata is to regard each vertex as representing a set of strings with appropriate properties. Thus, $\ominus$ represents the empty string ϵ as well as the initial vertex. In Figure 9.2.3, the vertices labeled "A" and "B" represent the strings "a" and "b", respectively. The vertex $\oplus$ represents the required set, and the rejected set of strings with length ≥ 2 is the sink we noted earlier.

In the same way, it is possible to show that any finite set of strings is the language of an automaton. This will be left to the exercises.

EXAMPLE 9.2.9

Construct an automaton whose language consists of all strings containing two or more b's.

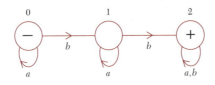

Figure 9.2.4

Method We think of the initial state as representing the strings with no b's so far. Similarly, state 1 of Figure 9.2.4 represents the strings with one "b" so far. Finally, the terminal state is the state with two or more b's. This makes the solution of Figure 9.2.4 evident.

In this example, it is natural to call the states q_0, q_1, and q_2. Here q_0 is initial and q_2 is terminal. The automaton function $f = f_A$ is given by

$$f(q_0, a) = q_0; \quad f(q_1, a) = q_1; \quad f(q_2, a) = q_2;$$
$$f(q_0, b) = q_1; \quad f(q_1, b) = q_2; \quad f(q_2, b) = q_2$$

where q_0 is the initial state and q_2 is the only terminal state. We usually omit such an explicit exhibition of the function, because it is more easily seen in the labeled digraph of the automaton.

EXAMPLE 9.2.10

Let L be the language consisting of strings with an even number of a's. Find an automaton A with $L(A) = L$.

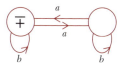

Figure 9.2.5

Method   Think of the two states "Even" and "Odd". This quickly leads to Figure 9.2.5. Be careful to include the loops about the vertices. This is needed because each vertex must have an edge leaving it for each of the letters of our alphabet.

EXAMPLE 9.2.11

Repeat Example 9.2.10 with an even number of a's *and* an odd number of b's.

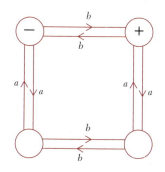

Figure 9.2.6

Method   Here, let i = the parity of the number of a's so far ($i = 0$ for even and $i = 1$ for odd), and let j be the similar parity for the number of b's. This leads to the four states (i, j), and the construction of Figure 9.2.6 is straightforward.

EXAMPLE 9.2.12

Find an automaton A whose language consists of all strings containing two consecutive a's or two consecutive b's.

Method   Think of the four states: (1) initial; (2) last letter scanned is "a", but no success so far; (3) last letter is "b," but no success so far; (4) success. Choosing vertices corresponding to these states, we construct the required automaton of Figure 9.2.7.

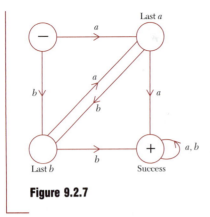

Figure 9.2.7

In the next section, we give an *algebraic* description of the languages that can be described by an automaton. We now show that there are relatively simple languages that are *not* the languages of an automaton. This is a more difficult problem because, as usual, it is often more difficult to show that an object doesn't exist than to construct it if it does.

EXAMPLE 9.2.13

Let R be the set of strings with an equal number of a's and b's. Then R is not the language of an automaton.

Method Suppose A were such an automaton, which had N vertices. Consider the string $\alpha = a^N b^N$, which is in R. (See Definition 9.2.14.) The path for α starts at $\ominus$ and must end at a terminal vertex labeled $\oplus$. Now, the path for the string a^N alone has N edges and uses $N + 1$ vertices (counting repetitions). Thus, since A is assumed to have N vertices, one vertex is repeated in the path for a^N, and we must have a **cycle** in the path for a^N as in Figure 9.2.8.

Suppose this cycle had length $K \geq 1$. Then, since the path for α starts with a^N, we can introduce a loop into this cycle before finishing the path for α, and we would still end at the same vertex. This means that the path $a^{N+K} b^N \in L(A)$, because it also begins at $\ominus$ and ends at $\oplus$. But this string does not have an equal number of a's and b's, and so is not in the language R. Hence, $L(A) \neq R$, and R cannot be the language of any automaton. This completes the proof.

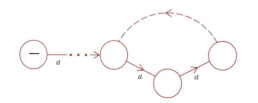

Figure 9.2.8

Remark   If we had tried to use our states for counting purposes, as in some of our previous examples, we might decide to let q_n represent the state where there was an excess of n a's over the b's. This would have led to the **infinite** automaton as in Figure 9.2.9. This failed attempt only suggests that we cannot do the job with finitely many states, as is required by an automaton. The foregoing proof gives the necessary demonstration.

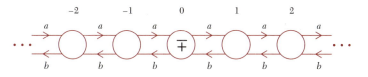

Figure 9.2.9

We used the notation a^N. This is the string "a" concatenated with itself N times. The formal definition is as follows.

DEFINITION 9.2.14

If α is any string and n a natural number, the string α^n is defined recursively as follows:

$$\alpha^0 = \epsilon$$
$$\alpha^{n+1} = \alpha^n\alpha \qquad \text{for } n \geq 0$$

Here ϵ is the empty string.

The method of Example 9.2.13 can be generalized.

THEOREM 9.2.15   The Pumping Lemma

Let A be an automaton with N states, and let α be in the language of A. Suppose length$(\alpha) \geq N$. Then α can be written $\alpha = \beta\gamma\delta$, where $\gamma \neq \epsilon$ and $\beta\gamma^i\delta$ is in the language of A for $i = 0, 1, 2, 3, \ldots.$ ■

The proof is similar to the one in Example 9.2.13, where α was taken as a^N and the loop was γ. See Exercise 9.2.11.

Example 9.2.14 is a relatively simple language that is not given by any automaton. Thus, it is not surprising that the language of propositional calculus is not the language of an automaton. The following example shows this. For simplicity, we choose one letter P to denote a variable or constant symbol and the letter B to denote a binary connective, and we retain ¬ and the two parentheses. Thus, in this example, we use an alphabet of five symbols: P, B, ¬, (, and).

EXAMPLE 9.2.16

The language of the propositional calculus is not the language of an automaton.

Method   As in the previous example, suppose that it were and that it had N states. Now find a formula $\alpha = ((\ldots (\beta$ that starts with N left parentheses. (This can be done by induction.) As in the previous example, if we follow the path for just the first N parentheses, we necessarily have a loop with $K > 1$ edges. Thus, by adding K left parentheses at the beginning of this formula, we obtain a string that terminates at the same state that α does. But this latter string is not a formula, because it has more left parentheses than right. But, by definition, α terminates at a terminal state, and hence the nonformula "$((\ldots (\alpha$", with K additional parentheses, will be accepted by the automaton. This is a contradiction.

Exercises

For Exercises 1–7 draw the graph of an automaton that has the given language.

1. All α starting with "a" and ending with "b".
2. All α containing the substring "aba".
3. All α of length 4 or 5.
4. All α except the empty string ϵ.
5. All α with no repetitions [no "aa" or "bb"].
6. All α where no letter is isolated. (Any "a" must be adjacent to another "a", and similarly for "b".)
7. All strings of the form $a^n b^m$ with $m, n \geq 0$.
8. Show that the set of strings of the form $a^n b$ $(n > 0)$ is an automaton language, as is the set of strings of the form $a^n b a^m$ $(m, n > 0)$, but the strings of the form $a^n b a^n$ $(m, n > 1)$ do not constitute the language of an automaton.
*9. Show that the strings of the form a^{n^2} are not the language of any automaton.
10. Let G be the set of strings with fewer a's than b's. Prove that G is not the language of an automaton.
*11. Prove the Pumping Lemma (Theorem 9.2.15). (*Hint:* Find the path of the string α and look for a repeated vertex.)
12. Prove: If an automaton has N states, then it accepts a string of length $\geq N$ if and only if the language of the automaton is infinite. (*Hint:* Use the Pumping Lemma.)
*13. Sharpen the previous exercise as follows. Prove that if an automaton has N states and its language is infinite, then there is a string α with

$N \le \text{length}(\alpha) < 2N$ that is in the language of the automaton. [*Hint:* Find a string α accepted by the automaton such that $\text{length}(\alpha) \ge 2N$. Let α be the string of least length satisfying this property. Now write $\alpha = \sigma\tau$, where $\text{length}(\sigma) = N$, and find a cycle in σ.]

14. Prove that if an automaton has N states and a nonempty language, then its language contains a string of length $\le N - 1$.

15. Let $G = L(A) \subseteq \Sigma^*$. Prove that the complement $\Sigma^* - G$ is also the language of some automaton.

16. Prove: If $S = L(A) \subseteq \Sigma^*$ and $T = L(B) \subseteq \Sigma^*$, then $S \cap T$ and $S \cup T$ are also languages of some automata.

17. Prove that any finite set of strings is the language of an automaton.

Describe the language of each of the following automata in the simplest way you can.

18.   21.

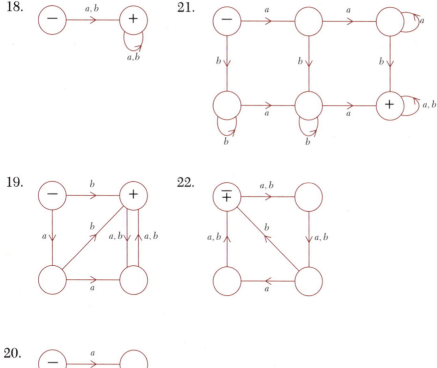

19.   22.

20.

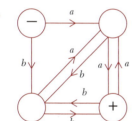

9.3 Regular Sets

We now introduce an algebra on sets of strings that is useful for defining a language on an alphabet Σ.

DEFINITION 9.3.1

Let Σ be an alphabet. We define three basic operations on sets of strings (subsets of Σ^*) as follows.

1. **Union**   If A and B are subsets of Σ^*, then $A + B$ denotes the union $A \cup B$.
2. **Concatenation**   If A and B are subsets of Σ^*, then $A \cdot B$ or AB denotes the set of *all* concatenations of a string in A with a string in B. Thus,

$$AB = \bigcup_{\substack{\alpha \in A \\ \beta \in B}} \{\alpha\beta\}$$

3. **Starring**   If A is a subset of Σ, A^* consists of all possible concatenations $\alpha_1 \alpha_2 \cdots \alpha_k$, where k is an arbitrary natural number and $\alpha_i \in A$. *This includes the empty string ϵ, which corresponds to $k = 0$.*

EXAMPLE 9.3.2

The following examples, using $\Sigma = \{a, b, c\}$, illustrate these operations.
   (a) $a + b$ = the set consisting of a and b. For convenience, we identify the element a and the singleton set $\{a\}$. Similarly, $\Sigma = a + b + c$.
   (b) $(a + b)^*$ = all strings of Σ that use only the letters a and b.
   (c) $(ab + ba + aa + bb)^*$ = all strings of a's and b's having even length.
   (d) $(b + c)^*a(b + c)^*$ = all strings with exactly one a.
   (e) $(aa + bc)(a + b + c)^*$ = all strings that begin with aa or with bc.
   (f) $(((ab + bc)^* + a)^* + (ab + ba^*)^*)^*$ = a set not easily described except by the formula itself.

DEFINITION 9.3.3

A **regular set** on an alphabet Σ is a set of strings that can be finitely expressed by using only the individual letters of Σ, the empty string ϵ, and the operators $+$, $\cdot$, and $*$. The null set $\varnothing$ is also defined as a regular set.

Examples 9.3.2(a)–(f) are regular sets.

EXAMPLE 9.3.4   *Recursive Definition of a Regular Set*

We can loosely describe the regular sets of Σ as the sets that have an algebraic description, using the operations of $+$, $\cdot$, and $*$. More formally, the definition is given recursively as follows.

1. $\varnothing$ is a regular set; $\{\epsilon\}$ is a regular set.
2. If $x \in \Sigma$, then $\{x\}$ is regular.
3. A is regular if and only if it is of type 1 or 2, or if
   (a) $A = B + C$, where B and C are regular, or
   (b) $A = BC$, where B and C are regular, or
   (c) $A = B*$, where B is regular.

In this definition, 1 and 2 are the basis. Definition 3(a)–(c) allows algebraic operations to be used and shows that any regular set is built up from such operations.

EXAMPLE 9.3.5

Show that the set of strings of the form $a^n b^m$, $m, n \geq 0$, is a regular set. Similarly, show that the set of strings (using the alphabet $\{a, b\}$) that contain two consecutive identical letters ("aa" or "bb") is a regular set.

Method   1. The set of strings of the form a^n is, by definition, $a*$. Similarly, the set of strings of the form b^m is $b*$. We need to concatenate any string in $a*$ with one in $b*$. But this is the regular set $a*b*$.
2. The set consisting of "aa" and "bb" is the union $aa + bb$. We now want to take any string of a's and b's, concatenate it with aa or bb and follow it by any other string of a's and b's. The set obtained in this way is precisely the set $(a + b)*(aa + bb)(a + b)*$, which is a regular set.

We shall show that the regular sets **are precisely those sets that are the language of some automaton.** But before we do this, we introduce two useful generalizations of an automaton.

DEFINITION 9.3.6

A **transition graph** is a directed general graph $\mathbf{G}$ with finitely many vertices, one *or more* of which are labeled $\ominus$, and some of which are labeled $\oplus$. There are finitely many directed edges of the graph $\mathbf{G}$, each of which is labeled with a string of $\Sigma*$.

Figure 9.3.1 is a simple example.
In contrast to an automaton, it is not necessary to have exactly $|\Sigma|$ directed edges leaving each vertex. Nor do the edges have to be labeled

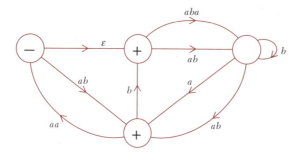

Figure 9.3.1

only with letters of Σ. Thus, a transition graph is more general. An automaton is perfectly **deterministic**—once a string is given, there is a unique path on its graph corresponding to it. A transitional graph is **nondeterministic**—there are possibly several, or possibly no, paths that correspond to a given string. Transition graphs represent nondeterministic automata.

Still, the *language* of a transition graph may be defined. Clearly, any path on the graph determines a unique string of Σ, obtained by concatenating the strings along the path.

DEFINITION 9.3.7

Each path on a transition graph **G** determines a string obtained by concatenating the labels of the edges of that path. The **language** $L(\mathbf{G})$ of a transition graph **G** is the set of strings that are determined by the paths joining a vertex labeled $\ominus$ to a vertex labeled $\oplus$.

We continue to use $\Sigma = \{a, b\}$ to illustrate these ideas. The generalization to any finite alphabet is straightforward.

Remark   It is possible for a path to start at a vertex labeled $\ominus$ and to end at a vertex that is *not* labeled $\oplus$, and yet the string α determined by that path might be in the language $L(\mathbf{G})$. The nature of

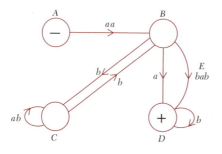

Figure 9.3.2

a nondeterministic automaton is that there may be several paths yielding the same string α. In Figure 9.3.2, the string $\alpha = aababb$ is in the language of the transition graph by virtue of the path $AB(E)DD$, yet the path $ABCCB$ also generates this string but does not end at a terminal vertex. The edge (E) is designated here because there were two edges joining B to D. It is not immediately clear that there is an algorithm to determine whether a string α is in $L(\mathbf{G})$. We shall soon show that there is.

We now generalize transition graphs.

DEFINITION 9.3.8

A **generalized transition graph G** is defined as in Definition 9.3.6, except that each directed edge is labeled with a regular set.

Figure 9.3.3 is an example.

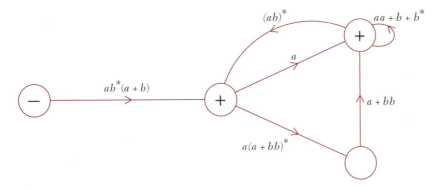

Figure 9.3.3

In an automaton or transition graph, any path determines a **string** obtained by concatenating the strings labeling the edges of the path. With generalized transition graphs, any path in the graph leads to a **set,** namely the one obtained by the **concatenation** (i.e., **multiplication**) **of the sets** along the edges of the path.

DEFINITION 9.3.9

The **language** $L(\mathbf{G})$ of a generalized transition graph $\mathbf{G}$ is defined as the union of all the sets corresponding to all directed paths joining a vertex labeled $\ominus$ to a vertex labeled $\oplus$.

We now prove that the language of an automaton is a regular set and, conversely, that any regular set is the language of an automaton. The proof is constructive. Along the way, we show how the nondeterministic automata (transition graphs) can be converted to (deterministic) automata.

THEOREM 9.3.10   (Kleene)

1. Let G be a generalized transition graph. Then $L(G)$ is a regular set. In particular, $L(A)$ is a regular set for any automaton A, since an automaton is a special case of a generalized transition graph.
2. Let R be a regular set. Then there is an automaton A such that $L(A) = R$.

Proof We prove part 1 in steps. In each step, we change the transition graph in such a way that the languages of the original and the changed graphs are the same.

Step 1 Replace all the labels $\ominus$ by only one labeled $\ominus$. We can do this by introducing a new $\ominus$ vertex. We then introduce directed edges from the new initial vertex $\ominus$ to all of the given vertices $\ominus$, labeling these with the empty string ϵ, and take away the $\ominus$ designation from the original set of initial vertices. This is illustrated in Figure 9.3.4. Note that the language of the resulting graph is clearly identical to the language of the original graph.

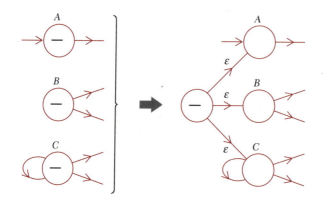

Figure 9.3.4

Step 2 Similarly, replace all terminal vertices (if any) with a single new vertex $\oplus$ as in Figure 9.3 5.

Step 3 If there is more than one edge connecting any two vertices v_1 and v_2, combine them into one edge, using the $+$ operation, as in Figure 9.3.6.

Once again, the language of the graph does not change when we use any of these operations. Thus, we have created a generalized transition graph with one initial vertex and one final vertex, such that there is at most one directed edge from any vertex to any other vertex.

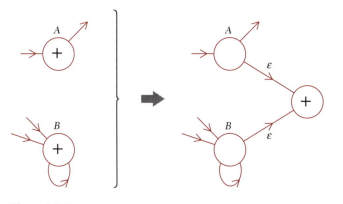

Figure 9.3.5

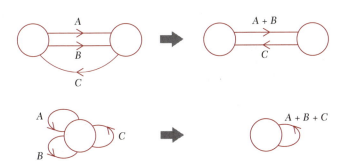

Figure 9.3.6

Step 4   We now show how to eliminate any vertex v that is not labeled $\ominus$ or $\oplus$. If there are no such vertices, go to step 6. Any string in the language of G that uses a vertex v must enter v on one directed edge, say from a vertex V, then loop about v for a finite number of steps (if there is a loop there), and finally exit along a directed edge to a vertex W. Such a situation is indicated in Figure 9.3.7. The one below is the case where $V = W$.

Our procedure in step 4 is as follows. Suppose v has a loop C: a directed edge from v to v. Further suppose that the edge Vv is labeled with the regular set A, and that edge vW is labeled with B. Then introduce a new edge from V to W labeled AC^*B as in Figure 9.3.7. If there is no loop at v, simply use the label AB for the new edge from V to W. Now repeat this step (for the fixed vertex v) using *every possible choice* of V and W. Finally, eliminate the vertex v from the graph altogether, along with any of the edges entering it and leaving it.

Step 5   Go to step 3. (Thus, a WHILE instruction is being used. Steps 3 and 4 are repeated for as long as there are vertices v not equal to the terminal or initial vertices. See the second sentence of step 4.)

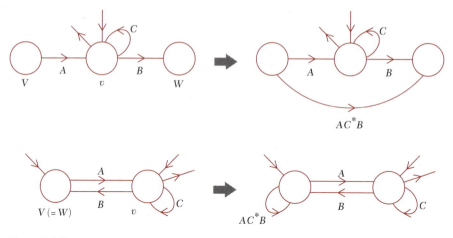

Figure 9.3.7

Step 6 The previous steps have eliminated all vertices except the two labeled ⊖ and ⊕. The process did not allow any new edges to enter ⊖ or to leave ⊕. Thus, we are left with the generalized transition graph of Figure 9.3.8. Its language is clearly the regular set R. ■

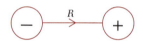

Figure 9.3.8

This completes part 1 of the theorem. In particular, it shows that the language of an automaton is necessarily a regular set.

An outline of the algorithm is as follows:

1. Introduce a single initial vertex.
2. Introduce a single terminal vertex.
3. WHILE there are vertices other than ⊖ and ⊕
    1. Combine multiple edges into one, using +.
    2. Eliminate a vertex other than ⊖ and ⊕.
4. Combine multiple edges into one, using +.
5. Find the regular set R as in Figure 9.3.8.

We interrupt the proof of Theorem 9.3.10 to illustrate this method with an example.

EXAMPLE 9.3.11

Consider the generalized transition graph of Figure 9.3.9. Express its language as a regular set.

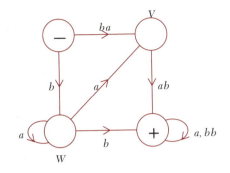

Figure 9.3.9

Method   We do not use steps 1 and 2 initially. Therefore, we arrive at a
diagram somewhat more complicated than Figure 9.3.8. Let us first eliminate the upper right vertex V. We first combine the *ba* and *ab* edges
from ⊖ through V to ⊕ and introduce one edge, labeled *baab*, from ⊖ to
⊕. Similarly, replace the edge from W to V to ⊕ by introducing one edge
from W to ⊕, labeled *aab*. This covers all paths through V, so we eliminate
V and all edges adjacent to it. This leads to the generalized transition
graph of Figure 9.3.10.

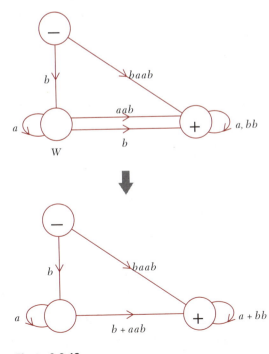

Figure 9.3.10

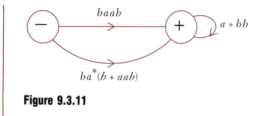

Figure 9.3.11

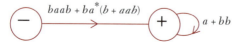

Figure 9.3.12

Using instruction 3.1 of the algorithm, we obtain the lower graph of Figure 9.3.10. We can now eliminate W in a similar manner to obtain Figure 9.3.11.

Again, using instruction 3.1, this finally simplifies to Figure 9.3.12. Steps 1 and 2 are not necessary here because the language represented by Figure 9.3.12 is clearly

$$R = [baab + ba*(b + aab)](a + bb)*$$

***Proof* of Theorem 9.3.10 Part 2**   Now suppose, conversely, that R is a regular set, given by some expression in a and b. We show how to unravel this expression to find an **automaton** that has R as its language. The method is completely algorithmic in character, much as part 1. Thus, we show how a nondeterministic automaton can be converted to a deterministic one. We do this in two steps.

Step 1   We first find a transition graph, with a single terminal and initial vertex, whose language is R and whose edges are labeled either a, b, or ϵ. We do this by reversing some of the steps in the proof of part 1. Once again, each step in the process keeps the language of the generalized transition graph unchanged. We first start with the simple transition graph of Figure 9.3.8, whose language is R.

We now show how to convert any edge labeled with a regular set other than ϵ, a, or b into new edges, labeled with simpler regular sets. Each such regular set R is, by definition, of the form $A + B$, AB, or $A*$, where A and B are simpler regular sets. The replacements are given in Figure 9.3.13.

Each of these replacements leaves the language of any transition graph unchanged. We can continue this process until we finally arrive at a transition graph whose edges are labeled a, b, or ϵ. This shows that any regular set is the language of such a transition graph.

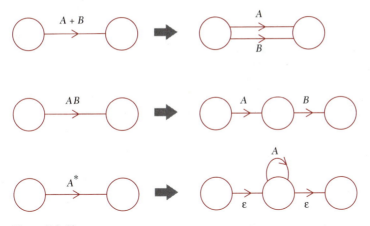

Figure 9.3.13

To illustrate this procedure, consider $R = b(aa + bb)^*(a + b)$. Figure 9.3.14 shows the series of conversions to reduce R to the language of a transition graph.

Step 2 This step takes a transition graph whose edges are labeled a, b, or ϵ and creates an **automaton** with the same language. The idea is to

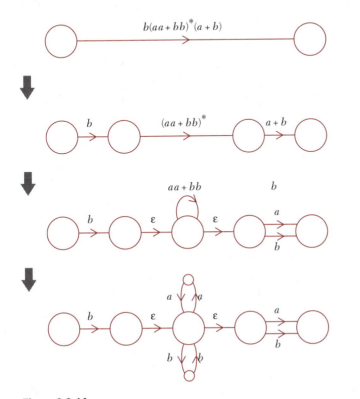

Figure 9.3.14

create an automaton whose **vertices** are **subsets** of the original set of vertices. If S is any subset of the vertices, we find the set Sa of vertices reachable from any of the vertices of S in one step along a path labeled "*a*". Regard a path labeled ϵ as a free move—that is, no steps are used to move along it. Similarly, define the set Sb.

Suppose vertices are labeled 1, 2, . . . , k, with $\ominus$ labeled 1. Now start a table with three columns labeled S, Sa, and Sb, and start with 1 in the S column:

S	Sa	Sb
1		

In the Sa column, we put the set of vertices reachable from a vertex of S by going one step along an "*a*" edge, allowing any number of free moves along an ϵ edge.

For example, in the transition graph of Figure 9.3.15, if S = {1}, then Sa = {2, 3, 4, 5} and Sb = {5}. Now once we find Sa and Sb, we put them in the S column and find the corresponding second and third columns. *Now* we *continue until no new subset appears.* The result is then a system of subsets S for which Sa and Sb are part of the system.

We illustrate with an example of this process. In what follows, we label $\ominus$ as 1. If no confusion occurs, we use 1 instead of {1}. Similarly, sets such as {1, 2, 4} are denoted 124.

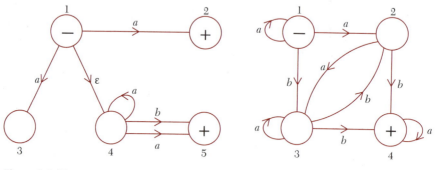

Figure 9.3.15   **Figure 9.3.16**

Consider the transition graph of Figure 9.3.16. Starting with S = 1, we find that Sa = 12 and Sb = 3. Thus, we have the start of the required table:

S	Sa	Sb
1	12	3
12		
3		

For $S = 12$, we find

$Sa = 123$. [Just check $\{2\}a$, since we know about $\{1\}a$.]
$Sb = 34$.

For $S = 3$, we find

$Sa = 3$.
$Sb = 24$.

This leads to the extension of the previous table:

S	Sa	Sb
1	12	3
12	123	34
3	3	24
123		
34		
24		

We continue the process and complete it as follows:

	S	Sa	Sb
⊖	1	12	3
	12	123	34
	3	3	24
	123	123	234
⊕	34	34	24
⊕	24	34	4
⊕	234	34	24
⊕	4	4	∅
	∅	∅	∅

This is an automaton! Specifically, its states are the subsets in column 1, and its transition function f is given by the two columns Sa and Sb. The terminal states are the subsets containing the vertex 4. The initial state continues to be 1. We write $f(S, a)$ as Sa, and similarly for Sb. Its graph is constructed from this table in Figure 9.3.17.

Why does this work? Any path in the constructed automaton that begins at ⊖ and ends at a vertex labeled ⊕ must have a corresponding path in the original transition graph ending at 4 (the original ⊕). For example, consider the string $aaab$ scanned by the automaton of Figure 9.3.17. It leads to the subsets indicated in Figure 9.3.18.

Now work backwards from 234. Any vertex of 234 is reachable by b (in the original graph) from one of 1, 2, or 3. This is what is meant by

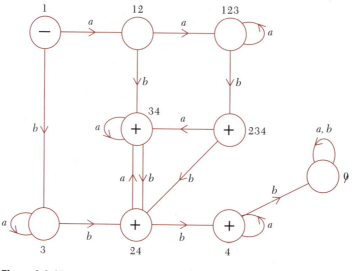

Figure 9.3.17

Figure 9.3.18

the formula $(123)b = (234)$. In particular, 4 is so reachable. Figure 9.3.16 shows that 4 is reachable by b from vertex 2 or 3, say 2. Now go backwards from 123. Vertex 2 is reachable from 123 along an edge labeled "a" (in fact, from vertex 1). Proceeding in this way, we can trace a path from 1 to 4 in the original graph, whose string is $aaab$. Conversely, any path from vertex 1 to vertex 4 in the original graph yields a path to one of the terminal vertices in the constructed automaton. The languages are thus the same, and this completes the proof.

The general proof, for any transition graph, follows the reasoning given in this illustration. ■

We conclude this section with a method of determining whether two automata have the same language. By Theorem 9.3.10, this is the same as determining whether two expressions for a regular set constitute the same set of strings. The idea is to keep track of the two automata at the same time. This is done by the device of building a superautomaton that contains within it all of the information in each automaton.

THEOREM 9.3.12   (Moore)

There is an algorithm to determine whether two automata have the same language.

Proof   We show how to determine whether two given automata have the same or different languages. Suppose the given automata are A and A'. Let v and v' denote vertices of A and A', respectively. We form a new automaton B as follows: Take the vertices of B to be the ordered couples (v, v'). Let va denote the vertex w in A such that v is joined to w by a directed edge labeled "a". A similar definition applies to vb. For all choices (v, v') in B we join (v, v') to $(va, v'a)$ by a directed segment labeled "a". We similarly form the directed segment from (v, v') to $(vb, v'b)$ with the label "b". In this manner, B is like an automaton except that we have not specified the initial and terminal vertices. If v_0 and v_0' are the initial vertices of A and A', we take (v_0, v_0') as the initial vertex of B.

Now we systematically keep track of all vertices (v, v') that can be reached from (v_0, v_0') by any string. (Note that these vertices are finite in number.) Now if the vertex (v, v') is reachable and if the languages of A and A' are the same, then v is terminal if and only if v' is terminal. We thus proceed as follows. Make a table giving the action of a and b on any vertex (v, v') as follows:

(v, v')	$(va, v'a)$	$(vb, v'b)$
(v_0, v_0')	...	...

Start the process with the initial vertex (v_0, v_0'), and each time a new ordered couple arises in column 2 or 3 put that ordered couple in column 1 and continue until no new pairs are introduced. Then

> The automata A and A' have the same language if for each ordered couple (v, v') that occurs in the foregoing construction, either both v and v' are terminal or both are not. If any pair (v, v') shows up with one coordinate terminal and the other not, then the languages are different. ∎

EXAMPLE 9.3.13

Determine whether the two automata in Figure 9.3.19 have the same language.

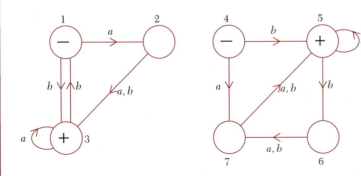

Figure 9.3.19

Method   We start with $(1, 4)$:

	v	v'	va	$v'a$	vb	$v'b$
	1	4	2	7	3	5
OK: both nonterminal	2	7				
OK: both terminal	3	5				

Continuing, we have

v	v'	va	$v'a$	vb	$v'b$
1	4	2	7	3	5
2	7	3	5	3	5
3	5	3	5	1	6
1	6	2	7	3	7
				↑	↑

The process stops here. The vertex 3 is terminal, whereas 7 is not. The languages are therefore not the same. It is now a simple matter to find a string in **A** that is not in **A'**. We arrived at $(3\ 7)$ via the route

$$(3\ 7) \xleftarrow{\ b\ } (1\ 6) \xleftarrow{\ b\ } (3\ 5) \xleftarrow{\ a\ } (2\ 7) \xleftarrow{\ a\ } (1\ 4)$$

Thus, $aabb$ is accepted by **A** but not by **A'**.

   If this process continued until no new vertices were reached, and there were no rejections, then the automata would necessarily have the same language.

Exercises

In Exercises 1–5, describe in English, as simply as you can, the regular set.

1. $(bb)^* + a^*$
2. $ab^*a + ba^*b$
3. $a^*ba^*ba^*$
4. $(a + b)^4 + (a + b)^3$
5. $a^*b + b^*a + (aa + bb)^*$

6. For any sets A, B, C, prove that
   (a) $A^{**} = A^*$.
   (b) If $A \subseteq B$, then $A^* \subseteq B^*$.
   (c) $A(B + C) = AB + AC$.

7. Show why the following are *not* in general true.
   (a) $(A + B)* = A* + B*$
   (b) $(AB)* = A*B*$
   (c) $AB + B = AB$
8. *By inspection*, express the language of the transition graph of Figure 9.3.20 as a regular set.

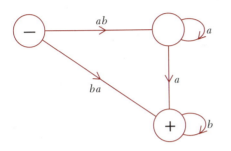

Figure 9.3.20

9. Using the algorithm of Theorem 9.3.10 part 1, find a formula (as a regular set) for the language of the automaton of Figure 9.3.21a.
10. Repeat Exercise 9 for Figure 9.3.21b.
11. Repeat Exercise 9 for Figure 9.3.21c.
12. Repeat Exercise 9 for Figure 9.3.21d.

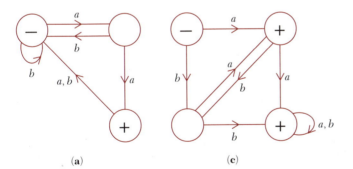

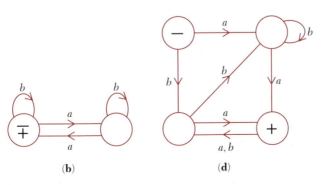

Figure 9.3.21

13. Find a transition graph whose edges are labeled a, b, or ϵ and whose language is the regular set $(aa + bb)^*$.
14. Find a transition graph whose edges are labeled a, b, or ϵ and whose language is the regular set $(ab^*a + b)^*$.
15. Find a transition graph whose edges are labeled a, b, or ϵ and whose language is the regular set $(a + b)(a + b)a^*$.
*16. Find an *automaton* whose language is the same as the transition graph of Figure 9.3.22a.

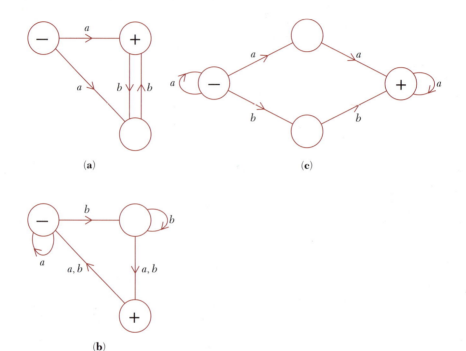

(a)

(c)

(b)

Figure 9.3.22

*17. Repeat Exercise 16 for Figure 9.3.22b.
*18. Repeat Exercise 16 for Figure 9.3.22c.
*19. Prove: If Σ is a finite alphabet and S is a regular set, then $\Sigma^* - S$ is also regular.
20. Find an automaton whose language is $(ab + ba)^*$.
21. Find an automaton whose language is $a + b^* + (ab)^*$.
22. Find an automaton whose language is $a^*b^* + b^*a^*$.
23. In the construction of the automaton B from A and A' of Theorem 9.3.12, an initial vertex was given, but no terminal ones were. Now suppose L is the language of A and L' is the language of A'.
    (a) Suppose (v, v') is defined to be terminal if either v or v' is terminal. What is the language of B?
    (b) Redo (a), but (v, v') is terminal if v and v' are.

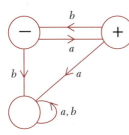
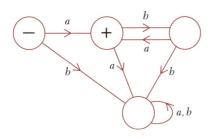

(a)

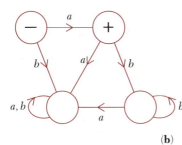
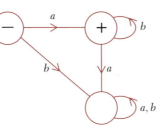

(b)

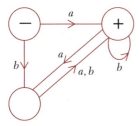
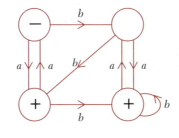

(c)

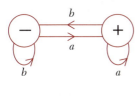
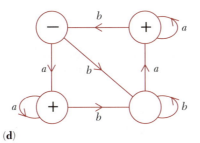

(d)

Figure 9.3.23

24. Using the method of Theorem 9.3.12, determine whether the pair of automata in Figure 9.3.23a have the same language.
25. Repeat Exercise 24 for the pair of Figure 9.3.23b.
26. Repeat Exercise 24 for the pair of Figure 9.3.23c.
27. Repeat Exercise 24 for the pair of Figure 9.3.23d.

Summary Outline

- The **alphabet** or the **symbols** of the propositional calculus:

$p, q, r, \ldots$	(**infinitely many variable symbols**)
T, F	(two **constant symbols**)
$\wedge, \vee, \rightarrow, \leftrightarrow$	(four binary **connective symbols**)
$\neg$	(a **unary connective symbol**)
$(\,,\,)$	(left and right **parentheses**)

- A **formula** (or a compound statement) is a string on the symbols of the propositional calculus that can be obtained by the use of the following rules:

  1. Any constant symbol or variable symbol is a formula.
  2. If α and β are formulas, so are

     (a) $(\alpha \wedge \beta)$;   (b) $(\alpha \vee \beta)$;   (c) $(\alpha \rightarrow \beta)$;   (d) $(\alpha \leftrightarrow \beta)$;   (e) $\neg\alpha$

- **Theorem**   Let $\mathbb{C}$ be a collection of formulas of the propositional calculus. Suppose that

  1. All variable or constant symbols are in $\mathbb{C}$.
  2. If α and $\beta \in \mathbb{C}$, and if λ is any of the four binary connectives, then $(\alpha\lambda\beta) \in \mathbb{C}$.
  3. If $\alpha \in \mathbb{C}$, then $\neg\alpha \in \mathbb{C}$.

  Then $\mathbb{C}$ consists of all formulas of the propositional calculus.

- **Alternative version of theorem**   If a statement about formulas is true for the variable or constant symbols, and if, whenever it is true for formulas α and β it is also true of $(\alpha\lambda\beta)$ for any binary connective λ, and it is also true for $\neg\alpha$, then the statement is true for all formulas.

- **Theorem**   Any formula has an equal number of left and right parentheses.

- If $\alpha = \sigma\tau$ (the concatenation of σ and τ), where σ and τ are nonempty strings, σ is a **proper initial segment** of α and τ is a **proper final segment.**

- **Lemma**   Let β be a proper initial segment of a formula α. Then β has more left parentheses than right, or β consists only of a string of $\neg$'s.

- **Theorem** (Unique Readability of Formulas)   Let α be a formula. Then α is precisely one and only one of the following types. In each case, α' and/or β' are uniquely determined formulas.

  1. α is a variable or constant symbol α'; or
  2. $\alpha = (\alpha' \wedge \beta')$; or
  3. $\alpha = (\alpha' \vee \beta')$; or
  4. $\alpha = (\alpha' \rightarrow \beta')$; or
  5. $\alpha = (\alpha' \leftrightarrow \beta')$; or
  6. $\alpha = \neg\alpha'$.

- Σ^* denotes **all strings** using the letters of Σ. Σ^+ denotes the **nonempty strings** on the alphabet Σ. The symbol ϵ denotes the **empty string.**

- If Σ is an alphabet, let V be a disjoint set, called the **variable set.** V is assumed to contain an element S, called the **starter.** We define $\Gamma = \Sigma \cup V$, the **extended alphabet.**

- A **production** is an ordered couple (α, β), written $\alpha \Rightarrow \beta$, where $\alpha \in \Gamma^+$ and $\beta \in \Gamma^*$. (Γ is the extended alphabet.) If σ is any string of Γ^*, we say that this production **transforms σ into τ** if $\sigma = \sigma_1\alpha\sigma_2$ and $\tau = \sigma_1\beta\sigma_2$. In this case, we write $\sigma \Rightarrow \tau$, or $\sigma_1\alpha\sigma_2 \Rightarrow \sigma_1\beta\sigma_2$.

- A **type-0 grammar** is given by the alphabet Σ, a variable set V, the starter S, and a given finite set of productions. The set of all strings α on the alphabet Σ produced using the productions of a type-0 grammar starting with the starter S is called a **type-0 language.**

- An **automaton** is a list $A = [\mathbf{Q}, \Sigma, q_1, T, f]$, where $\mathbf{Q}$ is a set of states $\{q_1, q_2, \ldots, q_k\}$ and Σ is a set of letters $\{a_1, \ldots, a_n\}$. The state q_1 is called the **initial state.** T is a subset of $\mathbf{Q}$ whose elements are called **terminal states.** f is a function $f: \mathbf{Q} \times \Sigma \to \mathbf{Q}$, called the **transition function** of the automaton.

- An automaton **scans** a string by first viewing the first letter in the initial state q_1. It then moves right and changes to a new state according to the state it is in and the letter being scanned, using the transition function f. Scanning ceases when there are no new letters to view. A string is **accepted** by the automaton if its final state is terminal. Otherwise, it is **rejected.**

- The **language $L(A)$** of an automaton A is the set of strings that are accepted by A.

- **String operations** on subsets of Σ^*:
    1. **Union.** If A and B are subsets of Σ^*, then $A + B$ denotes the union $A \cup B$.
    2. **Concatenation.** If A and B are subsets of Σ^*, then $A \cdot B$ or AB denotes the set of *all* concatenations of a string in A with a string in B. Thus,

    $$AB = \bigcup_{\substack{\alpha \in A \\ \beta \in B}} \{\alpha\beta\}$$

    3. **Starring.** If A is a subset of Σ, A^* consists of all possible concatenations $\alpha_1\alpha_2 \cdots \alpha_k$, where k is an arbitrary natural number and $\alpha_i \in A$. This includes the empty string ϵ, which corresponds to $k = 0$.

- A **regular set** on an alphabet Σ is a set of strings that can be finitely expressed using only the individual letters of Σ, the empty string ϵ, and the operators $+$, $\cdot$, and $*$. The null set $\varnothing$ is also defined as a regular set.

- A **transition graph** is a directed graph G with finitely many vertices, one *or more* of which are labeled $\ominus$, and some of which are labeled $\oplus$. There are finitely many directed edges of the graph **G,** each of which is labeled with a string of Σ^*.

- The **language** $L(G)$ of a transition graph **G** is the set of strings that are determined by the paths joining a vertex labeled $\ominus$ to a vertex labeled $\oplus$.

- A **generalized transition graph G** is defined as a transition graph, except that each directed edge is labeled with a regular set.

- **Theorem** (Kleene)

  1. Let **G** be a generalized transition graph. Then its language $L(G)$ is a regular set. In particular, $L(A)$ is a regular set for any automaton **A,** because an automaton is a special case of a generalized transition graph.
  2. Let R be a regular set. Then there is an automaton **A** such that $L(A) = R$

- **Theorem** (Moore)   There is an algorithm to determine whether two automata have the same language.

CHAPTER 10

Turing Machines

The Turing machine was one of the earliest conceived all-purpose computing machines. Turing machines are theoretical, in that they are not used to do practical computing. However, they are easy to understand, with a limited "language," and their computing power is equal to that of any computer available today. Their study leads to a theory of what is computable and what is not. In particular, certain problems are introduced, which we show are not computable. The study includes an excursion into Cantor's approach to counting sets and the Cantor diagonal method.

10.1 Turing Machines

A **Turing machine** (named after its inventor, Alan Turing) is a generalization of an automaton. We show that it has the capabilities of a full-fledged computer—in fact more, since there is no limit on the memory capacity of a Turing machine. However, Turing machines are theoretical in nature, and it is not expected that they would be used in practical situations.

The study of Turing machines leads to important results on the limitations of computers, as we see in Section 10.3. Programming a Turing machine, even to perform a simple-looking task, can be a long and tiresome job. Nevertheless, we do some of this in this section to introduce the subject and see some of its possibilities. In the next section, we learn a method to bypass this programming.

Like an automaton, a Turing machine takes as input a string of symbols. We use 0 and 1 as the symbols because we intend to compute. Similarly, the Turing machine starts reading at one of the symbols. As with an automaton, it has finitely many states, and it performs according to the symbol scanned and the state it is in. It differs in these following important respects:

1. A Turing machine can scan left or right as directed by its instructions. (The automaton scans only right.)
2. A Turing machine can overprint any symbol, thereby changing the tape on which the initial string is given. (An automaton does not overprint.)
3. The output for a Turing machine, if it stops, is a string. (The output for an automaton is either *yes*, for accepted, or *no*, for rejected.)

We can interpret 1 as yes and 0 as no, so a Turing machine generalizes an automaton with respect to output. Because of its extended power, it is possible for a Turing machine to go on forever. The formal definitions follow.

DEFINITION 10.1.1

The special symbols used to describe a Turing machine are as follows.

$q_1, q_2, q_3, \ldots$	(The states; q_1 is the initial state.)
L, R	(Symbols for moving left or right.)
$0, 1, B, S_3, S_4, \ldots$	(The letters of strings. We use S_0 for 0, S_1 for 1, S_2 for B (blank). S_3, S_4, and so on, are optional other letters that are sometimes convenient.)

Thus the alphabet is $\Sigma = \{S_0, S_1, S_2, \ldots\}$. The set of states is $Q = \{q_1, q_2, \ldots\}$.

DEFINITION 10.1.2

There are three kinds of instructions for a Turing machine. Each instruction is a **quadruple**—it has four components. The types are as follows, along with their meanings:

1. $q_i \, S_j \, L \, q_k$ (In state q_i, when viewing the symbol S_j, move left and go to state q_k.)
2. $q_i \, S_j \, R \, q_k$ [As in type 1, but move right.]
3. $q_i \, S_j \, S_k \, q_m$ (In state q_i, when viewing the symbol S_j, overprint the symbol S_k, and go to state q_m.

Note that the instructions for an automaton are all of type 2 in Definition 10.1.2. In type 3, the symbol S_j is lost once it is overprinted.

DEFINITION 10.1.3

A Turing machine is a finite nonempty set of instructions of type 1, 2, or 3 in Definition 10.1.2, with the added condition that no two instructions in this set start out with the same pair (q_i, S_j).

Remark   This last condition is needed in order for a Turing machine to act in a unique manner when it is in state q_i viewing the symbol S_j. If, during execution, a Turing machine is in state q_i, viewing the symbol S_j, and there is no instruction starting with (q_i, S_j), the machine stops. (See the halting procedure of Definition 10.1.4 for further details.) A Turing machine is a **program** using a rather restricted language. The instructions may be given in any order, because the program proceeds according to the state and the symbol read and does not proceed sequentially. With the exception of the restriction in the definition, *any* nonempty set of instructions constitutes a Turing machine.

We have discussed how an instruction is to be interpreted. We now consider this in a little more detail.

DEFINITION 10.1.4   *Input, Output, and Halting Procedures*

The **input** for a Turing machine is visualized as a two-way infinite tape with a finite string printed on it. (See Figure 10.1.1.) The input tape also includes a pointer or cursor on the tape (indicated by ↑) that points to the letter that the machine is scanning.

The unprinted places are understood to be blanks—the letter B. Thus, the tape of Figure 10.1.1 is understood to be the same as the one in Figure 10.1.2.

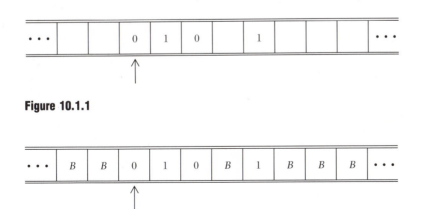

Figure 10.1.1

Figure 10.1.2

We make the convention that, unless otherwise stated, the pointer initially points to the first nonblank entry on the tape (reading from left to right). If the tape is totally blank, the pointer can point anywhere; it won't much matter where!

There is no formal **halting procedure**. The machine stops if it is in a state q_i, viewing the letter S_j, but it has no instruction that starts with the pair (q_i, S_j). We usually reserve a special state q_h as a state that never appears as the first part of any instruction. In effect, then, q_h may be viewed as a halt state, and we occasionally refer to it as the HALT state. For example, the instruction $q_7\ B\ B\ q_h$ will effect a halt if the machine views B while in state q_7.

The **output** occurs when and if the machine halts. The output is the string on the tape. Included as part of the output is the pointer location.

We now illustrate with a few examples.

EXAMPLE 10.1.5

Design a Turing machine that takes any string of 0's and 1's as input and adjoins a 1 at its right end.

Method   Keep moving right till you encounter a blank. Print 1 and stop. This is accomplished by a three-statement machine:

$$
\begin{array}{cccc}
q_1 & 1 & R & q_1 \\
q_1 & 0 & R & q_1 \\
q_1 & B & 1 & q_{100}
\end{array}
$$

In words, q_1 moves the pointer right, past all 0's and 1's, then overprints 1 on the first blank it encounters, and finally halts.

EXAMPLE 10.1.6

Design a Turing machine that concatenates two strings. The input consists of two nonempty strings separated by a blank. The pointer, as usual, points to the leftmost nonblank letter.

Method   The method is to scan right till the first blank is met. This is the blank that separates the strings. Then pull over the right string, letter by letter, and stop when the next blank is met. The following steps do this. They are designed so that each step corresponds to a state in the Turing machine. These steps are followed by the corresponding Turing machine instructions.

1. q_1:   Move right to the first blank. Move right once again.

2. q_2:   If 0 is scanned, overprint with a blank, and go to† step 3. If 1 is scanned, overprint with a blank, and go to 4. If B is scanned, then HALT.
3. q_{10}:   Move left and overprint the blank with 0. Go to step 1.
4. q_{20}:   Move left and overprint the blank with 1. Go to step 1.

The Turing machine (with corresponding step numbers) is

1. q_1 1 R q_1; q_1 0 R q_1; q_1 B R q_2;
2. q_2 0 B q_{10}; q_2 1 B q_{20}; q_2 B B HALT;
3. q_{10} B L q_{11}; q_{11} B 0 q_1;
4. q_{20} B L q_{21}; q_{21} B 1 q_1.

EXAMPLE 10.1.7

Let us illustrate how this Turing machine works by tracing its action on the input strings 11 and 01. The input is taken as the string 11B01. By our convention, this string of length 5 is preceded by blanks and followed by blanks. The entry in the "Next Instruction" column is obtained by scanning the instructions of the Turing machine and finding the instruction, if any, that starts with the current state, viewing the current letter. We take the first 1 as position 1 for the pointer. The action of the Turing machine is as follows:

String	Pointer	State	Viewing	Next Instruction			
11B01	1	q_1	1	q_1	1	R	q_1
11B01	2	q_1	1	q_1	1	R	q_1
11B01	3	q_1	B	q_1	B	R	q_2
11B01	4	q_2	0	q_2	0	B	q_{10}
11BB1	4	q_{10}	B	q_{10}	B	L	q_{11}
11BB1	3	q_{11}	B	q_{11}	B	0	q_1
110B1	3	q_1	0	q_1	0	R	q_1
110B1	4	q_1	B	q_1	B	R	q_2
110B1	5	q_2	1	q_2	1	B	q_{20}
110BB	5	q_{20}	B	q_{20}	B	L	q_{21}
110BB	4	q_{21}	B	q_{21}	B	1	q_1
1101B	4	q_1	1	q_1	1	R	q_1
1101B	5	q_1	B	q_1	B	R	q_2
1101B	6	q_2	B	q_2	B	B	HALT
1101B	6	HALT	B	None—machine stops			

---

† Because of the nature of Turing machine instructions, "go to" instructions are useful. Thus, the instruction q B B q' implicitly has a "go to" state q'.

Thus, after 14 steps, the machine stops, and the strings are concatenated.

EXAMPLE 10.1.8

Design a Turing machine that simulates an automaton **A.** Input should be a string of 0's and 1's. Output should be either 1, for accept, or 0, for reject.

Method   An instruction for an automaton is q_i S_j R q_k, where S_j is one of the symbols 0 or 1. The automaton stops when it views B, and then accepts (prints 1) if it is in a terminal state, or rejects (prints 0) if not. In either case, we erase the input. The required Turing machine has each of the instructions of the given automaton plus a few more to complete the output. Introduce the additional states a_1 (for accept), r_1 (for reject), $p_1 - p_3$ to blank out the input, and p_{100} to halt. Now adjoin the following instructions to the given ones used to define the automaton:

q_k B B a_1 for all q_k that are terminal states

q_k B B r_1 for all q_k that are not terminal states

a_1 B 1 p_1; r_1 B 0 p_1 (p_1 begins the erase procedure)

p_1 1 L p_2; p_1 0 L p_2 (move past the intended output)

p_2 0 B p_3; p_2 1 B p_3; p_2 B B HALT (blank initial inputs)

p_3 B L p_2 (loop to continue blanking the initial string)

These added instructions perform the required simulation.

EXAMPLE 10.1.9

Design a Turing machine that copies a string α. The input is the string α, and the output is $\alpha B \alpha$.

Method   Copy the letters of α from left to right, using the letter X as a place marker. A strategy follows. As in Example 10.1.6, the Turing machine instructions are labeled to correspond to the labels here. We leave the q_{20} routine (transporting the symbol 1) to the reader.

1. q_1:   Read the first symbol. If it is 0, then overprint X and go to step 2, corresponding to q_{10}. If the first symbol is 1, then overprint X and go to step 6, corresponding to state q_{20}. If the first symbol is B, then HALT.

2. q_{10}:   Move right to the first blank.

3. q_{11}:   Continue moving right to the next blank and overprint this blank with 0.

4. q_{12}:   Now move left until the symbol X is reached.
                          Overprint X with 0.
5. q_{13}:   Move right and go to step 1.
6. to 9. q_{20} through q_{23}:   Steps similar to 2 through 5.

The Turing machine instructions are

1. q_1 0 X q_{10}; q_1 1 X q_{20}; q_1 B B HALT;
2. q_{10} X R q_{10}; q_{10} 1 R q_{10}; q_{10} 0 R q_{10}; q_{10} B R q_{11};
3. q_{11} 1 R q_{11}; q_{11} 0 R q_{11}; q_{11} B 0 q_{12};
4. q_{12} 0 L q_{12}; q_{12} 1 L q_{12}; q_{12} B L q_{12}; q_{12} X 0 q_{13};
5. q_{13} 0 R q_1;

with similar instructions for q_{20} (transporting the letter l).

This example introduced the letter X for marking purposes. This is a convenience, but not strictly necessary. If we regard the input/output tape as consisting of bits rather than bytes, we can get away with two symbols: B and 1. For example, we can let BB stand for Blank, $B1$ for 1, $1B$ for 0, and 11 for X. Of course the programming would be more complicated, and we would have different conventions about input and output, but the principle is clear.

We want to show how a Turing machine can do computations. Thus, we need to have a convention regarding how numbers are to be represented as strings. We can, of course, use binary notation because we allow 0's and 1's as symbols. However, for ease in programming, we use the much more primitive way of expressing a positive integer by a string of 1's. Thus, the number 8 is represented by 11111111. This is not the most economical use of space, but with an infinite memory and no time problem it won't matter.

The possibility of copying by a Turing machine indicates that we can do computing on a Turing machine. The arithmetic operation of addition is little more than concatenation. We input any positive integer n as a string α of n 1's. Two numbers m and n are inputted as strings α and β, separated by a blank. Their sum is simply obtained as a concatenation of these strings (interpreted as a number). This was done in Example 10.1.6. Thus, the operation $+$ can be performed on a Turing machine. We now illustrate the product in a similar way.

EXAMPLE 10.1.10

Find a Turing machine that computes the product mn of two positive integers m and n.

Method   As noted, we code m and n as a string of m 1's, followed by a blank, followed by a string of n 1's. We indicate this by mBn. Starting at the first 1 in m, our procedure replicates n one time for each 1 appearing in m. A strategy is as follows:

1. If 1 is scanned, then overprint B and go to 2. (The tape reads $(m-1)Bn$ after the first step.) If B is scanned then HALT.
2. Go right to the first blank, and then move one space right.
3. Copy the string beginning at the pointer and ending at the next blank. (Example 10.1.9 shows how to do this.) The copy is placed on the second blank following this string. (After the first pass, the tape reads $(m-1)BnBn$, and the cursor is on the last nonblank symbol.)
4. Move left to the third blank, move right, and go to step 1.

In this way, m copies of the string n are formed and concatenated, leading to the string $nBmn$. This is the result. If desired, of course, m can first be duplicated and, hence, will show up on the tape. Similarly, n can be eliminated, and the tape just has output mn. We leave the details of the actual Turing machine instructions to the reader.

EXAMPLE 10.1.11 *Macros*

Devising a Turing machine to perform certain operations can be considerably simplified by using **macros**. Macros are sets of instructions performing some subtask, and may be regarded as subroutines or procedures.

For example, many of our examples called for moving right until the first blank is encountered. The Turing machine

$$\text{MoveRToBlank: } [p_1\ 0\ R\ p_1;\ p_1\ 1\ R\ p_1;\ p_1\ B\ B\ p_2]$$

does this when in state p_1 and exits in state p_2. The macro

$$\text{Print1: } [p_1\ 0\ 1\ p_2;\ p_1\ B\ 1\ p_2;\ p_1\ 1\ 1\ p_2]$$

overprints 1 when in state p_1 and exits in state p_2. Thus, the simple Turing machine of Example 10.1.5 can be realized as the succession of the macros MoveRToBlank and Print1. It is understood that the process starts with the initial state q_1, which must be used for p_1 in the macro MoveRToBlank, and that the exit state of this macro becomes the initial state of the next macro.

We now illustrate how macros can be used to create a Turing machine.

EXAMPLE 10.1.12

Using macros, create a Turing machine that takes a string of 0's and 1's, adjoins its first digit to the end of the string, and moves the cursor back to the first digit of the string.

Method We start by reading the first symbol. If it is 0, we move right to the first blank and print 0. Then we move back to the first blank and over one to place the cursor at the first symbol. If the first symbol is 1, we proceed similarly, except we print a 1. We can thus use the following macros:

1. MoveRToBlank: (Comment: Move right to first blank—enter in state p_1 and exit in state p_2.)
2. MoveLToBlank: p_1 0 L p_1; p_1 1 L p_1; p_1 B B p_2 (Comment: Move left to first blank—enter in state p_1 and exit in state p_2.)
3. MoveRFrB&Halt: p_1 B R HALT (Comment: Move right when viewing a Blank, and then halt.)
4. Print1onB: p_1 B 1 p_2 (Comment: Print 1 when viewing a Blank.)
5. Print0onB: p_1 B 0 p_2 (Comment: Print 0 when viewing a Blank.)

Using these macros we form a **pseudo Turing machine** as follows:

> If 0 go to A (Meaning: If viewing 0, go to step A)
> If 1 go to B (Meaning: If viewing 1, go to step B)

A: MoveRToBlank
   Print0onB
   Go to C

B: MoveRToBlank
   Print1onB
   Go to C

C: MoveLToBlank
   MoveRFrB&Halt

We now take this pseudo Turing machine and convert it to a Turing machine. Opposite the macro name, we put the macro, and then the Turing machine instruction. The entry states for A, B, and C are taken as q_{10}, q_{20}, and q_{30} respectively.

If 0 go to A	Macro: none	
	Instr: q_1 0 0 q_{10}	
If 1 go to B	Macro: none	
	Instr: q_1 1 1 q_{20}	
A: MoveRToBlank	Macro:	p_1 0 R p_1; p_1 1 R p_1; p_1 B B p_2
	Instr:	q_{10} 0 R q_{10}; q_{10} 1 R q_{10}; q_{10} B B q_{11}
Print0onB	Macro:	p_1 B 0 p_2
	Instr:	q_{11} B 0 q_{30}
B: MoveRToBlank	Macro:	p_1 0 R p_1; p_1 1 R p_1; p_1 B B p_2
	Instr:	q_{20} 0 R q_{20}; q_{20} 1 R q_{20}; q_{20} B B q_{21}
Print1onB	Macro:	p_1 B 1 p_2
	Instr:	q_{21} B 1 q_{30}
C: MoveLToBlank	Macro:	p_1 0 L p_1; p_1 1 L p_1; p_1 B B p_2
	Instr:	q_{30} 0 L q_{30}; q_{30} 1 L q_{30}; q_{30} B B q_{31}
MoveRFrB&Halt	Macro:	p_1 B R HALT
	Instr:	q_{31} B R HALT

The Turing machine is the list of instructions, labeled *Instr*. The reader should carefully note how the generic p_i in a macro converts to a specific q_j in an instruction.

EXAMPLE 10.1.13   Concatenation of Strings

Example 10.1.6 used a "go to" and an "IF" statement. We can now give a pseudo Turing machine for this example as follows:

```
     A:      MoveRToBlank
             MoveRFrBlank
             If 0 then go to Transf0
             If 1 then go to Transf1
             If B then HALT
Transf0:     MoveLFrBlank
             Print0onB
             go to A
Transf1:     MoveLFrBlank
             Print1onB
             go to A
```

This pseudo Turing machine is similar to Example 6.1.12 and uses many of the same macros. New macros used are MoveLFrBlank and MoveRFrBlank, which are cursor moves left and right when viewing blanks. We leave these definitions to the reader. Also, instead of labeling statements with A, B, and so on, we used more descriptive labels, which explain their use. The pseudoinstruction "If B then HALT" is described by the macro "p_1 B B HALT."

Of course, the listing is not the Turing machine itself. But as in Example 10.1.12, it is a short step from the macros constituting a pseudo Turing machine to the Turing machine itself.

As we noted in our introduction, it is not intended that Turing machines do serious real-life computing. However, we do want to show the theoretical possibilities of such a computing device. Therefore, we do not concern ourselves with shortcuts, time-saving maneuvers, and elegant programming techniques. With this in mind, let us now consider how computing can be done on a Turing machine.

We work with natural numbers only. As noted before, a positive integer can be given by a string of 1's. Let 0 be designated by the string "0". Since the tape is infinite, we have no problem listing any finite sequence of integers. Our convention is to use a single blank B to separate various variables. Inputting the variables V_1, V_2, V_3 as 3, 1, 0 respectively would lead to the tape of Figure 10.1.3.

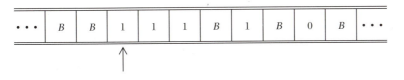

Figure 10.1.3

It is then a simple matter to position the pointer at any time at the beginning of any variable. For example, move left till two consecutive blanks occur. The first variable occurs to the right of these blanks; the second, after a move to the first blank to the right; and so forth.

Arithmetic operations on a variable are then possible, if care is taken to preserve the other variables. For example, incrementing the first in-putted variable V_1 (assumed >0) involves moving the rest of the variables over one, and then adjoining 1 to the end of the string V_1. More details are given as exercises.

At the time that Turing invented the Turing machine, the notion of "effective computability" or "algorithmic computability" was very much an open question mathematically. Turing machines were one of the first attempts to define this notion in a rigorous way. Turing himself noted that he tried to idealize how one computes with a pencil and paper and came up with his idea. The class of effectively computable functions is identified in Section 10.2 as the **recursive functions.** We indicate why this class is identical to the functions that are computable using a Turing machine. In this section, general methods are developed that help us avoid the tedious task of explicitly constructing a Turing machine for each such function.

Exercises

In Exercises 1–4, find a Turing machine that performs the following tasks. Except for special markings, the alphabet is taken to be $\{0, 1, B\}$.

1. Move left until a 1 is scanned.
2. Scan the letter scanned at the pointer (0, 1, or B); move right until X is scanned and overprint X with that letter. (Assume X is the only special marker used.)
3. Take the entire string starting at the pointer and ending at the first blank to the right and move it over one space to the left. The pointer is assumed to be scanning a nonblank entry.
4. Erase (blank out) the entire string starting at the pointer over to the first blank on the right. Return to the original pointer position.

In Exercises 5–10, strings α and β are nonempty strings on the alphabet $\{0, 1\}$. Design a Turing machine to give the correct output. In each case, give a pseudo Turing machine as well as the full set of instructions. You may use macros defined in this section, or create new ones, as needed.

5. Input: α; Output: $00\alpha 1$.
6. Input: α; Output: $\alpha\hat{\ } = \alpha$ written in reverse order.
7. Input: α; Output: $\alpha' = $ string with each 0 in α replaced by 1, and each 1 replaced by 0.
*8. Input: α; Output: αBn, where n is the length of α, coded as a string of 1's of length n.
*9. Input: $\alpha B\beta$; Output: $\beta\alpha$.
*10. Input: $\alpha B\beta$; Output: α or β, whichever has the greater length. In case of a tie, the output is α.

*11. Following the technique of Example 10.1.12, devise a pseudo Turing machine for the computation of the string $\alpha\alpha$ as output, when α is given as input. α is a string of 0's and 1's.

Describe, in words, the effect of each of the Turing machines of Exercises 12–17. Be sure to state when, if ever, the machine fails to stop. Assume as input a nonempty string of 0's and 1's, with the pointer pointing at the first character of that string.

12. $q_1\ 0\ B\ q_1; q_1\ B\ R\ q_1; q_1\ 1\ 1\ q_2$.
13. $q_1\ 0\ 1\ q_1$.
14. $q_1\ 1\ R\ q_1; q_1\ 0\ R\ q_1; q_1\ B\ L\ q_2;$
    $q_2\ 0\ B\ q_3; q_2\ 1\ B\ q_3;$
    $q_3\ 1\ B\ q_1; q_3\ B\ L\ q_2$.
15. $q_1\ 0\ 0\ q_1$.
16. $q_1\ 1\ R\ q_1; q_1\ 0\ R\ q_1; q_1\ B\ R\ q_2; q_2\ 1\ 0\ q_2; q_2\ B\ 1\ q_2$.
17. $q_1\ 1\ 0\ q_2; q_1\ 0\ 1\ q_2; q_1\ B\ B\ q_2;$
    $q_2\ 0\ R\ q_2; q_2\ 1\ R\ q_2; q_2\ B\ B\ q_1$.

18. Find a pseudo Turing machine and a Turing machine that compute $x + 1$. Input is $x \geq 0$. Output is $x + 1$.
*19. Find a pseudo Turing machine and a Turing machine that compute the sum $S = x + y$. Input is x, y, and output is S.
*20. Devise a macro that inserts the letter X into the tape at the current pointer position, moving the rest of the tape one unit to the right. Assume the tape consists of 0's, 1's, and B's, and that two consecutive B's are followed by nothing but B's.
*21. Devise a macro that inserts another copy of V_1 (the first inputted variable) after V_1. Thus, it converts the input tape $V_1, V_2, V_3, \ldots$ into the output tape $V_1, V_1, V_2, V_3, \ldots$.
*22. Devise a pseudo Turing machine to test whether m divides n. The input is mBn. The output is 1 if m divides n, and 0 if not. For example, the input $11B111111$ has output 1, whereas $11B111$ has output 0.
*23. Suppose natural numbers are represented in binary notation, both for input and for output purposes. Devise a pseudo Turing machine to compute the function $f(x) = x + 1$.
*24. Repeat Exercise 23 for $f(x, y) = x + y$.

10.2 Recursive Functions

Recursive functions are the **computable** functions. During the twentieth century, there were several approaches to a precise formulation of this idea, and all have been shown to be equivalent. Our approach does not use programming techniques to define computable functions, but we shall show that the approach in this chapter yields functions that can be computed by Turing machines.

The term *recursive function* should not be confused with the concept of a definition by recursion. We might use the term *computable* as perhaps a more descriptive term, but *recursive function* is now standard terminology. The concepts are related, however, because a definition by recursion is one component in the definition of a recursive function.

The functions f that we are concerned with are functions $f: \mathbb{N} \to \mathbb{N}$, where $\mathbb{N}$ is the set of natural numbers. More generally, we consider functions $f: \mathbb{N}^k \to \mathbb{N}$ of several variables. Thus, $y = f(x_1, \ldots, x_k)$ is a natural number whenever each $x_i \in \mathbb{N}$.

The idea of our definition is to start with some fixed functions (the *starter* functions) and allow three operations on functions: (i) substitution, (ii) definition by recursion, and (iii) searching. These operations are defined later. A recursive function is then defined as a function that can be defined by using the starter functions and any of these operations applied a finite number of times.

DEFINITION 10.2.1

The following functions are called **starter** functions:

1. $Z: \mathbb{N} \to \mathbb{N}$, the zero function. By definition,

$$Z(x) = 0 \qquad \text{for all } x \in \mathbb{N}$$

2. $S: \mathbb{N} \to \mathbb{N}$, the successor function. By definition,

$$S(x) = x + 1 \qquad \text{for all } x \in \mathbb{N}$$

3. $p_i^n: \mathbb{N}^n \to \mathbb{N}$, the projection functions. By definition,

$$p_i^n(x_1, \ldots, x_n) = x_i$$

Thus, p_i^n selects the ith variable from n variables. Thus, $p_1^2(x, y) = x$. The identify function I, defined by $I(x) = x$, is the function p_1^1, and is thus a starter function.

Each starter function can be computed on a Turing machine. Note also that in any reasonable definition of computable functions, we would expect to find the starters included. We now consider the three basic operations on functions that will be used to define recursive functions.

DEFINITION 10.2.2   Substitution

Suppose $f_0(y_1, \ldots, y_k)$ is a function of k variables ($f_0\colon \mathbb{N}^k \to \mathbb{N}$), and that

$$y_1 = f_1(x_1, \ldots, x_n)$$

$$\vdots$$

$$y_k = f_k(x_1, \ldots, x_n)$$

are k functions of n variables. Then the function

$$f(x_1, \ldots, x_n) = f_0(f_1(x_1, \ldots, x_n), \ldots, f_k(x_1, \ldots, x_n))$$

is said to be defined from $f_0; f_1, \ldots, f_k$ by **substitution.**

For example, the function $f(x) = x^2 + x^3$ is defined by substitution from the functions $f_0 = u + v, u = x^2$, and $v = x^3$. Intuitively, we would want recursive functions to be closed under substitutions. For if f_i are all computable, we can compute $y_1, \ldots, y_k$ and then $f_0(y_1, \ldots, y_k)$.

Turing-machine-computable functions are also closed under substitution. For if we have Turing machines that compute $f_1, f_2, \ldots, f_k$, and one that computes f_0, then we can piece the machines together to form a Turing machine that is defined from $f_0; f_1, \ldots, f_k$ by substitution. All we do is take $x_1, \ldots, x_n$ as input and compute $f_1, f_2, \ldots, f_k$. Then, using $f_1, \ldots, f_k$ as inputs, we compute $f_0(f_1, \ldots, f_k)$, and the result is $f(x_1, \ldots, x_n)$. The details are left to the reader.

DEFINITION 10.2.3   Recursion

Suppose $f_0(x_1, \ldots, x_k)$ and $f_1(x_1, \ldots, x_k, z, y)$ are functions of k and $k + 2$ variables, respectively. Now suppose $f(x_1, \ldots, x_k, z)$ satisfies the equations

$$f(x_1, \ldots, x_k, 0) = f_0(x_1, \ldots, x_k) \tag{10.2.1}$$

and

$$f(x_1, \ldots, x_k, z + 1) = f_1(x_1, \ldots, x_k, z, f(x_1, \ldots, x_k, z)) \tag{10.2.2}$$

Then f is said to be **defined recursively** from f_0 and f_1. Equations (10.2.1) and (10.2.2) are said to be the recursive definition of f. The most familiar case is $k = 0$. (A function of 0 variables is understood to be a constant.) Here we have

$$f(0) = a_0 \quad \text{(a constant)} \tag{10.2.3}$$

$$f(z + 1) = f_1(z, f(z)) \tag{10.2.4}$$

EXAMPLE 10.2.4

The function $\text{Fact}(x) = x!$ is recursively defined by the equations

$$\text{Fact}(0) = 1$$

$$\text{Fact}(z + 1) = (z + 1)\text{Fact}(z)$$

Here, $a_0 = 1$ and $f_1(z, y) = (z + 1)y$ in Equations (10.2.3) and (10.2.4).

By any reasonable definition of computability, we should expect f to be computable if it is recursively defined from computable f_0 and f_1. Equation (10.2.1) gives the value of f at $(x_1, \ldots, x_k, 0)$. Then for any value x_{k+1} in $\mathbb{N}$, Equation (10.2.2) may be used over and over again to compute f at $(x_1, \ldots, x_k, 1)$, then at $(x_1, \ldots, x_k, 2)$, and so on, until f is computed at $(x_1, \ldots, x_k, x_{k+1})$. We now use this idea to find a Turing machine for f, when we have one for f_0 and f_1.

EXAMPLE 10.2.5   *A Turing Machine for a Recursively Defined Function*

For convenience, we abbreviate the k variables $x_1, \ldots, x_k$ by x. Suppose $f_0(x)$ and $f_1(x, z, y)$ are Turing-machine computable and that $f(x, y)$ is recursively defined from f_0 and f_1. We now indicate how to construct a Turing machine to compute $f(x, z)$. Besides x and z, we introduce two new variables, t and v. The function $f(x, t)$ will be computed for $t = 0$, $1, \ldots, z$, at which point we have $f(x, z)$. The variable v is the current value of $f(x, t)$. Initially, we take $t = 0$ and $v = f_0(x)$, which is $f(x, 0)$ by (10.2.1). The following algorithm computes $f(x, z)$:

1. $t := 0$
2. $v := f(x, 0)$ [initializations]
3. IF $t = z$ THEN
    1. $f(x, z) := v$
    2. End
4. IF $t < z$ THEN
    1. $v := f_1(x, t, v)$
    2. $t := t + 1$
5. Go to step 3

The assignment $v := f_1(x, t, v)$ in step 4.1 gives $v = f(x, t + 1)$ by the recursion Equation (10.2.2). The assignment $t := t + 1$ updates t. Step 5 repeats this process until t has reached the value z. The algorithm can be written more easily using a WHILE or a FOR statement, but we use the "Go to" statement 5 because it can be easily simulated on a Turing machine. This algorithm can be used as a basis for constructing a Turing machine that computes f. We shall not give the details.

We now proceed to the final operation on functions that we use in our definition of recursive functions.

DEFINITION 10.2.6   *Searching*

Suppose $f(x_1, \ldots, x_k, y)$ is a function of $k + 1$ variables. Suppose for every $x_1, \ldots, x_k$, the equation $f(x_1, \ldots, x_k, y) = 0$ has a solution y in $\mathbb{N}$. Let $g(x_1, \ldots, x_k) =$ least value of y such that $f(x_1, \ldots, x_k, y) = 0$. Then g is said to be defined from f by a search, or by a search of y.
We write

$$g(x_1, \ldots, x_k) = (\mu y)(f(x_1, \ldots, x_k, y) = 0)$$

This process is also called a definition by minimization. Here, μy may be read "the least y such that."

Once again, a reasonable definition of a computable function would have g computable if f is. The way we would compute y for given $x = (x_1, \ldots, x_k)$ is to test $y = 0$ to see if $f(x, 0) = 0$. If it is not, we would test $y = 1$ to see if $f(x, 1) = 0$, and continue doing this until we find the first y making $f(x, y) = 0$.

If f is Turing-machine computable, then so is g. For we can construct a Turing machine that simulates the preceding description.

Remark   Care must be taken when defining a function by a search. For example, $(\mu y)(y^2 + x + 40 = 0)$ is not defined for any x. The search procedure used to find y will go on forever. Thus, the condition that for every x there is at least one y satisfying $f(x, y) = 0$ is a crucial one. This is unlike the substitution and the recursive procedures, where no special conditions on the functions were required.

DEFINITION 10.2.7

A function $f = f(x_1, \ldots, x_k)$ is called a **recursive function** if it can be defined, in a finite number of steps, from the starter functions using only the operations of substitution, recursion, and searching.

In brief, all functions used to define recursive functions must be either starter functions or recursive.

This definition can be given recursively. A recursive function is a function that can be obtained by the use of the following rules:

1. The starter functions are recursive functions.
2. f is recursive if it is defined by substitution from the recursive functions $f_0; f_1, \ldots, f_k$.
3. f is recursive if it is recursively defined from the recursive functions f_0 and f_1.
4. $g(x_1, \ldots, x_k)$ is recursive if it is defined by a search on y of the recursive function $f(x_1, \ldots, x_k, y)$.

THEOREM 10.2.8

A recursive function is Turing-machine computable.

Proof This follows from our previous remarks. All of the operations used to define a recursive function can be accomplished by using a Turing machine. ∎

The converse is also true, but we do not prove it here.

We illustrate the power of the definition by showing that certain familiar functions are recursive. An important consequence is that all of these functions are Turing-machine computable by Theorem 10.2.8. We now know this without going through the detail for constructing a Turing machine for these functions.

EXAMPLE 10.2.9 Some Recursive Functions

(a) The constant function defined by $C_1(x) = 1$ for all x is recursive. Here, $C_1(x) = S(Z(x))$ and, hence, is obtained from the starter functions Z and S by substitution.

(b) Similarly, defining $C_n(x) = n$, the constant function $C_n(x)$ is recursive, using n substitutions. For example, $C_5(x) = 5$, and $C_5(x) = S(S(S(S(S(Z(x))))))$. A proof may be given by induction on n.

(c) The sum function $x + y$ can be recursively defined, since

$$x + 0 = x \tag{10.2.5}$$

$$x + (z + 1) = (x + z) + 1 = S(x + z) \tag{10.2.6}$$

According to Equations (10.2.1) and (10.2.2) of Definition 10.2.3, what is required in order to recursively define the function $f(x, y) = x + y$ are functions $f_0(x)$ and $f_1(x, z, y)$ satisfying

$$x + 0 = f_0(x) \tag{10.2.7}$$

and

$$x + (z + 1) = f_1(x, z, x + z) \tag{10.2.8}$$

Comparing (10.2.5) and (10.2.7) shows that we need $f_0(x) = x$, or that f_0 is the identity function. But we already noted that the identity function is recursive because it is one of the starter functions. Comparing (10.2.6) and (10.2.8), we need f_1 to satisfy $f_1(x, z, x + z) = S(x + z)$. We thus take the function $f_1(x, z, y) = S(y)$, and (10.2.1) and (10.2.2) are thus satisfied for the sum function. To regard $S(y)$ as a function $f_1(x, z, y)$ of three variables, we have $S(y) = f_1(x, z, y) = S(p_3^3(x, z, y))$, using the starter functions S and p_3^3. (See Definition 10.2.1.) Thus the sum function is recursive because it can be defined solely in terms of starter functions, substitution, and recursion.

In the future, we shall not go into such explicit details for the construction of the functions f_0 and f_1.

(d) The product function $x \cdot y$ is recursive because the formulas

$$x \cdot 0 = 0$$

$$x \cdot (y + 1) = x \cdot y + y$$

define the product recursively. The sum function is recursive by part (c).

(e) The exponential function x^y is recursive:

$$x^0 = 1$$

$$x^{y+1} = x^y \cdot x$$

These formulas define x^y recursively. The product function is known to be recursive by part (d).

(f) $x!$ is recursive as noted.

EXAMPLE 10.2.10   More Recursive Functions

The following functions are needed in what follows.

(g) The predecessor function

$$p(x) = \text{predecessor of } x$$

$$= x - 1 \qquad \text{if } x > 0$$

$$= 0 \qquad \text{if } x = 0$$

is recursive, since the equations

$$p(0) = 0$$

$$p(x + 1) = x$$

define $p(x)$ recursively.

(h) *Natural Number Subtraction*   The function

$$x \div y = x - y \qquad \text{if } x \geq y$$

$$= 0 \qquad \text{if } x < y$$

is recursive, because the equations

$$x \div 0 = x$$

$$x \div (z + 1) = p(x \div z)$$

define $x \div y$ recursively.

(i) The function

$$|x - y| = (x \mathbin{\dot-} y) + (y \mathbin{\dot-} x)$$

is recursive.

(j) The function

$$\mathrm{pos}(x) = 0 \qquad \text{for } x = 0$$
$$= 1 \qquad \text{for } x > 0$$

is recursive. (Show this by recursion.)

(k) The function

$$\mathrm{zero}(x) = 1 \qquad \text{for } x = 0$$
$$= 0 \qquad \text{for } x > 0$$

is recursive, because $\mathrm{zero}(x) = 1 \mathbin{\dot-} \mathrm{pos}(x)$.

We shall see that $\mathrm{pos}(x)$ and $\mathrm{zero}(x)$ may be regarded as recursive *predicates*, expressing the predicates "x is positive" and "x is zero." It is convenient to use the μ operator on predicates. For example, to find the least number y whose square is greater than x, we would use the notation $(\mu y)(y^2 > x)$. Our approach is to code true and false as 1 and 0, respectively. Thus, a predicate in variables $x_1, \ldots, x_k$ is simply regarded as a **function** whose sole values are 0 and 1.

DEFINITION 10.2.11

A **predicate** $P(x_1, \ldots, x_k)$ is a function whose only values are 0 or 1. It is understood that each x_i is in $\mathbb{N}$. If $P = 1$ for the value $x = (x_1, \ldots, x_k)$ we say the predicate is true for that value of x; otherwise it is false.

DEFINITION 10.2.12

Suppose that $P(x_1, \ldots, x_k, y)$ is a predicate such that for any choice of $x_1, \ldots, x_k$ there is a value of y such that P is true. Then we define (μy) $P(x_1, \ldots, x_k, y)$ as the least value of y for which $P(x_1, \ldots, x_k, y)$ is true.

Let us pause for a moment and take stock. In the quest to discover what is meant by a computable function, the class of recursive functions was defined. Certain familiar numerical functions were then shown to be recursive. It was also shown that any computation of recursive functions could be simulated using a very primitive language, namely the language of a Turing machine. When we now talk about recursive predicates, we are asking for a computable method to decide whether a predicate is true or false. Therefore, we now bring some logic onto the scene.

Since we regard a predicate as a function, it now makes sense to speak of a **recursive predicate.** However, because predicates appear in natural context as statements involving variables (for example $x > y$), it is convenient to use the logical symbols $(\wedge, \vee, \neg, \ldots)$ as well as the quantifiers $\forall$ and $\exists$. We have the following results.

THEOREM 10.2.13

If $P(x_1, \ldots, x_k)$ and $Q(x_1, \ldots, x_k)$ are recursive predicates, then so are

$$(1)\ \neg P;\ (2)\ P \wedge Q;\ (3)\ P \vee Q;\ (4)\ P \to Q;\ \text{and}\ (5)\ P \leftrightarrow Q$$

Proof We convert the appropriate truth tables to numerical operations. Thus,
1. $\neg P = 1 \div P$
2. $P \wedge Q$ is true (1) if and only if both P and Q are. Therefore, $P \wedge Q$ is the same as PQ.
3. Similarly, $P \vee Q = P + Q \div PQ$.
The proofs for parts 4 and 5 are similar. ∎

It is easy to show that the familiar predicates of equality and order are recursive.

THEOREM 10.2.14

The predicates

$$EQ(x, y) \equiv x = y$$
$$LT(x, y) \equiv x < y$$
$$LE(x, y) \equiv x \le y$$

are recursive.

Proof $LT(x, y) = pos(y \div x)$. Thus, LT is obtained by substitution. [See Example 10.2.10, where $pos(x)$ was defined as the predicate "x is positive."]

$$LE(x, y) = \neg LT(y, x)$$
$$EQ(x, y) = zero(|\,y - x\,|)$$

We can also define $EQ(x, y) = LE(x, y) \wedge LE(y, x)$. ∎

The following theorem relates recursive functions to corresponding recursive predicates.

THEOREM 10.2.15

Let $f(x_1, \ldots, x_k)$ be a recursive function. Then the following are recursive predicates:

1. $f(x_1, \ldots, x_k) = 0$
2. $y = f(x_1, \ldots, x_k)$

Further, suppose $P(x_1, \ldots, x_k, y)$ is a recursive predicate and, for any $x_1, \ldots, x_k, P(x_1, \ldots, x_k, y)$ is true for at least one value of y. Then

3. $g(x_1, \ldots, x_k) = (\mu y)P(x_1, \ldots, x_k, y)$ is a recursive function.

Proof We take $x = (x_1, \ldots, x_k)$ in what follows.
1. The equation $f(x) = 0$ is equivalent to $\mathrm{EQ}(f(x), 0)$. This is recursive, by substitution. (Note that $0 \, [= Z(x)]$ is recursive because it is a starter function.)
2. $Q(x_1, \ldots, x_k, y) = \mathrm{EQ}(y, f(x_1, \ldots, x_k))$ is recursive.
3. $g(x_1, \ldots, x_k) = (\mu y)(1 \dotdiv P(x_1, \ldots, x_k, y) = 0)$. ∎

THEOREM 10.2.16   Definition by Cases

Let $f(x_1, \ldots, x_k)$ and $g(x_1, \ldots, x_k)$ be recursive functions, and let $P(x_1, \ldots, x_k)$ be a recursive predicate. Then the function

$$h(x_1, \ldots, x_k) = f(x_1, \ldots, x_k) \qquad \text{if } P(x_1, \ldots, x_k)$$
$$= g(x_1, \ldots, x_k) \qquad \text{otherwise}$$

is recursive.

Proof $h = fP + g(1 \dotdiv P)$. ∎

EXAMPLE 10.2.17

Show that the function defined by

$$f(x) = x \qquad \text{for } x \le 10$$
$$= 10 \qquad \text{for } 11 \le x \le 100$$
$$= x \dotdiv 90 \qquad \text{for } x > 100$$

is recursive.

Method Let $P(x)$ be the predicate $\mathrm{LE}(x, 10)$, let $Q(x)$ be the predicate $\mathrm{LE}(11, x) \wedge \mathrm{LE}(x, 100)$, and let $R(x)$ be the predicate $\mathrm{LT}(100, x)$. By Theorems 10.2.13 and 10.2.14, these are all recursive predicates. Now consider the function

$$g(x) = xP(x) + 10R(x) + (x \dotdiv 90)Q(x)$$

Then if $x \le 10$, $P(x) = 1$ and $R(x) = Q(x) = 0$, so $G(x) = x$. Similarly, if $11 \leq x \leq 100$, $Q(x) = 1$, and $P(x) = R(x) = 0$. Thus, in this case, $g(x) = 10 \cdot R(x) = 10$. Similarly, if $x > 100$, we have $g(x) = x \div 90$ or simply $x - 90$ in this case. Thus, $g(x) = f(x)$. We have taken advantage of our choice of the numbers 0 and 1 for the truth values F and T.

THEOREM 10.2.18   Sums and Products

Let $f(x_1, \ldots, x_k, y)$ be recursive. Then the sum and product

$$S(x_1, \ldots, x_k, z) = \sum_{y=0}^{z} f(x_1, \ldots, x_k, y)$$

$$P(x_1, \ldots, x_k, z) = \prod_{y=0}^{z} f(x_1, \ldots, x_k, y)$$

are recursive functions.

Proof   These are recursively defined. For example,

$$P(x_1, \ldots, x_k, 0) = f(x_1, \ldots, x_k, 0)$$

$$P(x_1, \ldots, x_k, z + 1) = P(x_1, \ldots, x_k, z)f(x_1, \ldots, x_k, z + 1)$$

with a similar recursive definition for the sum. ∎

Theorem 10.2.18 can be used to show that certain quantifier operations on recursive predicates may be performed, keeping the predicate recursive. Intuitively, we cannot expect this without restrictions for existential and universal quantifiers. For example, if $P(x, y)$ is computable, we would have no reason to believe that $R(y) = \exists x P(x, y)$ is recursive. To compute its value for fixed y, we would apparently have to compute whether $P(0, y)$ is true, and if not, then whether $P(1, y)$ is true, and so forth, through all the natural numbers. If $P(x, y)$ were false for all x, this process would not work, but would go on forever. Similarly, the universal quantifier applied to $P(x, y)$ would yield $\forall x P(x, y)$, whose truth would have to be checked for infinitely many x's, unless some shortcut method were available.

For example, Fermat's Last Theorem—that the equation $a^n + b^n = c^n$ has no nonzero solutions for $n > 2$—has not been proved to date. If this result were true, a direct computer search would be hopeless. We would only be able to verify that for finitely many attempts, the equation indeed has no solution. Fermat's Last Theorem is an example of a mathematical proposition that uses universal quantifiers. This is the statement:

$$\forall n \forall a \forall b \forall c (n > 2 \wedge abc \neq 0) \rightarrow (a^n + b^n \neq c^n)$$

Such a result can be disproved by one counterexample, but a proof using a direct computational verification is not possible unless further facts about number theory, so far unknown, are discovered.

One way of getting around this limitation on the use of quantifiers is with the use of a *bounded quantifier*. These quantifiers $(\forall x)_y$ and $(\exists x)_y$ are to be interpreted as "for all $x \leq y$" and "there is an $x \leq y$". The definitions are as follows:

DEFINITION 10.2.19

$$(\forall x)_y P(x_1, \ldots, x_k, x) \equiv \forall x(x \leq y \rightarrow P(x_1, \ldots, x_k, x))$$

$$(\exists x)_y P(x_1, \ldots, x_k, x) \equiv \exists x(x \leq y \wedge P(x_1, \ldots, x_k, x))$$

The quantifiers $(\forall x)_y$ and $(\exists x)_y$ are called **bounded quantifiers**, and y is called the **bound**.

Thus, for fixed $x_1, \ldots, x_k, y$, either of these predicates can be verified to be true or false in finitely many steps. Note that the prefix $(\forall x)_y$ and $(\exists x)_y$ makes x into a bound variable, but introduces y as a new free variable. Thus, if $P(x_1, \ldots, x_k, x)$ is a predicate in the $k + 1$ variables $x_1, \ldots, x_k, x$, then $(\forall x)_y P(x_1, \ldots, x_k, x)$ is a predicate in the variables $x_1, \ldots, x_k, y$, and similarly for the existential bounded predicate. The following theorem shows that recursive predicates stay recursive when bounded quantifiers are applied.

THEOREM 10.2.20

Let $P(x_1, \ldots, x_k, x)$ be a recursive predicate. Then

$$Q(x_1, \ldots, x_k, y) = (\forall x)_y P(x_1, \ldots, x_k, x)$$

$$R(x_1, \ldots, x_k, y) = (\exists x)_y P(x_1, \ldots, x_k, x)$$

are recursive.

Proof Q is true (1) if and only if $P(x_1, \ldots, x_k, k) = 1$ for all k between 0 and y inclusive. This gives

$$Q(x_1, \ldots, x_k, y) = \prod_{x=0}^{y} P(x_1, \ldots, x_k, x)$$

As for R, we may verify that, as for (unbounded) universal and existential quantifiers, $\neg(\forall x)_y$ is equivalent to $(\exists x)_y \neg$. Thus, $R(x_1, \ldots, x_k, y) \equiv \neg(\forall x)_y \neg P(x_1, \ldots, x_k, x)$. This is recursive by the previous result and Theorem 10.2.13. ∎

This theorem is very useful to show that predicates that are defined using quantifiers are recursive. We need only find a bound for the universal or existential quantifier. Some familiar examples are illustrated here. We remark again that all of the following functions and predicates can be computed on a Turing machine, since they will be shown to be recursive.

EXAMPLE 10.2.21   Recursive Predicates

1. $x \mid y$ is recursive.

For x divides y provided there exists z with $y = xz$. However, if $y > 0$, such a z will necessarily have to be less than or equal to y. This is the required bound. Thus,

$$x \mid y \equiv (y = 0) \lor (y > 0 \land (\exists z)_y (y = xz))$$

2. "x is composite" or $\text{comp}(x)$ is recursive.

For, x is composite if it can be written as a product of two numbers > 1. But this gives a bound—each must be smaller than x. Thus:

$$\text{comp}(x) \equiv (\exists u)_x (\exists v)_x (u > 1 \land v > 1 \land x = uv)$$

3. "x is prime" or $\text{prime}(x)$ is recursive.
For,

$$\text{prime}(x) \equiv x > 1 \land \neg \text{comp}(x)$$

4. The function $p_n = p(n) =$ the nth prime is recursive.
For we can recursively define $p(n)$:

$$p(0) = 0 \quad \text{[for convenience]}$$

$$p(n + 1) = (\mu y)(\text{prime}(y) \land y > p(n))$$

Note that we have used the search operator (μy) here. This is only possible because we know that there are infinitely many primes, and so there is always a prime $> p(n)$.

5. Let $\pi(x) =$ number of primes $\leq x$. Then $\pi(x)$ is recursive:

$$\pi(x) = \sum_{y=0}^{x} \text{prime}(y)$$

6. The relation "$x \equiv y \bmod z$" is recursive.
For,

$$x \equiv y \bmod z \quad \text{iff} \quad z \text{ divides } |x - y|$$

These examples illustrate how our results enable us to demonstrate why some familiar functions and predicates, all intuitively computable, are recursive. By Theorem 10.2.8, they are also computable on a Turing machine. In the next sections we show, with the help of Turing machines, some of the built-in limitations of computability.

Exercises

1. Suppose $f(x, y)$ is recursive. Let $g(y) = f(2, y)$. Give a proof, using the definition, that g is recursive.
2. Suppose $f(x, y, z)$ is recursive and $g(x, y) = f(y, x, x)$. Give a proof, using the definition, that g is recursive.
3. Suppose $f(x, y)$ is recursive. Define $g(x_1, x_2, x_3, x_4) = f(x_1, x_3)$. Give a proof, using the definition, that g is recursive.

*4. Define the *bounded* μ operator as follows:

$$(\mu y)_x P(x_1, \ldots, x_k, y) = \begin{cases} \text{least } y \le x \text{ such that } P(x_1, \ldots, x_k, y) \\ \quad \text{is true if such a } y \text{ exists} \\ 0 \qquad \text{otherwise} \end{cases}$$

Show that $f(x_1, \ldots, x_k, x) = (\mu y)_x P(x_1, \ldots, x_k, y)$ is a recursive function provided P is.

*5. Let

$$P(n) = \text{largest prime divisor of } n \qquad \text{if } n \ge 2$$
$$= 0 \qquad \text{if } n = 0 \text{ or } n = 1$$

Show that $P(n)$ is recursive.

*6. Let

$$E(n, p) = \text{largest exponent } r \text{ such that } p^r \text{ divides } n$$
$$\text{if } p \ge 2 \text{ and } n \ge 1$$
$$= 0 \qquad \text{if } p \le 1 \text{ or } n = 0$$

Show that $E(n, p)$ is recursive.

7. Let

$$(x)_n = \text{remainder when } x \text{ is divided by } n \qquad \text{if } n \ge 1$$
$$= x \qquad \text{if } n = 0$$

Show that $(x)_n$ is recursive.

8. Let

$$[x/y] = \text{greatest integer in } x/y \qquad \text{if } y \ge 1$$
$$= 0 \qquad \text{if } y = 0$$

Show that $[x/y]$ is recursive.

*9. Let

$$R(n) = \text{number of distinct prime divisors of } n \qquad \text{if } n \ge 1$$
$$= 0 \qquad \text{for } n = 0$$

Show that $R(n)$ is recursive.

*10. Let $P(x_1, \ldots, x_k, x)$ be a recursive predicate. Define the ν operator as follows:

$$(\nu y)_x P(x_1, \ldots, x_k, y) = \text{largest value of } y \leq x \text{ such that } P(x_1, \ldots,$$
$$x_k, y) \text{ is true if such a } y \text{ exists}$$
$$= 0 \quad \text{if no such } y \text{ exists}$$

Show that $(\nu y)_x P(x_1, \ldots, x_k, y)$ is recursive.

11. Show that $EQ(f(x), 0) = 1 \dot- f(x)$.

12. If $f(x_1, \ldots, x_n)$ is recursive, show that the predicate $f > 0$ is a recursive predicate.

13. Define the predicate sq(x) to mean that x is a square. (For example, sq(25) is true, but sq(2) is false. Show that sq(x) is recursive.

14. Show that the predicate even(x) (x is even) is recursive.

15. Show that the predicate twin(x), meaning that both x and $x + 2$ are primes, is recursive.

*16. Show that the predicate $D(m, n)$, meaning that m divides some power of n, is recursive. [For example, $D(8, 10)$ is true.]

17. Show that the function $[\sqrt{x}]$ (the greatest integer in $\sqrt{x}$) is recursive.

18. Show that the function $\max(x, y)$ is recursive.

19. The predicate $P(y) \equiv \exists x(y = x^2 + 2x)$ uses an unbounded quantifier. Is this predicate recursive? Explain.

*20. Is the predicate

$$P(y) \equiv \forall x(x \mid y \to \exists z(x = 2^z))$$

recursive? Explain.

21. Is the predicate

$$Q(m, n) \equiv \exists x \exists y(1 + mx = ny)$$

recursive?

22. Is the predicate

$$\text{RelPrime}(m, n) \equiv \forall d(d \mid n \wedge d \mid m \to d = 1)$$

recursive?

10.3 Enumeration Methods and the Nonexistence of Algorithms

In this section we discuss and apply some of the classical theory of infinite sets developed by G. Cantor in the nineteenth century. The purpose is to **arithmeticize** Turing machines in a way that will allow Turing machines to analyze Turing machines! We first want to show how we can arrange *all* of the Turing machines in a sequence $T_1, T_2, \ldots$. Thus, every Turing machine will be found in this list. Our method will show that this can be done *effectively*—it is possible to give an algorithm to find the nth Turing machine. (In fact there is a Turing machine that can do this.) The method we use is based on a few facts in number theory, which we review here.

1. A prime number is an integer > 1 whose only positive divisors are itself and 1. The first few primes are

$$2, 3, 5, 7, 11, 13, \ldots$$

2. There are infinitely many primes. Thus, the preceding list continues forever.

3. *The Unique Factorization Theorem* Any positive integer can be written as a product of prime numbers. Except for the order of the factors, this product is unique.

The Unique Factorization Theorem can be used to code any finite sequence of positive integers as *one* integer. We use the sequence as the exponents in a unique factorization. For example, the sequence [3, 2, 1] is coded as the single integer $2^3 \cdot 3^2 \cdot 5^1$, or 360. By unique factorization, the code number 360 can be uniquely decoded by first factoring and then picking off the exponents. In this example, we chose the first three primes. The general process codes a sequence $[a_1, \ldots, a_k]$ as the number

$$p_1^{a_1} p_2^{a_2} \cdots p_k^{a_k}$$

where p_i is the ith prime.

We now use this procedure to code each *Turing machine* as a single number. When we use the term *code*, we understand that a decoding system is also available. Thus, a code number for a Turing machine can be decoded to obtain the machine itself. We first code the symbols of a Turing machine.

DEFINITION 10.3.1 *Coding Turing Machine Symbols*

The code $c(S)$ for any symbol S used to define a Turing machine is given in Table 10.3.1. This coding device is arbitrary, and clearly one of many.

Table 10.3.1 Symbol Codes

State	q_1	q_2	q_3	$\cdots$	q_n	$\cdots$
Code	5	9	13	$\cdots$	$4n+1$	$\cdots$
Motion	L	R				
Code	3	7				
Symbol	0	1	B	S_3	$\cdots$	S_n	$\cdots$
Code	11	15	19	23	$\cdots$	$4n+11$	$\cdots$

We recall that symbols S_0, S_1, and S_2 are understood to be 0, 1, and B, respectively. We use the notation $c(S)$ to represent the code number of S, where S is a symbol, motion, or state. Thus, $c(q_7) = 29$, $c(R) = 7$, and so on. Similarly, we use c^{-1} for the decoding operation: $c^{-1}(9) = q_2$, $c^{-1}(15) = 1$ (or S_1), and $c^{-1}(4)$ is undefined.

A Turing machine is given as a sequence of instructions. Each instruction contains four items (states, symbols, or motions). Therefore, a sequence of k instructions will have $4k$ items, which represent either states, symbols, or motions. Not every such sequence is a Turing machine. For example, each instruction must, by definition, start and end with a state; its second component must be a symbol; and so on. If a Turing machine T is given, we can code each item in the sequence for T using Table 10.2.1. Thus, a Turing machine with k instructions is coded as a sequence of $4k$ positive integers. Finally, we can code this sequence, via unique factorization, *as a single integer C*, which we call the code number for T: $C = C(T)$.

DEFINITION 10.3.2

If T is given by the sequence of symbols

$$X_1, X_2, X_3, X_4; X_5, X_6, X_7, X_8, \ldots, X_{4k}$$

then the code $C(T)$ is defined as the number

$$C(T) = 2^{c(X_1)} \cdot 3^{c(X_2)} \cdot 5^{c(X_3)} \cdots p_{4k}^{c(X_{4k})}$$

If I is a Turing machine instruction given by the sequence of symbols X_1, X_2, X_3, X_4, then the code $C(I)$ is defined by the formula

$$C(I) = 2^{c(X_1)} \cdot 3^{c(X_2)} \cdot 5^{c(X_3)} \cdot 7^{c(X_4)}$$

Remark   A Turing machine was defined as a **set** of instructions rather than a sequence. Thus, we can rearrange the instructions of a given Turing machine to arrive at a different sequence of instructions, and so a different code number C for the same Turing machine. This will have no bearing on the results that follow, because, in any case, a code number C can be uniquely decoded to a Turing machine.

For example, the very simple Turing machine given by the single instruction $q_1\, 0\, L\, q_1$ has the code number $2^5 3^{11} 5^3 7^5$. This is an enormous number, but through it we encapsulate this Turing machine. And, conversely, if a number C is given, we can factor it into primes and decide whether C is the code number of a Turing machine and, if so, what the instructions of that machine are. This gives us a way of regarding a Turing machine as a single positive integer. This observation permits us to construct Turing machines that will investigate Turing machines, which will be coded as integers.

Because the coding process uses such high numbers, it is quite time-consuming and inefficient. However, the results we develop in this chapter are not dependent on efficiency or time, so a coding was chosen that was relatively straightforward in principle.

THEOREM 10.3.3

There is a Turing machine that can decide whether a number C is the code number of a Turing machine instruction.

Proof C is a code number of an instruction if $C = 2^a 3^b 5^c 7^d$, where a and d are the code numbers of states (i.e, numbers of the form $4n + 1$), b is the code number of a symbol (11, 15, or 19, as well as other possibilities, depending on what strings are allowed), and c is the code number of a symbol or a motion (3 or 7). This is a recursive predicate, and so there is a Turing machine that can compute this. ∎

Let us elaborate on this proof. Referring to Table 10.2.1, an integer is a code number of a state if it is of the form $4n + 1$ for $n > 0$. We may then define the recursive predicate

$$\text{State}(x) \equiv (\exists n)_x (x = 4n + 1 \wedge n > 0)$$

Thus $\text{State}(x)$ expresses the statement "x is the code number of a state." Similarly, we may define

$$\text{Symbol}(x) \equiv (\exists n)_x (x = 4n + 11)$$

$$\text{Motion}(x) \equiv x = 3 \vee x = 7$$

Finally,

$$\text{Instruction}(x) \equiv (\exists a)_x (\exists b)_x (\exists c)_x (\exists d)_x$$
$$(x = 2^a 3^b 5^c 7^d)$$
$$\wedge \text{State}(a) \wedge \text{Symbol}(b) \wedge (\text{Symbol}(c) \vee \text{Motion}(c)) \wedge \text{State}(d))$$

These definitions show that the predicate $\text{Instruction}(x)$ (meaning "x is the code number of an instruction") is recursive. Hence, it can be computed by a Turing machine.

We can go further. We can clearly reconstruct the instruction from its code. The output of the Turing machine of Theorem 10.2.1 can be the instruction itself, suitably coded. For example, it can be the four integers a, b, c, d, which appear as exponents in the expression for C. In what follows, if the code number of an instruction is $2^a 3^b 5^c 7^d$, it often will be convenient to refer to code numbers a, b, c, d as the instruction itself. A similar convention applies to Turing machines.

COROLLARY 10.3.4

There is a Turing machine that takes an integer C as input and gives 0 as output if C is not the code number of an instruction for a Turing machine. If C is the code number of an instruction, the output is the instruction itself.

The same procedures apply to code numbers of Turing machines rather than individual instructions. For in this case C must be of the form

$$p_1^{a_1} p_2^{a_2} \cdots p_{4k}^{a_{4k}}$$

and the methods of Theorem 10.3.3 and Corollary 10.3.4 apply to code numbers of Turing machines. It is also necessary to check that no two instructions start with the same pair. Thus, we have the next result.

THEOREM 10.3.5

The predicate Tur(C), meaning "C is the code number $C(T)$ of a Turing machine T" is recursive. There is a Turing machine that, for input C, computes the value (0 or 1) of Tur(C). There is also a Turing machine that, for input C, has output 0 if Tur(C) = 0, but lists the instructions of the Turing machine that has code number C if Tur(C) = 1. ■

We now wish to enumerate (i.e, list) all Turing machines. We first give some general definitions.

DEFINITION 10.3.6

A set X is **enumerable** if its elements can be listed in a sequence x_1, x_2, $\ldots, x_n, \ldots$. Every $x \in X$ must be at least one of the x_i, but not necessarily for a unique i, because there might be repetitions in this list. The sequence x_i is called an **enumeration**, or **listing**, of the set X. If $f(i) = x_i$ is a recursive function, then the set X is **recursively enumerable.** In this case, the elements of X must be natural numbers.

Remark Thus, a way of expressing that X is recursively enumerable is that we can find a computation that spews forth the elements of X, one at a time, in some order and possibly with repetition. The set of prime numbers is recursively enumerable because, as we saw in the previous section, the function $p(i) =$ the ith prime is a recursive function.

We can now enumerate Turing machines by enumerating the code numbers of the Turing machines. The first Turing machine is the one with the least code number; the second is the one with the next smallest code number; and so on. The enumeration is

$$T_1, T_2, \ldots, T_n, \ldots$$

THEOREM 10.3.7

The code numbers of Turing machines are recursively enumerable.

Proof Let $C(n) =$ code number of T_n. Then $C(n)$ may be defined by the recursion

$$C(1) = (\mu C)\mathrm{Tur}(C)$$

$$C(n + 1) = (\mu C)(\mathrm{Tur}(C) \wedge C > C(n))$$

This enumeration contains all Turing machines, because any Turing machine has a code number C, which will ultimately be reached in the definition of $C(n)$. ∎

By means of the enumeration T_n, we can regard any positive integer n as identifying a Turing machine, namely the nth Turing machine T_n. It is sometimes convenient to identify a Turing machine with its code number. When we do this, we may say that the set of Turing machines is recursively enumerable.

A Turing machine is an idealized computing program. It is idealized because, since it uses an infinite tape, it can be an arbitrarily large program with an arbitrarily large memory. Therefore, it may seem odd that they can be enumerated. But the same is also true for all BASIC programs. We can see this because each BASIC program is finite in length and uses a finite alphabet. Therefore, we can imagine all programs using exactly one letter, then two, then three, . . . , and list them one after the other to get a listing of all BASIC programs. The following theorem, however, is not so easy to visualize. We find *one* program (finite in length) to simulate *every* program, no matter how complicated or long!

THEOREM 10.3.8   *A Universal Turing Machine*

There is one Turing machine T that simulates any Turing machine. This means that if (n, α) is taken as the input string (n a positive integer, α a string), the action of T will be the same as the action of T_n with input string α. In brief, T can be made to act like the Turing machine T_n for any choice of n, and so can be made to act like any Turing machine.

Proof   First, find the Turing machine to compute the recursive function $C(n)$, the code number of T_n. Applied to n, we find the code number for T_n. As in Theorem 10.3.5, we can find a Turing machine to list the instructions of the Turing machine with this code. We can now use these instructions and operate with them on the string α. In brief, find T_n, erase n from the original input tape, and operate with T_n on the string α. Putting these machines together, we have the universal Turing machine. ∎

Remark   Thus, infinitely many (all!) Turing machines are encapsulated in *one* Turing machine. For example, this universal Turing machine is one of the T_n, and so can even simulate itself. The reader is referred to Penrose, *The Emperor's New Mind* (Oxford University Press, New York, 1989) for an explicit listing of a universal Turing machine. His coding and conventions differ from ours, but the idea is the same.

To further analyze recursive functions, we use a technique introduced by G. Cantor in the nineteenth century, called the *Cantor diagonal process* for reasons that will be seen in a moment. Cantor was a mathematician who became interested in enumeration techniques. He is best known today for his work on infinite (transfinite) counting numbers. He was the first to show that certain infinite sets cannot be enumerated. We now illustrate with a simple example of the Cantor diagonal method.

THEOREM 10.3.9

Let **B** be the set of all infinite sequences of bits (0's or 1's). Then the elements of **B** *cannot* be enumerated.

Proof   The proof is by contradiction. A typical element of **B** is a sequence

$$b_1, b_2, \ldots, b_n, \ldots \qquad (\text{each } b_i = 0 \text{ or } 1)$$

If **B** can be enumerated, we would have a sequence

$$B_1, B_2, \ldots, B_n, \ldots$$

where each B_i is itself a sequence of bits. This is naturally listed as a double array:

$$
\begin{aligned}
B_1: &\quad b_{11}, b_{12}, b_{13}, \ldots, b_{1n}, \ldots \\
B_2: &\quad b_{21}, b_{22}, b_{23}, \ldots, b_{2n}, \ldots \\
B_3: &\quad b_{31}, b_{32}, b_{33}, \ldots, b_{3n}, \ldots \\
&\quad \cdots
\end{aligned}
$$

This list $B_1, B_2, \ldots, B_n, \ldots$ is supposed to consist of *all* sequences of bits. We now use this list to construct a sequence d_n that does not appear anywhere in the list. That is, it does not appear as any row in this array. We go down the main diagonal as indicated and change each diagonal entry.

$$
\begin{aligned}
B_1: &\quad \underline{b_{11}}\; b_{12}\; b_{13} \ldots b_{1n} \ldots \\
B_2: &\quad b_{21}\; \underline{b_{22}}\; b_{23} \ldots b_{2n} \ldots \\
B_3: &\quad b_{31}\; b_{32}\; \underline{b_{33}} \ldots b_{3n} \ldots \\
&\quad \cdots\cdots\cdots\cdots \\
B: &\quad b_{n1} \ldots\ldots\ldots\ldots \underline{b_{nn}} \ldots \\
&\quad \cdots\cdots\cdots\cdots
\end{aligned}
$$

To change the diagonal elements, we define

$$d_i = 1 - b_{ii} \qquad \text{for all } i$$

We have defined a new sequence d_i to be different from each row. Thus, it differs from the first row because b_{11} was changed, and, similarly, it differs from each row. To see this more formally, we prove that the sequence D whose terms are d_i is not equal to any B_N (whose ith term is b_{Ni}). For if

$$D = B_N,$$

then $d_i = b_{Ni}$ for all i. Thus, taking $i = N$,

$$d_N = b_{NN}$$

By definition this leads to

$$1 - b_{NN} = b_{NN}$$

which is impossible for B_{NN}, because it is either 0 or 1. We have thus constructed a sequence D that is not in the original list, and this contradicts our assumption that the original list consisted of all sequences of bits. ■

Remark   It might seem unremarkable that no sequence $\{B_n\}$ can consist of all sequences of bits. On the other hand, we have seen that we can have a sequence consisting of all Turing machines, and this might seem equally unlikely. Another way of stating the preceding theorem is as follows.

THEOREM 10.3.10   (Cantor)

There are more real numbers than natural numbers.

Proof   If we put a decimal point in front of an infinite sequence of bits, we arrive at a real number. Thus, Theorem 10.3.3 shows that the real numbers x, $0 \le x < 1$, whose decimal expansion consists of 0's and 1's exclusively, cannot be listed as a sequence. It follows that the set of all reals cannot be listed as a sequence. Equivalently, there is no 1-1 correspondence between the natural numbers and the reals. As Cantor put it: there are more reals than natural numbers. ■

We now illustrate the power of the Cantor diagonal method with two theorems showing the nonexistence of algorithms.

DEFINITION 10.3.11

A Turing machine is a **function machine** if, for any input n, the machine has an output which is a natural number. (According to our convention, this means that the machine has output 0, 1, 11, 111, and so on.)

DEFINITION 10.3.12

The predicate $\text{Fcn}(n)$ is defined to be true (1) if the nth Turing machine T_n is a function machine, and false (0) if not.

For example, the Turing machine given by the set of instructions

$$q_1 \; 0 \; 1 \; q_{100}; \; q_1 \; 1 \; R \; q_1; \; q_1 \; B \; 1 \; q_{100}$$

is a function machine. Do you see which function it calculates?

THEOREM 10.3.13

The predicate $\text{Fcn}(n)$ is *not* recursive.

Proof (by contradiction)   Suppose it were. We first make a list of all the function machines $F_1, F_2, \ldots, F_n, \ldots$. This is done by finding the first number $n = F(1)$ such that T_n is function machine, then the second number $n = F(2)$ such that T_n is function machine, and so forth. We do this by the same process we used to list the code numbers of Turing machines. Namely, we recursively define $F(n)$ by the equations

$$F(1) = (\mu k)(\text{Fcn}(k) = 1)$$

$$F(n + 1) = (\mu k)(\text{Fcn}(k) = 1 \wedge k > F(n))$$

Because we are assuming that $\text{Fcn}(n)$ is recursive, it follows that $F(n)$ is recursive. Thus, the list of all the function machines in order is

$$T_{F(1)}, T_{F(2)}, \ldots, T_{F(n)}, \ldots$$

The function machines thus correspond to the Turing machines $T_{F(1)}$, $T_{F(2)}, \ldots, T_{F(n)}, \ldots$. Call the corresponding functions $f_1, f_2, \ldots$. Since F is recursive, it can be computed by means of a Turing machine. Hence, the Turing machine for the function f_k can be found and used to compute the values $f_k(n)$ for any n.

Thus, we can find *one* Turing machine that does the following: With input (k, n) it first computes $N = F(k)$, then brings up and simulates the Turing machine T_N with input n. The output will be the value of $f_k(n)$.

This gives a double array:

$$f_1(1), \; f_1(2), \; f_1(3), \ldots, \; f_1(n), \ldots$$
$$f_2(1), \; f_2(2), \; f_2(3), \ldots, \; f_2(n), \ldots$$
$$f_3(1), \; f_3(2), \; f_3(3), \ldots, \; f_3(n), \ldots$$
$$\cdots\cdots\cdots\cdots$$
$$f_m(1), \; f_m(2), \; f_m(3), \ldots, \; f_m(n), \ldots$$
$$\cdots\cdots\cdots\cdots$$

We now use the diagonal method. Use the foregoing Turing machine to create another Turing machine (call it S) that takes input n and has output $s(n) = f_n(n) + 1$. (We are altering the nth row of this double array in the nth column.) Now S is clearly a function machine, so the function it computes must be in the list $f_1, f_2, \ldots$ of all functions that come from function machines. Thus, we must have $s = f_N$ for some N. Evaluate s at N to find

$$s(N) = f_N(N), \quad \text{or}$$
$$f_N(N) + 1 = f_N(N)$$

a contradiction. This completes the proof. ■

Remark   Fcn is our first example of a **nonrecursive function.** Since recursive functions are viewed as those that can be computed algorithmically, the theorem can be restated: There can be no algorithm that looks at a (Turing machine) program and decides whether the program represents a function.

If someone offers you what is claimed to be a foolproof way of looking at any algorithm or program and deciding whether it represents a function, you can safely ignore that so-called method.

We end with a similar result concerning algorithms to decide whether a program will terminate.

DEFINITION 10.3.14

A Turing machine T with input α is written $\langle T, \alpha \rangle$. We say that $\langle T, \alpha \rangle$ **loops** if the program does not stop. Otherwise, it is said to **halt.** The predicate $\text{Halt}(n, m)$ is defined to be true if and only if $\langle T_n, m \rangle$ halts.

Briefly, $\text{Halt}(n, m)$ is true if and only if the nth Turing machine halts when the input tape contains the number m.

THEOREM 10.3.15   The Halting Problem

Halt(n, m) is not a recursive predicate.

Proof   Suppose it were. Then the predicate $P(n) \equiv$ Halt(n, n) would also be recursive. (Note: We use input string n for the nth Turing machine.) Thus there is a Turing machine S such that $\langle S, n \rangle$ determines whether $\langle T_n, n \rangle$ halts: S has output 1 if $\langle T_n, n \rangle$ halts, and 0 if it loops. Now construct a simple Turing machine T that loops for input 1 and halts for input 0. (See Exercise 1.) Finally, construct a Turing machine H combining S and T that takes the output of S and inputs it into T. Then

$$\langle H, n \rangle \text{ halts iff } \langle T_n, n \rangle \text{ loops} \tag{10.3.1}$$

Condition (10.3.1) yields the contradiction: H is itself a Turing machine, so $H = T_N$ for some N. We now let H test itself! Put $n = N$ in (10.3.1) to obtain

$$\langle H, N \rangle \text{ halts iff } \langle T_N, N \rangle \text{ loops} \tag{10.3.2}$$

But since $H = T_N$, this yields

$$\langle T_N, N \rangle \text{ halts iff } \langle T_N, N \rangle \text{ loops} \tag{10.3.3}$$

This is the contradiction.   ■

Remark   The proof uses the diagonal method once $\langle T_n, n \rangle$ is considered, but its use is hidden. We can see the double array and the diagonal more explicitly as follows. Halt(m, n) is itself a double array:

Halt(1, 1), Halt(1, 2), Halt(1, 3), . . . , Halt(1, n), . . .

Halt(2, 1), Halt(2, 2), Halt(2, 3), . . . , Halt(2, n), . . .

Halt(3, 1), Halt(3, 2), Halt(3, 3), . . . , Halt(3, n), . . .

. . . .

Halt(m, 1), Halt(m, 2), Halt(m, 3), . . . , Halt(m, n), . . .

. . . .

The diagonal Halt(n, n) is 1 if $T_n(n)$ halts, and 0 otherwise. The diagonal method suggests changing this diagonal—converting a halt to a loop, and a loop to a halt. This is the purpose of the Turing machine H.

Further Remarks   There are many mathematical and philosophical implications in this result. Consider what would happen if it were not true. The predicate Halt(n, m) would be recursive, and there would exist a Turing machine to determine whether T_n halts or loops with input m. This machine would effectively solve many mathematical problems for us by a simple computation.

For example, let us try to solve Fermat's Last Theorem on a computer. Let us systematically search for all solutions of the equation $a^n + b^n = c^n$, with $n > 2$, and a, b, c all positive. To do this, we construct a Fermat search Turing machine called FermatSearch. As soon as it finds a solution, it gives the integers a, b, c, and n as counterexamples for Fermat's Last Theorem, then writes "EU-REKA" and halts. If there are, in fact, no solutions to find, FermatSearch searches in vain and loops. Since FermatSearch is a Turing machine, it is equal to T_N for some N.

Now suppose Halt(n, m) were recursive. Then we would be able to compute the value of Halt(N, 1). If it were 0 (no halt), we would have finally proved Fermat's Last Theorem using the Halt Turing machine. But if Halt(N, 1) were 1, we could sit back and wait, assured that at some time in the future, FermatSearch would halt and we would find a counterexample. Theorem 10.3.15 shows that there can be no such universal shortcut to mathematical problems involving universal or existential quantifiers. The reader may wish to read Penrose's book, referred to after Theorem 10.3.8, for many more details.

Exercises

1. The proof of Theorem 10.3.15 (The Halting Problem) needed a Turing machine T that loops for input 1 and halts for input 0. Explicitly construct such a Turing machine.
*2. Using the Cantor diagonal method, show that the set of all positive integer-valued functions of a positive integer cannot be enumerated.
3. If Σ is a finite alphabet, show that Σ^* is enumerable.
*4. If Σ is an enumerable finite alphabet, show that Σ^* is enumerable.
5. Let $\alpha_1, \alpha_2, \ldots, \alpha_n$ be n strings on the alphabet $\{a, b\}$ each of length n. Show how the diagonal method (on a finite two-dimensional array) can be used to construct a string β of length n that is not one of the α_i. Hence, show that $2^n > n$.
6. Find T_1, the first Turing machine.
7. Find T_2, the second Turing machine.
8. Discuss the action of T_1 on any natural number input. Is it a function machine?
9. Discuss the action of T_2 on any natural number input. Is it a function machine?
10. Find the first Turing machine that has an overprint instruction.
11. Find the Turing machine with the smallest code number that is a function machine. Explain why your answer is correct.

12. Prove that the predicate

Left$(x) \equiv (x$ is the code number of an instruction that moves
the pointer left when viewing a certain symbol)

is recursive.

*13. Show that the predicate

$P(x, n) \equiv (x$ is the code number of a Turing machine
with n instructions)

is recursive.

*14. Show that the predicate

NoLeft$(x) \equiv (x$ is the code number of a Turing machine
that has no motions to the left)

is recursive.

*15. Show that the predicate

NoPrint$(x) \equiv (x$ is the code number of a Turing machine
that has no instruction that overprints)

is recursive.

*16. A set A of natural numbers is called **recursive** if its characteristic
function f_A is recursive. Prove that a recursive set is recursively
enumerable.

Summary Outline

- **Turing machine symbols**

$q_1, q_2, q_3, \ldots$	(The states. q_1 is the initial state.)
L, R	(Symbols for moving left or right.)
$0, 1, B, S_3, S_4, \ldots$	(The letters of strings. We use S_0 for 0, S_1 for 1, S_2 for B (blank). S_3, S_4, and so on are optional other letters that are sometimes convenient.)

  The **alphabet** is $\Sigma = \{S_0, S_1, S_2, \ldots\}$. The set $\{q_1, q_2, \ldots\}$ is the **set of states.**

- **Turing machine instructions** and interpretation:
  Each instruction is a **quadruple**—it has four components. The three types
  are as follows:

  1. $q_i\ S_j\ L\ q_k$ (In state q_i, when viewing the symbol S_j, move left and go to state q_k.)
  2. $q_i\ S_j\ R\ q_k$ [As in type 1, but move right.]
  3. $q_i\ S_j\ S_k\ q_m$ (In state q_i, when viewing the symbol S_j, overprint the symbol S_k and go to state q_m.)

- A **Turing machine** is a finite nonempty set of instructions of type 1, 2, or 3 with the condition that no two instructions in this set start out with the same pair (q_i, S_j).

- A Turing machine **halts** if it is in a state q_i, viewing the letter S_j, but it has no instruction that starts with the pair (q_i, S_j).

- Turing machine **macros** are sets of instructions performing some subtask. They may be regarded as subroutines.

- The following functions are called **starter** functions:

  1. $Z: \mathbb{N} \to \mathbb{N}$, the zero function. By definition, $Z(x) = 0$ for all $x \in \mathbb{N}$.
  2. $S: \mathbb{N} \to \mathbb{N}$, the successor function. By definition, $S(x) = x + 1$ for all $x \in \mathbb{N}$.
  3. $p_i^n: \mathbb{N}^n \to \mathbb{N}$, the projection functions. By definition, $p_i^n(x_1, \ldots, x_n) = x_i$.

- **Definition by Substitution** Suppose $f_0(y_1, \ldots, y_k)$ is a function of k variables ($f_0: \mathbb{N}^k \to \mathbb{N}$) and

$$y_1 = f_1(x_1, \ldots, x_n),$$

$$\cdots$$

$$y_k = f_k(x_1, \ldots, x_n).$$

are k functions of n variables. Then the function

$$f(x_1, \ldots, x_n) = f_0(f_1(x_1, \ldots, x_n), \ldots, f_k(x_1, \ldots, x_n))$$

is said to be defined from $f_0; f_1, \ldots, f_k$ by substitution.

- **Definition by Recursion** Let $f_0(x_1, \ldots, x_k)$ and $f_1(x_1, \ldots, x_k, z, y)$ be functions of k and $k + 2$ variables, respectively. Now suppose that $f(x_1, \ldots, x_k, z)$ satisfies the equations

$$f(x_1, \ldots, x_k, 0) = f_0(x_1, \ldots, x_k)$$

and

$$f(x_1, \ldots, x_k, z + 1) = f_1(x_1, \ldots, x_k, z, f(x_1, \ldots, x_k, z))$$

Then f is said to be **defined recursively** from f_0 and f_1.

- **Definition by a Search** Let $f(x_1, \ldots, x_k, y)$ be a function of $k + 1$ variables. Suppose for every $x_1, \ldots, x_k$, the equation $f(x_1, \ldots, x_k, y) = 0$ has a solution y in $\mathbb{N}$. Let $g(x_1, \ldots, x_k) =$ the least value of y such that $f(x_1, \ldots, x_k, y) = 0$. Then g is said to be defined from f by a search, or by a search of y. We write

$$g(x_1, \ldots, x_k) = (\mu y)(f(x_1, \ldots, x_k, y) = 0)$$

(μy may be read "the least y such that").

- **A recursive function** is one that can be defined, in a finite number of steps, from the starter functions using only the operations of substitution, recursion, and searching.

- **Theorem**   A recursive function is Turing-machine computable.

- **A recursive predicate** $P(x_1, x_2, \ldots, x_k)$ is a recursive function whose only values are 0 or 1. If $P = 1$ for the value $x = (x_1, x_2, \ldots, x_k)$, we say that the predicate is true for that value of x; otherwise it is false.

- **Theorem**   If $P(x_1, x_2, \ldots, x_k)$ and $Q(x_1, x_2, \ldots, x_k)$ are recursive predicates, then so are

$$(1) \; \neg P; \; (2) \; P \wedge Q; \; (3) \; P \vee Q; \; (4) \; P \to Q; \text{ and } (5) \; P \leftrightarrow Q$$

- **Bounded quantifiers** are defined by the formulas

$$(\forall x)_y P(x_1, \ldots, x_k, x) \equiv (\forall x)(x \le y \to P(x_1, \ldots, x_k, x))$$

$$(\exists x)_y P(x_1, \ldots, x_k, x) \equiv (\exists x)(x \le y \wedge P(x_1, \ldots, x_k, x))$$

- **Theorem**   Let $P(x_1, \ldots, x_k, x)$ be a recursive predicate. Then

$$Q(x_1, \ldots, x_k, y) = (\forall x)_y P(x_1, \ldots, x_k, x)$$

$$R(x_1, \ldots, x_k, y) = (\exists x)_y P(x_1, \ldots, x_k, x)$$

are recursive.

- **Code numbers** for Turing machine symbols:

State	q_1	q_2	q_3	$\cdots$	q_n	$\cdots$
Code	5	9	13	$\cdots$ $4n + 1$		$\cdots$

Motion	L	R
Code	3	7

Symbol	0	1	B	S_3	$\cdots$	S_n	$\cdots$
Code	11	15	19	23	$\cdots$	$4n + 11$	$\cdots$

- **Code numbers** for a Turing machine:
  If T is given by the sequence

$$X_1, X_2, X_3, X_4; \; X_5, X_6, X_7, X_8; \; \ldots; \; \ldots, X_{4k}$$

of Turing machine symbols, then the code $C(T)$ is defined by

$$C(T) = 2^{c(X_1)} \cdot 3^{c(X_2)} \cdot 5^{c(X_3)} \cdots p_{4k}^{c(X_{4k})}$$

If I is a Turing machine instruction given by the sequence of symbols X_1, X_2, X_3, X_4, then the code $C(I)$ is defined as the number

$$C(I) = 2^{c(X_1)} \cdot 3^{c(X_2)} \cdot 5^{c(X_3)} \cdot 7^{c(X_4)}$$

● **Theorem**   There is a Turing machine that can decide whether a number C is the code number of a Turing machine instruction.

● **Corollary**   There is a Turing machine that takes an integer C as input and gives 0 as output if C is not the code number of an instruction for a Turing machine. If C is the code number of an instruction, the output is the instruction itself.

● **Theorem**   The predicate Tur(C), meaning "C is the code number $C(\textbf{\textit{T}})$ of a Turing machine $\textbf{\textit{T}}$," is recursive. There is a Turing machine that, for input C, computes the value (0 or 1) of Tur(C). There is also a Turing machine that, for input C, has output 0 if Tur$(C) = 0$, but if Tur$(C) = 1$ it will list the instructions of the Turing machine that has code number C.

● A set X is **enumerable** if its elements can be listed in a sequence x_1, x_2, $\ldots, x_n, \ldots$. Every $x \in X$ must be at least one of the x_i, but not necessarily for a unique i. The sequence x_i is called an **enumeration**, or **listing**, of the set X. If $f(i) = x_i$ is a recursive function, then the set X is **recursively enumerable.**

● The set of all infinite sequences of bits (0's or 1's) *cannot* be enumerated.

● **Theorem** (A Universal Turing Machine)   There is one Turing machine $\textbf{\textit{T}}$ that simulates any Turing machine. This means that if (n, α) is taken as the input string (n a positive integer, α a string), the action of $\textbf{\textit{T}}$ is the same as the action of $\textbf{\textit{T}}_n$ with input string α.

● A Turing machine is a **function machine** if for any input n, the action of the machine is such that the machine has an output that is a natural number.

● **Theorem**   The predicate Fcn(n) is *not recursive*. [The predicate Fcn(n) is defined to be true (1) if the nth Turing machine $\textbf{\textit{T}}_n$ is a function machine, and false (0) if not.]

● A Turing machine $\textbf{\textit{T}}$ with input α is written $\langle \textbf{\textit{T}}, \alpha \rangle$. We say that $\langle \textbf{\textit{T}}, \alpha \rangle$ **loops** if the program does not stop. Otherwise, it is said to **halt.** The predicate Halt(n, m) is defined to be true if and only if $\langle \textbf{\textit{T}}_n, m \rangle$ halts.

● **Theorem** (The Halting Problem)   Halt(n, m) is not a recursive predicate.

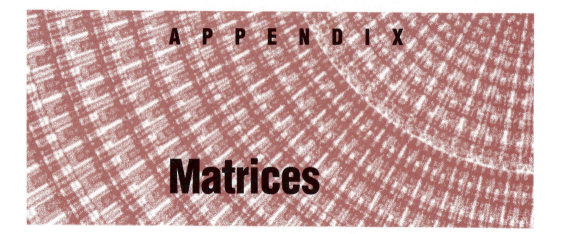

APPENDIX

Matrices

This appendix covers those properties of matrices needed for the material in this text.

DEFINITION A.1 Matrices

If m and n are positive integers, we define a **matrix A** as a function A from $\mathbb{I}_m \times \mathbb{I}_n \to \mathbb{R}$. The value $A(i, j)$ is also written a_{ij}. Here, $1 \leq i \leq m$ and $1 \leq j \leq n$. A is also called an **m by n matrix**. The ordered pair (m, n) is called the **dimension** of the matrix. For fixed i and j, a_{ij} is called an **entry** of the matrix.

We visualize the numbers a_{ij} as a rectangular array of numbers with m rows and n columns. Simply put, the number a_{ij} is in the ith row and jth column of the rectangle.

For example, if

$$a_{11} = 0,\, a_{12} = 3, \quad a_{13} = -8$$

$$a_{21} = 1,\, a_{22} = 4, \quad a_{23} = 5$$

$$a_{31} = 4,\, a_{32} = -1,\, a_{33} = -6$$

$$a_{41} = 9,\, a_{42} = 2, \quad a_{43} = 10$$

then the matrix A is a 4 by 3 matrix, and we can put the numbers a_{ij} in a rectangular array.

$$A = \begin{bmatrix} 0 & 3 & -8 \\ 1 & 4 & 5 \\ 4 & -1 & -6 \\ 9 & 2 & 10 \end{bmatrix}$$

A matrix will be represented by a capital letter, like X. Its entries will then be written x_{ij}, using the lowercase version of the letter.

DEFINITION A.2

A **square** matrix is an n by n matrix for some n. The entries a_{ii} are called the **diagonal entries**. If $i \neq j$, a_{ij} is called off diagonal.

DEFINITION A.3

A square matrix is called **symmetric** if $a_{ij} = a_{ji}$ for all i and j. If this equality does not hold for all i and j, the matrix is called nonsymmetric.

EXAMPLE A.4

The matrix

$$C = \begin{bmatrix} 2 & 3 & 1 \\ 3 & 0 & -9 \\ 1 & -9 & 5 \end{bmatrix}$$

is a square, symmetric, 3 by 3 matrix. Its diagonal entries are 2, 0, 5. An easy way to see whether a square matrix is symmetric is to see whether each row is the same as the corresponding column. Another way is to check that each off diagonal entry is equal to its mirror image about the diagonal.

DEFINITION A.5

A matrix is a (0, 1) matrix if its entries are either 0 or 1. Thus,

$$A = \begin{bmatrix} 0 & 1 & 0 & 0 & 1 \\ 1 & 1 & 1 & 1 & 1 \\ 0 & 1 & 0 & 1 & 1 \\ 0 & 0 & 1 & 0 & 1 \\ 1 & 0 & 1 & 0 & 0 \end{bmatrix}$$

is a (0, 1), nonsymmetric square matrix.

Such matrices are used in the study of relations, and in the study of general directed graphs.

There is an important *algebra* of matrices.

DEFINITION A.6

Let A and B be m by n matrices. Their sum

$$C = A + B$$

is defined by the formula

$$c_{ij} = a_{ij} + b_{ij} \qquad\qquad \text{(A.1)}$$

for all i and j, with $1 \leq i \leq m$, $1 \leq j \leq n$.
This is simply an entry by entry addition. Thus,

$$\begin{bmatrix} 1 & 3 \\ 2 & -4 \end{bmatrix} + \begin{bmatrix} 6 & -1 \\ 2 & 7 \end{bmatrix} = \begin{bmatrix} 7 & 2 \\ 4 & 3 \end{bmatrix}$$

Remark   Only matrices of the same dimension can be added.

DEFINITION A.7

The negative $-A$ of a matrix is the matrix N where $n_{ij} = -a_{ij}$. The zero matrix O is the matrix whose entries are all 0.

Strictly, there are O matrices for each choice of (m, n). However, the context usually makes it clear which dimension O has. The following result shows that there are no surprises as far as matrix addition is concerned.

THEOREM A.8

If A and B are matrices of the same dimension, then

$$A + B = B + A$$
$$A + O = A$$
$$-A + A = O \quad \blacksquare$$

Matrix multiplication is more complicated.

DEFINITION A.9

If A is an m by n matrix and B is an n by p matrix, then the product $D = AB$ is an m by p matrix defined by the formula

$$d_{ij} = a_{i1}b_{1j} + a_{i2}b_{2j} + \cdots + a_{in}b_{nj}$$

or

$$d_{ij} = \sum_{k=1}^{n} a_{ik}b_{kj} \qquad\qquad (A.2)$$

for $1 \le i \le m$, and $1 \le j \le p$.

Multiplication is "row by column" multiplication, since the ith row, jth column of the product is obtained by taking the ith row of A, the jth column of B, multiplying the corresponding entries, and summing the results. Note the new dimensions of the product: (m by n) (n by p) = (m by p).

EXAMPLE A.10

An example of a (3, 2) matrix times a (2, 3) is as follows:

$$\begin{bmatrix} 1 & 2 \\ 2 & 0 \\ 4 & 6 \end{bmatrix} \begin{bmatrix} 1 & 2 & 5 \\ 1 & 1 & 1 \end{bmatrix} = \begin{bmatrix} 3 & 4 & 7 \\ 2 & 4 & 10 \\ 10 & 14 & 26 \end{bmatrix}$$

Here, for example, if the product is D, to find d_{23}, take the 2nd row [2 0] of the first matrix, the 3rd column [5 1] of the second matrix, and multiply corresponding entries and add:

$$d_{23} = 2 \cdot 5 + 0 \cdot 1 = 10$$

This does take time, and people who do intensive matrix multiplications on a computer know this. For a multiplication of two n by n matrices, each of the n^2 entries of the answer is a sum of n numbers, each of which is a product. The sum of n numbers takes $O(n)$ time, and since this must be done n^2 times, the time for a computation of a product is $O(n^3)$, not even counting the time to compute products.

Whenever we write the product AB, we are assuming that multiplication is possible; namely that the number of columns of A is equal to the number of rows of B. Similarly, when matrices are added, we assume that they have the same dimensions.

The following algebraic laws hold for matrix multiplication.

THEOREM A.11

For any matrices A, B, and C, if the additions and multiplications are possible, then

$$A(B + C) = AB + AC$$

$$(A + B)C = AC + BC$$

$$OA = O$$

$$AO = O$$

$$A(BC) = (AB)C \quad \blacksquare$$

Thus, the first of these equations is true if A is m by n and B and C are n by p. Similar dimensional restrictions apply to the other equations.

We have unrestricted addition and multiplication if the matrices under consideration are all square, n by n matrices, so Theorem A.11 holds unrestrictedly for all square, n by n matrices.

We also have an equivalent of a "1" for matrices.

DEFINITION A.12

The n by n matrix I is defined as the matrix whose diagonal entries a_{ii} are all 1, and whose off diagonal entries are all 0. For example (for 3 by 3 matrices), we have

$$I = \begin{bmatrix} 1 & 0 & 0 \\ 0 & 1 & 0 \\ 0 & 0 & 1 \end{bmatrix}$$

As for O, there is an identity for each size square matrix. But usually the context makes it clear which size I is. For square matrices, we have the following additional laws:

THEOREM A.13

For any n by n matrix,

$$IA = AI = A \quad \blacksquare$$

We should point out the laws of algebra which fail.

First, it is possible to define an **inverse matrix** A^{-1} for a wide class of matrices A (these are called nonsingular matrices). For these matrices, we do have $AA^{-1} = A^{-1}A = I$. But it is not true that any nonzero matrix has an inverse. We do not use inverses in this text.

Secondly, the commutative law fails for multiplication of matrices. In general, we can expect $AB \neq BA$, though specific matrices may commute.

Powers of matrices can be defined in a way similar to powers of ordinary numbers.

DEFINITION A.14

If A is any n by n matrix, the matrix A^n is defined for any natural number n by recursion as follows:

$$A^0 = I$$
$$A^{n+1} = A^n \cdot A$$

The following are consequences of this definition.

THEOREM A.15

If A is any n by n matrix,

$$A^m A^n = A^{m+n}$$
$$(A^m)^n = A^{mn} \quad \blacksquare$$

We do not prove this here, but it is a good exercise in induction to prove these. All that is needed is the associative law, and Theorem A.13.

A consequence of this theorem is that the powers of A do commute: $A^m A^n = A^n A^m = A^{m+n}$.

A definition similar to Definition A.14 can be used in any system with an associative multiplication and an identity. For example, it is used to define powers of a string α, where the null string ϵ plays the role of I, and the multiplication is concatenation.

Answers to Odd-Numbered Exercises

Chapter 1

Section 1.1

1. (a) {1, 3};   (b) {1, 2, 3, 4, 6, 8, 10};   (c) {1, 2, 3};   (d) {1, 3}
3. See Figure A1.1 for parts (a) and (b).

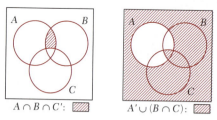

$A \cap B \cap C'$: $A' \cup (B \cap C)$:

Figure A1.1

5. (a) $\emptyset$;   (b) the three-digit even numbers;   (c) integers smaller than 50 or three-digit positive integers;   (d) even positive integers other than the positive three-digit integers.
7. (a) O';   (b) $O \cap H'$;   (c) $O' \cup H$
9. 15
11. j must divide i.
13. A* consists of all strings of 1's, including the empty string. $x =$ "11111".

15. (a) $\{x \mid 0 < x \text{ and } x < 1\}$. The universe is the set of reals $\mathbb{R}$.
    (b) $\{p/q \mid p \text{ is an odd integer, and } q \text{ is an even integer}\}$. The universe is the set $\mathbb{Q}$ of rational numbers.
    (c) $\{X \mid X \subseteq U \text{ and } \mid X \mid \text{ is even}\}$. The universe is the power set $P(U)$.
17. $f(f(n)) = n + 2$, $f(g(n)) = 3n + 1$, $g(f(n)) = 3n + 3$, $g(g(n)) = 9n$. The range of f is $\mathbb{N} - \{0\}$; the range of g consists of the multiples of 3: $\{3n \mid n \in \mathbb{N}\}$.

19.

n	0	1	2	3	4	5	6
$f(n)$	2	5	1	4	0	3	6

21. (a) 0, 1, 3, 7, 15, 31
    (b) $b_n = 2^{2n} - 1$; the first six terms are 0, 3, 15, 63, 255, 1023.
    (c) $c_n = 2^{n+2} - 1$; the first six terms are 3, 7, 15, 31, 63, 127.
    (d) The b_n sequence gives every other term of the a_n sequence. The c_n sequence is the a_n sequence with the first two terms omitted.
23. (a) PROD(EXP(a, b), EXP(a, c)) = EXP(a, SUM(b, c)).
    (b) EXP(PROD(x, y), z)) = PROD(EXP(x, z), EXP(y, z)).
25. (a) g is the reflection of the plane about the line $y = x$.
    (b) If $P = (x, y)$, $g(g(P)) = g(g(x, y)) = g(y, x) = (x, y) = P$.
    (c) All points (x, x); or all points on the line $y = x$.
27. They are in 1-1 correspondence because each string may be regarded as the characteristic function of a subset of $\mathbb{I}_{10}$, as in Example 1.1.36.
29. For $a_0 = 11$, $a_{14} = 1$, $a_i > 1$ for $i < 14$, so 14 computations were made.
31. 01110101110101110101
33. Write $U = \{0000, 0001, 0010, 0011, 0100, 0101, 0110, 0111, 1000, 1001, 1010, 1011, 1100, 1101, 1110, 1111\}$. Then the required set consists of the fourth, sixth, seventh, tenth, eleventh, and thirteenth terms. Its characteristic function is 0001011001101000.
35. (a) $A = \{x \mid x \in \mathbb{I}_{10} \text{ and } x \text{ is even}\}$; $B = \{x \mid x \in \mathbb{I}_{10} \text{ and } x \text{ is a perfect square}\}$; $C = \{x \mid x \in \mathbb{I}_{10} \text{ and } x \neq 6\}$.
    (b) $A = \{2, 4, 6, 8, 10\}$; $B = \{1, 4, 9\}$; $C = \{1, 2, 3, 4, 5, 7, 8, 9, 10\}$.

37. $$\begin{bmatrix} 0 & 1 & 2 & 3 \\ 1 & 2 & 3 & 0 \\ 2 & 3 & 0 & 1 \\ 3 & 0 & 1 & 2 \end{bmatrix}$$

Section 1.2

1. p: You take care; q: You get into trouble; $p \rightarrow \neg q$.
3. p: That jewelry is fake; q: I'm a monkey's uncle; $p \vee q$.
5. p: Jane takes the algebra course; q: The algebra course is taught by Mr. Smith; r: Jane will fail the algebra course; $p \wedge q \rightarrow r$.
7. p: John is always happy; q: Peter watches too much television; r: Peter gets unhappy; s: Peter must go for a long walk. $p \wedge (q \rightarrow r \wedge s)$.

9. $(x < y) \rightarrow (x^2 < y^2)$. Sometimes true: true for $x = 1$, $y = 2$; false for $x = -2$, $y = 1$.

11. $(x < 0) \wedge (x > 0) \rightarrow (x$ is even$)$. Always true because $(x < 0) \wedge (x > 0)$ is always false.

13. $(x^2 = 5) \rightarrow (x$ is even$)$. Always true because $x^2 = 5$ is always false when x is an integer.

15. $(x^2 = 4) \rightarrow (x = 2) \vee (x = -2)$. Always true, by algebra. It is true for $x = 2$ and for $x = -2$, because both sides of the implication are true. Otherwise, $x^2 = 4$ is false, so the implication is true.

17. $(x^2 + 1 = 0) \rightarrow (x = 2)$. Always true because $x^2 + 1$ is never 0 for a real number x.

19. $(x > y) \rightarrow (x > 2y)$. Sometimes true: true if $x = 1$, $y = 0$; false if $x = 2$, $y = 1$.

21. (a) This movie is very long and interesting.
    (b) If this movie is about insects, it is not interesting.
    (c) This movie is very long and interesting, and not about insects.

23. (a) $x = 3$ if and only if $x = -3$ or $x^2 = 9$. Sometimes true.
    (b) If $x = 3$, then $x \neq -3$ and $x^2 = 9$. Always true.
    (c) $x^2 = 9$ and $x = 3$ and $x = -3$. Always false, since $x = 3$ and $x = -3$ cannot both be true.

25. (a) If it's raining, I'll stay home. [Or: I'll stay home if it's raining.]
    (b) It's not true that John either walks or sleeps.
    (c) If Frederick goes to the movies, then Denise goes too, but Elise doesn't.
    (d) If Beth studies math, then she won't watch TV or she'll eat a lot of popcorn.

27. (a) If x is divisible by 6, then it's even.   (b) If the square of an integer x is 5, then x must be 7.

29. (a) p and q are both false.   (b) p is true and q is false.

31.

p	q	$p \wedge q \rightarrow \neg q$
F	F	T
F	T	T
T	F	T
T	T	F

33.

p	q	$p \rightarrow p \rightarrow p \rightarrow q$
F	F	T
F	T	T
T	F	F
T	T	T

35.

p	q	r	$(p \rightarrow q) \rightarrow ((q \rightarrow r) \rightarrow (p \rightarrow r))$
F	F	F	T
F	F	T	T
F	T	F	T
F	T	T	T
T	F	F	T
T	F	T	T
T	T	F	T
T	T	T	T

37.

p	q	r	$\neg(p \to r) \vee (q \wedge \neg p)$
F	F	F	F
F	F	T	F
F	T	F	T
F	T	T	T
T	F	F	T
T	F	T	F
T	T	F	T
T	T	T	F

39.

p	q	r	$p \wedge (q \to (r \wedge \neg p))$
F	F	F	F
F	F	T	F
F	T	F	F
F	T	T	F
T	F	F	T
T	F	T	T
T	T	F	F
T	T	T	F

41.

s	t	u	$s \to \neg(t \to (s \vee u))$
F	F	F	T
F	F	T	T
F	T	F	T
F	T	T	T
T	F	F	F
T	F	T	F
T	T	F	F
T	T	T	F

43. $\neg(p \vee q)$ or $\neg p \wedge \neg q$

45. True for $x = 1$ because $x < 3$ and $x < 9$ are true in this case. True for $x = 5$ and $x = 10$, because $x < 3$ is false in this case. It is true for all x. When $x < 3$ is true, then, since $3 < 9$, $x < 9$ is also true. When $x < 3$ is false, the implication is automatically true.

47. Since $1 \in A$, we have $\neg(1 \in B)$. Thus, $1 \notin B$. $2 \in A$ is false. Therefore $\neg(2 \in B)$ is false, so $2 \in B$ is true. Similarly, $4 \in B$, $5 \in B$, $3 \notin B$. Thus $B = \{2, 4, 5\}$.

49. If $(p \wedge q \wedge \neg r) \to (r \vee s \vee t)$ is false, $(r \vee s \vee t)$ is false, so r, s, and t must be false. Also, $(p \wedge q \wedge \neg r)$ must be true, so p, q, and $\neg r$ must be true. Thus, the implication is false only when p and q are true and r, s, and t are false.

Section 1.3

7. Interchange all $\cap$ and $\cup$ symbols, and $\wedge$ and $\vee$ symbols, in the proof of Theorem 1.3.2.

9. A tautology.

11. Not a tautology. For example, assign truth values T to p and F to q and r.

13. Not a tautology. Assign truth value F to p.

15. Take $m =$ Mary is cutting the chemistry lab; $j =$ John is cutting the chemistry lab. The hypotheses are m, $\neg j \to \neg m$. Using the contrapositive law, we have $m \to j$. Since we have m, we conclude j, using modus ponens.

17. Take i = The book is interesting; j = Joan will finish reading the book; b = The book is about baseball. The hypotheses are $i \rightarrow j$, $\neg b \rightarrow i$, $j \rightarrow b$. Proof of b by contradiction: If $\neg b$, then by modus ponens we obtain i, then j, then b. This contradicts the assumption $\neg b$. Therefore b follows from the hypotheses.

19. Take d = Henry will dance all night; p = Henry goes to the party; c = Henry will be cross. The hypotheses: $p \rightarrow d$, $\neg p \rightarrow c$, $d \rightarrow c$. Proof of c by contradiction: Assume $\neg c$. From $\neg p \rightarrow c$, we obtain p, using the contrapositive. From $p \rightarrow d$ we obtain d, and from $d \rightarrow c$ we obtain c by modus ponens. This is the contradiction.

21. Take h = I have homework; m = I go to the movies; e = Bio class is easy. The hypotheses: $h \rightarrow m$; $\neg e \rightarrow h$; $e \rightarrow m$. Proof by cases: If e, then m follows from $e \rightarrow m$ by modus ponens. If $\neg e$, then modus ponens gives h, then m. Thus m is proved by a case analysis.

23. The conclusion is sound. Take z = Zookies are crunchy; b = Balzoes are malleable. The hypotheses are $\neg z \rightarrow b$, $b \rightarrow z$. If $\neg z$, then applying modus ponens gives b and then z. This is a contradiction, so we must have z.

25. Unsound reasoning. Take a = John gets angry; L = Mary is late; g = Mary forgets gloves. The hypotheses are $L \rightarrow a$, $L \rightarrow g$. Then $g \rightarrow a$ is not a consequence because it is possible to assign truth values F to a and L, and T to g, making the hypotheses true and the conclusion false.

27. Sound reasoning. The conclusion is a restatement of one of the hypotheses.

Section 1.4

1. Some examples: $\neg(p \vee q) \equiv \neg p \wedge \neg q$, and $(p \rightarrow q) \equiv (\neg q \rightarrow \neg p)$
3. Some examples: $p \wedge (q \wedge r) \equiv (p \wedge q) \wedge r$, $p \wedge (q \vee r) \equiv (p \wedge q) \vee (p \wedge r)$
5. Some examples: $p \wedge p \equiv p$, $p \vee p \equiv p$, $\neg\neg p \equiv p$
7. (a) $\neg(\neg p \wedge \neg q)$
   (b) Replace any occurrence of $P \vee Q$ by $\neg(\neg P \wedge \neg Q)$. In this way all occurrences of $\vee$ may be eliminated.
   (c) $p \rightarrow q \equiv \neg p \vee q \equiv \neg(p \wedge \neg q)$; $p \leftrightarrow q \equiv (p \rightarrow q) \wedge (q \rightarrow p)$. Now use the previous equivalence to obtain the equivalence: $p \leftrightarrow q \equiv \neg(p \wedge \neg q) \wedge \neg(q \wedge \neg p)$.
   (d) Using parts (b) and (c), we can eliminate all occurrences of $\leftrightarrow$, $\rightarrow$, and $\vee$ in favor of the connectives $\neg$ and $\wedge$.
9. $p \vee (q \rightarrow r) \equiv \neg(\neg p \wedge \neg(q \rightarrow r)) \equiv \neg(\neg p \wedge q \wedge \neg r)$
11. (a) Both have the same truth table:

p	q	$\cdots$
F	F	T
F	T	F
T	F	F
T	T	F

(b) $p \downarrow p \equiv \neg(p \vee p) \equiv \neg p$
(c) $(p \downarrow q) \downarrow (p \downarrow q) \equiv \neg(p \downarrow q) \equiv \neg\neg(p \vee q) \equiv p \vee q$

13. $p \vee q \equiv \neg(\neg p \wedge \neg q) \equiv \neg(p \mid p \wedge q \mid q) \equiv (p \mid p) \mid (q \mid q)$

15. $P \to Q$ is a formula of the propositional calculus when P and Q are. P logically implies Q is a relation between formulas P and Q. P logically implies Q if and only if $P \to Q$ is a tautology.

17. If P and Q logically imply R, then $P \wedge Q \to R$ is a tautology. But $P \wedge Q \to R \equiv P \to (Q \to R)$ (Exportation Law). Thus, $P \to (Q \to R)$ is a tautology. Therefore, P logically implies $Q \to R$. The converse is similarly proven by using the Exportation Law.

19. 1. p Hyp
    2. $p \to r$ Hyp
    3. r T 1, 2
    4. $r \vee s$ T 3

21. 1. p Assumption
    2. $p \to q$ Hyp
    3. q T 1, 2
    4. $q \to r$ Hyp
    5. r T 3, 4
    6. $r \to \neg p$ Hyp
    7. $\neg p$ T 5, 6
    8. $\neg p$ Discharge Assumption 1, 7

23. 1. p Assumption
    2. $p \to q \wedge r$ Hyp
    3. $q \wedge r$ T 1, 2
    4. q T 3
    5. r T 3
    6. $r \to \neg q$ Hyp
    7. $\neg r$ T 4, 6
    8. $\neg p$ Discharge Assumption 5, 7

25. 1. p Assumption
    2. $p \to q \wedge s$ Hyp
    3. $q \wedge s$ T 1, 2
    4. $s \to t$ Hyp
    5. s T 3
    6. t T 4, 5
    7. $q \to r$ Hyp
    8. q T 3
    9. r T 7, 8
    10. $t \wedge r$ T 6, 9
    11. $p \to t \wedge r$ Discharge Ded 1, 10

27. 1. $\neg p \wedge q$ Assumption
    2. q T 1
    3. $p \vee q$ T 2
    4. $p \vee q \to r$ Hyp
    5. r T 3, 4
    6. $\neg p \wedge q \to r$ Discharge Ded 1, 5

Note that only one hypothesis was needed for this deduction.

29. 1. q Assumption
    2. $p \lor q$ T 1
    3. $p \lor q \to s$ Hyp
    4. s MP 2, 3
    5. $\neg r \to \neg s$ Hyp
    6. r T 4, 5
    7. $q \to r$ Discharge Ded 1, 6

31. (Proof by cases)
    1. t Assumption
    2. $t \to p$ Hyp
    3. p MP 1, 2
    4. $p \lor q$ T 3
    5. $t \to p \lor q$ Discharge Ded 1, 4
    6. $\neg t$ Assumption
    7. $\neg t \to q$ Hyp
    8. q MP 6, 7
    9. $p \lor q$ T 8
   10. $\neg t \to p \lor q$ Discharge Ded 6, 9
   11. $p \lor q$ T 5, 10 (Cases)
   12. $p \lor q \to r$ Hyp
   13. r T 11, 12

33. Valid conclusion:

    1. $p \to q$ Hyp
    2. $\neg p \lor q$ T 1

35. Not valid. Take r, p, and q false.
37. Not valid. Assign truth value F to all variables.
39. Not valid. Take p, q, r, and s false.

Section 1.5

1. $\text{Bro}(x, y) \equiv \text{Sib}(x, y) \land M(x) \land M(y)$
3. $\text{Gramma}(x, y) \equiv F(x) \land \exists z(P(x, z) \land P(z, y))$
5. $\text{Cou}(x, y) \equiv \exists w \exists z(P(w, x) \land P(z, y) \land \text{Sib}(w, z))$
7. $\exists x(\text{Gramma}(x, \text{Mary}) \land \text{Gramma}(x, \text{Alice}))$, using the definition in Exercise 3.
9. Define $\text{Nephew}(x, y) \equiv \exists z(\text{Sib}(y, z) \land P(z, x)) \land M(x)$. The required answer is $\text{Nephew}(\text{Robert}, \text{Matilda}) \land \neg \exists w P(\text{Robert}, w)$.
11. $x < y \lor x = y$
13. $(y = x \lor y = -x) \land (0 = y \lor 0 < y)$
15. $\exists x(0 < x \land x^2 = 2)$
17. $\forall m \forall n((\text{Int}(m) \land \text{Int}(n) \land 0 < n) \to \neg(x = m/n))$
19. $\forall x \forall y(x < y \to \exists m \exists n(\text{Int}(m) \land \text{Int}(n) \land n > 0 \land x < m/n \land m/n < y))$
21. $\exists y(x = 2y)$
23. $\forall x(0 < x \to \exists r \exists s \exists t \exists u(x = r^2 + s^2 + t^2 + u^2))$

25. $\exists x(x^2 = -1)$. Negation: $\forall x(\neg(x^2 = -1))$. The square of any number is not equal to -1.

27. $\exists x \forall y(y \neq x \rightarrow P(x, y))$. Negation: $\forall x \exists y(y \neq x \land \neg P(x, y))$. Any person is not the parent of someone.

29. $\exists x \forall y(y \leq x)$. Negation $\forall x \exists y(x < y)$. Any number is less than some other number.

31. $\exists x \exists y(x \neq y \land x^2 = 9 \land y^2 = 9)$. Negation: $\forall x \forall y(x = y \lor x^2 \neq 9 \lor y^2 \neq 9)$. Either any two numbers are equal or one of them has square unequal to 9.

33. $\exists a \exists b \exists c \forall x(ax^2 + bx + c \neq 0)$. Negation: $\forall a \forall b \forall c \exists x(ax^2 + bx + c = 0)$. Any quadratic equation has a solution.

35. Both statements are true, since $1 + 1 = 2$ regardless of what x you choose.

37. $\forall x(2 \cdot x \neq 1)$ is true for $\mathbb{Z}$, but not for $\mathbb{R}$ or $\mathbb{Q}$.
$\exists x(x \cdot x = 2)$ is true for $\mathbb{R}$, but not for $\mathbb{Q}$ or $\mathbb{Z}$.
$\forall x(x \cdot x \neq 2) \land \exists y(2 \cdot y = 1)$ is true for $\mathbb{Q}$ but not for $\mathbb{Z}$ or $\mathbb{R}$.

39. $\exists M \forall x(x \in A \rightarrow f(x) \leq M)$

41. $\exists t \forall x(x \in A \rightarrow f(x) \leq f(t))$

43. $\mathrm{SQ2}(x) \equiv \exists r \exists s(x = r \cdot r + s \cdot s)$

45. At least one term in the matrix a_{ij} is T.

47. At least one row of the matrix a_{ij} consists only of T's.

49. $\mathrm{Smarter}(x, y) \equiv x$ is smarter than y. $\forall x(\mathrm{Smarter}(\mathrm{John}, x))$. Negation: $\exists x(\neg \mathrm{Smarter}(\mathrm{John}, x))$. John is not smarter than a certain person.

51. $\forall x(\neg \mathrm{Smarter}(x, \mathrm{Jane}))$. Negation: $\exists x(\mathrm{Smarter}(x, \mathrm{Jane}))$. Somebody is smarter than Jane.

53. $\mathrm{Likes}(x, y) \equiv x$ likes y. $\forall x(\mathrm{Likes}(x, \mathrm{Judy}) \land \neg \mathrm{Likes}(x, \mathrm{Karen}) \rightarrow \mathrm{Likes}(\mathrm{Judy}, x))$. Negation: $\exists x(\mathrm{Likes}(x, \mathrm{Judy}) \land \neg \mathrm{Likes}(x, \mathrm{Karen}) \land \neg \mathrm{Likes}(\mathrm{Judy}, x))$. There is someone who likes Judy, and doesn't like Karen, whom Judy doesn't like.

55. (a) $\exists r \exists s \forall t(\neg P(r, s, t) \land \neg Q(r, s, t))$
(b) $\exists x \forall y \exists z(x \geq y \land y \geq z)$
(c) $\forall x \forall y \forall z(x^4 + y^4 \neq z^4)$

57. False. No positive integer is divided by all positive integers. For example, x is not divisible by $2x$.

59. True. 1 divides all integers.

61. False. For example, 3 does not divide 4.

Chapter 2

Section 2.1

1. Not reflexive or symmetric. Is transitive. Not a poset or an equivalence relation.

3. Symmetric. Not reflexive or transitive. Not a poset or an equivalence relation.

5. Not reflexive. Is symmetric and transitive (vacuously). Not a poset or an equivalence relation.

7. The graph is given in Figure A2.1. In the transitive closure R' of the given relation R, $xR'y$ for all x and y. Proof: If $x < y$, then the sequence $x, x + 1, \ldots, y$ shows that $xR'y$. Similarly, if $y < x$, we form the sequence $y, y + 1, \ldots, x$. If $x = y < 10$, $xR(x + 1)$ and $(x + 1)Rx$ shows that $xR'x$. If $x = 10$, the sequence $10, 9, 10$ shows that $10R'10$.

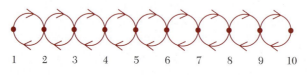

1   2   3   4   5   6   7   8   9   10

Figure A2.1

9. Not reflexive. The graph is given in Figure A2.2. The transitive closure R' is an equivalence relation. $xR'y$ if and only if x and y are an even number of seats away from each other. There are two equivalence classes.

Figure A2.2   **Figure A2.3**

11. $x \text{ PARENT}' y$ if and only if y is a *descendent* of x.

13. Suppose $xR'y$. Then there is a sequence $x, x_1, x_2, \ldots, x_{n-1}, y$ in which successive terms are related by R. The reversed sequence $y, x_{n-1}, \ldots, x_1, x$ has consecutive terms related by R, since R is symmetric. Therefore $yR'x$.

15. See Figure A2.3. No skeleton.

17. See the left graph in Figure A2.4. The symmetric skeleton is on the right.

19. Symmetric, reflexive, and transitive. The full graph is on the left in Figure A2.5. The reflexive-symmetric-transitive skeleton is on the right.

21. Neither symmetric, reflexive, nor transitive. The graph is given in Figure A2.6.

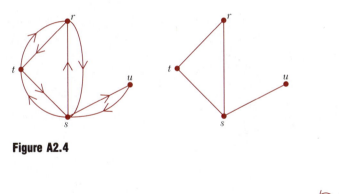

Figure A2.4

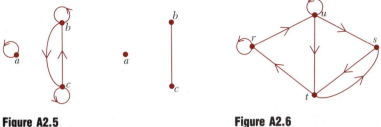

Figure A2.5   **Figure A2.6**

23. If xRy and yRz, then $y = 3^m x$ and $z = 3^n y$, where $n, m \in \mathbb{N}$. Therefore $z = 3^{m+n}x$ and hence xRz. If xRy and yRx, then $z = x$ in this calculation and, hence, $x = 3^{m+n}x$, giving $m = n = 0$. But this shows that $x = y$. Therefore $(\mathbb{N}^+, R)$ is a poset. xCy if and only if $y = 3x$. A graph of C on $\mathbb{I}_{12}$ is given in Figure A2.7.

25. The vertices of the square are the points (x, y), where $x = 0$ or 1, and $y = 0$ or 1. In algebraic terms, $(x, y)R(x', y')$ if and only if $x \leq x'$ and $y \leq y'$. Transitivity: Suppose $p_1 R p_2$ and $p_2 R p_3$. In coordinates this is $x_1 \leq x_2 \leq x_3$ and $y_1 \leq y_2 \leq y_3$. Thus $p_1 R p_2$. Similarly, if $p_1 R p_2 R p_1$, we have $x_1 \leq x_2 \leq x_1$, so $x_1 = x_2$, with similar reasoning for the y coordinates of the points. The graph is indicated in Figure A2.8.

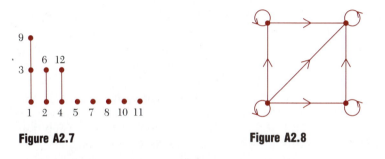

Figure A2.7   **Figure A2.8**

27. As in Exercise 25, except the coordinates (x, y) of a point are arbitrary.

29. The proof that this is a partial ordering is as in Exercise 25 for three coordinates (x, y, z). If $p = (x, y, z)$ and $q = (x', y', z')$, then pCq if and only if $x \le x'$, $y \le y'$, and $z \le z'$, with equality for two of these conditions and strict inequality for the third. For example $(0, 1, 0)C$ $(1, 1, 0)$. The graph is in Figure A2.9.

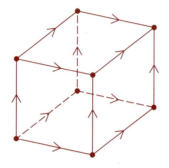

Figure A2.9

31. An equivalence class consists of the people living in a particular state. There are 50 equivalence classes, one for each state. An alternative answer is 51, if the District of Columbia is included.

33. Suppose $x < y$ and $y < z$. Then $x \le y$ and $y \le z$. Therefore $x \le z$. Now suppose $x = z$. Then from $x \le y$ and $y \le z$, we would have $x = y$. But this contradicts $x < y$. Therefore $x \ne z$ and so $x < z$.

Section 2.2

1. f is a function, since a unique value of $\mathbb{Z}$ is assigned for each value of n in $\mathbb{Z}$. It is onto since $n - 1$ is in $\mathbb{Z}$ when n is and $f(n - 1) = n$. It is 1-1 because if $n + 1 = m + 1$, we must have $m = n$.

3. f is a function. It is onto because if $n \in \mathbb{N}^+$, then $n - 1 \in \mathbb{N}$, and $n = f(n - 1)$. It is 1-1 because if $n + 1 = m + 1$, we must have $m = n$.

5. g is 1-1 but not onto. It is 1-1 because if $2n = 2m$, we must have $m = n$. It is not onto because 7, for example, is not in the range of g. h is not 1-1 because, for example, $h(2) = h(3)$. It is onto because for any n, $n = h(2n)$, so n is in the range of h. We have $h(g(n)) = [2n/2] = n$. However, $g(h(n)) = 2[n/2]$. This is not n when n is an odd integer. g and h are not inverses because neither is 1-1 and onto.

7. Computing $f(n)$ for each $n \in \mathbb{Z}_7$, we discover $f(1) = f(2) = 3$. Therefore f is not 1-1. Since we need all seven values for $f(n)$, and 3 is repeated, f is not onto.

9. Let $f: A \rightarrow B$ be onto. For $y \in B$, choose an element $x \in A$ such that $f(x) = y$ and call this element $g(y)$. By definition, $f(x) = y$ or $f(g(y)) = y$. This is the result, except that we used y for a variable element in B.

11. f is onto but not 1-1: $|A| > |B|$; f is 1-1 but not onto: $|A| < |B|$.

13. g must be onto. For if $c \in C$, we have $c = (g \circ f)(a)$ for some $a \in A$. Thus, $c = g(f(a))$ and hence $c = g(b)$ for $b = f(a)$. f need not be onto. To show this, we need an example. Take $A = I_1, B = I_2, C = I_1$. Define $f: A \to B$ and $g: B \to C$ by the formulas $f(x) = 1; g(x) = 1$ for all x. Then $g \circ f$ is onto, but f is not.

15. $\sigma\tau = (124935867); \tau\sigma^{-1} = (179368542); \tau^{-1}\sigma\tau = (1249)(78); \sigma\tau\sigma^{-1} = (3649)(58)$

17. $\tau\sigma\tau^{-1} = (2431)(45)$

19. Let $\rho =$ reverse permutation $= (1\ 100)(2\ 99) \cdots (50\ 51)$. The required permutation is $\rho^{-1}(2\ 43\ 48\ 34\ 51)(4\ 87\ 31)\rho$. This is the permutation $(99\ 58\ 53\ 67\ 50)(97\ 14\ 70)$.

21. $\sigma = (123)^{-1}(123456)(5432)^{-1} = (156)(234)$

23. $I = \sigma_1\sigma_2\sigma_3$, where $\sigma_1 = (13245); \sigma_2 = (1563)(24); \sigma_3 = (364)$. The identity is $I = (13245)(1563)(24)(364)$.

25. $(145867)(345) = (15867)(34) = (15)(18)(16)(17)(34)$. This leads to the sequence

$$\text{ABCDEFGH} \to \text{EBCDAFGH} \to \text{HBCDAFGE}$$

$$\to \text{FBCDAHGE} \to \text{GBCDAHFE} \to \text{GBDCAHFE}$$

27. (a) $(1\ 10)(2\ 9)(3\ 8)(4\ 7)(5\ 6)$
    (b) $(5\ 1\ 2\ 3\ 4)$
    (c) $(7\ 10\ 9\ 8)$
    (d) $(1\ 7\ 3\ 9\ 5)(2\ 8\ 4\ 10\ 6)$
    (e) $(10\ 3\ 4\ 5\ 6\ 7\ 8\ 9)$

29. $F^{-1}\alpha F$. Here, $\alpha = (13), F = (1467)$

Section 2.3

1. $S_n = \sum_{k=1}^{n} \frac{1}{k(k+1)} = \frac{n}{n+1}$. It is true for $n = 0$, since $0 = 0/1$. Assuming the result for n we have, using some algebra,

$$S_{n+1} = S_n + \frac{1}{(n+1)(n+2)}$$

$$= \frac{n}{n+1} + \frac{1}{(n+1)(n+2)} = \frac{n+1}{n+2}$$

This is the result for $n + 1$, proving the result by induction.

3. Let $S_n = \sum_{k=1}^{n} k$. Then $S_0 = 0 = 0 \cdot 1/2$, so the result is true for $n = 0$. Assuming the result for n, we have $S_{n+1} = S_n + (n + 1) = n(n+1)/2 + (n + 1) = (n + 1)(n + 2)/2$, by algebra. This is the result for $n + 1$, and this proves the result by induction.

5. Let $S_n = \sum_{k=1}^{n} k(k + 1)$. Then $S_0 = 0 = 0 \cdot 1 \cdot 2/3$, so the result is true

for $n = 0$. Assuming the result for n, we have

$$S_{n+1} = S_n + (n+1)(n+2) = n(n+1)(n+2)/3 + (n+1)(n+2)$$

$$= (n+1)(n+2)(n/3+1) = (n+1)(n+2)(n+3)/3$$

This is the result for $n + 1$, and this proves the result by induction.

7. We prove that $n(n+1)$ is even for all natural numbers. The result is true for $n = 0$ because $0 \cdot 1 = 0$ is even. Now assume the result for n. By algebra, we have

$$(n+1)(n+2) - n(n+1) = 2(n+1)$$

Therefore $(n+1)(n+2) = n(n+1) + 2(n+1)$, the sum of two even numbers. Therefore $(n+1)(n+2)$ is even. This is the result for $n + 1$, and this proves the result by induction.

9. We prove this by induction on r. It is true for $r = 1$ because all integers are divisible by 1. This is the basis. Now assume that this result is true for r, and let us consider the product $n \cdot (n+1) \cdots (n+r)$ of $r + 1$ consecutive integers. We prove this is divisible by $(r+1)!$ by induction on n (an induction within an induction). It is true for $n = 0$ because 0 is divisible by any integer. Assume that this result is true for some n. Then

$$(n+1) \cdot (n+2) \cdots (n+r)(n+r+1)$$

$$= n \cdot (n+1) \cdots (n+r) + (r+1)(n+1) \cdot (n+2) \cdots (n+r)$$

The first of these summands is divisible by $(r+1)!$ by our induction hypothesis on n. The second of these summands is $(r+1)$ times the product of r consecutive integers. But the product of r consecutive integers is divisible by $r!$, by our induction hypothesis on r. Thus the second summand is also divisible by $(r+1)!$. Therefore the product $(n+1) \cdot (n+2) \cdots (n+r)(n+r+1)$ is itself divisible by $(r+1)!$ because it is the sum of two terms divisible by $(r+1)!$. This proves the result by induction. (The proof is an example of a *double induction*, in which an induction was used within an induction.)

11. The result is true for $n = 0$, as in this case it is simply the statement $1 \geq 1$. Assuming the result for n, we have

$$(1+c)^n \geq 1 + nc \qquad (n \in \mathbb{N})$$

Multiply this inequality by $1 + c$, which is positive, to obtain

$$(1+c)^{n+1} \geq (1+nc)(1+c)$$

$$= 1 + (n+1)c + nc^2$$

$$\geq 1 + (n+1)c$$

This is the result for $n + 1$, proving the result by induction.

13. The result is true for $n = 0$ because both sides are equal to 1. Assuming the result for n, we have

$$(1 - c)^n \le 1 - nc + n(n - 1)c^2/2 \qquad (n \in \mathbb{N})$$

Multiply by $(1 - c)$, which is positive, to obtain

$$(1 - c)^{n+1} \le (1 - nc + n(n - 1)c^2/2)(1 - c)$$
$$= 1 - (n + 1)c + (n + 1)nc^2/2 - n(n - 1)c^3/2$$
$$\le 1 - (n + 1)c + (n + 1)nc^2/2$$

since $c > 0$. This is the result for $n + 1$, and this proves the result by induction.

15. $a_n = n + 1$. True for $n = 0$ because $a_0 = 1 = 0 + 1$. Assuming the result for n, we have $a_{n+1} = a_n + 1 = n + 1 + 1$, which is the result for $n + 1$. This proves the result by induction.

17. $a_n = (3^{n+1} + 1)/2$. Since $a_0 = 2 = (3^1 + 1)/2$, the result is true for $n = 0$. If it is true for n, then

$$a_{n+1} = 3a_n - 1 = (3^{n+2} + 3)/2 - 1 = (3^{n+2} + 1)/2$$

This is the result for $n + 1$, proving the result by induction.

19. (a) By a case analysis, depending on whether n is in the set or not, we have

$$F_n = F_{n-1} + P_{n-2}; P_n = P_{n-1} + F_{n-2} \qquad (n \ge 2)$$

$$F_0 = 1, F_1 = 1; P_0 = 0, P_1 = 1$$

(b) $F_n + P_n$ is the number of subsets of $\mathbb{I}_n$ such that no two elements are consecutive. This has f_{n+2} elements, by Example 2.3.19.

(c) $F_n = P_n = f_{n+2}/2$ for n of the form $6k + 1$ and $6k + 4$ $(n = 1, 4, 7, \ldots)$.
$F_n = (f_{n+2} - 1)/2$, $P_n = (f_{n+2} + 1)/2$, for n of the form $6k + 2$ and $6k + 3$.
$F_n = (f_{n+2} + 1)/2$, $P_n = (f_{n+2} - 1)/2$, for n of the form $6k$ and $6k + 5$.

(d) The result is verified for $n = 0$ and $n = 1$. A detailed induction gives the result for all n. In the induction, six cases are considered according to the remainder of n when divided by 6. For example, suppose $n = 6k + 5$. Then $n - 1 = 6k + 4$ and $n - 2 = 6k + 3$. Thus, using the result for all values less than n, we have, by part (a),

$$F_n = F_{n-1} + P_{n-2} = f_{n+1}/2 + (f_n + 1)/2 = (f_{n+2} + 1)/2$$

$$P_n = P_{n-1} + F_{n-2} = f_{n+1}/2 + (f_n - 1)/2 = (f_{n+2} - 1)/2$$

The other five cases are similar.

21. We prove the inequality for $n \ge 6$. Since $60 < 64 = 2^6$, the result is true for $n = 6$. Assume the result for some n $(n \ge 6)$. Then, since

$1 < n$, we have $n + 1 < 2n$. By assumption, $10n < 2^n$. Multiplying by 2, we obtain $20n < 2^{n+1}$. But $10(n + 1) < 20n$. Thus, $10(n + 1) < 2^{n+1}$. This is the result for $n + 1$, proving the result by induction.

23. (Sketch) By induction. If true for n, adjoin 1 to the set. If 1 is already in it, replace it with 2. Any time two consecutive Fibonacci numbers are in the resulting set, replace them with their sum, the next Fibonacci number. The result is a set of the required type. For example, since $17 = 1 + 3 + 13$, we obtain $18 = 1 + 1 + 3 + 13 = 2 + 3 + 13 = 5 + 13$. Thus, the result will be true for $n + 1$ when it is true for n.

25. $f_2 = 1 \geq \Psi^0$. This is the result for $n = 2$. Similarly, $f_3 = 2 \geq \Psi^1 = 1.6\ldots$. Now suppose $n \geq 4$, and assume the result for all values $m < n$. Then, since $\Psi^2 = \Psi + 1$, we have

$$f_n = f_{n-1} + f_{n-2} \geq \Psi^{n-3} + \Psi^{n-4}$$

$$= \Psi^{n-4}(\Psi + 1) = \Psi^{n-4}\Psi^2 = \Psi^{n-2}$$

This is the result for n.

27. Basis: $0! = 1$. Recursion: $(n + 1)! = (n + 1) \cdot n!$. Using the recursion for $n = 0$, we obtain $1! = 1 \cdot 0! = 1$. Continuing, $2! = 2 \cdot 1! = 2$; $3! = 3 \cdot 2! = 3 \cdot 2 = 6$, and $4! = 4 \cdot 3! = 4 \cdot 6 = 24$.

29. One definition of f^n is given recursively by $f^0 = i$, and $f^{n+1} = f^n \circ f$. We prove $f^{m+n} = f^m \circ f^n$ by induction on n. We have $f^{m+0} = f^m = f^m \circ i = f^m \circ f^0$. Thus the result is true for $n = 0$. If it is true for n, then

$$f^{m+(n+1)} = f^{(m+n)+1} = f^{m+n} \circ f = f^m \circ f^n \circ f = f^m \circ f^{n+1}$$

This is the result for $n + 1$, and thus the result is proved by induction.

31. $SQ^k(n) = n^{P(k)}$, where $P(k) = 2^k$. We shall use $P(k + 1) = 2P(k)$. We prove this formula by induction on k. We have $SQ^0(n) = i(n) = n = n^1 = n^{P(0)}$. This proves the case $k = 0$. Assume the result for k. Then

$$SQ^{k+1}(n) = SQ^k \circ SQ(n) = SQ^k(n^2) = n^{2P(k)} = n^{P(k+1)}$$

This is the result for $k + 1$, and thus the formula is proved by induction.

33. Yes. If $P(n)$ is true, then $P(2n)$ is true. Then, by repeatedly using condition (c), we can prove that $P(n + 1)$ is true. Thus, ordinary induction shows that $P(n)$ is true for all n.

35. By induction on the number of transformations starting with the string "p".
   1. For $n = 1$, the two possibilities both begin with "L" and end with "R", which is the result for $n = 1$. Assuming the result for n, then the $(n + 1)$st step transforms p, which is internal to the string at stage n. Thus, the initial and terminal letters remain the same. This is the inductive step.

2. True for $n = 1$, since each possibility has exactly one "L" and one "R". Assuming the result for n, each of T1 and T2 will introduce one extra "L" and one extra "R". Thus the number of L's is equal to the number of R's at stage $n + 1$.

3. For $n = 0$, $\alpha =$ "p", and we have $x = 1$, $y = z = 0$, so the result is true. Suppose at stage n there are x_n, y_n, and z_n occurrences of "p", "C", and "N", respectively. Assume that $x_n - y_n - z_n = 1$. If T1 is used to get to stage $n + 1$, then $x_{n+1} = x_n + 1$. In this case, $y_{n+1} = y_n + 1$, and $z_{n+1} = z_n$. Therefore $x_{n+1} - y_{n+1} - z_{n+1} = x_n - y_n - z_n = 1$. A similar argument gives the result if the $(n + 1)$st transformation is T2. This proves the result by induction.

37. We prove by induction that if a path starts and ends with the same element, with n internal elements, then all these elements are equal. This is true for $n = 0$, since the path is simply $x \leqslant x$. Assume this for n. If we now have a sequence $x \leqslant x_1 \leqslant \cdots \leqslant x_n \leqslant x_{n+1} \leqslant x$, then $x \leqslant x_{n+1} \leqslant x$. Therefore, by antisymmetry, $x_{n+1} = x$, so the part of the path preceding the last x also begins and ends with x. Therefore all the other terms are x, by induction.

39. If A were not well-ordered, there would be an infinite decreasing sequence x_n ($n = 1, 2, \ldots$) where $x_{n+1} < x_n$ for all n. The set $A = \{x_n \mid n \geq 1\}$ is then a set with no least element, which contradicts the hypothesis. Thus A is well-ordered.

41. Suppose $A \times B$ were not well-ordered. Then there would be an infinite decreasing sequence (a_n, b_n). Since $a_{n+1} \leqslant a_n$, the a_n must be a finite set. Thus, there is some number N such that $a_n = a_N$ for $n > N$. If we now consider the sequence b_n for $n \geq N$, this must be an infinite decreasing sequence in B, contradicting the hypothesis that B is well-ordered. We omit the relatively straightforward proof that $A \times B$ is a linearly ordered poset.

Section 2.4

1. (a) If $2n + 1$ objects are placed in n boxes, then one box must contain at least three objects.
   (b) For $rn + 1$ objects, one box must contain at least $r + 1$ objects.
   (c) For infinitely many objects in finitely many boxes, one box must contain infinitely many objects.

3. If r red chips, g green chips, and b blue chips are placed into n boxes so that no box contains two chips of the same color, and $r + g + b > 2n$, then one box will be occupied by three chips of different colors.

5. By breaking up the square into 20 rectangles, size $\frac{1}{4}$ by $\frac{1}{5}$, we can show that $x = \sqrt{41}/20 \approx 0.32$. This is not necessarily the best possible x.

7. Choose x to form an equilateral triangle as in Figure A2.10. Here we have $x^2 + x^2 = 1 + (1 - x)^2$, and so $x = \sqrt{3} - 1$. The answer is $y = x\sqrt{2} = \sqrt{6} - \sqrt{2} \approx 1.03528$.

9. Two of the points must be in the same rectangle of Figure A2.11. The diagonal of this rectangle is $\sqrt{13}/6 \approx 0.6009$.

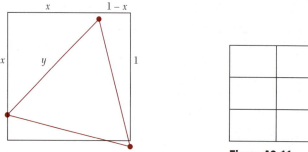

Figure A2.10   **Figure A2.11**

11. (a) Let the numbers $a_1, \ldots, a_{17}$ be called red. Define $b_i = a_i + 3$, and call these green. The a_i's and the b_i's are between 1 and 33 inclusive. Put each of the numbers into a box labeled with its own number. Since there are $34 = 17 + 17$ numbers, one box N contains the same number as red and as green. This means that $N = a_i = b_j = a_j + 3$.

   (b) Call the first three boxes $\{1, 4\}, \{2, 5\}, \{3, 6\}$. Similarly, the next three boxes use the numbers from 7 through 12. These are $\{7, 10\}, \{8, 11\}, \{9, 12\}$. Continuing in this way, the numbers from 1 through 30 can be partitioned into 15 sets, each of which contains two integers differing by 3. Therefore, if there are 16 numbers from 1 through 30, and each is put into the box containing its own number, two will be in the same box, and these will differ by 3.

   (c) The 15 even numbers $2, 4, 6, \ldots, 30$ have the property that no two differ by 3. The result is false for this case.

13. A midpoint of (a, b) and (c, d) is a lattice point if a and c are both even (E) or both odd (O), and similarly for b and d. Any lattice point is one of four types: $(E, E), (E, O), (O, E)$, and (O, O). If each of the five lattice points is assigned such a type, two must have the same type by the Box Principle. These two will have a lattice point as midpoint.

15. Any integer n may be written $2^k N$, where N is odd. We call N the type of n. Then for any two integers of the same type, one must divide the other. For integers between 1 and 100, there are exactly 50 types—the odd numbers in this range. Therefore, for 51 integers, two will have the same type, and one must divide the other.

17. No. Separate the points into two classes A and B of size n, and join any point in A to a point in B. The resulting system has n^2 segments, but no triangle is formed.

19. Let a_i $(i = 1, \ldots, n^2 + 1)$ be a sequence of distinct numbers. Then there is a subsequence of length $n + 1$ that is ascending or is descending. Proof: For each i $(1 \leq i \leq n^2 + 1)$, let r be the length of the largest subsequence that starts with the term a_i and is ascending. Similarly, let s be the largest descending subsequence starting with a_i. Thus, each i is assigned the numbers (r, s). Assuming there are no subsequences of the required type, we must have $1 \leq r, s \leq n$. Therefore there are n^2 possible pairs (r, s). Since there are $n^2 + 1$ possible values of i, it follows that two terms a_i and a_j $(i < j)$ have the same type (r, s). But this is not possible, since a subsequence starting with a_j can have a_i tacked on in front to form a larger ascending, or descending, subsequence starting at a_i.

21. $f(n) = n^2$.

Section 2.5

1. $8352 = (20240)_8 = (231402)_5 = 20A0H$

3. $(10011110110011011)_2 = (10\,011\,110\,110\,011\,011)_2 = (236633)_8$
$$= (1\,0011\,1101\,1001\,1011)_2 = 13D9BH$$
$$= (1\,00\,11\,11\,01\,10\,01\,10\,11)_2 = (103312123)_4$$

5. $4EAH = 1258$

7. $(3452)_7 = (1262) = (1652)_9$

9. $4^7 < x = (10011001)_4 < 4^8$. Thus $\log_{10} x$ is between $7 \log_{10} 4$ and $8 \log_{10} 4$, or between 4.2 and 4.9. The number of decimal digits is thus 5. $D(157654, 2) = 1 + [\log 157654/\log 2] = 18$.

11. Any integer amount up to 121 $(1 + 3 + 9 + 27 + 81)$ can be weighed. We must show that if $1 \leq n \leq 121 = (11111)_3$, n can be written

$$n = a_0 + a_1 \cdot 3 + a_2 \cdot 3^2 + a_3 \cdot 3^3 + a_4 \cdot 3^4, \qquad a_i = -1, 0, \text{ or } 1$$

In the ternary representation of n, we show how to replace all 2's so that the resulting answer has digits 0, 1, or -1. Writing $2 = 3 - 1$, the term $2 \cdot 3^k$ can be written $(3 - 1)3^k = 3^{k+1} - 3^k$. Thus, replace the 2 with -1 and raise the next digit by 1. An illustration is

$$15 = (1\,\underline{2}\,0)_3 \rightarrow (\underline{2}\,-1\,0)_3 \rightarrow (1\,-1\,-1\,0)_3$$

showing that $15 = 27 - 9 - 3$. Similarly,

$$25 = (2\,\underline{2}\,1) \rightarrow (3\,-1\,1) \rightarrow (1\,0\,-1\,1)$$

leading to $25 = 27 - 3 + 1$. The general situation is similar.

13. (All numbers in the computation are understood to be in binary notation.)

Let $\qquad\qquad\qquad\qquad x = 1010 \ldots 10$

Then $\qquad\qquad\qquad 100x = 1010 \ldots 1000$

$$100x + 10 = 1010 \ldots 1010$$

Since $\qquad\qquad\qquad\qquad\qquad\qquad\qquad\qquad x = 1010\ldots10$

we subtract to get $\qquad (100 - 1)x + 10 = 100000\ldots00$

and so $\qquad\qquad\qquad\qquad\qquad\qquad x = (100\ldots00 - 10)/11$
$$(2k + 1 \text{ zeros})$$

In decimal, this is $x = (2^{2k+1} - 2)/3$.

17. E: 127, 0, 73, 0, 73, 0, 73, 0, 65, 0, 0

b: 127, 0, 72, 0, 72, 0, 72, 48, 0, 0, 0

α: 56, 68, 0, 68, 0, 68, 40, 16, 40, 0, 0

19. $(0.1111\ldots)_b = 1/(b - 1)$; $(0.121212\ldots)_3 = 5/8$.

21. $\frac{5}{26}$

23. $(0.1\underline{0110}01100\ldots)_2$

25. $\pm 2^{-33}$

27. If $N = bQ_1 + R_1 = bQ_2 + R_2$, then $b(Q_1 - Q_2) = R_2 - R_1$, so b divides $R_2 - R_1$. But $0 \le R_2 < b$ and $0 \le R_1 < b$. Now multiply this latter inequality by -1 and add to the former. This yields $-b < -R_1 \le 0$, and $-b < R_2 - R_1 < b$. Therefore, $|R_2 - R_1| < b$. Since b divides $R_2 - R_1$, this implies that $R_1 = R_2$. From the equation $bQ_1 + R_1 = bQ_2 + R_2$, we obtain $bQ_1 = bQ_2$. Dividing by b, we obtain $Q_1 = Q_2$. This shows that the R's and the Q's are unique.

Chapter 3

Section 3.1

1. x and z are assigned the initial value of y; y is assigned the initial value of z.

3. 1. Min $:= x$

   2. Loc $:= 1$ [temporarily setting minimum $= x$, located at the first position]

   3. IF $y <$ Min THEN
      1. Min $:= y$
      2. Loc $:= 2$

   4. IF $z <$ Min THEN
      1. Min $:= z$
      2. Loc $:= 3$
      [Now put Min as new value of x]

   5. IF Loc $= 2$ THEN
      1. $y := x$ [moving x to the position of the minimum]
      2. $x :=$ Min

   6. IF Loc $= 3$ THEN
      1. $z := x$ [as in step 5.1]
      2. $x :=$ Min
      [Now switch y and z if necessary]

   7. IF $z < y$ THEN
      1. Min $:= z$

  2. $z := y$

  3. $y :=$ Min [which was the old value of z]

5. $x = 4; y = 1$

7. $n = 5$: 1, 4, 5, 5, 5. $n = 10$: 1, 4, 9, 10, 10, 10, 10, 10, 10, 10.

9. All negative values are changed to 0, and all values larger than 2 are changed to 2.

11. No. The first replaces both a_{ij} and a_{ji} by their minimum. In the second, no check is made on a_{ij} when $j < i$. For example, if $a_{21} > a_{12}$, the first algorithm changes a_{21} to a_{12}. The second doesn't change it.

13. 1. Loc $:= 0$ [initializing, assuming all $a_i \neq 0$]

 2. FOR $i := 1$ to n

  1. IF $a_i = 0$ THEN

   1. Loc $:= i$

   2. End

15. 1. Save $:= a_i$

 2. $a_i := a_j$

 3. $a_j :=$ Save

17. 1. FOR $i := 1$ to $[n/2]$ [swap the term a_i with a_{n+1-i} for $i = 1$ to $[n/2]$]

  1. Save $:= a_i$

  2. $a_i := a_{n+1-i}$

  3. $a_{n+1-i} :=$ Save

19. 1. Min $:= a_i$

 2. Loc $:= 1$ [temporary minimum and location]

 3. FOR $i := 2$ to n

  1. IF $a_i <$ Min THEN [a better candidate]

   1. Min $:= a_i$

   2. Loc $:= i$ [the new location]

 4. FOR $i :=$ Loc -1 down to 1 [push up the beginning of the sequence]

  1. $a_{i+1} := a_i$

 5. $a_1 :=$ Min [OK, since we made room for this term in step 4]

21. 1. FOR $i := 1$ to n

  1. FOR $j := 1$ to n

   1. IF NOT $(a_{ij} = 1$ or $a_{ij} = 0)$ THEN

    1. Output "Not a matrix of a relation"

    2. End [check to see if it is a 0-1 matrix]

 2. FOR $i := 1$ to n

  1. IF $a_{ii} = 0$ THEN

   1. Output "Not a reflexive relation"

   2. End

 3. Output "Relation is reflexive" [matrix passed all of our checks]

23. 1. IF $i = j$ THEN End [not necessary but saves a little time in this case]

 2. FOR $k := 1$ to n [swap a_{ik} and a_{jk}]

  1. Save $:= a_{ik}$

 2. $a_{ik} := a_{jk}$
 3. $a_{jk} :=$ Save
25. 1. FOR $k := 1$ to n [add a_{ik} to a_{jk}]
    1. $a_{jk} := a_{jk} + a_{ik}$

Section 3.2

 1. By Theorem 3.2.8, $f(n) = O(n)$. Since $g(n) = O(n)$, $f(n) + g(n) = O(n)$ by Theorem 3.2.13.
 3. $O(n)$
 5. $O(\sqrt{n})$
 7. $O(\log n)$
 9. Suppose $f \preccurlyeq g$ and $g \preccurlyeq h$. Then $f = O(g)$ and $g = O(h)$. This means that for large n, $f(n) \leq Kg(n)$ and $g(n) \leq Lh(n)$ for certain constants K and L. This gives $f(n) \leq KLh(n)$ for large n, so $f = O(h)$. Thus, $f \preccurlyeq h$, and transitivity is proved. Reflexivity is true since $f = O(f)$.
11. The transitivity of $\prec$ is similar to the proof in Exercise 9 and the proof is omitted. If $f = o(g)$ and $g = o(f)$, then $f = o(f)$ by transitivity. This means that $f(n) < 0.5f(n)$ for large n (choosing $\epsilon = 0.5$ in the definition). This is impossible since f is positive.
13. (a) Since $f \leq f$, we have $f \approx f$.
    (b) Since $f \approx g$ means that $f \preccurlyeq g$ and $g \preccurlyeq f$, it follows that if $f \approx g$ then $g \approx f$.
    (c) If $f \approx g$ and $g \approx h$, then $f \preccurlyeq g$ and $g \preccurlyeq h$, so $f \preccurlyeq h$. Similarly, $h \preccurlyeq f$, so $f \approx h$.
15. No. If we take an oscillatory f such as $f = \sin^2 n + 1/n$, then $f = O(1)$. But f approaches 0 as n goes through the values $k\pi$ ($k = 1, 2, \ldots$). If 1 were $O(f)$, then we would have $1 \leq Kf$, or $f \geq 1/K$ for large n. This is not possible for $n = k\pi$ for sufficiently large k. On the other hand, f is not $o(1)$ since $f \geq 1$ for the values $k\pi/2$ where $k = 1, 2, \ldots$.
17. n^5
19. $\log n$
21. 1
23. n
25. $\log n$
27. $\log(n!) = \log(n(n - 1) \cdots 2 \cdot 1)$
    $= \log n + \log(n - 1) + \cdots + \log 2 + \log 1$
    $\leq \log n + \log n \quad\quad + \cdots + \log n + \log n$
    $= n \log n$
29. Replace the first $n - 1$ terms by $(n - 1)^{n-1}$, to obtain

$$1 + 2^2 + 3^3 + \cdots + n^n \leq (n - 1)(n - 1)^{n-1} + n^n$$

$$\leq n \cdot n^{n-1} + n^n = n^n + n^n = 2n^n$$

Thus the sum is $O(n^n)$.
31. We first show that $\log n < n^p$ for large n. This inequality is equivalent to $\log(\log n) < p \log n$ for large n. But $\log t = o(t)$ by (3.2.10). Thus,

$\log t < pt$ for large t. Putting $t = \log n$, we get the result. This shows that $\log n = O(n^p)$ for any fixed $p > 0$. Therefore, using $p/2$ instead of p, we obtain $\log n = O(n^{p/2}) = o(n^p)$.

33. $2^n/n! = (2/1)(2/2)(2/3)(2/4) \cdots (2/n) \le (2/1)(2/n) = 4/n$ if $n \ge 2$. This shows that $2^n/n! = o(1)$, proving the result. A similar method works for any exponential a^n instead of 2^n.

35. (a) $n^k < a^n$ if and only if $k \log n < n \log a$ or $\log n < (n \log a)/k$. But this follows immediately for large n from $\log n = o(n)$, by choosing $\epsilon = (\log a)/k$ in the definition of little Oh.

   (b) $n^{k+1} < a^n$ for large n, so $n^{k+1} = O(a^n)$. Thus, since $n^k = o(n^{k+1})$, we have the result.

37. For $k = 0$, $n = 2^0 = 1$, and the statement is $1 \ge 1$, which is true. Letting

$$S_k = 1 + \frac{1}{2} + \frac{1}{3} + \cdots + \frac{1}{2^k}$$

we assume the result $S_k \ge 1 + k/2$. Then

$$S_{k+1} = S_k + \frac{1}{1 + 2^k} + \frac{1}{2 + 2^k} + \cdots + \frac{1}{2^{k+1}}$$

$$\ge S_k + \frac{1}{2^{k+1}} + \frac{1}{2^{k+1}} + \cdots + \frac{1}{2^{k+1}}$$

$$= S_k + \frac{2^k}{2^{k+1}} = S_k + \frac{1}{2}$$

$$\ge 1 + \frac{k}{2} + \frac{1}{2} = 1 + \frac{k+1}{2}$$

This is the result for $k + 1$, proving the inequality by induction.

39. Let $S(n) = 1 + 1/2 + \cdots + 1/n$. If $2^k \le n < 2^{k+1}$, we have

$$S(2^k) \le S(n) \le S(2^{k+1})$$

The inequality on n implies

$$k \log 2 \le \log n < (k + 1) \log 2$$

The inequalities in Exercises 38 and 37 show that

$$S(2^k) = O(k) \quad \text{and} \quad k = O(S(2^k))$$

Thus,

$$S(n) \le S(2^{k+1}) = O(k + 1) = O(k) = O(\log n)$$

and

$$\log n = O(k) = O(S(2^k)) = O(S(n))$$

Thus $S(n)$ and $\log n$ have the same order of magnitude.

41. If n has $d = d(n)$ binary digits, then $2^{d-1} \le n < 2^d$. Taking logarithms to the base 2, we get $d - 1 \le \lg n < d$. Therefore $0 < d(n) - \lg n \le 1$. Thus, $d(n) - \lg n = O(1)$ or $d(n) = \lg n + O(1)$.

Section 3.3

1. Setting $n/b^r = \sqrt{n}$, we have $b^r = \sqrt{n}$, so $r = (\log n)/(2 \log b)$. We now take $R = [(\log n)/(2 \log b)]$ terms in this series, underestimate the sum by replacing the first R terms by the Rth term, and ignore the rest. We then obtain $S(n) \geq R \log(n/b^R)$. In this latter term, R has order of magnitude $\log n$, and n/b^R has order of magnitude $\sqrt{n}$. Therefore, the term $R \log(n/b^R)$ has order of magnitude $(\log n)(\log \sqrt{n}) = (\log n)^2/2$, which has order of magnitude $(\log n)^2$.

3. The variables in this algorithm are Q, X, and d. The table gives the value of these variables after the indicated step.

Step	Q	X	d
1	—	—	—
2	13	—	—
3	13	ϵ	—
4.1	13	ϵ	1
4.2	3	ϵ	1
4.3	3	B	1
4.1	3	B	3
4.2	0	B	3
4.3	0	DB	3

5.

Step	Q	i	d_0	d_1	k
1	—	—	—	—	—
2	53	—	—	—	—
3	53	0	—	—	—
4.1	53	0	5	—	—
4.2	6	0	5	—	—
4.3	6	1	5	—	—
4.1	6	1	5	6	—
4.2	0	1	5	6	—
4.3	0	2	5	6	—
5	0	2	5	6	1

7. Algorithm SWITCH(i)
   1. IF $a_i < a_{i+1}$ THEN
       1. Save $:= a_i$
       2. $a_i := a_{i+1}$
       3. $a_{i+1} :=$ Save

The outlined BubbleSort algorithm switches from $i = 1$ to $n - 1$, then from 1 to $n - 2$, and so on. The algorithm is

Algorithm BubbleSort
1. FOR $j := n - 1$ down to 1
    1. FOR $i := 1$ to j
        1. SWITCH(i)

9. $O(n^2)$. Each loop takes $O(n)$ time, and there are n loops.
11. The algorithm CheckPlace (S, n) has the sorted sequence $S = \{a_i, 1 \le i \le n\}$ as input, and a number x to be placed into the sequence. The output is the new sequence.

Algorithm CheckPlace(x, S, n)
1. IF $n = 1$ THEN [the basis for this algorithm]
    1. IF $x \le a_i$ THEN
        1. Put x before a_i
    OTHERWISE
        2. Put x after a_i
    2. End
2. Split the sequence into two nonempty parts,
    the lower part L of length l,
    and the upper part U of length u.
3. IF $x \le$ the first (and least) element of U THEN
    1. CheckPlace(x, L, l) [recursion used here]
    OTHERWISE
    1. CheckPlace(x, U, u) [recursion used here]
4. Put the U sequence after the L sequence [reconstituting the upper and lower sequences]
5. End

13. One method is to search from the beginning to find and remove the term x that is out of order [time $O(n)$]. Then consolidate the sequence [time $O(n)$]. Then locate the position where x goes into the sorted sequence [time $O(\log n)$], and insert x into this sequence [time $O(n)$ to push up the terms after x]. The total time is $O(n)$.
15. We define MERGE(A, B, C). A, B, and C are *sorted* sequences of length a, b, and c, and are inputs. The output is the revised sequence C, which is the sequence obtained when A and B are merged and placed at the end of the original C. For convenience, we include the empty sequence $\varnothing$. If the first term x is removed from a sequence X, we call the resulting sequence $X - \{x\}$. If x is put at the end of a sequence X, we call the resulting sequence $X + \{x\}$. The algorithm is

Algorithm MERGE(A, B, C)
1. IF $A = \varnothing$ and $B = \varnothing$ THEN End
2. IF $A = \varnothing$ THEN [a basis step; now adjoin B to the end of C]
    1. WHILE $B \ne \varnothing$
        1. $C := C + \{b_1\}$ [pile B at the end of C]
        2. $B := B - \{b_1\}$
        3. End
3. IF $B = \varnothing$ THEN
    1. MERGE(B, A, C) [recursive call]
4. IF $a_1 \le b_1$ THEN [deciding which term to add to C]
    1. $C = C + \{a_1\}$

   2. $A = A - \{a_1\}$
   3. MERGE(A, B, C) [recursive call]
   OTHERWISE $[a_1 > b_1]$
   4. $C = C + \{b_1\}$
   5. $B = B - \{b_1\}$
   6. MERGE(A, B, C) [recursive call]

17. Algorithm DigitSum(n)
   1. DigitSum$(n) := 0$
   2. FOR $i := 0$ to k
       1. DigitSum$(n) :=$ DigitSum$(n) + d_i$

   Algorithm Divide9(n)
   1. WHILE $n \geq 10$
       1. $n :=$ DigitSum(n)
   2. IF $n = 0$ or $n = 9$ THEN
       1. Divide9$(n) =$ True
   OTHERWISE
       2. Divide9$(n) =$ False

19. Take Palindrome(α) as a function whose values are "Yes" or "No."
The algorithm is

   Algorithm Palindrome(α)
   1. Palindrome$(\alpha) =$ "Yes" [assume true until proven otherwise]
   2. $m = [n/2]$
   3. FOR $i := 1$ to m [now check condition]
       1. IF $\alpha_i \neq \alpha_{n+1-i}$ THEN [α_i is the ith letter in α]
           1. Palindrome$(\alpha) =$ "No"
           2. End

21. Continue halving as indicated:

```
6 56   73   19    4   37   17 28    99   30   23   21   42   34
6 56   73   19    4   37   17 ■ 28   99   30   23   21   42   34
6 56 73 ■■ 19    4   37   17 ■ 28   99   30   23 ■■ 21   42   34
6 56 73 ■■ 19    4 | 37   17 ■ 28   99 | 30   23 ■■ 21   42   34
```

   Now sort the pieces:

```
6   56   73 ■■ 4   19 | 17   37 ■ 28   99 | 23   30 ■■ 21   34   42
6   56   73 ■■    4 17   19   37 ■ 23   28   30   99 ■■ 21   34   42
    4   6   17   19   37   56   73 ■ 21   23   28   30   34   42   99
    4   6   17   19   21   23   28 30   34   37   42   56   73   99
```

23. 1. FOR $i := 1$ to n
       1. IF $a_i = 1$ and $b_i = 1$ THEN [a common element i]
       2. Output "The sets are not disjoint"
       3. End
   2. Output "The sets are disjoint" [passed all tests]

25. 1. FOR $i := 1$ to n
    1. IF $a_i = 1$ and $b_i = 0$ THEN
    2. Output "A is not included in B"
    3. End
   2. Output "A is included in B" [passed all tests]

Chapter 4

Section 4.1

1. 300
3. 81
5. 22,464; 6097
7. 75
9. 120; 200 additional
11. 180; 190 if a number can begin with 0
13. $|A \cup B \cup C| = |(A \cup B) \cup C| = |A \cup B| + |C| - |(A \cup B) \cap C|$
   But $|A \cup B| = |A| + |B| - |A \cap B|$ and
   $$|(A \cup B) \cap C| = |(A \cap C) \cup (B \cap C)|$$
   $$= |A \cap C| + |B \cap C| - |(A \cap C) \cap (B \cap C)|$$
   $$= |A \cap C| + |B \cap C| - |A \cap B \cap C|$$
   Therefore,
   $$|A \cup B \cup C| = (|A| + |B| - |A \cap B|) + |C|$$
   $$- (|A \cap C| + |B \cap C| - |A \cap B \cap C|)$$
   $$= |A| + |B| + |C| - |A \cap B|$$
   $$- |A \cap C| - |B \cap C| + |A \cap B \cap C|$$
15. 5
17. 44
19. 4
21. Let A_1 be the set of strings with a vowel in the first place, and similarly let A_2 and A_3 be the sets with a vowel in the second and third places, respectively. Then
   $$|A_1| = |A_2| = |A_3| = 5 \cdot 26^2 = 3380$$
   $$|A_1 \cap A_2| = |A_1 \cap A_3| = |A_2 \cap A_3| = 5^2 \cdot 26 = 650$$
   $$|A_1 \cap A_2 \cap A_3| = 5^3 = 125$$
   The inclusion-exclusion equality gives
   $$|A_1 \cup A_2 \cup A_3| = 3 \cdot 3380 - 3 \cdot 650 + 125 = 8315$$
23. 1560. In general, $4^n - 4 \cdot 3^n + 6 \cdot 2^n - 4$.
25. $2^{20} = 1,048,576$
27. 5

29. 6

31. $f_{A\Delta B} = f_A + f_B - 2f_A f_B$. Therefore,

$$f_{(A\Delta B)\Delta C} = f_{A\Delta B} + f_C - 2f_{A\Delta B}f_C$$
$$= f_A + f_B - 2f_A f_B + f_C - 2(f_A + f_B - 2f_A f_B)f_C$$
$$= f_A + f_B + f_C - 2f_A f_B - 2f_A f_C - 2f_B f_C + 4f_A f_B f_C$$

with a similar derivation for $f_{A\Delta(B\Delta C)}$.

33. $|A\Delta B| = \Sigma f_{A\Delta B} = \Sigma(f_A + f_B - 2f_A f_B) = \Sigma(f_A + f_B - 2f_{A\cap B})$
$= |A| + |B| - 2|A \cap B|$

35. 96; 768

37. If n is odd, $2n$ has an additional prime factor 2. The Euler formula for $\varphi(2n)$ contains an addition factor $1 - \frac{1}{2} = \frac{1}{2}$ in addition to the factor $F = (1 - 1/p_1) \cdots (1 - 1/p_k)$. Since $\varphi(n) = nF$, we have $\varphi(2n) = 2n(\frac{1}{2})F = nF = \varphi(n)$.

39. $n = 343 = 7^3$ is the only number in the given range.

Section 4.2

1. $C(9, 3) = 84$; $P(9, 3) = 504$

3. 18/7

5. (a) 45;   (b) 25

7. 518

9. 75

11. 10! or 3,628,800; 4!7! or 120,960

13. 5-3-2-2 is more likely

15. $(4C(13, 5) - 40)/C(52, 5) \approx 0.0019654$, or about 1 chance in 500

17. $(9/19)^5 \approx 0.02385$

19. 1191

21. 9

23. 45; 120

25. $C(36, 6) = 1{,}947{,}792$

27. $180/C(36, 6) \approx 0.000{,}092$, or about 1 chance in 10,000

29. $1/C(40, 6) \approx 0.000{,}000{,}261$, or about 1 in four million. The improvement factor is $C(40, 6)/C(36, 6) \approx 1.97$. Your chances are about twice as good in a 36-number lottery compared with a 40-number lottery.

31. Using Equation (4.2.2), we write

$$C(n, r) + C(n, r - 1) = \frac{n!}{r!(n - r)!} + \frac{n!}{(r - 1)!(n - r + 1)!}$$

$$= \frac{n!}{(r - 1)!(n - r)!}\left(\frac{1}{r} + \frac{1}{n - r + 1}\right)$$

$$= \frac{n!}{(r - 1)!(n - r)!} \cdot \frac{n + 1}{r(n - r + 1)}$$

$$= \frac{(n + 1)!}{r!(n - r + 1)!} = C(n + 1, r)$$

33. (a) $C(20, 13)$ or 77,520;   (b) $C(32, 13)$ or over 347 million
35. $C(11, 8)$ or 165 different solutions
37. $C(15, 6)$ or 5005
39. $C(18, 10) - 10 = 43{,}748$
41. 36
43. 40

Section 4.3

1. $a_n = 2^n + 3^n$
3. $a_n = [3^n + 3(-1)^n]/2$
5. $a_n = -1 + [2 \cdot 5^n + 5(-2)^n]/7$
7. $a_n = -2 \cdot 2^n + [5^n - (-2)^n]/7$
9. $a_n = (n - 1)5^{n+1} + 6(-2)^n$
11. $a_n = 7 \cdot 2^n$ if n is even; $a_n = 11 \cdot 2^{n-1}$ if n is odd. An equivalent formula is $a_n = [25 \cdot 2^n + 3(-2)^n]/4$.
13. $a_n = (2 + n)2^{n-1}$
15. $a_n = (2n - 3)3^n + 2^{n+2}$
17. The sequence f_{n+1} starts 1, 1, 2, ... , and the sequence $2f_n$ starts 0, 2, 2, ... Therefore, by inspection, the sequence $f_{n+1} + 2f_n$ starts with 1, 3, 4, ... But this latter sequence is a linear combination of two sequences that satisfy the difference equation $a_n = a_{n-1} + a_{n-2}$. Therefore it satisfies the same difference equation. Since it has the same initial values as g_n, it is equal to it: $g_n = 2f_n + f_{n+1}$.
19. The sequence starts $-6, 9, -3, 15, 9$. It satisfies the difference equation $a_n - a_{n-1} - 2a_{n-2} = 0$ for $n \geq 2$, and its initial conditions are $a_0 = -6$, $a_1 = 9$.
21. $a_n = (n^2 - 3n + 2)/2$
23. $a_n = (1 + n)3^n$
25. $b_n = (An^2 + Bn + C) - (A(n - 1)^2 + B(n - 1) + C)$ is linear in n. Similarly, c_n is the constant $2A$, and $d_n = 2A - 2A = 0$.
27. By algebra, if $a_n = n^k$, then $\Delta a_n = kn^{k-1} +$ lower powers of n. Thus $\Delta^k a_n = k!$, so $\Delta^{k+1} a_n = 0$.
29. \$139,365.07. This uses Equation (4.3.17), with $A = \$1{,}000{,}000$, $r = 0.09$, $N = 10$, and $F = \$250{,}000$. In this case P is negative.
31. The solution of $a_n - 0a_{n-1} = 0$ $(n \geq 1)$ is the sequence $c, 0, 0, 0, \ldots$. This is given by the formula $a_n = c \cdot 0^n$ by our convention that $0^0 = 1$.
33. $(3^{16} + 1)/2 = 21{,}523{,}361$
35. $(5 \cdot 3^{15} + 1)/4 = 17{,}936{,}134$
37. Since $b_n = 2b_{n-1}$ has solution $b_n = b_0(2^n)$, the inequality satisfies $b_n \leq b_0(2^n)$. Since $b_n \geq 0$, this implies $b_n = O(2^n)$.
39. The equality has solution $a(3^n) + b(-1)^n + Kn$ for some constants a, b, and K. Since 1 and n are $O(3^n)$, and $a_n \geq 0$, we have the result.
41. $b_n = O(5^n)$
43. Let $K = \text{Max}(a_0, a_1/2, a_2/4)$. By induction, we can show that $a_n \leq K(2^n)$. This is proved for $n = 3r$, $3r + 1$, and $3r + 2$.

45. $a_n = n + Kn \lg n = n(1 + K \lg n)$ when n is a power of 2.

47. By induction. It is true for $n = 1$. If $n > 1$, then

$$b_n \leq 2b_{\lceil n/2 \rceil} + Kn \leq 2a_{\lceil n/2 \rceil} + Kn = a_n$$

Section 4.4

1. If $f = f(x)$ is the generating function, then

$$f = 2xf + 3/(1 - x) - 2$$

Solving, the generating function is

$$f(x) = \frac{1 + 2x}{(1 - x)(1 - 2x)} = \frac{-3}{1 - x} + \frac{4}{1 - 2x}$$

This gives $a_n = -3 + 4(2^n)$.

3. The generating function f satisfies

$$f = xf + 2x^2 f + x$$

Solving for $f(x)$, we obtain

$$f(x) = \frac{x}{1 - x - 2x^2} = \frac{x}{(1 - 2x)(1 + x)} = \frac{1}{3}\left[\frac{1}{1 - 2x} - \frac{1}{1 - x}\right]$$

This gives $a_n = [2^n - (-1)^n]/3$.

5. The generating function f satisfies

$$f - 5xf + 6x^2 f + 1/(1 - x) = 2 - 3x$$

Solving for $f(x)$, we obtain

$$f(x) = \frac{2 - 3x - 1/(1 - x)}{(1 - 2x)(1 - 3x)} = \frac{1 - 5x + 3x^2}{(1 - x)(1 - 2x)(1 - 3x)}$$

$$= \frac{1}{2}\left[\frac{-1}{1 - x} + \frac{6}{1 - 2x} + \frac{-3}{1 - 3x}\right]$$

This gives $a_n = [-1 + 6(2^n) - 3(3^n)]/2$.

7. $f(x) = \dfrac{x}{(1 - x)^2}$, $g(x) = \dfrac{1}{1 + x}$, and $h(x) = \dfrac{x}{(1 - x)(1 - x^2)}$.

9. $1, -8, 3, 1, -8, 3, 1, -8, 3, \ldots$. The sequence repeats after every three terms: $a_n = a_{n-3}$ $(n \geq 3)$.

11. $C(n + 2, r + 2) = C(n, r) + 2C(n, r + 1) + C(n, r + 2)$

13. To choose $r + k$ coins from n nickels and k pennies, divide into cases, according to the number y of pennies chosen. If y pennies are chosen, then $r + k - y$ nickels are chosen. In this case, there are $C(k, y)$ choices for the pennies, and $C(n, r + k - y)$ choices for the nickels, for a total of $C(k, y)C(n, r + k - y)$ choices. To find the total for all possible cases, sum from $y = 0$ to $y = k$. The result is $C(n + k, r + k)$, the number of ways of choosing $r + k$ coins from among $n + k$. Thus, $C(n + k, r + k) = \displaystyle\sum_{y=0}^{k} C(k, y)C(n, r + k - y)$.

15. Since $(1 + x)^{2n} = (1 + x)^n(1 + x)^n$, we can equate the coefficients of x^n on both sides. On the left, it is $C(2n, n)$. On the right, it is the sum of products $C(n, r)C(n, n - r)$, corresponding to a multiplication of x^r and x^{n-r}. But $C(n, n - r) = C(n, r)$, so $C(n, r)C(n, n - r) = [C(n, r)]^2$.

17. $1/(1 - x)^2$

19. $1/(1 + 2x^2)$

21. $1/(1 + x)^2$

23. The generating function f satisfies the equation

$$f = x/(1 - x)^2 + 2xf - x^2f + 1 - 2x$$

Therefore

$$f(x) = \frac{1 - 2x + [x/(1 - x)^2]}{1 - 2x + x^2} = \frac{1 - 2x}{(1 - x)^2} + \frac{x}{(1 - x)^4}$$

By Equation (4.4.17), this gives $a_n = n + 1 - 2n + C(n + 2, 3)$, or $a_n = (n^3 + 3n^2 - 4n + 6)/6$.

25. We write the coefficients of f, and difference:

f:	0	1	16	81	256	625	1296 ...
$(1 - x) \ f$:	0	1	15	65	175	369	671
$(1 - x)^2 \ f$:	0	1	14	50	110	194	302
$(1 - x)^3 \ f$:	0	1	13	36	60	84	108
$(1 - x)^4 \ f$:	0	1	12	23	24	24	24 ...
$(1 - x)^5 \ f$:	0	1	11	11	1	0	0

Therefore $f(x) = (x + 11x^2 + 11x^3 + x^4)/(1 - x)^5$. As in the proof of Equation (4.4.5), the fifth difference vanishes from $n = 5$ on.

27. Start with $(1 + x)(1 + x^2)(1 + x^4)(1 + x^8) \ldots = 1/(1 - x)$. If m is any odd number, substitute x^m for x to get

$$(1 + x^m)(1 + x^{2m})(1 + x^{4m})(1 + x^{8m}) \ldots = 1/(1 - x^m)$$

Multiply these for all odd m. The left-hand side consists of products $1 + x^t$, where t is of the form $2^k \cdot$odd. But any number is the unique product of an odd number and a power of 2. Thus the left side consists of the product factors $(1 + x^k)$, where $k \geq 1$.

29. $C(30, 4) - 5C(24, 4) + 10C(18, 4) - 10C(12, 4) + 5C(6, 4) = 0$. More generally, if $t \geq 26$,

$$\sum_{k=0}^{5} (-1)^k C(5, k)C(4 + t - 6k, 4) = 0$$

31. The coefficient of x^5 in the expansion of $(1 + x + x^2 + x^3 + x^4)^{13}$. This expression is $(1 - x^5)^{13}(1 - x)^{-13}$, and the required coefficient is $C(17, 12) - 13 = 6175$

Chapter 5

Section 5.1

1. See Figure A5.1.
3. See Figure A5.2.
5. Directed multigraph, as in Figure A5.3.
7. General digraph as in Figure A5.4.

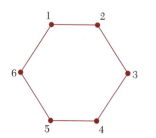

Figure A5.1

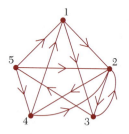

Figure A5.2

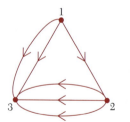

Figure A5.3

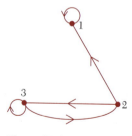

Figure A5.4

9. $$\begin{bmatrix} 0 & 1 & 0 & 1 & 0 & 0 & 0 & 0 & 0 \\ 1 & 0 & 1 & 0 & 1 & 0 & 0 & 0 & 0 \\ 0 & 1 & 0 & 0 & 0 & 1 & 0 & 0 & 0 \\ 1 & 0 & 0 & 0 & 1 & 0 & 1 & 0 & 0 \\ 0 & 1 & 0 & 1 & 0 & 1 & 0 & 1 & 0 \\ 0 & 0 & 1 & 0 & 1 & 0 & 0 & 0 & 1 \\ 0 & 0 & 0 & 1 & 0 & 0 & 0 & 1 & 0 \\ 0 & 0 & 0 & 0 & 1 & 0 & 1 & 0 & 1 \\ 0 & 0 & 0 & 0 & 0 & 1 & 0 & 1 & 0 \end{bmatrix}$$

11. $$\begin{bmatrix} 0 & 0 & 1 & 1 \\ 2 & 1 & 0 & 0 \\ 1 & 1 & 0 & 0 \\ 0 & 0 & 3 & 0 \end{bmatrix}$$

13. $C(2n, 2)$; n^2

15.

$$C + C^2 + C^3 + C^4 = \begin{bmatrix} 1 & 1 & 1 & 1 \\ 1 & 1 & 1 & 1 \\ 1 & 1 & 1 & 1 \\ 1 & 1 & 1 & 1 \end{bmatrix}; \quad C^4 = \begin{bmatrix} 1 & 0 & 0 & 0 \\ 0 & 1 & 0 & 0 \\ 0 & 0 & 1 & 0 \\ 0 & 0 & 0 & 1 \end{bmatrix}$$

17. 17
19. 8
21. Not symmetric
23. Let Σ be a set of vertices in a digraph G. A vertex v is said to be **adjacent** to Σ if it is not in Σ, but $C(w, v) = 1$ for some vertex w of Σ. Let Σ be a nonempty set of vertices in a digraph. Let w be a vertex not in Σ, but $v_0 \approx w$ for some vertex v_0 in Σ. Then there is a vertex v adjacent to Σ. The argument is the same as for digraphs.
25. Suppose G is connected. If i and j are any vertices, there is a path of length $<n$ joining them. Thus the number of such paths is greater than 0. But this number is precisely the ith row and jth column of the matrix $I + C + C^2 + \cdots + C^{n-1}$. Conversely, if there is a path joining any two vertices, the graph is connected.
27. (a) See Figure A5.5.
    (b) Yes. $A \rightarrow F \rightarrow E \rightarrow D \rightarrow C$.
    (c) $A \rightarrow B \rightarrow C \rightarrow D \rightarrow G \rightarrow E \rightarrow F \rightarrow A$
    (d) C

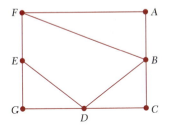

Figure A5.5

Section 5.2

1. At least 75 edges
3. 2
5. S is twice the number of edges. $S \leq v(v - 1)$, where v is the number of vertices.
7. Let G be a directed multigraph such that the in-degree and the out-degree of each vertex of G are the same. Further, assume that if v and w are any two vertices of G, then there is a path joining v to w. Then there is an Euler cycle on G. The proof closely follows the proof of Theorem 5.2.12.

9. Let G be a connected directed multigraph such that the in-degree and the out-degree of each vertex of G are the same, except for vertices v_0 and v_1. For v_0 the out-degree exceeds the in-degree by 1, and for v_1 the in-degree exceeds the out-degree by 1. Then there is an Euler path on G joining v_0 to v_1. The proof closely follows the proof of Theorem 5.2.13.

11. See Figure A5.6.

13. See Figure A5.7.

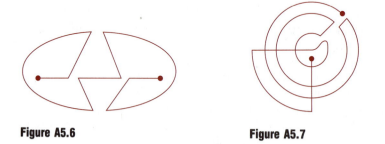

Figure A5.6   **Figure A5.7**

15. Not possible. One vertex has in-degree 1 and out-degree 3.

17. One approach is as in Figure A5.8. The streets are all made one way as indicated. This is consistent with the given directions. The light path can be made one way in either direction. The result is a directed digraph in which the in-degree and out-degree are equal at each vertex. Thus an Euler cycle can be formed by using the techniques of Theorem 5.2.13.

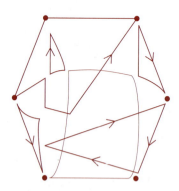

Figure A5.8

19. None is possible. We can label the vertices A and B as in Exercise 21 in such a way that no A vertex is adjacent to a B vertex, and there will be two more A's than B's.

21. Each vertex of a Hamilton cycle alternates between an A vertex and

a B vertex. Since we start and end with an A vertex, the number is the same. A similar argument proves the result for paths.

23. In Figure A5.9, the vertices P and Q each have degree 2. Therefore any Hamiltonian cycle must necessarily include the edges with these vertices and end vertices. The resulting cycle would have to include the indicated edges, which is impossible.

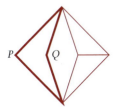

Figure A5.9

25. Make a graph out of the director and all of the parents by putting an edge between any two people who speak the same language. A telephone tree starting and ending with the director is then a Hamiltonian path on this graph. Since the number of vertices is 30, and the degree of each vertex is at least 15, there is a telephone tree by Ore's Theorem.

27. The ith row of D corresponds to the vertex v_i. The row sum is the number of entries d_{ij} that are 1. But each such entry corresponds to an edge e_j that has v_i as end vertex. Thus, the row sum is the number of edges that have v_i as end vertex. By definition, this is the degree of v_i.

29. Choose $\mathbb{I}_m$ as the set V of vertices. For each column k of D, find the two rows i and j for which d_{ik} and d_{ij} are 1. Join i and j by an edge. By definition, the resulting graph G has D as its incidence matrix. To find the matrix C of G, suppose i and j are in V with $i \neq j$. Then define c_{ij} as the number of columns in which the ith and the jth rows are equal to 1. Further define $c_{ii} = 0$ for all $i \in V$. Then C is the matrix of the graph of G.

31. See Figure A5.10.

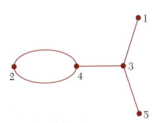

Figure A5.10

Section 5.3

1. 2^{n^2} bipartite graphs and $2^{C(2n,2)}$ graphs
3. Graphs with no edges
5. The graph is bipartite with one more of one color than the other. Hence there is no Hamiltonian cycle on it. A Hamiltonian path is indicated in Figure A5.11.

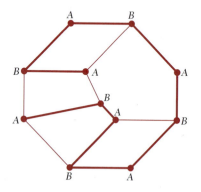

Figure A5.11

7. No Hamiltonian path or cycle exists. The graph is bipartite with two more vertices of one color than the other.
9. $v - e + r = 1 + k$. To form new components, a new operation (d) is needed: Adjoin a new vertex without any new edge. The proof is again by induction. Operation (d) increases v by 1, but it also increases k by 1. Operation (a) increases v and e by 1. Operation (c) increases e and r by 1. Operation (b) has two cases. If the vertices are in different components, e is increased by 1, but k is decreased. If the vertices are in the same component, e and r are both increased by 1. Thus, the equation remains true when any of the operations (a) through (d) are done. In the proof of Theorem 5.3.14, new components were never introduced, because operation (d) was not used.
11. $e = 3v - 3 - p$
13. The Four-Color Theorem is stronger. If you can color a graph with four colors, then the 4-coloring is by definition a 5-coloring.
15. See Figure A5.12. The graph is not bipartite.
17. See Figure A5.13. If the center vertex is colored 1, then the pentagon surrounding it must be colored with at least three colors, all different from 1.
19. See Figure A5.14.
21. See Figure A5.15.
23. It is convenient to introduce a procedure to determine whether a vertex i is adjacent to the set Σ. Define the function $\text{Adj}(i) = 1$ if i

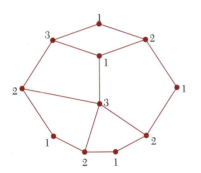

Figure A5.12

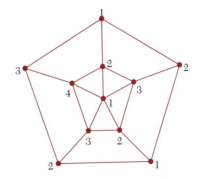

Figure A5.13

Figure A5.14

Figure A5.15

is adjacent to Σ, and 0 if it is not. Then we have the following procedure to compute Adj(i):

Adj(i) Algorithm
1. IF $\sigma(i) = 1$ THEN [i is in Σ]
    1. Adj(i) := 0
    2. End
2. FOR $j := 1$ to n [check each vertex j of the graph]
    1. IF ($c_{ij} = 1$ and $\sigma(j) = 1$) THEN
        1. Adj(i) := 1 [found an adjacent vertex]
        2. End
3. Adj(i) := 0 [could not find an adjacent vertex]

We can now give the required algorithm:
1. FOR $i := 1$ to n [check each vertex of the graph]
    1. IF Adj(i) = 1 THEN
        1. Output "There is an adjacent vertex"
        2. End
2. Output "There is no adjacent vertex" [we checked all vertices, and couldn't find any]

25. Define Consistent(Σ, f) to be true if the condition holds, and false if not. The algorithm is as follows:
    1. FOR $u := 1$ to n [check all vertices]
        1. IF ($\sigma(u) = 1$ and $\sigma(w) = 0$ and $c_{uw} = 1$ and $f(u) \neq 1 - f(w)$) THEN
            1. Consistent(Σ, f) is false
            2. End
    2. Consistent(Σ, f) is true

27. Yes. See Figure A5.16 for a planar representation.
29. Not planar. Here, $e = 13$ and $v = 6$, and $3v - 6 = 12$. Since $e > 3v - 6$, the graph is not planar by Theorem 5.3.16.
31. See Figure A5.17, where a subgraph is not even planar.

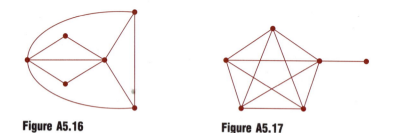

Figure A5.16   **Figure A5.17**

Section 5.4

1. See Figure A5.18.
3. See Figure A5.19. Structurally, the vertex with degree 4 is necessarily connected to all the other vertices. The subgraph spanned by the other four vertices is a graph in which each of the four vertices has degree 2. If vertex A is joined to vertices B and C, as in Figure A5.19, then vertex D cannot be joined to A, so it must be joined to B and C also. Thus the full graph is as indicated in Figure A5.19.

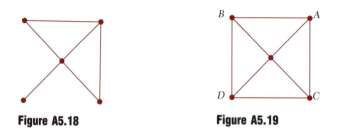

Figure A5.18   **Figure A5.19**

5. Call a vertex B **between** A and C if A, B, and C are all different and ABC is a path. Since all vertices have degree 2, it follows that any vertex is between exactly two others. Therefore, starting with any vertex A_1, we can form a path $A_1 A_2 A_3 \ldots$, where each internal vertex

is between the vertices to its left and its right. Because the number of vertices is finite, this must close up to form a cycle. Then all vertices must be included in this cycle. For if not, since the graph is connected, there would be another vertex adjacent to the vertices in this cycle, and this would imply that some vertex has degree >2. (See Figure A5.20.) This is a contradiction, and hence the graph must be isomorphic to the graph of a polygon as in Figure A5.21.

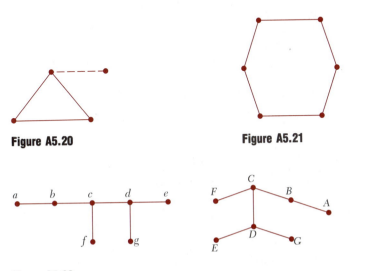

Figure A5.20

Figure A5.21

Figure A5.22

7. Isomorphic. The isomorphism is exhibited in Figure A5.22.
9. Not isomorphic. The graph on the right has two adjacent vertices of degree 2. This is not so for the graph on the left.
11. Figure A5.23 gives two nonisomorphic graphs with the same degree list, 2, 2, 2, 2, 3, 3.
13. In the left graph of Figure A5.23, there is a vertex of degree 2 whose neighbors have degree list 3, 3. But in the right graph, if a vertex has degree 2, the degree list of the neighbors of v is 2, 3. Thus the graphs are not isomorphic by the results of Exercise 12.

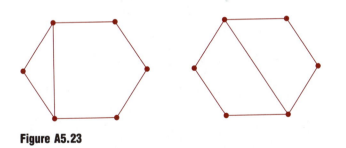

Figure A5.23

15. One possibility is to relabel $4 \to 1, 5 \to 4$, and $1 \to 5$. The resulting graph has matrix

$$\begin{bmatrix} 0 & 0 & 1 & 0 & 0 \\ 0 & 0 & 0 & 1 & 1 \\ 1 & 0 & 0 & 0 & 1 \\ 0 & 1 & 0 & 0 & 1 \\ 0 & 1 & 1 & 1 & 0 \end{bmatrix}$$

It is still possible to relabel vertices 2, 3, and 4, so other matrices are possible.

17. The most economical route is $M \to I \to G \to C$ for cost 205.

19. A possible path is $RDDRDR$, where R and D represent right and down, respectively, for a total of 90.

21. 1. IF $r = s$ THEN End [nothing to do]
    2. FOR $i := 1$ to n [interchange columns r and s]
        1. Save $:= a_{ir}$
        2. $a_{ir} := a_{is}$
        3. $a_{is} :=$ Save
    3. FOR $j := 1$ to n [now the column interchanges]
        1. Save $:= a_{rj}$
        2. $a_{rj} := a_{sj}$
        3. $a_{sj} :=$ Save

    The time for this algorithm is $O(n)$. There is a total of $8n$ assignments in the loops counting loop variables.

23. 1. FOR $i := 1$ to n [find the degree sum for each row i]
        1. $d_i := 0$ [initialize a sum]
        2. FOR $j := 1$ to n [now sum for each entry in the ith row]
            1. $d_i := d_i + c_{ij}$
    2. MERGESORT(d_i) [Now arrange in ascending order]

    The time for this algorithm is $O(n^2)$.

25. Yes. If care is taken with adjacency and if there is a path joining vertex A to vertex B, then the same algorithm works.

Section 5.5

1. A game graph and a strategic labeling are given in Figure A5.24. The first player loses. The second player moves to 3, then 6, then 10.

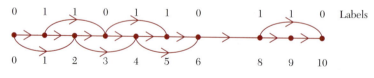

Figure A5.24

3. The first player wins by moving to 2, then 7 (regardless of what the second player does), then 11, and then 16, at which point the second player loses.

5. Change the 9-pile to a 6-pile.

7. 3

9. The (1, 2) position can be won by the usual Nim method of changing the 2-pile to a 1-pile. Similarly for the (1, 3) position. The (2, 3) also can be won by changing the 3 to a 2 and then playing copycat.

11. If each pile has one chip in it, there are no options: You win if there are an even number of chips and lose if there are an odd number. Otherwise, if you are in a winning Nim position, play the usual winning move on the largest pile possible, unless this leaves only one-chip piles. In that case, remove all, or all but one, to leave an odd number of 1-piles, and you then win the reversed game.

13. Since (11, 18) is a 0-position, change the 43-pile to 18.

15. The position (4, 7) is winning, with exactly two possible winning moves into (4, 4) or (2, 5). [The graphical method indicates a symmetry in the 0-positions. These consist of all (m, n) with either (a) $n \equiv 0$ and $m \equiv 0$ mod 4, (b) $n \equiv 1$ and $m \equiv 2$ mod 4, or (c) $n \equiv 2$ and $m \equiv 1$ mod 4.]

17. If the piles are $(n, 0)$, then the rules allow subtracting either 1 or 2 from n, with the object of reaching 0. This is equivalent to the game of ten. The direction of motion is changed, but the games are equivalent.

19. Label the 0-vertices $(-1, 1)$ and the 1-vertices $(1, -1)$. Then the method of Example 5.5.26 will yield a strategic labeling according to Definition 5.5.12.

21. The valued game graph is in Figure A5.25. The best move for A is either to the left or to the right. B's best move is to the right.

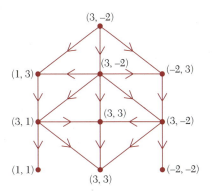

Figure A5.25

23. See Figure A5.26.

25. See Figure A5.27.

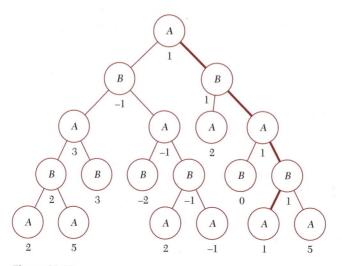

Figure A5.26

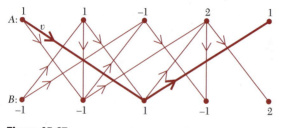

Figure A5.27

27. The Grundy labeling $g(v)$ is defined on vertices as follows:

1. $\{v_1, v_2, \ldots, v_k\} :=$ set of end vertices
2. FOR $i := 1$ to k
   1. $g(v_i) := 0$ [initializing g on the end vertices]
3. $L := \{v_1, v_2, \ldots, v_k\}$
4. WHILE $L \neq G$ [enlarge L and define g on the enlargement]
   1. $v :=$ a vertex in $G - L$ such that $L \cup \{v\}$ is a lower set
   2. $g(v) :=$ least natural number n such that $g(w) \neq n$ for all w such that vw is an edge of G
   3. $L := L \cup \{v\}$

29. See Figure A5.28.

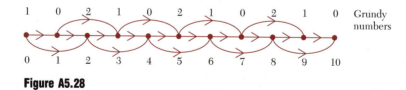

Figure A5.28

31. If the Grundy label of a vertex v is >0, there is an edge vw such that the Grundy label of w is 0. If the Grundy label of a vertex v is 0, then for each edge vw, the Grundy label of w is >0. The Grundy label of an end vertex is 0. Thus, if all positive labels are converted to 1, the resulting labeling is a strategic labeling by Definition 5.5.12.

33. Treat the Grundy numbers as Nim piles. Using the strategy in Nim, decide which number to reduce, as in Nim. Go to the corresponding game and move to a position with that Grundy number. This is possible by the definition of the Grundy numbers.

35. A Nim pile of size n can be changed to one of size x, where $0 \leq x \leq n - 2$, except that if $n = 1$, $x = 0$ is possible. A little experimentation shows that the Grundy numbers of piles of sizes 1, 2, 3, 4 are, respectively, 1, 1, 2, 2. This suggests the general formula $g(n) = [(n + 1)/2]$, which can be proved by induction. The given piles of sizes 4, 6, 12 had Grundy numbers 2, 3, 6. The Nim strategy indicates that Grundy number 6 should be changed to 1. This means that, for the original pile of chips, you must remove 10 or 11 from the 12-pile.

Section 5.6

1. See Figure A5.29. The maximum flow is 12.
3. See Figure A5.30. The maximum flow is 11.
5. Labeling the vertices as indicated, an a-path is $CbcK$, as in Figure A5.31.
7. There is no a-path.

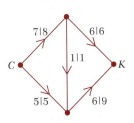

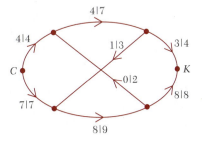

Figure A5.29

Figure A5.30

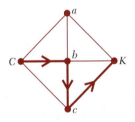

Figure A5.31

9. The sum of the outflows at the sources is equal to the sum of the inflows at the sinks. The proof is similar to the proof of Theorem 5.6.4.

11. This allows any $f(e) \geq 0$ in a flow. An edge e is allowable if it is in N, or if e' is in N with $f(e') > 0$.

13. If v_0 is any vertex and f is an algebraic flow, use the notation Inflow(v, f) to denote the inflow of f at the vertex v. By definition Inflow$(v, f) = \Sigma f(wv)$, the sum over all w. Then we have

$$\text{Inflow}(v, f + g) = \Sigma(f(wv) + g(wv)) = \Sigma f(wv) + \Sigma g(wv)$$

$$= \text{Inflow}(v, f) + \text{Inflow}(v, g)$$

with a similar definition and proof for the outflow. From this it follows that if v is not the source or the sink, then

$$\text{Inflow}(v, f + g) = \text{Inflow}(v, f) + \text{Inflow}(v, g)$$

$$= \text{Outflow}(v, f) + \text{Outflow}(v, g)$$

$$= \text{Outflow}(v, f + g)$$

Similarly, Outflow$(C, f + g) =$ Outflow$(C, f) +$ Outflow(C, g), so the value of $f + g$ is the sum of the values of f and g.

15. Introduce the function Lee(v) for labeled vertices. Add the following instructions immediately after step 1.1.1:

2. IF $v = C$ THEN Lee$(v_i) := L$
3. IF $v \neq C$ THEN
   1. IF Lee$(v) \leq L$ THEN
      1. Lee$(v_i) = $ Lee(v)
      OTHERWISE
      2. Lee$(v_i) := L$ [Thus, Lee$(v_i) = \min($Lee$(v), L)$]

It is necessary to renumber the original step 1.1.1.2 (IF $v_i = K$ THEN) as step 1.1.1.4.

17. Assume the vertices are labeled $v_0, v_1, \ldots, v_n$.

Increase(f) Algorithm
1. APath$(f; \alpha)$
2. IF there is no a-path THEN
   1. OUTPUT "No increase is possible"
   2. End
3. $b := $ Lee(K) [as defined in Exercise 15]
4. $g := $ Path(α, b) [using the algorithm of Exercise 16] [now add $g = b \cdot \alpha$ to f]
5. FOR $i := 0$ to n
   1. FOR $j := 0$ to n [through all pairs of vertices]
      1. IF $v_i v_j$ is an edge of N THEN
         1. $f(v_i v_j) := f(v_i v_j) + g(v_i v_j)$

19. *CbdeK* and *CcbdeK* as in Figure A5.32.

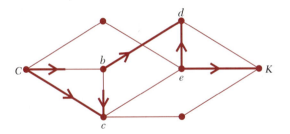

Figure A5.32

Chapter 6

Section 6.1

1. See Figure A6.1. There are seven leaves.

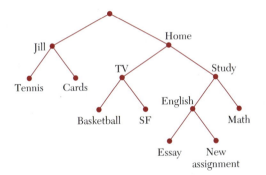

Figure A6.1

3. No. The married pair becomes identified as one vertex that will have two different paths leading to it.
5. See Figure A6.2. There are 13 such strings of length 3. There are 2745 paths of length 10. Use a recursion for a_n = the number of such paths of length n. The recursion is $a_n = a_{n-1} + 2a_{n-2} + a_{n-3}$.

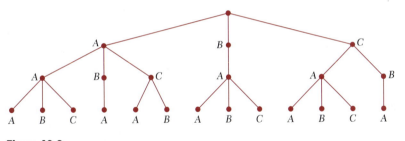

Figure A6.2

7. $((x + y)*y) + z$
9. See Figure A6.3.
11. See Figure A6.4.
13. $p \leftrightarrow \neg\neg p$
15. See Figure A6.5.
17. See Figure A6.6.

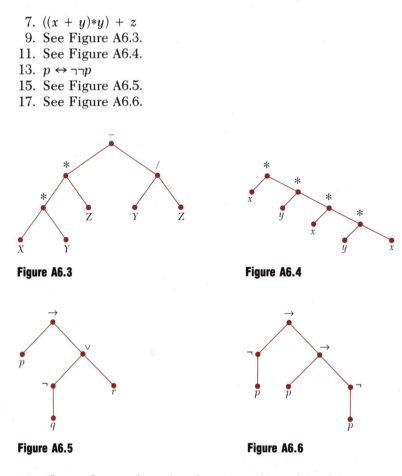

Figure A6.3   **Figure A6.4**

Figure A6.5   **Figure A6.6**

19. If g_n is the number of such strings of length n, then a case analysis, depending on whether the string starts with A or with B, yields the recursion $g_n = g_{n-1} + g_{n-2}$ for $n \geq 2$. Initial conditions are verified to be $g_0 = 1$ and $g_1 = 2$. An induction now shows $g_n = f_{n+2}$.

Section 6.2

1. No for each case. See Figure A6.7 for counterexamples.

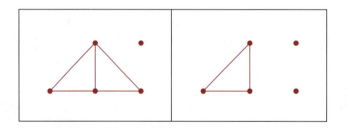

Figure A6.7

3. $n - k$
5. Suppose the cycle is $P_1 P_2 \cdots P_n P_1$ and edge $P_1 P_2$ is removed. Let P and Q be any two vertices in the graph. By hypothesis, they can be connected in the original graph. If the connection uses edge $P_1 P_2$, replace that edge by $P_1 P_n P_{n-1} \cdots P_2$, as in Figure A6.8. Similarly, replace $P_2 P_1$ by $P_2 P_3 \cdots P_n P_1$. The result is a connecting path (with possible repeating vertices) not using edge $P_1 P_2$.

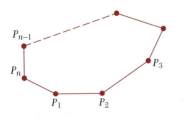

Figure A6.8

7. We first find whether C is the matrix of a connected graph G. If so, we find the number e of edges of that graph and check whether $e = n - 1$. We compute e by the formula $2e = S =$ the sum of the degrees of the vertices $=$ the sum of all the entries in the matrix. This latter sum is directly computed by two simple FOR loops. The algorithm is

   1. IF G is not connected THEN
      1. Output is "no tree"
      2. End [here we've used the Connected Algorithm]
   2. $S := 0$ [initializing S]
      1. FOR $i := 1$ to n
         1. FOR $j := 1$ to n
            1. $S := S + c_{ij}$ [$S = 2e$]
   3. IF $S = 2(n - 1)$ THEN
      1. Output is "tree"
     OTHERWISE
      2. Output is "no tree"

9. Each component C_i of G is a tree. If C_i has e_i edges and v_i vertices, then $e_i = v_i - 1$ by Theorem 6.2.10. We now sum these equations and use $\Sigma e_i = e$ and $\Sigma v_i = v$. This gives the result $e = v - k$, where k is the number of components of G. Thus, $e \le v - 1$, since $k \ge 1$.

11. Suppose an edge $f = PQ$ is removed and replaced by e as suggested, forming a subgraph H'. We can see that H' still spans, because any connecting path in H that used the edge PQ can be changed into a connection in H' that uses e. We simply replace PQ with the path α joining P to Q as in Figure A6.9. Thus H' is connected and still has all the vertices of G. Since it still has $n - 1$ edges, it is a tree.

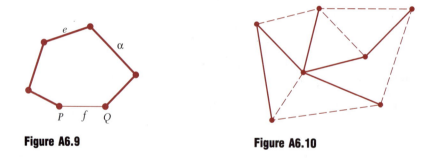

Figure A6.9   **Figure A6.10**

13. See Figure A6.10. The minimal weight is 53.
15. 1. List the edges of G in order of decreasing weights. In case of a tie, any order of edges with the same weight is permissible.
    2. Starting with the full graph G, remove edges in the order fixed by step 1 to form a subgraph H. (Do not remove any vertices.) However, if removing an edge causes the resulting graph to be disconnected, do not remove that edge, and go on to the next edge.
    3. Proceed with the removals of step 2 until all edges have been tested.

    By Theorem 6.2.13, the resulting graph H is a spanning graph. The proof that it is a spanning graph with least weight is similar to the proof of Kruskal's Algorithm.
17. The edges are eliminated in the order shown in Figure A6.11a. The resulting minimal spanning graph is in Figure A6.11b.

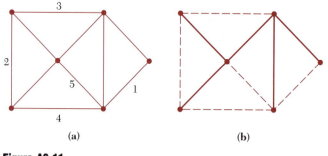

(a)   (b)

Figure A6.11

19. By construction, the resulting graph $H = (\Sigma, E)$ is, at all stages, connected. Since G is connected, there are always adjacent vertices to H, so the process continues until all vertices of G are reached. Initially, there is one vertex and no edges. Since one edge is adjoined each time a vertex is adjoined, the number of edges is one less than the number of vertices. This proves that H is a spanning tree.

To show that H is a spanning tree with minimal weight, we show that at each step of the procedure, $H = (\Sigma, E)$ constitutes a **subgraph** of some spanning tree M of least weight. This is true in the beginning, since H is then a 1-vertex, no-edge graph. We now show that this continues to be true after we adjoin a new edge and vertex. Suppose, then, that E is a subset of the edges of some spanning graph M of minimal weight. In Prim's Algorithm, suppose edge uv is adjoined, where $u \in \Sigma$ and v is adjacent to Σ.

If uv is an edge of M, then the extension of H is still a subgraph of M. If not, the path in M starting from u to v must exit at some vertex u_0 of Σ to some vertex v_0 adjacent to Σ, as in Figure A6.12. But Prim's Algorithm chose uv and not $u_0 v_0$. This means that the weight $w(uv) \leq w(u_0 v_0)$. By replacing uv with $u_0 v_0$ in M, we obtain a spanning tree M' whose weight is not larger than M. M' is therefore a spanning tree of minimal weight. H continues to be a subgraph of a minimal spanning tree M'. This proves the result, since at the final step H is itself a spanning tree, and if H is a subgraph of the spanning tree M, we must necessarily have $H = M$.

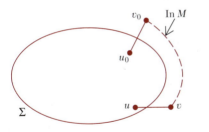

Figure A6.12

21. Starting at the vertex P of Figure A6.13a, the edges are adjoined in the indicated order to form the spanning tree of minimal weight in Figure A6.13b.

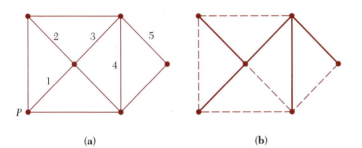

(a)   (b)

Figure A6.13

23. The greedy strategy gives the string $CACACA$ with value 36. For the best strategy, note that BB has value 14, and so is better than CA in a string. The strategy is to avoid using C until the end, yielding $BBBBBC$ for a maximum value of 45.

25. In H, we have $v \approx w \approx v'$. Since $u \approx v$ for any element u of G, and $u' \approx v'$ for any element u' of G', it follows that H is connected, since any elements can be connected through w. To show that H is a tree, suppose there were a cycle α on H, where $\alpha = P_0 P_1 \cdots P_n$.

Suppose, first, that w is *not* a vertex of α. Then if P_0 were in G, each P_i would be in G (by induction on n), so α would be a cycle in G, which is impossible since G is a tree. Similar reasoning would apply if P_0 were in G'.

On the other hand, if w *were* a vertex of α, then, by reversing the direction of α if necessary, we could obtain a cycle of the form $\beta = \ldots P_2 P_1 v w v' P_1' P_2' \ldots$, where the P_i are in G and the P_i' are in G'. (See Figure A6.14.) But then β cannot be a cycle in this case, too, since it can't close up. The initial vertex of β, in G, is not its final vertex, which is in G'.

A less direct proof is possible using Theorem 6.2.10.

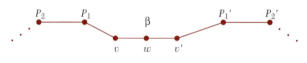

Figure A6.14

Section 6.3

1. (a) $E + xy - x/yz$;   (b) $xy + xyz/ - E$

3. $\leftrightarrow \vee \rightarrow pqq \rightarrow pq$

5. Algebraic: $(((((a + b)/c) + d)/e) + f)$; reverse Polish notation is: $ab + c/d + e/f +$

7. $(p \rightarrow q \vee \neg s) \vee \neg p$

9. We shall show, by induction, that if w is an nth-generation descendent of v, then $v \approx w$. It is true for $n = 0$ since $v \approx v$. Assuming it true for n, suppose w is an $(n + 1)$st-generation descendent of v. By definition w is the child of some w', where w is an nth-generation descendent. Therefore $v \approx w' \approx w$. Thus the descendents of v are connected.

No cycles are possible in the graph spanned by the descendents since any such cycle would be a cycle in the full tree, which is impossible. Thus the descendents form a subtree of the original tree.

11. Defining the weight of a term a_n to be $1 - a_n$, the resulting sequence must give a tree by the List/Tree Algorithm.

13. Let T_i be the tree associated with the tree sequence S_i $(1 \le i \le k)$. Introduce a new vertex v and graft the trees T_i onto v. The resulting tree will then have the tree sequence $k, S_1, \ldots, S_k$.

15. Basis: 0 is a tree sequence. Recursion: A sequence is a tree sequence if it is a sequence $k, S_1, \ldots, S_k$, where $S_1, \ldots, S_k$ are tree sequences. In Exercise 11, choose $k = 3$ and (writing the sequences as strings) $S_1 = 2010$; $S_2 = 2103000$; $S_3 = 0$. Here S_1 is a tree sequence since 0 and 10 are. S_2 is a tree sequence since 10 and 3000 are, and S_3 is the basis tree sequence.

17. See Figure A6.15.

19. Not a preorder traversal for a tree. The sum of the weights is not 1.

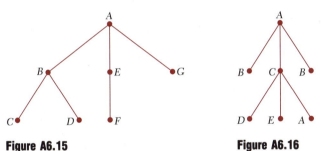

Figure A6.15   **Figure A6.16**

21. See Figure A6.16.

23. $\vee \to p \,\neg\, q \to p \vee q \, s$

25. $acbfehigd$

27. A recursive algorithm is as follows. The input is a tree T; the output is a list L of its vertices.

ReversePolish(T; L) Algorithm
1. IF T has only one vertex v THEN
    1. $L := v$
    2. End [the basis]
2. IF T has root v THEN
    1. $k :=$ the number of children of v
    2. FOR $i := 1$ to k
        1. $v_i :=$ ith child of v
        2. $T_i :=$ the tree of descendents of v_i
        3. ReversePolish(T_i; L_i) [a recursive call; now tie the lists together]
    3. $L := L_1$ [initialize]
    4. FOR $i := 2$ to k
        1. $L := L, L_i$ [adjoining the other lists]
    5. $L := L, v$ [tossing v at the end, the idea of reverse Polish notation]

Section 6.4

1. $g_n = -\dfrac{1}{3} + \dfrac{2\sqrt{3} - 3}{3\sqrt{3}}(1 - \sqrt{3})^n + \dfrac{3 + 2\sqrt{3}}{3\sqrt{3}}(1 + \sqrt{3})^n$
3. 822,678,869
5. $2^{2^n} - 1$
7. $2^{3^n} - 1$
9. 866. The recursion is $G_n = G_{n-1}^2 + G_{n-2}^2$, where G_n is the number of such graphs with leaves on level n.
11. 16 leaves; 39 vertices
13. The best such cluster is $A \oplus (B \oplus [E \oplus (D \oplus C)])$, whose graph is given in Figure A6.17. The weight for this coding is 226. The code developed in Example 6.4.15 has weight 192, so the hunch was wrong, as predicted by the theory. The text coding has a lesser weight.

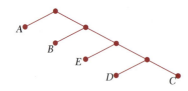

Figure A6.17

15. 2
17. The table gives a Huffman coding.

Character	Code	Character	Code
A	0110	E	10
R	00	I	010
C	01111	S	111
T	110	X	01110

The average number of bits used per character is 2.75. For a fixed length code, 3 bits can be used per character.
19. If the characters have frequency n at levels L and M, the contribution of the variables to the weight is $n(L + M)$. When interchanged, their contribution is still $n(L + M)$. A similar argument is used for different characters at the same level.
21. Suppose cluster C_1 has weight W_1, frequency a, and is at level L, and that cluster C_2 has weight W_2, frequency b, and is at level M. Suppose $L > M$, and $a > b$. Then switching the locations of C_1 and C_2 yields a cluster of smaller weight.

Proof: Using the previous exercise, the contributions to the weights before and after the switch are $W_1 + W_2 + aL + bM$ and

$W_1 + W_2 + aM + bL$, respectively. The difference is

$$(aL + bM) - (aM + bL) = a(L - M) + b(M - L)$$
$$= (a - b)(L - M) > 0$$

Thus a switch yields a smaller weight. The proof of Lemma 6.4.14 is similar to the proof of Lemma 6.4.11, except that clusters, rather than variables, are moved.

23. $(C \oplus C) \oplus (C \oplus C)$

25.
Leonard:	Root	Bill:	000	Frieda:	001
George:	0	Tom:	1	Ellen:	0010
Carol:	00	Alice:	0000	Marie:	10
Harry:	01	John:	011	Anthony:	00001

27. A recursive algorithm Compare(α, β) is as follows. We use ϵ as the empty string—the string identifying the root. The input consists of strings α and β. The output is one of the statements $\alpha < \beta$, $\alpha = \beta$, or $\beta < \alpha$, where $\alpha < \beta$ means that the name corresponding to α alphabetically precedes the name corresponding to β.

Compare(α, β)
1. IF $\alpha = \epsilon$ THEN [3 cases for β]
   1. IF $\beta = \epsilon$ THEN
      1. $\alpha = \beta$
      2. End
   2. IF β starts with 0 THEN
      1. $\beta < \alpha$
      2. End
   3. IF β starts with 1 THEN
      1. $\alpha < \beta$
      2. End [this completes the basis for the recursion]
2. IF $\beta = \epsilon$ THEN
   1. Compare(β, α) [a recursive call; already covered in instruction 1]
   2. End
3. IF α starts with 0 and β starts with 1 THEN
   1. $\alpha < \beta$
   2. End
4. IF α starts with 1 and β starts with 0 THEN
   1. $\beta < \alpha$
   2. END
5. IF α and β both start with the same digit x so that $\alpha = x\alpha'$ and $\beta = x\beta'$ THEN
   1. Compare(α', β') [a recursive call]
   2. IF $\alpha' < \beta'$ THEN $\alpha < \beta$
   3. IF $\beta' < \alpha'$ THEN $\beta < \alpha$
   4. IF $\alpha' = \beta'$ THEN $\alpha = \beta$

Chapter 7

Section 7.1

1. $pq' \vee p'q \vee p'q'$
3. $pr' \vee p'r$
5. $pqr \vee pqr' \vee pq'r \vee pq'r' \vee p'qr \vee p'qr' \vee p'q'r$
7. $\neg(\neg[\neg p \wedge \neg q] \wedge \neg r)$
9. $\neg(\neg p \vee \neg q \vee \neg r) \vee \neg(\neg[\neg p \vee \neg q] \vee r)$
11. We show by induction that any statement P using $\vee$ as the only connective must necessarily be true when all of the atomic sentences $p, q, \ldots$ used in P are true. This is true for the atomic statements $p, q, \ldots$. If it is true for P and Q, then it is true for $P \vee Q$, since by induction P and Q are true when the atomic statements used in their construction are all true. This proves the result. But the statement $\neg p$ is false when p is true. Therefore, it cannot be built up out of atomic statements using $\vee$ as the only connective.
13. $\neg p \wedge q \equiv \neg(p \vee \neg q) \equiv (p \downarrow [q \downarrow q])$
15. $q \vee pr'$. See Figure A7.1.

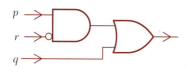

Figure A7.1   **Figure A7.2**

17. $p' \vee r'$. See Figure A7.2.
19. $p's' \vee pq' \vee pr's$
21. $p'q' \vee p'rs \vee prs' \vee pr's$
23. $p'q'r \vee prs'$
25. $p'r \vee q'r$
27. $pq' \vee p'q$. See Figure A7.3.

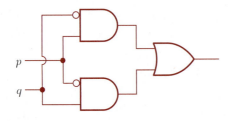

Figure A7.3

29. $pqr \vee pqr' \vee pq'r \vee pq'r'$
31. $pq' \vee p'q$

33. Not associative. For example, $(T \mid T) \mid F = F \mid F = T$, but $T \mid (T \mid F)$
    $= T \mid T = F$.

Section 7.2

1. The table for both sides of this equivalence is as follows:

p	q	r	Statement
F	F	F	F
F	F	T	F
F	T	F	F
F	T	T	F
T	F	F	F
T	F	T	T
T	T	F	T
T	T	T	F

3. $p(1 + p) + 1 = 1$

5. $a(b \vee c) = a(b + c + bc) = ab + ac + abc$

   $ab \vee ac = ab + ac + abac = ab + ac + abc$

   Thus both sides are equal.

7. $a \vee (b \vee c) = a \vee (b + c + bc)$

   $$= a + (b + c + bc) + a(b + c + bc)$$

   $$= a + b + c + ab + ac + bc + abc$$

   A similar calculation gives the same result for $(a \vee b) \vee c$.

9. Using $p \rightarrow q \equiv \neg p \vee q$, we have the equivalence

   $$p \rightarrow q = (1 + p) \vee q = 1 + p + q + (1 + p)q = 1 + p + pq$$

11. Since $p \leftrightarrow q$ is tautologically equivalent to $(p \wedge q) \vee (\neg p \wedge \neg q)$, we have the equivalence

   $$p \leftrightarrow q = pq \vee (1 + p)(1 + q)$$

   $$= pq + (1 + p)(1 + q) + pq(1 + p)(1 + q)$$

   $$= pq + 1 + p + q + pq + 0$$

   $$= 1 + p + q$$

13. All humming-birds are small. (Answer courtesy Lewis Carroll.)

15. No kitten with green eyes will play with a gorilla. (Answer courtesy Lewis Carroll.)

17. Write $B + C = B \cap C' \cup B' \cap C$. Then

   $$A \cap (B + C) = A \cap B \cap C' \cup A \cap B' \cap C$$

Similarly,

$$A \cap B + A \cap C = A \cap B \cap (A \cap C)' \cup (A \cap B)' \cap (A \cap C)$$
$$= A \cap B \cap (A' \cup C') \cup (A' \cup B') \cap (A \cap C)$$
$$= A \cap B \cap C' \cup A \cap B' \cap C$$

proving the result.

19. $x \in (A + B) + C$ if and only if x is in $A \cap B \cap C$ or x is in only one of the three sets A, B, and C. This condition is symmetric in A, B, and C. Thus,

$$x \in (A + B) + C \quad \text{iff} \quad x \in (B + C) + A$$

Therefore, $(A + B) + C = (B + C) + A = A + (B + C)$, since the operation $+$ is commutative.

21. $(ab)a = aba = aab = ab$. Therefore $ab \subseteq a$ by definition.

23. $xy'z' + xyz + x'yz + x'y'z$

25. $xyz + xyz' + xy'z + xy'z' + x'y'z + x'yz$

27. Since $ab = a$, we have

$$a \vee (a + b) = a + a + b + a(a + b)$$
$$= a + a + b + aa + ab$$
$$= a + a + b + a + a = b$$

29. $0x = (x + x)x = xx + xx = 0$

31. Since $\subseteq$ is transitive, it is only necessary to show that if $a \subseteq b$ and $b \subseteq a$ then $a = b$. But if $ab = a$ and $ba = b$, it follows by commutativity that $a = b$.

33. Take $d = ab$. Then $d \subseteq a$ and $d \subseteq b$. Also if $x \subseteq a$ and $x \subseteq b$, then $x \subseteq d$. To see this latter fact, we have $xa = x$ and $xb = x$. Multiplying, we get $xab = x$. Thus $x \subseteq ab = d$.

35. In order to show that $0 < a + b$, we must show that $0 \neq a + b$, and that if $0 \subseteq x \subseteq a + b$, then $x = 0$ or $x = a + b$. Now suppose $x \neq 0$ and $x \neq a + b$. (See Figure A7.4 for a set-theoretic motivation here.) Let $y = x + a$. Then we claim that $y \neq a$, $y \neq b$, yet $a \subseteq y$ and $y \subseteq b$. The inequalities are straightforward. The inclusions can be obtained as follows. From $x \subseteq a + b$, multiply by a to get $ax \subseteq a + ab = a + a = 0$. Therefore, $ax = 0$. Since $x \subseteq a + b$, we have $x = x(a + b) = xa + xb = xb$. Thus, $bx = x$.

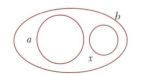

Figure A7.4

We can now show that $a \subseteq y$:

$$ay = a(a + x) = a + ax = a$$

Similarly, we show that $y \subseteq b$:

$$by = b(a + x) = ab + bx = a + x = y$$

Thus, we have found an element $y = a + x$ strictly between a and b, contradicting $a < b$.

37. We always have $0 \subseteq ab \subseteq a$ for any b. If a is small, then $ab = 0$ or $ab = a$. Similarly, if b is small, $ab = 0$ or $ab = b$. Therefore, if $ab \neq 0$, we have $a = ab = b$.

39. We must first show that any element x of B is the sum of small elements. Because B is finite, there is a chain

$$0 = x_0, x_1, x_2, \ldots, x_k = x$$

in which each term is covered by the one after it. (See Example 2.3.16.) Using the results of Exercise 35, the elements $x_i + x_{i+1}$ are all small $(i = 0, \ldots, k - 1)$. Then

$$x = (0 + x_1) + (x_1 + x_2) + \cdots + (x_{k-1} + x_k)$$

expresses x as the sum of small elements. See Figure A7.5 for a set-theoretic motivation of this proof.

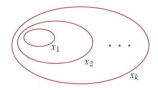

Figure A7.5

Using the results of Exercise 38, if there are n small elements in B, there will be 2^n elements in B. These correspond in a 1-1 way with the subsets of the small elements.

Chapter 8

Section 8.1

1. $138 = 1 \cdot 72 + 66$; $72 = 1 \cdot 66 + 6$; $66 = 11 \cdot 6$. The GCD is 6. A linear combination is: $6 = 2 \cdot 72 - 1 \cdot 138$.

3. By induction. It is true for $f_0 = 0$ and $f_1 = 1$. Assume the result for f_n and f_{n+1}. Now suppose d is a positive common divisor of f_{n+1}

and f_{n+2}. Then $d \mid (f_{n+2} - f_{n+1})$ or $d \mid f_n$. Thus d is a common divisor of f_n and f_{n+1}, so $d = 1$ by induction.

5. For $n = 0$, we have $0 = b \cdot 0 + 0$. Assuming the result for n, we have $n = bq + r$. Therefore, $n + 1 = bq + r + 1$. There are two cases. If $r < b - 1$, this is the required representation of $n + 1$. If $r = b - 1$, then $r + 1 = b$, so

$$n + 1 = bq + b = b(q + 1) + 0$$

This is again of the required form. If $n = -m$, write $m = b \cdot q + r$. Then $n = -m = -bq - r$. There are two cases. If $r = 0$, this is the required form: $n = (-b)q$. If $0 < r < b$, then $0 < b - r < b$, so $n = -bq - b + b - r = (-q - 1)b + (b - r)$, which is of the required form.

7. If $a = bq + r = bq_1 + r_1$, then $b(q - q_1) = r_1 - r$. Therefore b divides $r_1 - r$. But because of the size of the remainders, we have $|r_1 - r| < b$. Since b divides $r_1 - r$, we have (by Lemma 8.1.2, part 8) $r_1 - r = 0$, or $r_1 = r$. It follows that $bq = bq_1$. Dividing by b, we obtain $q = q_1$. This proves uniqueness.

9. Let $d = \text{GCD}(a, b)$. Write the full Euclidean algorithm to compute d. Now multiply each of the equations in this algorithm by x. The result is the Euclidean algorithm for the GCD of xa and xb, and it gives xd as their GCD.

11. If $\text{GCD}(a, b) = \text{GCD}(a, c) = 1$, then, by Theorem 8.1.15,

$$ax + by = 1 \qquad au + cv = 1$$

Multiplying and combining terms, we obtain $aX + bcY = 1$ where $X = axu + buy + cxv$, and $Y = yv$.

13. Suppose that d is the largest proper divisor of n. Then $n = de$, where $2 \le e \le d$. Therefore $d = n/e \le n/2$. To use this fact in the LongSearch Algorithm, change the FOR statement 3 of this algorithm to

$$3. \text{ FOR } i := 2 \text{ to } [b/2]$$

15. (a) $288 = 2^5 3^2$;   (b) $448 = 2^6 7$
17. (a) $1001 = 7 \cdot 11 \cdot 13$;   (b) $221 = 13 \cdot 17$
19. The code names for $(2, 3)$ and $(5, 0)$ are 108 and 32, respectively.
21. The code names for $(0, 1, 1)$ and $(1, 2, 2)$ are 15 and 450, respectively.
23. (a) $(3, 2)$;   (b) $(0, 4)$
25. Neither can be decoded since each has 7 as a divisor.
27. If $\sqrt{7} = m/n$, then squaring and simplifying, we have $7n^2 = m^2$. If n has 7 to the power r in its unique factorization, and m has 7 to the power s in its unique factorization, the equation yields, by the unique factorization theorem, $2r + 1 = 2s$. But this gives $s - r = \frac{1}{2}$, which is impossible.
29. The primes are

2 3 5 7 11 13 17 19 23 29 31 37 41 43 47 53 59 61 67 71 73 79 83 89
97 101 103 107 109 113 127 131 137 139 149 151 157 163 167 173
179 181 191 193 197 199

31. The following algorithm has input $n \geq 1$. The output is either "prime" or "no prime."

    1. IF $n = 1$ THEN
        1. Output "no prime"
        2. End
    2. IF $n = 2$ THEN
        1. Output "prime"
        2. End
    3. IF $2 \mid n$ THEN
        1. Output "no prime" [$n > 2$ here, and n is even]
        2. End
    4. $d := 3$ [initializing the first odd divisor]
    5. WHILE $d^2 \leq n$ [only these need be tested]
        1. IF $d \mid n$ THEN
            1. Output "no prime"
            2. End
        2. $d := d + 2$ [try the next odd divisor]
    6. Output "prime" [passed all divisor tests]

33. If m divisions are required, and $m \geq n$, then $b \geq f_{m+1} \geq f_{n+1}$ since the Fibonacci numbers are increasing.

35. Transitivity is stated in Lemma 8.1.2, part 4. Also, if $a \mid b$ and $b \mid a$, then $|a| \leq |b| \leq |a|$ by Lemma 8.1.2, part 8. Therefore $a = b$ since a and b are positive.

Section 8.2

1. $(3^{81})_7 = 6$; $(3^{81})_{19} = 18$
3. $x \equiv 5 \bmod 33$
5. $x \equiv 7 \bmod 35$
7. Take the numbers $x_1, x_2, \ldots, x_m, -x_1, -x_2, \ldots, -x_m$ modulo n. There are more than n of these numbers. Therefore, by the Box Principle, two must be equal modulo n. By hypotheses, none of the x_i are congruent mod n, and similarly none of the $-x_i$ are congruent mod n. Therefore, we must have $x_i \equiv -x_j \bmod n$ for some i and j. This is the result.
9. $n \equiv S(n) \bmod 7$, where $S(n) = \Sigma a_i$, the sum of the octal digits of n. $n \equiv A(n) \bmod 9$, where $\Sigma(-1)^i a_i$, the alternating sum of the octal digits of n. Finally, $n \equiv a_0 \bmod 8$. The proofs of these results are similar to the analogous results for base 10. Thus, we have $(543612)_8$

$\equiv 5 + 4 + 3 + 6 + 1 + 2 = 21 \equiv 0$ mod 7. Similarly $(543612)_8 \equiv -5 + 4 - 3 + 6 - 1 + 2 = 3$ mod 9, and finally, $(543612)_8 \equiv 2$ mod 8.

11. Converting to base 8, the number is $(3247163)_8 \equiv 26 \equiv 5$ mod 7, as in Exercise 9. Therefore it is not divisible by 7.

13. If $S(n)$ is the digit sum of n, then $n \equiv S(n)$ mod 3, since the congruence is valid mod 9. Thus, a simple rule is to take repeated digit sums until the answer is a single digit. The number is divisible by 3 if and only if that digit is 0, 3, 6, or 9.

15. We have $a \equiv 1$ mod $(a - 1)$. Raising both sides to the power n, we get $a^n \equiv 1$ mod $(a - 1)$. Thus $a^n - 1$ is divisible by $a - 1$. Similarly, from $a \equiv -1$ mod $(a + 1)$, we have $a^n \equiv -1$ when n is an odd number, and hence $a^n + 1$ is divisible by $a + 1$.

17. Suppose n were not a power of 2. Then n would have an odd factor m: $n = mp$. Then $2^n + 1 = (2^p)^m + 1$ is divisible by $2^p + 1$ by the result in Exercise 15.

19. $x \equiv 1$ and $y \equiv 9$ mod 13

21. $x \equiv 8$, $y \equiv 7$, and $z \equiv 6$ mod 9

23. Write $a = da'$, $b = db'$, $n = dn'$. Then if $a \equiv b$ mod n, we have $n \,|\, (a - b)$. Hence $dn' \,|\, (da' - db')$, and $n' \,|\, (a' - b')$. Thus $a' \equiv b'$ mod n' or $a/d \equiv b/d$ mod n/d. The converse is obtained similarly.

25. (a) EHCHG EHPCF YHXJV HWN;   (b) $m' = 21$ and $b' = 3$

27. LETTER: A B C D E F G H I J K L M N O P Q R S T U V W X Y Z
    CODE: K Z O D S H W L A P E T I X M B Q F U J Y N C R G V

    CODE: A B C D E F G H I J K L M N O P Q R S T U V W X Y Z
    DECODE: I P W D K R Y F M T A H O V C J Q X E L S Z G N U B

29. (a) 105

    (b) 7

    (c) 50, 16, 37, 34, 99, 9, 38, 50, 16, ... The sequence has period 7, as predicted by theory.

    (d) Only 7 of the 112 possibilities are reached since the sequence repeats after 7 different terms.

31. 56,746

33. Since $999 \equiv 0$ mod 37, we have $1000 \equiv 1$ mod 37. Therefore, if a number n is written in base 1000, so that $n = a_0 + 1000a_1 + 1000^2 a_2 + \cdots$, we have $n \equiv a_0 + a_1 + a_2 + \cdots$, the digit sum using base 10^3. Thus $12{,}409{,}362 \equiv 12 + 409 + 362 = 783 \equiv 6$ mod 37.

35. $89 \equiv -21$ mod 55

    $55 \equiv -8$ mod 21

    $21 \equiv -3$ mod 8

    $8 \equiv -1$ mod 3

    Thus, four divisions are required. If positive remainders are needed, eight divisions are necessary.

Section 8.3

1.

+	0 1 2 3 4 5 6
0	0 1 2 3 4 5 6
1	1 2 3 4 5 6 0
2	2 3 4 5 6 0 1
3	3 4 5 6 0 1 2
4	4 5 6 0 1 2 3
5	5 6 0 1 2 3 4
6	6 0 1 2 3 4 5

×	0 1 2 3 4 5 6
0	0 0 0 0 0 0 0
1	0 1 2 3 4 5 6
2	0 2 4 6 1 3 5
3	0 3 6 2 5 1 4
4	0 4 1 5 2 6 3
5	0 5 3 1 6 4 2
6	0 6 5 4 3 2 1

The equation $a + x = b$ has a unique solution x for each a and b. Therefore in the row opposite a in the addition table there is exactly one column (x) in which b occurs. Similar reasoning applies for the equations $x + a = b$, $ax = b$, and $xa = b$.

3. $x \equiv 15 \bmod 73$

5. 26

7. Using the Euclidean Algorithm, we find $1 = 7 \cdot 47 - 8 \cdot 41$. Therefore the inverse of 41 mod 47 is -8 or 39. We can trace through the QuotientModp(1, 41, c) Algorithm as follows:

a	b	$q := [47/b]$	
1	41	1	
-1	6	7	[here, $a := (-aq)_{47}$ and $b := 47 - bq$]
7	5	9	
31	2	23	
-8	1		[here, $c := -8$, so the reciprocal is -8]

9. $\mathbb{Z}_8^* = \{1, 3, 5, 7\}$; $\mathbb{Z}_{15}^* = \{1, 2, 4, 7, 8, 11, 13, 14\}$. The multiplication table for $\mathbb{Z}_8^*$ is as follows:

·	1 3 5 7
1	1 3 5 7
3	3 1 7 5
5	5 7 1 3
7	7 5 3 1

The equation $ax = b$ has a unique solution x for each a and b. Therefore in the row opposite a there will be exactly one column (x) in which b occurs. By commutativity, the ith column is the same as the ith row.

11. $x \equiv 23 \bmod 81$

13. If x does not have an inverse mod n, then x and n would not be relatively prime. Therefore $p \mid x$. If $n = p^k$, then $p^k \mid x^k$, so $n \mid x^k$ and

$x^k = 0$ in $\mathbb{Z}_n$. If n were not the power of a prime, then n would have two different prime factors p and q. Then p would not have an inverse in $\mathbb{Z}_n$, but no power of p would be 0 since q is not a factor of a power of p.

15. $1/(\alpha + \beta) = 1$; $\alpha/(1 + \beta) = 1$; $\beta^2 + \alpha^2 = 1$

17. $1^3 = 1$. $\alpha^3 = \alpha^2 \cdot \alpha = \beta\alpha = 1$, and similarly, $\beta^3 = 1$.

19. The fixed point is -7. Thus, $f(x) = 3(x + 7) - 7$. Using Equation (8.3.1), $f^{14}(x) = 3^{14}(x + 7) - 7$ in $\mathbb{Z}_{19}$. But $3^{14} = 4$ in $\mathbb{Z}_{19}$. Therefore $f^{14}(x) = 4(x + 7) - 7 = 4x + 2$ in $\mathbb{Z}_{19}$.

21. If the elements of $\mathbb{Z}_n^*$ are $a_1, a_2, \ldots, a_{\varphi(n)}$, then the elements aa_1, $aa_2, \ldots, aa_{\varphi(n)}$ are a rearrangement of these elements. This is so because the equation $ax = a_i$ has a unique solution in $\mathbb{Z}_n$. Therefore

$$a_1 a_2 \cdots a_{\varphi(n)} = aa_1 \cdot aa_2 \cdots aa_{\varphi(n)}$$

Now, dividing by $a_1 a_2 \cdots a_{\varphi(n)}$, we obtain $a^{\varphi(n)} = 1$ in $\mathbb{Z}_n$.

23. To find the inverse of 7 mod 12, we solve $7e \equiv 1$ mod 12. The solution is $e \equiv 7$ mod 12. The solution of $x^7 \equiv 2$ mod 13 is therefore $x \equiv 2^7 \equiv 11$ mod 13.

25. The equation is equivalent to $x^2 \equiv 9$ or $(x - 3)(x + 3) \equiv 0$ mod 7. The complete solution is $x \equiv 3$ or $x \equiv -3 \equiv 4$ mod 7.

27. If $x \equiv 0$ mod 29, then $a \equiv 0$. If not, then $x^{28} \equiv 1$ by Fermat's Little Theorem. Raising the given congruence $x^4 \equiv a$ mod 29 to the seventh power, we obtain $1 \equiv x^{28} \equiv a^7$ mod 29. Therefore the order of a must divide 7. Therefore a has order 7 or order 1, in which case $a = 1$.

29. Suppose that a has order $2r$. Then $a^{2r} = 1$. Thus $(a^2)^r = 1$ and a^2 has order $\leq r$. If it had order $s < r$, then we would have $a^{2s} = 1$ contrary to the hypothesis that a has order $2r$. This proves the result.

31. Suppose a^n has order $s < p - 1$. Then $a^{ns} \equiv 1$ mod p. Since n is relatively prime to $p - 1$, find n' such that $nn' \equiv 1$ mod $p - 1$. Then, by raising the congruence $a^{ns} \equiv 1$ mod p to the power n', we arrive at $a^s \equiv 1$ mod p. But this is impossible because the order of a is given as $p - 1$. For $p = 11$ and $a = 2$, we can verify that 2 has order 10. The elements of $\mathbb{Z}_{10}^*$ are $\{1, 3, 7, 9\}$. Therefore, other elements of order 10 in $\mathbb{Z}_{11}$ are $2^3 = 8$, $2^7 \equiv 7$, and $2^9 = 512 \equiv 6$. Thus, 2, 6, 7, and 8 are all elements of order 10 in $\mathbb{Z}_{11}$.

33.

n	2^n
1	2
2	4
4	16
8	256
16	544
32	87
64	122
128	667
256	100
512	522

35. Since $133 = 10000101$, starting with 2^1, we use the operations SSSS SM S SM. Here, "S" is the square operation and "M" multiplies by 2. The resulting table gives the intermediate results:

Operation	Power	Computation	Exponent (binary)
	2^1	2	1
S	2^2	4	10
S	2^4	16	100
S	2^8	256	1000
S	2^{16}	544	10000
S	2^{32}	87	100000
M	2^{33}	174	100001
S	2^{66}	488	1000010
S	2^{132}	517	10000100
M	2^{133}	357	10000101

Thus, $2^{133} \equiv 357 \bmod 677$.

37. After instruction 2.3 add the instructions

2.4. IF $b > p/2$ THEN
    1. $b := p - b$
    2. $a := p - a$

Section 8.4

1. $x \equiv 6 \bmod 7$
3. $x \equiv 1211 \bmod 1430$
5. 550, 891, and 540 mod 990
7. Basis 288, 441, and 280 mod 504. Solutions: (a) $x \equiv 275 \bmod 504$; (b) $x \equiv 463 \bmod 504$.
9. 628 coconuts. The number must be congruent to $-2 \bmod 630$.
11. (a) $x \equiv 31 \cdot 61 - 45 \cdot 60 = -809 \equiv 2851 \bmod 3660$
    (b) $x \equiv 4 \cdot 61 - 30 \cdot 60 = -1556 \equiv 2104 \bmod 3660$
13. Two consecutive integers are relatively prime, since if d is a positive common divisor, it must divide their difference, which is 1. Hence $d = 1$. Similarly, if d divides $r - 1$ and $r + 1$, it must divide their difference, 2. But if the middle term r is even, d cannot be 2 since $r - 1$ and $r + 1$ are odd. So d is 1 in this case, too, and the numbers are relatively prime in pairs. If the middle term is odd, then 2 is a common divisor of $r - 1$ and $r + 1$, so the numbers are not relatively prime in pairs.
15. A basis is $\alpha_1 = 240,318$; $\alpha_2 = -6083$; $\alpha_3 = 240,240$. The solution is $x \equiv 474,427 \bmod 474,474$.
17. Any four consecutive integers will have at least two even numbers.
19. 410 has components (5, 0, 3). Since $5^{89} \equiv 2 \bmod 9$, and $3^{89} \equiv 4 \bmod 11$, it follows that 410^{89} has components (2, 0, 4) mod 990. A basis is 550, -99, 540. The numbers (2, 0, 4) are also the components of $2(550) + 4(540) \equiv 290 \bmod 990$. Therefore we have the congruence $410^{89} \equiv 290 \bmod 990$.

21. Since the basis elements α_1, α_2, and α_3 have components $(1, 0, 0)$, $(0, 1, 0)$, and $(0, 0, 1)$, respectively, the sum $\alpha_1 + \alpha_2 + \alpha_3$ has components $(1, 1, 1)$. But 1 has components $(1, 1, 1)$ also. Therefore $\alpha_1 + \alpha_2 + \alpha_3 \equiv 1 \bmod mnp$.

23. Let $(a)_n = r$. Set $A = b^a - 1$, $N = b^n - 1$, $R = b^r - 1$. Then $(A)_N = R$. Proof: Write $a = nq + r$. From $b^n \equiv 1 \bmod N$, we get $b^{nq} \equiv 1 \bmod N$, and so $b^a \equiv b^{nq+r} \equiv b^r \bmod N$. Subtracting 1, we get $A \equiv R \bmod N$. This gives the result since $0 < R < N$.

25. Using the result of Exercise 22 or 23, if the remainder when r is divided by s is p, then the remainder when $2^r - 1$ is divided by $2^s - 1$ is $2^p - 1$. This shows that the sequence of divisions given by the Euclidean Algorithm starting with r and s that terminates in 1 gives rise to a sequence of divisions starting with $2^r - 1$ and $2^s - 1$ that terminates with $2^1 - 1 = 1$. Thus, these numbers are relatively prime.

27. $a = 2^{15} - 1$ and $b = 2^6 - 1$. Since $\mathrm{GCD}(6, 15) = 3$, it follows by Exercise 24 that $\mathrm{GCD}(a, b) = 2^3 - 1 = 7$.

Chapter 9

Section 9.1

1. (a) $(p \to \lnot(p \land r))$;   (b) $((s \to \lnot t) \lor (q \land (r \leftrightarrow s)))$
3. (a) See Figure A9.1.
   (b) Not a formula.
   (c) See Figure A9.2.

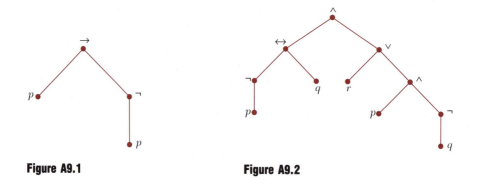

Figure A9.1   **Figure A9.2**

5. See Figure A9.3.
7. Not a formula because it has more left parentheses than right.
9. $P = C + 1$. It is true for variable or constant symbols, since $P = 1$ and $C = 0$ in this case. Suppose it is true for formulas α and β, and

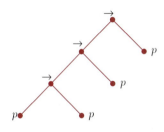

Figure A9.3

that α has C_1 binary connectives and P_1 symbols, and β has C_2 binary connectives and P_2 symbols. Then we have $P_1 = C_1 + 1$ and $P_2 = C_2 + 1$. If λ is any binary connective, then the formula $(\alpha\lambda\beta)$ has $C = C_1 + C_2 + 1$ binary connectives and $P = P_1 + P_2$ symbols. But by induction $P_1 = C_1 + 1$ and $P_2 = C_2 + 1$. Therefore $P = P_1 + P_2 = C_1 + C_2 + 2 = C + 1$, which is the result for $(\alpha\lambda\beta)$. The introduction of the negation symbol has no effect on C or P. Thus the result is proved by induction.

11. By induction we can first show that a formula can never begin or end with a binary connective. Assuming this fact, the proof of the result is by induction as follows. It is true in the base case. If it is true for α, it remains true for $\neg\alpha$, because no binary connective is introduced. If it is true for α and β, and if λ is any binary connective, then it is true for $(\alpha\lambda\beta)$, since there are no consecutive binary connectives within α and within β, and λ cannot be part of a consecutive pair since α cannot end with a binary connective and β cannot begin with one. This proves the result by induction.

13. This is true for the base case when the formula is a variable or constant symbol. If it is true for α and β, and λ is a binary connective, it remains true for $(\alpha\lambda\beta)$, since the introduction of "(" at the beginning and ")" at the end cannot give rise to the string ")(". Similarly, if it is true for α, it remains true for $\neg\alpha$, because no parentheses are added.

15. To prove (b), note that all productions either keep the same number of a's and b's or increase their number by 1. Thus, at any stage, the number of a's is equal to the number of b's. To show (a), we can arrive at $a^n b^n$ by n applications of the production $S \Rightarrow aSb$ and a final application of $S \Rightarrow \epsilon$. (This can be proved by induction.) Finally, the production $ab \Rightarrow ba$ allows us to "pull the letter a past a letter b." It is "evident" that any string with an equal number of a's and b's can be obtained by using this process over and over. A detailed proof is possible by induction on the number of a's.

17. Strings on a and b that (a) are palindromes (that is, they read the same backward and forward), (b) have an odd number of letters, and (c) have a as the middle letter.

19. Starting with S, the first move must be to $aaabbb$. We show that any

string in the language must be of the form αab^n, with $n \geq 3$. This is true after the first step, and we shall show that this structure persists for any further production. If the production $aba \Rightarrow ba$ is used on this form and affects the last a, the string must be of the form $\cdots abab^n$, which gets converted to $\cdots bab^n$ and has the required form. If the production $ab \Rightarrow bab$ is used, affecting the last a, then we have $\alpha ab^n \Rightarrow \alpha abab^n$, again of the required form. Finally, the last production, if it affects the last a, must be of the form $\cdots bab^n \Rightarrow \cdots bab^{n+1}$, which is again of the required form. This proves parts (a) and (b) of the exercise.

21. (a) The first production used must be $S \Rightarrow ab$, since it is the only one with an S. The other production increases the length of the string. Thus ab, which has length 2, cannot be attained.
    (b) $S \Rightarrow \underline{ab} \Rightarrow a\underline{ba} \Rightarrow a\underline{ba}b \Rightarrow aba\underline{bb} \Rightarrow aba\underline{bab} \Rightarrow abababb$ is a sequence of productions that yields the string.
    (c) If one production is used, it must be the one with S, so the string is ab. This is the basis case. Assume now that after n productions we arrive at the string $ab \cdots$. If the $(n + 1)$st production is $ab \Rightarrow aba$ and it affects the initial string ab, the $(n + 1)$st string is $aba \cdots$. If the $(n + 1)$st production is $ba \Rightarrow bab$ and it affects the initial string, the string must be of the form $aba \cdots$, and the $(n + 1)$st string is $abab \cdots$. In all cases, the $(n + 1)$st string starts with ab, and the result is proved by induction.

23. $S \Rightarrow a^3F$; $F \Rightarrow bF$; $F \Rightarrow \epsilon$; $ab \Rightarrow ba$. [The first production forces us to a^3F. The next allows as many b's as we wish, and the route $F \Rightarrow \epsilon$ is the only way to eliminate the variable F. Finally, the last production allows us to pull a's to the right of b's.

25. $S \Rightarrow aSa$; $S \Rightarrow bSb$; $S \Rightarrow a$; $S \Rightarrow b$; $S \Rightarrow \epsilon$

27. $S \Rightarrow aE$; $E \Rightarrow aaE$; $E \Rightarrow bE$; $E \Rightarrow \epsilon$; $ab \Rightarrow ba$

29. We think of constructing the strings in three stages: 1. Put down an equal number of a's, A's, and b's, with the b's at the extreme right. The A's are destined to move to the extreme right where they will become a's. Put a marker R at the extreme right. 2. Move the A's to the extreme right. 3. Convert A to a when A is at the extreme right. The productions are

    1. $S \Rightarrow aAXbR$; $X \Rightarrow aAXb$; $X \Rightarrow \epsilon$ [puts down the letters]
    2. $Aa \Rightarrow aA$; $Ab \Rightarrow bA$ [allows A to move to the right]
    3. $AR \Rightarrow aR$; $R \Rightarrow \epsilon$ [converts A to a and terminates the procedure]

31. We can show by induction that any formula α is the language generated by these productions. Using production (9.1.3), the variable and constant symbols are seen to be in the language. If α and β are in the language, we can produce $\neg\alpha$ using (9.1.2) and then producing α from S. If λ is a binary connective, we can produce $(\alpha\lambda\beta)$ by using (9.1.1), then producing α on the first occurrence of S, β on the second occurrence of S, and replacing B with λ using one of the productions (9.1.4).

To show that any production from this set is a formula, we proceed by induction on the number of steps used to produce a string of the language. If $n = 1$, we must use (9.1.3), and we arrive at a formula that is a constant or variable symbol. Now assume the result for all $m < n$. If the first step used is (9.1.1) $S \Rightarrow (SBS)$, then the next $n - 1$ steps must eliminate the two S's. By induction, this elimination will lead to formulas α and β, respectively. Further, B must be replaced by a binary connective λ as in the productions (9.1.4). Thus, we end with the formula $(\alpha\lambda\beta)$. Similar reasoning applies if we start with production (9.1.2) $S \Rightarrow \neg S$.

33. A recursive algorithm Tree(α; T) is as follows. α is the input: a string in the language of the propositional calculus. T is the tree representing the formula, or else a statement that no tree is possible.

Tree(α; T) Algorithm
1. IF α has length 1 THEN [the basis case]
    1. IF α is a constant or variable symbol p THEN
        1. $T :=$ tree consisting of single vertex p
        2. End
    OTHERWISE
        3. Output: "No tree is possible"
        4. End
2. IF $\alpha = \neg\alpha'$ THEN
    1. Tree(α'; S) [recursive call]
    2. IF S is a tree THEN
        1. $T :=$ tree with root labeled $\neg$, which has one child to which S is attached at its root
        2. End
    OTHERWISE
        3. Output: "No tree is possible"
        4. End
3. Scan(α; α', λ, β')
4. IF Scan procedure fails THEN
    1. Output: "No tree is possible"
    2. End
5. Tree(α'; R)
6. Tree(β'; S) [recursive calls]
7. IF R and S are trees THEN
    1. $T :=$ tree with root labeled λ grafting the trees R and S
    OTHERWISE
    2. Output: "No tree is possible"

Section 9.2

1. See Figure A9.4.
3. See Figure A9.5.
5. See Figure A9.6.
7. See Figure A9.7.

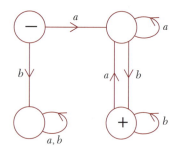

Figure A9.4

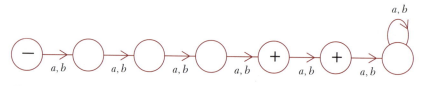

Figure A9.5

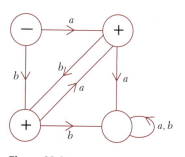

Figure A9.6

Figure A9.7

9. Suppose A were such an automation and it had N states. The path α for the string a^{N^2} ends at a terminal state. As in Example 13, this path has a cycle in it of length K, where $0 < K \le N$. By introducing this cycle into the path α, we arrive at the string a^{N^2+K}, which also ends at the same terminal vertex. But since $0 < K \le N < 2N$, $N^2 + K$ is strictly between N^2 and $(N + 1)^2 = N^2 + 2N + 1$. Therefore $N^2 + K$ is not a perfect square. This is a contradiction, because we have found a string not in the given language but which is in the language of A.

11. Since α has length $\ge N$, its path passes through $N + 1$ states. By the Box Principle, there is some state that is used twice in this path. This yields a cycle within this path that corresponds to a string γ. Then $\alpha = \beta\gamma\delta$ for some strings β and δ. By its construction, γ can be removed, or introduced freely, and the resulting string will still be in the language of A since it terminates at the same state as α.

This means that $\beta\gamma^i\delta$ is in the language of A for $i = 0, 1, 2, 3, \ldots,$ proving the theorem.

13. As in the hint, suppose $\alpha = \sigma\tau$ is the indicated string. Since σ has length N, its path passes through $N + 1$ vertices and so must contain a loop. Removing this loop, we obtain a string σ' whose length is smaller than N, the length of σ. Then $\alpha' = \sigma'\tau$ is the required string. Since length$(\tau) \geq N$, α' has length $\geq N$. Since α' has length smaller than α, and α was chosen to be the string of smallest length $\geq 2N$, it follows that length$(\alpha') < 2N$. This proves the result.

15. Interchange the terminal and nonterminal states. This has the effect of changing a string accepted by the automation to one that is not accepted, and vice versa.

17. Suppose the maximum length of the set of strings is N. Now consider the complete binary tree with all of its leaves on level N. Put the label "a" on edges going to the left and "b" on edges going to the right. Finally, introduce a sink and join all leaves to it with edges labeled "a, b." Now label the root $\ominus$ and label the vertices corresponding to the strings in the given set as $\oplus$. The result is the required automation. Figure A9.8 illustrates the procedure for the set $\{a, b, ab\}$.

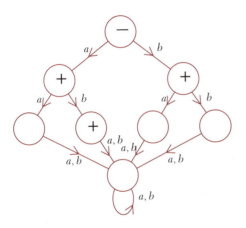

Figure A9.8

19. Strings of odd length that start with either "b" or "aa", together with strings of even length starting with "ab".

21. Strings containing at least two a's and one b.

Section 9.3

1. The strings consisting of an even number of b's, with no a's, or any number of a's with no b's.

3. Strings containing exactly two b's.

5. Strings beginning with any number of a's and ending with b, or strings beginning with any number of b's and ending with a, or strings consisting of consecutive pairs of the form aa or bb.

7. (a) If $A = a$, and $B = b$, then ab is in $(a + b)^*$, but ab is not in $a^* + b^*$.

   (b) If $A = a$, and $B = b$, then $abab$ is in $(ab)^*$ but is not in a^*b^*.

   (c) If $A = a$, and $B = b$, then b is in $ab + b$, but not in ab.

9. $(b + ab)^*aa[(a + b)(b + ab)^*aa]^*$

11. $[bb + (a + ba)(ba)^*(a + bb)](a + b)^* + (a + ba)(ba)^*$

13. See Figure A9.9.

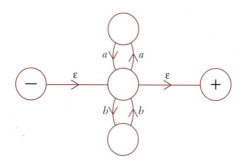

Figure A9.9

15. See Figure A9.10.

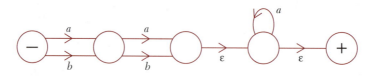

Figure A9.10

17. See Figure A9.11. The original states are labeled 1, 2, and 3.

19. S is the language of an automaton. If the terminal vertices are interchanged with the nonterminal vertices, then the resulting automaton will have language $\Sigma^* - S$. But as the language of an automaton, it is regular.

21. See Figure A9.12.

23. (a) $L \cup L'$;   (b) $L \cap L'$

25. Not the same language. One accepts ab, the other rejects it.

27. The same language.

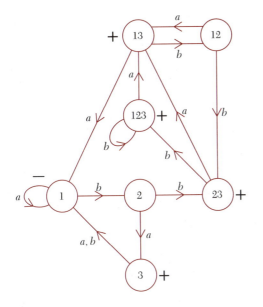

Figure A9.11

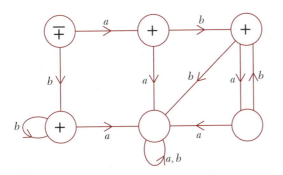

Figure A9.12

Chapter 10

Section 10.1

1. $q_1 B L q_1; q_1 0 L q_1$
3. $q_1 0 B q_{10}; q_1 1 B q_{20}; q_1 B B q_{100};$
   $q_{10} B L q_{11}; q_{11} B 0 q_{12}; q_{11} 1 0 q_{12}; q_{11} 0 0 q_{12};$
   $q_{12} 0 R q_{13}; q_{13} B R q_1;$
   $q_{20} B L q_{21}; q_{21} B 1 q_{22}; q_{21} 1 1 q_{22}; q_{21} 0 1 q_{22};$
   $q_{22} 1 R q_{23}; q_{23} B R q_1.$
5. MoveLToBlank; Print0onB; MoveLFrom0; Print0onB;
   MoveRToBlank; Print1onB.

Turing machine instructions reflecting this sequence of macros are

$q_1 \, 0 \, L \, q_1; \, q_1 \, 1 \, L \, q_1; \, q_1 \, B \, 0 \, q_2; \, q_2 \, 0 \, L \, q_3;$
$q_3 \, B \, 0 \, q_4; \, q_4 \, 0 \, R \, q_4; \, q_4 \, 1 \, R \, q_4; \, q_4 \, B \, 1 \, q_5.$

7. X: If 0 then go to P
   If 1 then go to Q
   If B then Halt
   P: Print1on0
   MoveRFrom1
   go to X
   Q: Print0on1
   MoveRFrom0
   go to X

Turing machine instructions based on these macros are

$q_1 \, 0 \, 0 \, q_{10}; \, q_1 \, 1 \, 1 \, q_{20}; \, q_1 \, B \, B \, q_{30};$
$q_{10} \, 0 \, 1 \, q_{11}; \, q_{11} \, 1 \, R \, q_1; \, q_{20} \, 1 \, 0 \, q_{21}; \, q_{21} \, 0 \, R \, q_1.$

9. X: If 0 then go to P; If 1 then go to Q; If B then Halt
   P: MoveRToBlank; MoveRFromB; MoveRToBlank;
   Print0onB; go to S;
   Q: Move RToBlank; MoveRFrBlank; MoveRToBlank;
   Print1onB; go to S;
   S: MoveLToBlank; MoveLFrBlank; MoveLToBlank;
   MoveRFrB; PrintBlank; MoveRFrBlank; go to X;

Turing machine instructions are

(X) $q_1 \, 0 \, 0 \, q_{10}; \, q_1 \, 1 \, 1 \, q_{20}; \, q_1 \, B \, B \, q_{100};$
(P) $q_{10} \, 0 \, R \, q_{10}; \, q_{10} \, 1 \, R \, q_{10}; \, q_{10} \, B \, R \, q_{11};$
  $q_{11} \, 0 \, R \, q_{11}; \, q_{11} \, 1 \, R \, q_{11}; \, q_{11} \, B \, 0 \, q_{30};$
(Q) $q_{20} \, 0 \, R \, q_{20}; \, q_{20} \, 1 \, R \, q_{20}; \, q_{20} \, B \, R \, q_{21};$
  $q_{21} \, 0 \, R \, q_{21}; \, q_{21} \, 1 \, R \, q_{21}; \, q_{21} \, B \, 1 \, q_{30};$
(S) $q_{30} \, 0 \, L \, q_{30}; \, q_{30} \, 1 \, L \, q_{30}; \, q_{30} \, B \, L \, q_{31};$
  $q_{31} \, 0 \, L \, q_{31}; \, q_{31} \, 1 \, L \, q_{31}; \, q_{31} \, B \, R \, q_{32};$
  $q_{32} \, 0 \, B \, q_{33}; \, q_{32} \, 1 \, B \, q_{33}; \, q_{33} \, B \, R \, q_1;$

11. We duplicate α as α', with each 0 replaced by Y and each 1 replaced by Z. In this way we keep track of the location where the duplicated string begins. After the duplication, we change all Y's to 0 and Z's to 1. (This is the procedure given by S.) X continues to be our marker. An algorithm is
    A: If 0 then go to P; If 1 then go to Q; If B then Halt;
    If Y then go to S; If Z then go to S;
    P: PrintXon0; MoveRToBlank; PrintYonB; MoveLToX;
    Print0onX; MoveRFr0; go to A;
    Q: PrintXon1; MoveRToBlank; PrintZonB; MoveLToX;
    Print1onX; MoveRFr1; go to A;
    S: If Y then Print0onY; If Z then Print1onZ; If 0 then MoveRFr0;
    If 1 then MoveRFr1; If B then Halt.

13. Changes the first character of the string to 1.

15. If the first character is 0, the machine never stops. If the first character is 1, nothing is done and the machine halts.

17. This machine never stops.

19. The method is to close up the blank between x and y. This is done by removing the last 1 in the variable y (given as a string of 1's) and putting it into the blank between x and y. Special procedures are used if $x = 0$ or $y = 0$. The pseudo Turing machine:

If 0 then go to Z;
MoveRToBlank;
MoveRFrBlank;
If 0 then go to Z; [$y = 0$ in this case]
MoveLFr1; [x and y are not 0 if this step is reached]
Print1onB; MoveRToBlank; MoveLFrBlank; PrintBon1; Halt;
Z: PrintBon0; Halt.

A Turing machine based on this pseudomachine is
q_1 0 B q_{100};
q_1 1 1 q_2; q_2 1 R q_2; q_2 B R q_3;
q_3 0 B q_{100}; q_3 1 L q_4; q_4 B 1 q_5;
q_5 1 R q_5; q_5 B L q_6; q_6 1 B q_{100}

21. We replicate V_1 before V_1. Note that the variable is either a string of 1's or a single 0. The pseudo Turing machine:
X: If 0 then go to P; If 1 then go to Q; If B then Halt;
P: MoveLFr0; MoveLFrB; Print0onB; Halt
Q: PrintXon1; MoveLToBlank; MoveLFrB;
    If B then go to S; MoveLToBlank; go to S;
S: Print1onB; MoveRToX; Print1onX; MoveRFr1; go to X;
A Turing machine based on this pseudomachine is
(X) q_1 0 0 q_{10}; q_1 1 1 q_{20}; q_1 B B q_{100};
(P) q_{10} 0 L q_{11}; q_{11} B L q_{12}; q_{12} B 0 q_{100};
(Q) q_{20} 1 X q_{21}; q_{21} X L q_{21}; q_{21} 1 L q_{21}; q_{21} B L q_{22}
    q_{22} B B q_{30}; q_{22} 1 L q_{23}; q_{23} 1 L q_{23}; q_{23} B B q_{30}
(S) q_{30} B 1 q_{31}; q_{31} 1 R q_{31}; q_{31} B R q_{31};
    q_{31} X 1 q_{32}; q_{32} 1 R q_1.

23. Starting from the right, we change 1's to 0's. As soon as we reach a 0 or a blank, we change it to 1 and halt. A pseudo Turing machine is as follows:
    MoveRToBlank; MoveLFrB;
X: If 1 then go to C;
    If 0 then go to D;
    If B then go to E;
C: Print0on1; MoveLFr0; go to X;
D: Print1on0; Halt;
E: Print1onB; Halt;

Section 10.2

1. The function $T(x) = S(S(Z(x)))$ is recursive by two substitutions from the starter Z function. The function $I(x) = x$ is recursive as the starter function $p_1^1(x)$. Since f is recursive, so is $f(2, x) = f(T(x), I(x))$ by substitution.

3. $g(x_1, x_2, x_3, x_4) = f(p_1^4(x_1, x_2, x_3, x_4), p_3^4(x_1, x_2, x_3, x_4))$ is recursive by substitution.

5. We can find the largest prime divisor of n by searching through x for divisors of the form $n - x$. Therefore we define

$$X(n) = (\mu x)[((n \div x) \mid n \wedge \text{prime}(n \div x)) \vee n < 2]$$

   Now define

$$P(n) = n \div X(n) \qquad \text{if } n \geq 2$$
$$= 0 \qquad \text{otherwise}$$

   Then $P(n)$ is recursive by Theorem 9.2.16. The condition $n < 2$ in the definition of $X(n)$ is needed to ensure that $X(n)$ is defined for all n.

7. Define $R(x, n) = (\mu y)[(\exists q)_{\leq n}(x = nq + y \wedge y < n) \vee n = 0]$. Then we have

$$(x)_n = R(x, n) \qquad \text{if } n > 0$$
$$= x \qquad \text{otherwise}$$

   Thus $(x)_n$ is recursive by Theorem 9.2.16.

9. Define the predicate $\text{PDiv}(x, n)$ [x is a prime divisor of n] by the formula

$$\text{PDiv}(x, n) \equiv x \mid n \wedge \text{prime}(x)$$

   Then

$$R(n) = \sum_{x=1}^{n} \text{PDiv}(x, n)$$

11. If $f(x) = 0$, then $\text{EQ}(f(x), 0) = 1$, and $1 \div f(x) = 1 \div 0 = 1$. Therefore $\text{EQ}(f(x), 0) = 1 \div f(x)$ in this case. Similarly, if $f(x) > 0$, we find $\text{EQ}(f(x), 0) = 1 \div f(x) = 0$. Therefore $\text{EQ}(f(x), 0) = 1 \div f(x)$ for all x.

13. $\text{sq}(x) = (\exists y)_{\leq x} (x = y^2)$

15. $\text{twin}(x) = \text{prime}(x) \wedge \text{prime}(x + 2)$

17. $[\sqrt{x}] = (\mu y)[x < y^2] \div 1$

19. Recursive, since the predicate $(\exists x)$ can be replaced by the bounded quantifier $(\exists x)_{x \leq y}$

21. Yes. $Q(m, n)$ is true if and only if m and n are relatively prime. Thus we can redefine Q by the formula

$$Q(m, n) = (\forall y)_{\leq m+n}(y \mid m \wedge y \mid n \rightarrow y = 1)$$

Section 10.3

1. $q_1\ 1\ 1\ q_1$.

3. Let $s = |\Sigma|$. Then there are s^n elements of Σ^n, and they can be arranged in a finite list L_n. All of the elements in Σ^* will then appear somewhere in one of the lists $L_0, L_1, \ldots, L_n, \ldots$. Since each of these lists is finite, we can write the elements of Σ^* as a sequence, by first listing the strings of L_0, then those of L_1, and so on. This shows that Σ is enumerable.

5. As in the infinite case, the finite sequence α_i yields a square matrix M consisting of a's and b's. The ith row of M is the string α_i. If we now define d_{ii} to be different from m_{ii} (changing a's to b's, and vice versa), we arrive at the string $d_{11} \ldots d_{nn}$, which differs from each of the α_i. This shows that there are more than n strings of length n on the letters a and b. Since there are 2^n strings of length n, we have $n < 2^n$.

7. $q_2\ 0\ L\ q_1$.

9. Nothing happens. The machine just halts. It is a function machine corresponding to the identity function.

11. $q_1\ 0\ L\ q_1$. This Turing machine has the smallest code number since the code number for each of its symbols is the smallest possible. It acts as the identity function as it executes no overprints and has at most one cursor movement.

13. Our coding shows that $P(x, n)$ is true if and only if x is the code number of some Turing machine, is divisible by $p(4n)$, but not divisible by $p(4n + 1)$. Here, $p(n)$ is the nth prime, which has been shown to be a recursive function. Thus,

$$P(x, n) \equiv \mathrm{Tur}(x) \wedge p(4n) \,|\, x \wedge \neg(p(4n + 1) \,|\, x)$$

$\mathrm{Tur}(x)$ is the recursive predicate that states that x is the code number of a Turing machine. (See Theorem 10.3.5.)

15. An overprint symbol has code number 11, 15, ... (See Table 10.3.1.) These symbols can only occur as the third Turing machine symbol in an overprint instruction quadruple. Therefore, if $p(n)$ denotes the nth prime, we can define $\mathrm{NoPrint}(x)$ as follows:

$$\mathrm{NoPrint}(x) \equiv \mathrm{Tur}(x) \wedge (\forall n)_{\leq x}\neg(p(4n + 3)^{11} \,|\, x)$$

Here, the bound "$\leq x$" on the quantifier $(\forall n)$ is much larger than it has to be, and can be improved if desired. All that is needed is that $n \leq$ the number of Turing machine instructions in the Turing machine represented by x.

Index

Boldface number refers to the page on which the definition appears.

LIST OF SYMBOLS

Topic	Symbol	Meaning	Page
NUMBER THEORY AND COUNTING	$a \mid b$	a divides b	480
	$a \nmid b$	a does not divide b	488
	$a \equiv b \bmod n$	a is congruent to b modulo n	495
	$(a)_n$	remainder when a is divided by n	497
	$C(n,r)$	the number of r-combinations of n objects	231
	$(d_k \ldots d_1 d_0)_b$	base b notation	142
	$D(N,b)$	number of digits in the base b expansion of N	151
	ψ	$(1 + \sqrt{5})/2$	124
	F_4	the four element field	515
	f_n	nth Fibonacci number	123
	$\mathrm{GCD}(a,b)$	greatest common divisor of a and b	481
	$P(n,r)$	the number of r-permutations of n objects	231
	$\mathbb{Z}_p$	$\mathbb{Z}_p$ as finite field	515
	$\mathbb{Z}_n$	$\mathbb{Z}_n$ as commutative ring	513
	$\mathbb{Z}_n^*$	invertible elements of $\mathbb{Z}_n$	525

Topic	Symbol	Meaning	Page
RECURSIVE FUNCTION THEORY	$(\forall x)_y$	bounded universal quantifier	611
	EQ	equality predicate	608
	$(\exists x)_y$	bounded existential quantifier	611
	LE	less than or equal predicate	608
	LT	less than predicate	608
	μy	the least y such that	604
	$p(x)$	predecessor function	606
	p_i^n	starter projection function	601
	$S(x)$	starter successor function $x + 1$	601
	$x \dotminus y$	natural number subtraction	606
	$Z(x)$	starter zero function	601